P9-DNZ-973

THE NEW OXFORD BOOK OF SIXTEENTH-CENTURY VERSE

THE NEW OXFORD BOOK OF SIXTEENTH CENTURY VERSE

Chosen and edited by
EMRYS JONES

Oxford New York
OXFORD UNIVERSITY PRESS
1991

Oxford University Press, Walton Street, Oxford OX2 6DP

Oxford New York Toronto
Delhi Bombay Calcutta Madras Karachi
Petaling Jaya Singapore Hong Kong Tokyo
Nairobi Dar es Salaam Cape Town
Melbourne Auckland

and associated companies in
Berlin Ibadan

Oxford is a trade mark of Oxford University Press

British Library Cataloguing in Publication Data
The New Oxford book of sixteenth century verse.
1. Poetry in english, 1400–1625—Anthologies
I. Jones, Emrys 1931–
821.208
ISBN 0–19–214126–0

Library of Congress Cataloging in Publication Data
The new Oxford book of sixteenth century verse/chosen and edited by Emrys Jones.
p. cm.
1. English poetry—Early modern, 1500–1700. I. Jones, Emrys, 1931–
PR1205.N49 1991 821'.308—dc20 90–40570
ISBN 0–19–214126–0

Typeset by Wyvern Typesetting Ltd, Bristol
Printed in Great Britain by
Courier International Ltd.
Tiptree, Essex

CONTENTS

CONTENTS

CONTENTS

INTRODUCTION

(i)

THE original *Oxford Book of Sixteenth Century Verse*, edited by Sir Edmund Chambers, appeared in 1932. By this time Chambers, born in 1866, had long established himself as an authority on the sixteenth century, and in fact his *Oxford Book* was to be the last of his contributions to the scholarship of that period. It was an admirable volume and has often been reprinted. Sixty years later, however, in a changed world—after the Second World War and a post-war social revolution, and after Modernism, Structuralism, Post-Structuralism, and much else—the sixteenth century and its poetry have come to look rather different. And as Chambers's volume has receded further into the past, it has become more obviously not just a product but also an expression of its own time. But even when it appeared it must have seemed, at least to some readers, old-fashioned. Its scholarship might be up-to-date, but the literary taste and cultural preferences that dictated the actual choice of poems were not so much Thirties as Edwardian or even Late Victorian. Reading it now, we are likely to find it even further away from us than the sixty-year interval might suggest.

Of course a fair number of the poems Chambers selected must be included in any big sixteenth-century anthology. Some of Shakespeare's sonnets and songs, Spenser's wedding odes, and one or two episodes from *The Faerie Queene*, this or that poem by Wyatt, Sidney, Ralegh, and Campion, and others again from the much smaller poets of the century (Vaux, Tichborne, Nashe, Barnfield), not to speak of those mysterious survivals whose authors are unknown ('Western wind' perhaps the most famous)—all these will claim a place in a book of this kind irrespective of the editor's personal tastes and interests. But once we have granted that the core of the selection is settled in advance, we are still left with the bulky remainder undecided, well more than half the number of poems chosen, and it is here that sharp differences in emphasis are likely to occur between anthologists working in different periods.

Behind Chambers's *Oxford Book* stands Palgrave's *Golden Treasury*, first published in 1861 but still in the 1920s and 1930s exerting a considerable influence. That influence was a matter of inducing readers to adopt a certain idea of what poetry was. Palgrave 'lyricized' the expectations of English readers of poetry. He helped to make such readers identify the very idea of poetry with short personal poems, what he defined as 'lyrical' poems. Palgrave originally meant his book to be a collection of strictly lyrical pieces. He called it *The Golden Treasury of the Best Songs and Lyrical Poems in the English Language*. But as his book went through its numerous editions and took its place as the leading anthology of English poetry, its full title became shortened to simply *The Golden Treasury*, as if it were, without qualification, a collection of the best poems of any kind—at least the best short poems—in English. Palgrave himself was initially much clearer as to what he was doing. In his Preface he states that only 'lyrical' poems qualify for inclusion; other kinds he considered 'foreign to the idea of the book'. And so, as he goes on to explain, most narrative, descriptive, didactic, and humorous poetry, along with 'occasional' and religious poetry, as well as poems written in blank verse and the heroic couplet, were not what he was looking for. It would be wrong to assume that Palgrave necessarily disliked or despised these other kinds and forms—he referred elsewhere, for instance, to 'the wonderful genius of Pope'. But whether or not he intended it, one of the long-term effects of his enormously popular volume was to downgrade what seemed to him these non-lyrical modes as well as the verse forms which were associated with them.

Palgrave's fairly small anthology, sensitively chosen and skilfully arranged, had the influence it did because he was fully in sympathy with the main direction of Romantic literary theory. Palgrave himself was not an original critic, nor did he present himself as one. Most of the views he helped to spread he had inherited from the Romantic poets and critics, chiefly Wordsworth and Keats. Another, more immediate, influence was Tennyson, a close friend, with whom he had many times discussed *The Golden Treasury* and to whom he finally dedicated it. So the process whereby 'Lyrical Poetry' (to use Palgrave's emphatically capitalized term) acquired its dominant position was already well advanced and would no doubt have been accomplished without the stimulus given by his anthology; what Palgrave in the event did was to hasten this process by institutionalizing in book form the beliefs which underpinned it. Though his tone is modest and unassuming, his strategy—preface followed by carefully selected texts—has almost the air of a manifesto, though a manifesto for a cause that was already won. The great Romantics were now firmly part of the literary establishment. In 1880 Matthew Arnold was to announce: 'Dryden and Pope are not classics of our poetry, they are classics of our prose.'

It was no doubt because he took for granted these lyricizing preferences of the late nineteenth century that Chambers admitted to his *Oxford Book* such a large number of short poems not just of the first rank but of the second or even third. Or rather, making allowances for the element of relativism in all literary judgements, perhaps one should say that he valued these poems—not the very good ones which we can agree about, but the less good which may now seem mediocre—in ways that we can no longer share. I am thinking, for example, of the inordinate space he gives to some of the lesser late Elizabethans and of the many pages devoted to a rather second-rate anthology, Davison's *Poetical Rhapsody*. In expressing his tastes in this way, Chambers was acquiescing in the tendency of his time to see the Elizabethans largely in terms of their 'dainty' pastorals and their pretty love songs. He was to some extent seeing the period through Palgravian spectacles, giving full attention to anything that qualified as 'lyrical' and allowing less weight to what did not so qualify. The justification for such an approach is of course that the age of Elizabeth was undeniably a great age of lyric, one of the greatest. To that extent Chambers's emphasis is understandable or even proper. The objection to such a way of looking is not just that it confirms a prettified and slightly trivializing image of the Elizabethan period but that it fails to find space for a whole range of writing that ought to be brought to the attention of any reader of such a collection as this. Chambers himself, as we know from some of his late essays, would have agreed with the view that much important verse was excluded from his book on the grounds of unsuitability. He was, after all, a very distinguished scholar who had familiarized himself with all the by-ways of the century; he was exceptionally well informed about its poetry. In any case—to be fair to him—he does not, in his *Oxford Book*, absolutely confine himself to poetry of a lyrical character: the capacious dimensions of his volume allowed him, perhaps compelled him, to admit a much wider range of writing, including some satirical and didactic verse. But the norm, so to speak, remains a lyrical one, and those kinds of poem which Palgrave explicitly rejected from his *Golden Treasury* are allowed by Chambers into *his* selection, if at all, only with what may be felt as a slight reluctance or, at most, a conscious broad-mindedness. Accordingly Chambers does indeed include Wyatt's first satire and two extracts from Daniel's long reflective poem *Musophilus*. But such pieces are uncharacteristic of the volume, and they are in any case poems warmed by personal feeling and coloured by the poet's personality, thereby approaching the Palgravian criteria; moreover Wyatt's satire is in *terza rima*, *Musophilus* in six-line stanzas, verse-forms which are, so to speak, potentially lyrical. Even these pieces are not, therefore, too far removed from the lyric norm.

The outer limits of Chambers's range as an anthologist, the extent of his editorial tolerance, can be viewed in the case of a major early Tudor poet, John Skelton. Chambers represents Skelton by two extracts from *Philip Sparrow* and three of the short lyrics addressed to high-born ladies from *The Garden of Laurel*. There is nothing wrong with the choice of these pieces; the short poems especially are fresh and delightful lyric utterances and show to advantage one side of Skelton's poetic talent. The point at issue is not what Chambers includes but what he leaves out. Much of Skelton's poetic effort went into satire of various kinds, political, moral, social. Some of his weightiest writing is in the form of the ten-syllable line. But none of his satirical or didactic writing is included by Chambers, all of whose items are in the short-lined Skeltonic. A Skelton without *The Bouge of Court* and *Speak Parrot* is a poet deprived of some of his most powerful notes. Nor is *Eleanor Rumming* allowed in, for obvious reasons—for, though one of Skelton's finest poems in the Skeltonic metre, it deals with low-born people engaged in somewhat sordid activities.

Chambers's late Victorian and Edwardian assumptions show themselves not just in a tacit partiality for the pretty and the genteel; they emerge also in an avoidance of the grotesque, the ugly, the difficult, the obscure, and—

to use a more general term—the historically circumstantiated. The preference for lyrical modes goes with a distaste for history. Chambers himself was a great literary historian; there can be no suggestion that he of all people was lacking in historical knowledge. But on the evidence of his *Oxford Book*, his preferences were for poems which were as far as possible unimplicated in historical circumstance—as if the finest poetry floated free of such contact and spoke to us with a transcendent directness across the centuries. Such a belief in the power of poetry to annihilate historical distance is not of course absurd; it may even have a measure of truth. But one consequence of holding such a belief is that poems which enthusiastically embrace the messy and perhaps obscure specificity of transient historical situations may be thought unsuitable for a general readership, despite the fact that such poems may very well be of high literary quality. Skelton's *Speak Parrot*, unsurprisingly rejected by Chambers, is one of the most difficult English poems of the sixteenth century. Yet to post-Modernist readers undaunted by *The Waste Land* and *Ulysses*, it should not, especially if accompanied by minimal annotation, prove too inaccessible. I have accordingly included a couple of extracts from it in this volume; it is, after all, one of Skelton's best poems.

On this, as on other matters, I have broken with Chambers's practice. Difficult poems, or obscure or troublesomely 'topical' or personal poems, like other poetic kinds which earlier anthologists might have considered unsuitable, qualify for a place here if their literary qualities justify it. And if difficult, etc. poetry is admitted, it follows that annotation must also be supplied, even at the risk of disfiguring an otherwise handsome page with the paraphernalia of scholarly aids and glosses. But the gains are considerable. The sixteenth century was a time of unparalleled linguistic innovation. Great numbers of words were taken into the language for the first time, some of which understandably failed to keep their place; such words need glossing. Many other words may have been in common use, inherited from earlier centuries, yet for some unknown reason they faded out: Tusser's word 'camping' for football is one instance; another is the term for a marsh-marigold in Barnfield's 'Affectionate Shepherd'—'boots'—which is not recorded in the *OED* but is explained in Gerard's *Herbal* of 1597. Something a reader of this volume will inescapably notice is the quite formidable—wonderful—expressive resources of the English language during this period, even though prose writings on one side and the verse drama on the other are not here represented. In order to increase appreciation of the richness of these linguistic resources and in order to demonstrate that poets were using words exactly, often more precisely and even delicately than they may seem to at a first reading, fuller glossing than in previous *Oxford Books* has been thought necessary. Even so, I have often been aware of the constraints of space; in many places even fuller glossing would have been desirable.

(ii)

No doubt all ages feel as if they were ages of transition, yet the sixteenth century in England was probably the most disorientating age of transition we have had, at least since the Norman Conquest. It saw England change from feudalism to early capitalism, from Catholicism to Protestantism, and from a fairly small country to an emergent nation-state soon to take its place in a select club of European great powers. In the course of the century London grew from a large town of about 60,000 to a metropolis of about 200,000. By the middle of the seventeenth century, London may well have been, after Constantinople, the biggest city in Europe. We often think of the sixteenth century as the first modern century; it might be just as well to think of it as the last medieval one, since 'medieval' and 'modern' elements, old established ways, and starkly innovative practices, fought it out throughout these hundred years. England was making itself anew. There is everywhere an exciting spirit of incipience, of beginning again, of building an entire culture on new foundations and of seeking to catch up with other nations which had started earlier in the race for pre-eminence. There is also at times a piercing sense of nostalgia, a passionate regret for well-remembered yet recently lost ways of life—and this is not confined to Catholics or recusants but is common to a wide range of people who felt at least a pang at the passing of what seemed in retrospect an irrecoverable 'merry England'.

In the realm of poetry, forms and styles change so much and so radically that there seems little point in trying to think of this century as a single literary period. Indeed the very notion of sixteenth-century verse as an entity, as something having unity, must be rather questionable and problematic. 'Sixteenth-century verse'—it has a confident ring, but the reality is much more untidy, unevenly distributed, ragged even, than the phrase might suggest. Taken as a whole, the sixteenth century in literature is a period with a hole in it—not just a hole either, rather a gaping crater. The century takes whatever cultural homogeneity it has not from literature, certainly not from poetry, but from politics. It is the century of the Tudors; and of these hundred years Henry VIII and Elizabeth I between them account for eighty. These two historical facts—the single dynasty and the long duration of the two chief Tudor reigns—are enough in themselves to create an impression of a cohesive temporal sequence which also happens to be the sixteenth century. But of course to think of this century as in some sense closely knit is not to deny those momentous events which disrupted it and which permanently changed the fabric of English life. The most far-reaching of these was unquestionably the Reformation. However, the crucial years which saw the break with Rome—the 1530s—fell well within the reign of Henry VIII; the reign itself was not interrupted. This was quite unlike the Cromwellian regime of the following century, which violently discontinued the rule of the Stuarts. The Tudor years which came closest to that Interregnum were the five-year reign of the Catholic

Mary. But the changes in religious practice which she authorized were in part mitigated by the dynastic continuity which she also embodied. Queen Mary was at least Henry's daughter. There was no interregnum, no break in the unfolding of the Tudor dynasty. It is in this sense—the continuity of its chief political institutions (the monarchy, the Tudor court administration as established by Henry VII, Parliament)—that we can speak of the relative cohesiveness of the sixteenth century.

To turn from political history to the history of poetry is to move from a measure of real continuity to something that is almost the reverse. At the beginning of the century, Skelton looks back to Chaucer as if to a recently deceased master; at the end of it, poets are writing social satires in heroic couplets which anticipate Dryden and Pope. In between, poetry develops—if such a term can be used—in fits and starts; sometimes it disappears, or seems to disappear, altogether. Viewed as a whole, what we witness in the course of the sixteenth century is the recession, the cessation indeed, of one great literary system or order (the medieval) and the initiation of a new one (Renaissance or 'early modern').

Nothing about this long process of change-over is tidy. The medieval literary system persisted well into the sixteenth century, alongside what we now think of as the 'Renaissance' experiments of the Petrarchan sonnet-poets Wyatt and Surrey. (To complicate the picture, the fourteenth-century Chaucer seems to us more and more to qualify as a 'Renaissance' poet on the strength of the multi-vocal 'dialogical' *Troilus and Criseyde* and the *Canterbury Tales*.) At some time in the 1530s, probably, an unknown 'Cambridge clerk' (as he calls himself) wrote a poem of over 1,400 lines now entitled 'The Court of Love'. The author was no doubt prompted to write it by Thynne's edition of Chaucer (1532); but for a long time afterwards the poem was included in editions of Chaucer and believed to be by Chaucer himself. 'The Court of Love' is undoubtedly a Tudor poem, post-Skelton and contemporary with Wyatt and Surrey, yet in everything except the linguistic accidentals which only philologists are equipped to notice it is also a 'medieval' poem written by a poet who may not even have known that he was writing in an 'obsolete' style—for him, it was not obsolete. For another instance of an ambitious medieval-type poem being written in Tudor England, we need to go back a little earlier. At some time after the battle of Flodden (1513), an unknown poet writing in Cheshire, probably a member of an old-established gentry household, wrote a poem in celebration of the English victory. He cast it in the alliterative metre once associated with his own part of the country; its greatest medieval achievement had been the fourteenth-century *Sir Gawayn and the Green Knight*. *Scottish Field*, as it is now called, is in fact the last surviving poem to use alliteration as a metrical system in the old way. Its outlook is uncompromisingly feudal: the poet writes, or sings, his epic story in the role of an ancient family's minstrel or bard. Yet his poem is undeniably a Henrician work, and is as much a product of the century of the Tudors as anything by Sidney or Spenser.

If certain things in the early part of the century seem to be more at home in the 'medieval' fifteenth, the century's last decade was in a state of such furious upheaval that literary historians have often tried to deny that some of its products belong to the sixteenth century at all. This scholarly reluctance to admit that the late Elizabethan period was, in literary terms, not entirely given over to pastorals, romances and sonnet-sequences but was in its way as 'dark' as the Jacobean decades that followed may no doubt be explained by invoking deep-seated, possibly unconscious, political motives: it seemed important to such scholars to maintain the fiction of an innocent, brightly tinted, upstanding, wholesome Elizabethan England. But the 1590s were an Age of Anxiety, the decade of Fulke Greville's fathomlessly gloomy Calvinist lyrics as well as Marston's queasily satiric diatribes. This is also the decade in which Donne made his unmistakable mark. Yet, although a fair number of Donne's poems can be dated to the 1590s, some critics persist even now in regarding him as a 'seventeenth-century poet' (i.e. someone who fits into their idea of the seventeenth century) rather than siting him where as a secular poet he historically belongs, firmly straddling the last Elizabethan decade and the first Jacobean one. Chambers in his *Oxford Book* illustrated the usual trend: 'Only for chronology, indeed, can Donne be an Elizabethan.' So Donne is not admitted to his *Oxford Book*. But Donne can be denied to be an 'Elizabethan' only if history—and not just 'chronology'—is set aside, and only if qualifications for 'Elizabethan' status are deemed to be identical with those 'golden' stylistic virtues which C. S. Lewis found supremely exemplified in Sidney and Spenser. The plain historical fact is that the verse satire of the 1590s which Donne pioneered and of which he was by far the most distinguished exponent was exclusively an Elizabethan phenomenon. Verse satire—together with the collections of epigrams which were associated with it—was banned by the Archbishop of Canterbury and the Bishop of London in 1599. The whole fashion for formal verse satire was over before the end of the century. It follows that a volume such as this *New Oxford Book* cannot omit Donne, since Donne's was one of the two or three leading poetic voices of the 1590s. I have therefore included a number of those poems which either can be dated to the 90s or cannot be shown to be later than about 1600. Since we do not know the dates of the *Songs and Sonnets*, this is a contentious area; but it is overwhelmingly likely, as all Donne experts agree, that some at least of these love poems were written during Donne's early or middle twenties.

Rather than using the term 'Elizabethan' to designate some special set of qualities, it seems better simply to accept as Elizabethan whatever was written during the reign of Elizabeth; only in this way will we avoid a perhaps unconsciously tendentious interpretation of the age. In the context of literary history, the late Elizabethan period has long been given—or so it can be argued—a rather distorted reading. Too much attention has focused on the Court, while not enough has been given to that rapidly growing part of fashionable London which in the following century became known as 'the Town'. It is this newly developing urban scene which distinguishes the 1590s and is carried over into the seventeenth century. This is where Donne belongs—the Donne whom Barbara Everett has described as a 'London poet'—together with the other verse satirists, Hall, Marston and the rest. But there are other poets too who, while not using the specific genre of verse-satire, none the less share something of the satirists' new urban outlook and in some cases aspirations to urbanity. Notable names here are Sir John Harington and Sir John Davies, both of whom have claims to be thought of as social poets of a new and forward-looking kind. These poets have been inadequately noticed by critics and perhaps undervalued. I have given them, in their role as poet-observers of society, a slightly fuller representation than they usually receive.

I have been saying that the sixteenth century in poetry does not have much unity; it certainly does not constitute a single literary period. (But then, why should it? A century is a fairly arbitrary way of chopping up the seamless sequence of history.) Its literary beginnings are indistinguishable from a lot of fifteenth-century work, while the 1590s are rich in work of a kind we associate with much later seventeenth-century developments. In between these two extremes comes that period of cultural striving which we think of as characteristically sixteenth century and which we traditionally identify as the English Renaissance. But these middle years of effortful imitative emulation come no closer to forming a coherent literary period than the century itself. They separate, rather, into two phases, or even three. For the process of bringing about a cultural rebirth, a new poetic order, needed to be attempted again and again. Much was accomplished, in their utterly distinct ways, by Wyatt and Surrey (and the full fineness of Wyatt's achievement is perhaps only recently being recognized after a long period of neglect). Their work belongs to the 30s and 40s, but before the end of the 40s both were prematurely dead, and those very minor poets associated with them were not strong enough to carry further the renewing impulse. Only after a twenty-or-more year gap does a successor not unworthy of Wyatt appear in Gascoigne, who in turn hands on to Ralegh some remnant of Wyatt's special gravity and seriousness.

But the decisive moment came, as contemporaries clearly recognized, with Sidney and Spenser in the late 70s and early 80s. What they did was at once to raise literary standards and establish new models both through the force of their own example and by putting English writers in touch with the most recent developments in Italy and France. Earlier literary historians sometimes chose to play it down, but the truth is that the Elizabethan literary Renaissance owed an immense inspirational debt to the group of French poets known as the Pléiade. Ronsard, Du Bellay and the rest had revitalized French poetry by deliberately fomenting a literary revolution, complete with carefully thought-out programmes and manifestos, not to speak of their own considerable talents. They revived old genres, structures, stanza forms, and metres. They enormously enlarged the vocabulary of poetry and enriched its diction. And they rejuvenated and reinvigorated the poetic psyche by rediscovering the natural world. One major difference between early Tudor and late Elizabethan poetry is that Wyatt and Surrey were pre-Pléiade poets, whereas Sidney and Spenser were post-Pléiade in that they took full advantage of the French example, particularly in its rediscovery of Nature. Early Tudor poetry is relatively poor in natural imagery: Wyatt, for instance, inherits from the late medieval French love lyric its narrow, abstract, and stereotyped diction and, though his achievement is nothing less than very impressive, he can be felt to be working within certain linguistic constraints and disadvantages. The post-Pléiade Elizabethan poets, on the other hand, make themselves free of a freshly apprehended world of living things—'every thing that grows', as Shakespeare put it in one of his sonnets. This is a world of transient natural phenomena: sunlight, moonlight, starlight, seasons, trees and flowers, the movement of air and water: 'With how sad steps, O moon, thou climb'st the skies . . .', 'Look, Delia, how we steem the half-blown rose . . .', 'Full many a glorious morning have I seen . . .', 'Sweet Thames, run softly till I end my song . . .' If in the earlier part of the century the presiding poetic deity was Fortune, in its closing years Time and Mutability take over as dominant presences, each of them given powerful imaginative treatment by Spenser and Shakespeare, the leading poets of the age. This move from Fortune to Time shows the main direction of the century's poetry: the poet now wants to write of his own experience, his own subjective sense, of the changing world of Nature. And in this respect, though they treat their subjects with totally different means and effects, Spenser and Shakespeare—and Donne—are recognizably alike.

(iii)

One of the interests of reading sixteenth-century poetry is in observing a modern corpus of texts, a national literature, in the process of being created. Some of the verse-writing of the mid-century has a quality of clumsy primitiveness as if we were in fact at the beginning, the opening stages, of a civilization: poems are unambitious, overemphatic, heavily moralistic, deadeningly regular in metre. Yet in a few decades we have passed even the achievements of Sidney and Spenser and reached Shakespeare, Donne, and the earlier poems of Jonson. A literary historian can make it seem as if the story is one of uninterrupted steep ascent, progress all the way. And it is true that, in terms of the acquisition of necessary techniques, small and often minute adjustments affecting diction, syntax, rhythm, and style in general, the story is indeed one of rapid improvement, even though perhaps no poet included in this volume was to acquire the flawless all-round technical mastery shown by Herbert and Milton in the following century. None the less such a smoothed-out progressivist narrative would give a misleading, because partial and distorted, impression of the period. For although poetry is in the process of becoming professionalized, it is not yet in the hands of the professionals.

One of the distinguishing features of the sixteenth century is the large number of people—people, not poets—who write poems without otherwise having any serious literary ambitions. Quite a few of these poem-writing persons seem to be remembered for only one poem—like Chidiock Tichborne, the young Catholic gentleman who was executed for his part in the Babington Plot and who is said to have written just before his death 'My prime of youth is but a frost of cares'. Another one-poem man is Lord Burghley, the Queen's chief counsellor, and certainly no lover of poetry if tradition is to be believed. The one poem of his that has survived, heavy, prosaic, and suitably old-fashioned in style, was written to accompany a present to his daughter. The poem—which I have not included—ends: 'To set you on work, some thrift to feel, / I send you now a spinning wheel.' It has no pretensions to be anything other than a private and domestic social gesture. For many writers, poetry, or rather verse, was not separated by any self-conscious literary theories from real life; like a good many other people of the time, Burghley had learned how to write verse and could turn out something that looked like a poem when required.

A reader of this *New Oxford Book* will often be brought up against abrupt discontinuities of style and literary sophistication. But this diversity, however momentarily startling, may be one of the uses a large anthology can have; it puts one on one's guard against the blandnesses of a certain kind of literary history, some of which is not very different from historical fiction. Few accounts of Elizabethan poetry have much to say, for instance, about one of the bestsellers of the time, Thomas Tusser's *One Hundred Points of Good Husbandry* (in later enlarged editions *Five Hundred Points* . . .). Its pleasantly detailed, and no doubt useful, instructions on how to be a good farmer, or a good housewife, are cast into amphibrachs whose vigorous galloping or jouncing is unflaggingly maintained throughout the entire volume. Far from being unacceptably incongruous, however, the effect of obtrusive metre and utilitarian, not to say prosaically down-to-earth, substance is oddly satisfying. Of course there is little point in reading Tusser for anything other than his subject-matter, which is in any case of some interest; even so, the literary-minded reader will notice that Tusser uses words exactly and that he rescues from oblivion a number of rare terms; while reading him, moreover, one may be alerted not only to possible uses of verse but to certain expressive resources of the English language which are now neglected. Another compilation, very popular at the time, also usually thought now to be below the level of true 'literature', is Warner's *Albion's England*. Like Tusser, Warner too can still entertain, and surprise, if one is prepared to take him at his own jogtrot tempo. His enormous poem is a hotchpotch of 'British' and English history, romance episodes, folklore, travel lore, and much else, and despite its often homely style did not appeal only to 'burgher'-type readers. The Cambridge academic Gabriel Harvey noted that the Earl of Essex 'much commends *Albion's England*'.

Too often anthologies of sixteenth-century poetry give the impression that poets wrote about little else than love—and an especially exalted, strainedly idealistic, literary kind of love at that. But as Dr Johnson remarked on another occasion, 'love is only one of many passions': much of life as lived by anyone has nothing to do with love. And sixteenth-century writers knew this as well as he did. Accordingly I have tried to show authors of poems responding to a wide range of other activities—in public and private life, in country and in town, at home and abroad, and on the part of woman as well as man, child as well as adult. One aim has been to evoke, however faintly, a sense of the resistant, unassimilable disorderliness of the period's actual life, as opposed to what usually gets into the historian's tidied narrative. This has meant trying to find poems which give voice to outlooks different from those of the governing or dominant classes, though inevitably *their* voices remain dominant. A search for poems written by women did not get very far: only four women could be found who qualified for inclusion—and one of those was Queen Elizabeth. Poems by recusants have been given, I hope, fair representation. This is no more than justice, since England was officially a Catholic country for a full forty years of the sixteenth century, and even after Elizabeth's accession there were many avowed recusants. (How many unacknowledged Catholics persisted in their beliefs, historians are still undecided.) As for those who refused belief in any form of religion, we are sometimes told that such total scepticism was beyond the mental 'set' to be expected of sixteenth-century people. As a way of questioning that view, I have included the remarkable broadside ballad by Thomas Gilbart on the execution of John Lewes. The two words spoken by that courageous heretic, though presented by the author with predictable disapproval, strike a refreshing and heartening note—for this is otherwise a period in which the repeated expression of orthodox religious beliefs can become somewhat oppressive. Still, some of the deepest, most searching poetry of the time is religious in subject and is here given ample representation. Fulke Greville's powerful religious lyrics, for instance, which are usually held over for the seventeenth century (in keeping with the prejudice already mentioned in connection with Donne) are here claimed for the 1590s.

Good poems written by ordinary commoners, who were of course the vast majority of the English nation, are not easy to come by, though there are a few. Most of the notable writers who were not already members of the gentry were to some extent absorbed into it, as were (in theory) the University Wits (Lyly, Peele, Greene, even Marlowe the son of a shoemaker), who could claim gentility on the strength of their university degree. The lowest ranks of Tudor society leave no first-hand poetic account of themselves—nor, of course, would anyone expect them to. However, the destitute, the true down-and-outs, receive some attention in Copland's neglected poem *The High Way to the Spital*. The early and mid-Tudor period was one in which Langland's great poem, *Piers Plowman*, was held in high esteem and printed for the first time. It contains some unforgettable accounts of the sufferings of the poor. Copland's poem is usually grouped with others in a *Ship of Fools* tradition only recently inaugurated; in his indignation and compassion, however, Copland can be placed just as plausibly in a native line descending from *Piers Plowman*.

A prejudice which persists from the nineteenth century is the view that verse translations are by definition an inferior form of writing. Whatever merits this notion might have had in the context of Victorian poetry, it makes no sense at all for the sixteenth century. Much of what we think of as 'original' poetry is in fact some form of translation, whether close paraphrase or free imitation. This applies not only to many of the sonnets of Wyatt, Surrey, Ralegh, Daniel, Drayton, and others but to dozens of songs, epigrams, short moral poems, and even longer narrative works. To take one example: Bryskett's elaborate pastoral elegy on the death of Sir Philip Sidney, which furnished Milton with an important model for *Lycidas*, was itself closely based on an eclogue by Bernardo Tasso yet was published with no suggestion that it derived from another poem. But if many seemingly original poems are in fact—or should we say, also—translations, the converse is equally true. Many translations are also original poems. And this is the justification for including so many translations in the present volume. No account of sixteenth-century poetry could leave out Wyatt's translations (not only of Petrarch but of Horace, Seneca, and the Penitential Psalms), Surrey's Virgil, Golding's *Metamorphoses*, Marlowe's *Amores* and Lucan, Harington's Ariosto, Chapman's Homer, and Fairfax's Tasso. To list these works is not to imply that they are all of the same poetic quality. But all of them are important in their own right as English poems.

In the case of the narrative poems just mentioned, Golding's Ovid and Harington's Ariosto especially, I have allowed them plenty of space, since in order to appreciate what these English poets are capable of, we need to assess them over a fair distance. There is no avoiding the fact that many sixteenth-century poets take time to make their effects; they are unhurried in their movement; they seem to live according to a rustic rather than an urban clock. Thomas Churchyard, for instance, is a poet of rather lowly literary status. During a long writing career (over 50 years) he churned out a mass of mediocre near-doggerel verses. I have included his little-known *Friar's Tale* (I think it may be his best poem) and despite its length have given it complete, since nothing less than the whole would do it justice. One needs to see that within its leisurely amplitude there is in fact a spareness and precision of statement which save the poem, giving it at times a kind of almost-witty terseness. Humble though it is, this Tudor fabliau takes its place in a Chaucerian, or rather sub-Chaucerian, tradition, and like Copland's *High Way* makes another Tudor poem which looks back to an English master of the fourteenth century. As a boy Churchyard had served as a page to the Earl of Surrey; but there could scarcely be a more extreme contrast between the two poets. Surrey's economy, elegance, and taste established a touchstone for English poets for several decades, until he was eclipsed, or at least dimmed, by Sidney. In his time, Surrey's concision must have seemed remarkable; his Virgilian translations can be best savoured if one has been put off by the common verbosity and otiosity of late medieval writers (Lydgate, Hawes). Comparably, much later, Marlowe's line-for-line heroic-couplet version of Ovid's *Amores* must be measured against the prevailing Elizabethan preference for a well-rounded redundancy of expression. But this is not to argue that a tight compression is the only road to poetic pre-eminence. On the contrary, it must be accepted that some good poets of the period are expansive and even diffuse; if their qualities are to be displayed, these poets must be given more space than concise writers. It does not follow, therefore, as quantity-surveying readers might be tempted to infer, that a poet given twenty pages is twice as good as another with only ten. I have had more regard to the poem than to the poet, as Christopher Ricks puts it in his *New Oxford Book of Victorian Verse*. It would seem a good anthological principle that each poem be taken on its own merits and that nice questions concerning the comparative status of poets be left in abeyance.

Some of the items in this volume are long because they are complete poems or at least complete narrative episodes, and in each case the poet must be allowed to finish his story. Others are long because, though working in a non-narrative genre, the poet equally needs time and scope to complete his argument or present his case. One instance is Sir Thomas More's poem on Fortune, here given complete; another is Daniel's *Musophilus*, represented by several extracts. In each of these, the poet seems to become more weighty, more discriminating, in general more impressive (or so I find), the more extensively one reads him. Brief snippets will not convey what is special about him in this particular genre. The poem from Drayton's *Shepherd's Garland* is given complete, because only thus can its form—its cunning assemblage of shapes—be recognized. What we have is one story inserted into another in a manner proper to pastoral, with its characteristic arrangement of recessed planes or enclosed places and its artful symmetries (120 lines for each of the two main sections). The old *Oxford Book* included only part—the inset part—of this poem; but in doing so, it conveyed a misleading sense of what Drayton was doing, for the effect of one episode contained within, but also countering, another is essential.

Given the peculiar 'shape', or distribution, of sixteenth-century poetry as a whole, the real difficulty for an anthologist is to avoid over-representing the 1590s. Such a volume as this could easily turn into an Elizabethan rather than a truly sixteenth-century selection. If the early and middle parts of the century are somewhat under-populated, the century's end suffers, if anything, from a sense of congestion and overcrowding. There is almost too much that competes for our attention. I have tried, up to a point, to resist the internal pressure set up by the texts themselves by drawing on a wider than usual range of work from the early years and from the 1560s and 1570s. In any case these pre-Elizabethan and early Elizabethan poems have qualities that fully justify their inclusion. On the other hand, it would be absurd to seek to diminish, or flatten, or in any way 'contain', the extraordinary explosion of poetic talent which occurred during the century's closing years. The 1590s have a claim to be considered the most remarkable decade in English literary history. This is not simply because they see the arrival of Shakespeare—though that might be thought distinction enough. Shakespeare, however, is only one of many new voices. To put it at its soberest, and avoiding merely inflationary rhetoric: during the 1590s well over thirty poets of at least some talent were known to be writing—and this does not include those anonymous poets who wrote poems of substantial merit. Two or three of these named poets were writers of genius: Spenser, Donne, and—though his writing career had only just started near the end of the decade—Ben Jonson. There are besides Greville, Ralegh, Marlowe, Daniel, Drayton, Chapman, and (not least) Campion—poets who at their best make a unique and irreplaceable contribution to writing in the English language. And to them must be added other smaller, widely diverse, talents who are still, it seems, very little known. So little known indeed are some of them that this culminating Elizabethan decade, despite its conventionally acknowledged achievements, might in reality be called one of the least explored regions of English poetry.

EDITORIAL PRINCIPLES

Inclusion. In a small number of cases I have overrun the strict limits of the sixteenth century. I have included, for example, some of Skelton's earliest poems, which belong to the very late fifteenth century, as well as 'The Nutbrown Maid', published 1503 (?) but probably written before 1500—the first in order to give a fuller sense of Skelton's development and range, the second because it has not been included in recent large anthologies and it seems a pity that it should not be made accessible to the sorts of readers for whom these anthologies are meant. At the other end of the century, I have tried to stop at around 1600, but in a couple of instances (Shakespeare, Ralegh), I have included later examples of their work.

Texts. Whenever possible, I have used recent scholarly editions. I have modernized texts, normalizing spelling and punctuation, though often allowing the poet to keep his or her distinctive spelling-variant when it seemed to contribute to meaning. I have tried to save as much as possible of Skelton's linguistic personality; I have followed editorial tradition in not modernizing *The Faerie Queene*; and I have allowed a stylistic eccentric like Stanyhurst his outlandish spellings. I have made use of modern editors' explanatory notes where these were available; otherwise, the glosses are my own. The editors concerned are acknowledged in the Notes and References.

Procedure. Poets are arranged in an approximately chronological order, according to the date of composition, when known, of the earliest poem included. The date of first printing is supplied after each poem; but in those cases where the date of composition is known to have been markedly earlier than the date of publication this other date is also supplied. Many sixteenth-century poems, particularly those written by court poets, circulated in manuscript for decades, or even centuries, before being printed. A glance at the dates of publication will show, for example, how slowly the full poetic achievement of Wyatt was made available to the ordinary book-buying reader. Some of what now seem Wyatt's best-known poems (e.g. 'Whoso list to hunt') were not published until 1815; while the Blage manuscript in Trinity College, Dublin, which contains unique versions of a number of fine poems by Wyatt, was not printed until 1961. Many good poems of the period are anonymous; in these cases, dating may be very difficult and must often remain conjectural. Some poems, including one or two famous ones, have traditionally been attributed to persons who, according to recent scholarship, cannot have been their authors. 'The Passionate Man's Pilgrimage', usually given to Ralegh, is one such misattributed poem; the quatrain beginning 'Christ was the Word that spake it', traditionally given to Queen Elizabeth, is another. On these matters I have tried to represent the most recent scholarly consensus, but in some cases I have no doubt failed to keep track of recent scholarly moves. Titles of poems in square brackets are the editor's; titles comprising the first lines of poems or their opening words are also editorial. Other titles are those given in the earliest texts, whether in manuscript or in printed editions.

ACKNOWLEDGEMENTS

I am grateful to Dr L. G. Black of Oriel College, Oxford, for allowing me to make use of his unpublished Oxford D.Phil. thesis on Elizabethan manuscript anthologies. The Librarian of Oscott College, Birmingham, kindly sent me a photocopy of a manuscript recusant poem. Professor E. G. Stanley supplied me with useful information about 'The Court of Love'. Otherwise I am strongly aware of my debts to earlier editors, scholars, critics, and anthologists who have not only made texts available but in many cases have drawn attention to neglected or misunderstood or unknown poets and poems.

JOHN SKELTON
*c.*1460–1529

from *The Garland of Laurel*

1 *To Mistress Isabel Pennell*

By saint Mary, my lady,
Your mammy and your daddy
Brought forth a goodly baby!

My maiden Isabel,
Reflaring rosabel,
The flagrant camomel;

The ruddy rosary,
The sovereign rosemary,
The pretty strawberry;

The columbine, the nept, 10
The jelofer well set,
The proper violet;

Ennewed your colour
Is like the daisy flower
After the April shower.

Star of the morning gray,
The blossom on the spray,
The freshest flower of May,

Maidenly demure,
Of womanhood the lure; 20
Wherefore I make you sure,

It were an heavenly health,
It were an endless wealth,
A life for God himself,

5 Reflaring] odorous
rosabel] beautiful rose
6 camomel] camomile, a creeping herb
10 nept] mint
11 jelofer] gillyflower
12 proper] pretty
13 Ennewed] made fresh

To hear this nightingale
Among the birdes smale,
Warbling in the vale

'Dug, dug,
Jug, jug!
Good year and good luck!' 30
With 'Chuck, chuck, chuck, chuck!'

2 *To Mistress Margaret Hussey*

MERRY Margaret
As midsummer flower,
Gentle as falcon
Or hawk of the tower;

With solace and gladness,
Much mirth and no madness,
All good and no badness;
So joyously,
So maidenly,
So womanly 10
Her demeaning
In every thing
Far far passing
That I can endite
Or suffice to write
Of merry Margaret
As midsummer flower,
Gentle as falcon
Or hawk of the tower.

As patient and as still, 20
And as full of good will
As fair Isyphill;
Coliander,
Sweet pomander,
Good Cassander;
Steadfast of thought,
Well made, well wrought.
Far may be sought

2
3 Gentle] of high breeding
4 hawk of the tower] hawk which towers high in the air
22 Isyphill] Hypsipyle, daughter of Thoas, King of Lemnos
23 Coliander] coriander
24 pomander] aromatic ball

Erst that ye can find 30
So courteous, so kind
As merry Margaret,
This midsummer flower,
Gentle as falcon
Or hawk of the tower.

(Wr. before 1495; pub. 1523)

3 [*My darling dear, my daisy flower*]

WITH lullay, lullay, like a child,
Thou sleepest too long, thou art beguiled.

'My darling dear, my daisy flower,
Let me,' quod he, 'lie in your lap.'
'Lie still,' quod she, 'my paramour,
Lie still hardely, and take a nap.'
His head was heavy, such was his hap,
All drowsy dreaming, drowned in sleep,
That of his love he took no keep,
With hey lullay, lullay, like a child, 10
Thou sleepest too long, thou art beguiled.

With 'Ba, ba, ba' and 'bas, bas, bas'
She cherished him both cheek and chin,
That he wist never where he was:
He had forgotten all deadly sin.
He wanted wit her love to win:
He trusted her payment and lost all his prey;
She left him sleeping and stale away.
With hey lullay, lullay, like a child,
Thou sleepest too long, thou art beguiled. 20

The rivers rough, the waters wan,
She spared not to wet her feet;
She waded over, she found a man
That halsed her heartily and kissed her sweet.
Thus after her cold she caught a heat.

3
6 hardely] confidently
9 keep] care
17 prey] booty, loot
24 halsed] embraced

'My lefe,' she said, 'routeth in his bed;
Ywis he hath an heavy head,
With hey lullay, lullay, like a child,
Thou sleepest too long, thou art beguiled.

What dreamest thou, drunkard, drowsy pate? 30
Thy lust and liking is from thee gone;
Thou blinkard blowbowl, thou wakest too late,
Behold thou liest, luggard, alone!
Well may thou sigh, well may thou groan,
To deal with her so cowardly:
Ywis, pole hatchet, she bleared thine eye.

(Wr. 1495–1500; pub. *c.*1527)

from *The Bouge of Court*

DREDE

4

THE sail is up, Fortune ruleth our helm,
We want no wind to pass now over all;
Favour we have tougher than any elm,
That will abide and never from us fall.
But under honey oft time lieth bitter gall:
For, as methought, in our ship I did see
Full subtil persons, in number four and three.

The first was Favell, full of flattery,
With fables false that well could feign a tale;
The second was Suspect, which that daily 10
Misdeemed each man, with face deadly and pale;
And Harvy Hafter, that well could pick a male,
With other four of their affinity,
Disdain, Riot, Dissimuler, Subtilty.

26 lefe] love
routeth] snores
27 Ywis] indeed
30 What] why
32 blowbowl] drunkard
36 pole hatchet] soldier who uses a pole-axe
bleared thine eye] deceived you

4
Bouge of Court] i.e. 'bouche de cour', the free rations provided in the royal household
8 Favell] Duplicity, Flattery
9 fables] lies
12 Hafter] Trickster
male] purse, bag

Fortune their friend, with whom oft she did dance;
They could not fail, they thought, they were so sure;
And oftentimes I would myself advance
With them to make solace and pleasure.
But my disport they could not well endure:
They said they hated for to deal with Drede. 20
Then Favell gan with fair speech me to feed.

FAVELL

'No thing earthly that I wonder so sore
As of your conning, that is so excellent;
Deyntee to have with us such one in store,
So virtuously that hath his dayes spent;
Fortune to you gifts of grace hath lent:
Lo, what it is a man to have conning!
All earthly treasure it is surmounting.

'Ye be an apt man, as any can be found,
To dwell with us, and serve my lady's grace; 30
Ye be to her, yea, worth a thousand pound!
I heard her speak of you within short space,
When there were divers that sore did you menace;
And, though I say it, I was myself your friend,
For here be divers to you that be unkind.

'But this one thing—ye may be sure of me;
For, by that Lord that bought dear all mankind,
I cannot flatter, I must be plain to thee!
An ye need ought, man, shew to me your mind,
For ye have me whom faithful ye shall find; 40
Whiles I have ought, by God, thou shalt not lack,
And if need be, a bold word I dare crack!

'Nay, nay, be sure, whiles I am on your side
Ye may not fall, trust me, ye may not fail.
Ye stand in favour, and Fortune is your guide,
And, as she will, so shall our great ship sail:
These lewd cockwats shall nevermore prevail
Against you hardely, therefore be not afraid.
Farewell till soon. But no word that I said!'

23 conning] learning
24 Deyntee] pleasure
in store] so much, in abundance
47 lewd] ignorant
cockwats] rascals
48 hardely] boldly, certainly

DREDE

Then thanked I him for his great gentleness. 50
But, as methought, he ware on him a cloak
That lined was with doubtful doubleness;
Methought, of words that he had full a poke;
His stomach stuffed oft times did reboke.
Suspicion, methought, met him at a braid,
And I drew near to hark what they two said.

'In faith,' quod Suspect, 'spake Drede no word of me?'
'Why? what then? wilt thou let men to speak?
He saith he can not well accord with thee.'
'Twyste,' quod Suspect, 'go play! him I ne reke!' 60
'By Christ,' quod Favell, 'Drede is sullen freke.
What, let us hold him up, man, for a while.'
'Yea so,' quod Suspect, 'he may us both beguile.'

And when he came walking soberly,
With hum and ha, and with a crooked look,
Methought his head was full of jealousy,
His eyen rolling, his handes fast they quoke;
And to meward the straight way he took.
'God speed, brother!' to me quod he than,
And thus to talke with me he began. 70

SUSPICION

'Ye remember the gentleman right now
That communed with you, methought, a pretty space?
Beware of him, for, I make God avow,
He will beguile you and speak fair to your face.
Ye never dwelt in such another place,
For here is none that dare well other trust—
But I would tell you a thing, an I durst!

'Spake he, i'faith, no word to you of me?
I wot, an he did, ye would me tell.
I have a favour to you, whereof it be 80
That I must shew you much of my counsel.
But I wonder what the devil of hell

53 poke] bag
54 reboke] belch
55 at a braid] suddenly
58 let] prevent
60 Twyste] (exclamation of contempt)
reke] care about
61 freke] man
62 hold him up] leave him alone
66 jealousy] suspicion
67 quoke] quaked, trembled
69 quod] said
72 communed] talked

He said of me, when he with you did talk!
By mine advice use not with him to walk.

'The sovereign'st thing that any man may have
  Is little to say, and much to hear and see;
For, but I trusted you, so God me save,
  I woulde nothing so plaine be:
  To you only, methink, I durst shrive me,
For now am I plenarly disposed 90
To show you things that may not be disclosed.'

DREDE

Then I assured him my fidelity
  His counsel secret never to discure,
If he could find in heart to truste me;
  Else I prayed him, with all my busy cure,
  To keep it himself, for then he might be sure
That no man earthly could him bewray,
Whiles of his mind it were locked with the key.

'By God,' quod he, 'this and thus it is.'
  And of his mind he showed me all and some. 100
'Farewell,' quod he, 'we will talk more of this.'
  So he departed. Where he would be come,
  I dare not speak, I promised to be dumb.
But, as I stood musing in my mind,
Harvy Hafter came leaping, light as lynde.

Upon his breast he bare a versing-box;
  His throat was clear, and lustily could fain.
Methought his gown was all furred with fox,
  And ever he sang, 'Sith I am nothing plain.'
  To keep him from picking it was a great pain. 110
He gazed on me with his goatish beard;
When I looked on him, my purse was half afeard.

90 plenarly] fully, completely
93 discure] reveal
95 busy cure] diligence
97 bewray] expose by betraying his secrets
105 lynde] lime tree, i.e. nimbly
106 versing-box] dicing-box
107 fain] sing
110 picking] stealing, picking my pocket

HARVY HAFTER

'Sir, God you save! why look ye so sad?
What thing is that I may do for you?
A wonder thing that ye wax not mad!
For, an I study should as ye do now,
My wit would waste, I make God avow!
Tell me your mind: methink ye make a verse;
I could it scan, an ye would it rehearse!

'But to the point shortly to proceed, 120
Where hath your dwelling been ere ye came here?
For, as I trow, I have seen you indeed
Ere this, when that ye made me royal cheer.
Hold up the helm, look up, and let God steer:
I would be merry, what wind that ever blow!
"Heave and ho rumbelow, row the boat, Norman, row!"

"Princes of Youth" can ye sing by rote?
Or "Shall I sail with you" a fellowship assay?
For on the book I cannot sing a note.
Would to God, it would please you some day 130
A ballad book before me for to lay,
And learn me to sing "re mi fa sol"!
And, when I fail, bob me on the noll.

'Lo, what is to you a pleasure great
To have that conning and ways that ye have!
By Goddes soul, I wonder how ye gate
So great pleasure, or who to you it gave.
Sir, pardon me, I am an homely knave,
To be with you thus pert and thus bold:
But ye be welcome to our household. 140

'And, I dare say, there is no man herein
But would be glad of your company.
I wist never man that so soon could win
The favour that ye have with my lady.
I pray to God that it may never die.
It is your fortune for to have that grace—
As I be saved, it is a wonder case!

126 'Heave . . . row!'] (a boatmen's song)
128 a fellowship assay] join in a part-song
133 bob me on the noll] hit me on the head

'For, as for me, I served here many a day
And yet unneth I can have my living:
But, I require you, no worde that I say! 150
For, an I know any earthly thing
That is again you, ye shall have weeting.
And ye be welcome, sir, so God me save!
I hope hereafter a friend of you to have.'

(Wr. 1498; pub. 1499)

5 from *Philip Sparrow*

Pla ce bo!
Who is there, who?
Di le xi!
Dame Margery.
Fa, re, my, my.
Wherefore and why, why?
For the soul of Philip Sparrow
That was late slain at Carrow,
Among the Nunnes Black.
For that sweet soules sake, 10
And for all sparrows' souls
Set in our bead-rolls,
Pater noster qui,
With an *Ave Mari*,
And with the corner of a Creed,
The more shall be your meed.

When I remember again
How my Philip was slain,
Never half the pain
Was between you twain, 20
Pyramus and Thisbe,
As then befell to me.
I wept and I wailed,
The teares down hailed,
But nothing it availed
To call Philip again,
Whom Gib, our cat, hath slain.

149 unneth] scarcely
150 require] ask, beseech
152 ye shall have weeting] you shall know about it

5
1 *Pla ce bo*] beginning of the Office for the Dead
8 Carrow] Carrow Abbey, near Norwich

Gib, I say, our cat
Worrowed her on that
Which I loved best. 30
It cannot be exprest
My sorrowful heaviness,
But all without redress!
For within that stound,
Half slumbering, in a sound
I fell downe to the ground.

Unneth I cast mine eyes
Toward the cloudy skies.
But when I did behold
My sparrow dead and cold, 40
No creature but that wold
Have rued upon me,
To behold and see
What heaviness did me pang:
Wherewith my hands I wrang,
That my sinews cracked,
As though I had been racked,
So pained and so strained
That no life wellnigh remained.

I sighed and I sobbed, 50
For that I was robbed
Of my sparrow's life.
Of maiden, widow, and wife,
Of what estate ye be,
Of high or low degree,
Great sorrow then ye might see,
And learn to weep at me!
Such paines did me frete
That mine heart did beat,
My visage pale and dead, 60
Wan, and blue as lead:
The pangs of hateful death
Wellnigh had stopped my breath.

Heu, heu, me,
That I am woe for thee!
Ad Dominum, cum tribularer, clamavi.
Of God nothing else crave I

29 Worrowed] bit
34 stound] moment
35 sound] swoon
37 Unneth] scarcely
64 *Heu, heu, me*] Woe, woe is me
66 *Ad Dominum . . . clamavi*] 'In my distress, I cried unto the Lord' (Ps. 120:1)

But Philip's soul to keep
From the marees deep
Of Acheronte's well, 70
That is a flood of hell;
And from the great Pluto,
The prince of endless woe;
And from foul Alecto,
With visage black and blo;
And from Medusa, that mare,
That like a fiend doth stare;
And from Megaera's adders
For ruffling of Philip's feathers,
And from her fiery sparklings 80
For burning of his wings;
And from the smokes sour
Of Proserpina's bower;
And from the dennes dark
Where Cerberus doth bark,
Whom Theseus did affray,
Whom Hercules did outray,
As famous poetes say;
From that hell-hound
That lieth in chaines bound, 90
With ghastly heades three;
To Jupiter pray we
That Philip preserved may be!
Amen, say ye with me!

Do mi nus,
Help now, sweet Jesus!
Levavi oculos meos in montes.
Would God I had Zenophontes,
Or Socrates the wise,
To shew me their device 100
Moderately to take
This sorrow that I make
For Philip Sparrow's sake!
So fervently I shake,
I feel my body quake;
So urgently I am brought
Into careful thought.

69 marees] marsh
87 outray] vanquish
97 *Levavi . . . montes*] 'I lifted up mine eyes unto the hills' (Ps. 121:1)
98 Zenophontes] Xenophon

Like Andromach, Hector's wife,
Was weary of her life,
When she had lost her joy, 110
Noble Hector of Troy;
In like manner also
Increaseth my deadly woe,
For my sparrow is go.

It was so pretty a fool,
It would sit on a stool,
And learned after my school
For to keep his cut,
With 'Philip, keep your cut!'

It had a velvet cap, 120
And would sit upon my lap,
And seek after small wormes,
And sometime white bread-crumbes;
And many times and oft
Between my breastes soft
It woulde lie and rest;
It was proper and prest.

Sometime he would gasp
When he saw a wasp;
A fly or a gnat, 130
He would fly at that;
And prettily he would pant
When he saw an ant.
Lord, how he would pry
After the butterfly!
Lord, how he would hop
After the gressop!
And when I said, 'Phip, Phip!'
Then he would leap and skip,
And take me by the lip. 140
Alas, it will me slo
That Philip is gone me fro! . . .

Lauda, anima mea, Dominum!
To weep with me look that ye come
All manner of birdes in your kind;
See none be left behind.

118 keep his cut] behave properly
127 prest] quick, sprightly
137 gressop] grasshopper
141 slo] slay
143 *Lauda . . . Dominum*] 'Praise the Lord, O my soul' (Ps. 145:1)

To mourning looke that ye fall
With dolorous songes funerall,
Some to sing, and some to say,
Some to weep, and some to pray, 150
Every birde in his lay.
The goldfinch, the wagtail;
The jangling jay to rail,
The flecked pie to chatter
Of this dolorous matter;
And Robin Redbreast,
He shall be the priest
The requiem mass to sing,
Softly warbeling,
With help of the red sparrow 160
And the chattering swallow,
This hearse for to hallow.
The lark with his long toe;
The spink, and the martinet also;
The shovelar with his broad beak;
The dotterel, that foolish peke,
And also the mad coot,
With balde face to toot;
The fieldfare and the snite;
The crow and the kite; 170
The raven, called Rolfe,
His plain-song to sol-fa;
The partridge, the quail;
The plover with us to wail;
The woodhack, that singeth 'chur'
Hoarsely, as he had the mur;
The lusty chanting nightingale;
The popinjay to tell her tale,
That tooteth oft in a glass,
Shall read the Gospel at mass; 180
The mavis with her whistle
Shall read there the Epistle.
But with a large and a long
To keepe just plain-song,

154 pie] magpie
164 spink] chaffinch
martinet] martin
165 shovelar] spoonbill
166 dotterel] a species of plover
peke] dolt
168 toot] peer, pry
169 snite] snipe
175 woodhack] woodpecker(?), nightjar(?)
176 mur] catarrh
178 popinjay] parrot
179 tooteth] peers
181 mavis] song-thrush

Our chanters shall be the cuckoo,
The culver, the stockdoo.
With 'peewit' the lapwing,
The Versicles shall sing.

The bittern with his bumpe,
The crane with his trumpe, 190
The swan of Maeander,
The goose and the gander,
The duck and the drake,
Shall watch at this wake;
The peacock so proud,
Because his voice is loud,
And hath a glorious tail,
He shall sing the Grail;
The owl, that is so foul,
Must help us to howl; 200
The heron so gaunce,
And the cormorance,
With the pheasant,
And the gaggling gant,
And the churlish chough;
The knot and the ruff;
The barnacle, the buzzard,
With the wild mallard;
The divendop to sleep;
The water-hen to weep; 210
The puffin and the teal
Money they shall deal
To poore folk at large,
That shall be their charge;
The seamew and the titmouse;
The woodcock with the longe nose;
The threstle with her warbling;
The starling with her brabling;
The rook, with the osprey
That putteth fishes to a fray; 220
And the dainty curlew,
With the turtle most true.

At this *Placebo*
We may not well forgo

186 culver] dove
stockdoo] wild pigeon
189 bumpe] booming voice
198 Grail] gradual
204 gaggling gant] cackling gannet
207 barnacle] species of wild goose
209 divendop] dabchick
211 teal] small member of duck family
215 seamew] common gull
217 threstle] song-thrush

The countering of the coe;
The stork also,
That maketh his nest
In chimneys to rest;
Within those walls
No broken galls 230
May there abide
Of cuckoldry side,
Or else philosophy
Maketh a great lie.

The ostrich, that will eat
An horseshoe so great,
In the stead of meat,
Such fervent heat
His stomach doth frete;
He cannot well fly, 240
Nor sing tunably,
Yet at a brayd
He hath well assayed
To sol-fa above E-la.
Fa, lorell, fa, fa!
Ne quando
Male contando,
The best that we can,
To make him our bell-man,
And let him ring the bells. 250
He can do nothing else.

Chanticleer, our cock,
Must tell what is of the clock
By the astrology
That he hath naturally
Conceived and caught,
And was never taught
By Albumazer
The astronomer,
Nor by Ptolomy 260
Prince of astronomy,
Nor yet by Haly;
And yet he croweth daily

225 countering] improvisation below plainsong
coe] jackdaw
230 galls] sores
239 frete] consume
242 at a brayd] suddenly
246–7 *Ne . . . cantando*] lest ever by singing badly

And nightly the tides
That no man abides,
With Partlot his hen,
Whom now and then
He plucketh by the head
When he doth her tread.

The bird of Araby, 270
That potentially
May never die,
And yet there is none
But one alone;
A phoenix it is
This hearse that must bless
With aromatic gums
That cost great sums,
The way of thurification
To make a fumigation, 280
Sweete of reflare,
And redolent of aire,
This corse for to cense
With greate reverence,
As patriarch or pope
In a blacke cope.
Whiles he censeth the hearse,
He shall sing the verse,
Libera me,
In de la, sol, re, 290
Softly B molle
For my sparrow's soul.
Pliny sheweth all
In his *Story Natural*
What he doth find
Of the phoenix kind;
Of whose incineration
There riseth a new creation
Of the same fashion
Without alteration, 300
Saving that olde age
Is turned into corage
Of freshe youth again;
This matter true and plain,

279 thurification] burning of incense
280 fumigation] odorous smoke
281 reflare] scent
289 *Libera me*] 'Deliver me' (opening of the Responsory)

Plain matter indeed,
Who so list to read.

But for the eagle doth fly
Highest in the sky,
He shall be the sub-dean,
The choir to demean, 310
As provost principal,
To teach them their Ordinal;
Also the noble falcon,
With the ger-falcon,
The tarsel gentil,
They shall mourn soft and still
In their amice of gray;
The saker with them shall say
Dirige for Philip's soul;
The goshawk shall have a roll 320
The choristers to control;
The lanners and the marlions
Shall stand in their mourning-gowns;
The hobby and the musket
The censers and the cross shall fet;
The kestrel in all this wark
Shall be holy water clerk.

And now the dark cloudy night
Chaseth away Phoebus bright,
Taking his course toward the west, 330
God send my sparrow's soul good rest!
Requiem aeternum dona eis, Domine!
Fa, fa, fa, mi, re, re,
A por ta in fe ri,
Fa, fa, fa, mi, mi.

Credo videre bona Domini,
I pray God, Philip to heaven may fly:
Domine, exaudi orationem meam
To heaven he shall, from heaven he came:

310 demean] supervise
314 ger-falcon] large falcon
315 tarsel gentil] male peregrine falcon
318 saker] large lanner falcon
319 *Dirige*] 'Direct [my steps]'
322 lanners] species of falcon
marlions] merlins, small swift falcons
324 hobby] small falcon
musket] male of the sparrowhawk
325 fet] fetch
332 *Requiem . . . Domine*] 'Grant them eternal rest, O Lord!'
334 *A por ta in fe ri*] 'From the gates of hell . . .'
336 *Credo . . . Domini*] 'I had thought to see the goodness of the Lord' (Ps. 26: 13)
338 *Domine . . . meam*] 'Lord, hear my prayer' (Ps. 102: 2)

Do mi nus vo bis cum! 340
Of all good prayers God send him some!
Oremus,
Deus, cui proprium est misereri et parcere,
On Philip's soul have pity!

(Wr. before 1505; pub. 1545)

from *Magnificence*

6

[*Fancy's song and speech*]

Fancy

STOW, birde, stow, stow!
It is best I feed my hawke now.
There is many evil favoured, an thou be foul.
Each thing is fair when it is young. All hail, owl!

Lo, this is
My fancy ywis:
Now Christ it blesse!
It is, by Jesse,

A bird full sweet,
For me full meet: 10
She is furred for the heat
All to the feet;

Her browes bent,
Her eyen glent:
From Tyne to Trent,
From Stroud to Kent,

A man shall find
Many of her kind.
How standeth the wind—
Before or behind? 20

Barbed like a nun,
For burning of the sun;
Her feathers dun,
Well-favoured, bonne!

343 *Deus . . . parcere*] 'O God, whose nature it is to be merciful and to spare'

6

1 stow] a call used by falconers to bring down the hawk
6 ywis] indeed, certainly
8 Jesse] Jesus
10 meet] fitting
14 glent] gleaming
21 Barbed] wearing a 'barb', a white linen headdress
24 bonne] good girl

Now, let me see about
In all this rout
If I can find out
So seemly a snout

Among this press:
Even a whole mess— 30
Peace, man, peace!
I rede we cease.

So farly fair as it looks,
And her beak so comely crooks,
Her nailes sharp as tenter-hooks!
I have not kept her yet three wooks.

And how still she doth sit!
Tewit, tewit! Where is my wit?
The devil speed whit!

That was before, I set behind: 40
Now too courteous, forthwith unkind,
Sometime too sober, sometime too sad,
Sometime too merry, sometime too mad;
Sometime I sit as I were solemn proud,
Sometime I laugh over-loud,
Sometime I weep for a gee-gaw,
Sometime I laugh at wagging of a straw;
With a pear my love you may win,
And ye may lose it for a pin.
I have a thing for to say, 50
And I may tend thereto for play;
But in faith I am so occupied
On this half and on every side,
That I wot not where I may rest.
First to tell you what were best,
Frantic Fancy-Service I hight;
My wits be weak, my brains are light.
For it is I that other while
Pluck down lead, and theke with tile;
Now will I this, and now will I that, 60

30 mess] company of diners
32 rede] advise
33 farly] marvellously
39 the devil . . . whit!] May the devil not prosper a jot!
40 that] what
42 sometime . . . sometime] on one occasion . . . on another
46 gee-gaw] trifle
56 hight] am called
59 theke] cover, thatch

Make a windmill of a mat;
Now I would, and I wist what—
Where is my cap? I have lost my hat!
And within an hour after
Pluck down a house, and set up a rafter.
Hither and thither, I wot not whither:
Do and undo, both together.
Of a spindle I will make a spar:
All that I make forthwith I mar!
I blunder, I bluster, I blow, and I blother, 70
I make on the one day, and I mar on the other.
Busy, busy, and ever busy,
I dance up and down till I am dizzy.
I can find fantasies where none is:
I will not have it so, I will have it this!

7 [*The conclusion of the play*]

Redress. UNTO this process briefly compiled,
Comprehending the world casual and transitory,
Who list to consider shall never be beguiled,
If it be registered well in memory;
A plain example of worldly vainglory,
How in this world there is no sickerness,
But fallible flattery enmixed with bitterness.

Now well, now woe, now high, now low degree;
Now rich, now poor, now whole, now in disease;
Now pleasure at large, now in captivity; 10
Now lief, now loth, now please, now displease;
Now ebb, now flow, now increase, now discrease;
So in this world there is no sickerness,
But fallible flattery enmixed with bitterness.

Circumspection. A mirror encleared is this interlude,
This life inconstant for to behold and see;
Suddenly advanced, and suddenly subdued,
Suddenly riches, and suddenly poverty,
Suddenly comfort, and suddenly adversity;
Suddenly thus Fortune can both smile and frown, 20
Suddenly set up, and suddenly cast down.

68 spar] rafter or pole
70 blother] babble

7
1 process] narrative
6 sickerness] security
7 fallible] deceitful
15 encleared] made clear or bright
17 advanced] raised

Suddenly promoted, and suddenly put back,
  Suddenly cherished, and suddenly cast aside,
Suddenly commended, and suddenly find a lack,
  Suddenly granted, and suddenly denied,
  Suddenly hid, and suddenly espied;
Suddenly thus Fortune can both smile and frown,
Suddenly set up, and suddenly cast down.

Perseverance. This treatise, devised to make you disport,
  Showeth nowadays how the world cumbered is, 30
To the pith of the matter who list to resort;
  Today it is well, tomorrow it is all amiss,
  Today in delight, tomorrow bare of bliss,
Today a lord, tomorrow lie in the dust,
Thus in this world there is no earthly trust.

Today fair weather, tomorrow a stormy rage;
  Today hot, tomorrow outrageous cold;
Today a yeoman, tomorrow made of page;
  Today in surety, tomorrow bought and sold;
  Today masterfast, tomorrow he hath no hold; 40
Today a man, tomorrow he lieth in the dust:
Thus in this world there is no earthly trust.

Magnificence. This matter we have moved, you mirthful to make,
  Pressly purposed under pretence of play,
Showeth wisdom to them that wisdom can take,
  How suddenly worldly wealth doth decay,
  How wisdom through wantonness vanisheth away;
How none estate living of himself can be sure,
For the wealth of this worlde cannot endure.

Of the terrestre treachery we fall in the flood, 50
  Beaten with storms of many a froward blast,
Ensorbed with the waves savage and wood;
  Without our ship be sure, it is likely to brast,
  Yet of magnificence oft made is the mast;
Thus none estate living of him can be sure,
For the wealth of this worlde cannot endure.

29 treatise] story
38 made of page] considered as a page
39 bought and sold] betrayed for a bribe
40 masterfast] bound to a master
44 pressly] briefly
48 estate] person of worldly standing
52 ensorbed] swallowed up
  wood] fierce

Redress. Now seemeth us fitting that ye then resort
Home to your palace with joy and royalty.
Circumspection. Where everything is ordained after your noble port.
Perseverance. There to endure with all felicity. 60
Magnificence. I am content, my friends, that it so be.
Redress. And ye that have heard this disport and game,
Jesus preserve you from endless woe and shame!

(Wr. 1515–16; pub. 1530)

from *Elinour Rumming*

8

[*Visitors to the ale-house*]

(i)

INSTEAD of coin and money
Some bringe her a coney,
And some a pot with honey,
Some a salt, and some a spoon,
Some their hose, some their shoon;
Some ran a good trot
With a skillet or a pot;
Some fill their pot full
Of good Lemster wool.
An hussif of trust 10
When she is athrust
Such a web can spin,
Her thrift is full thin.
Some go straight thither,
Be it slaty or slither:
They hold the highway,
They care not what men say!
Be that as be may,
Some, loth to be espied,
Some start in at the back-side 20
Over the hedge and pale,
And all for the good ale.
Some runne till they sweat,
Bring with them malt or wheat,
And Dame Elinour entreat
To birle them of the best.

2 coney] rabbit
7 skillet] pan, kettle
9 Lemster] Leominster (Herefordshire)
10 hussif] housewife, mistress of a household
11 athrust] thirsty
15 slaty] stony
slither] slippery
26 birle] pour out drink

Then cometh another guest.
She sweareth by the Rood of Rest
Her lippes are so dry
Without drink she must die; 30
Therefore, 'fill it by and by,
And have here a peck of rye.'
Anon cometh another,
As dry as the other,
And with her doth bring
Meal, salt, or other thing,
Her harnessed girdle, her wedding ring,
To pay for her scot
As cometh to her lot.
One bringeth her husband's hood 40
Because the ale is good;
Another brought her his cap
To offer to the ale-tap,
With flax and with tow;
And some brought sour dough
With 'Hey!' and with 'Ho!
Sit we down a row
And drink till we blow,
And pipe "tirly-tirlow!" '
Some laid to pledge 50
Their hatchet and their wedge,
Their heckle and their reel,
Their rock, their spinning-wheel;
And some went so narrow
They laid to pledge their wharrow,
Their ribskin and their spindle,
Their needle and their thimble.
Here was scant thrift
When they made such shift.
Their thirst was so great 60
They asked never for meat,
But 'Drink!' still 'Drink!
And let the cat wink!
Let us wash our gums
From the dry crumbs!'

31 by and by] at once
37 harnessed] ornamented
38 scot] reckoning
44 tow] fibre of flax for spinning
51 wedge] metal tool for splitting wood
52 heckle] instrument for combing flax or hemp
53 rock] distaff
54 narrow] close to their financial limits
55 wharrow] pulley on the spindle of a spinning-wheel
56 ribskin] leather apron (worn by women when 'rubbing' flax)
63 wink] close its eyes (i.e. connive at it)

(ii)

Some for very need
Laid down a skein of thread,
And some a skein of yarn.
Some brought from the barn
Both beans and pease; 70
Small chaffer doth ease
Sometime, now and than.
Another there was that ran
With a good brass-pan—
Her colour was full wan—
She ran in all the haste,
Unbraced and unlaced,
Tawny, swart and sallow
Like a cake of tallow.
I swear by all hallow 80
It was a stale to take
The devil in a brake!
And then came halting Joan,
And brought a gambone
Of bacon that was resty;
But, Lord, she was testy,
Angry as a waspy!
She began to gape and gaspy,
And bade Elinour go bet
And fill in good met; 90
It was dear that was far-fet.
Another brought a spick
Of a bacon flick;
Her tongue was very quick,
But she spake somewhat thick.
Her fellow did stammer and stut,
But she was a foul slut,
For her mouth foamed
And her belly groaned:
Joan said she had eaten a fiest. 100
'By Christ', said she, 'thou liest.

71 chaffer] trade, trading
77 Unbraced] with the fastenings of clothes undone
80 all hallow] All saints
81 stale] bait, lure
82 brake] trap
83 halting] limping
84 gambone] ham, bottom part of a flitch of bacon
85 resty] rancid
88 gape] crave
gaspy] desire earnestly
89 go bet] go more quickly, hurry
90 met] measure
91 far-fet] fetched from afar, a rarity
92 spick] piece (of fat meat or bacon)
93 flick] flitch
96 stut] stutter
100 fiest] foul smell, stink

I have as sweet a breath
As thou, with shameful death!'
  Then Elinour said, 'Ye callets,
I shall break your palates,
Without ye now cease!'
And so was made the peace.
  Then thither came drunken Alice,
And she was full of tales,
Of tidings in Wales, 110
And of Saint James in Gales,
And of the Portingales,
With 'Lo, gossip, ywis,
Thus and thus it is:
There hath been great war
Between Temple Bar
And the Cross in Cheap,
And thither came an heap
Of millstones in a rout . . .'
She spake thus in her snout, 120
Snivelling in her nose
As though she had the pose.
'Lo, here is an old tippet,
An ye will give me a sippet
Of your stale ale,
God send you good sale!'
And as she was drinking,
She fell in a winking
With a barlyhood;
She pissed where she stood. 130
Then she began to weep,
And forthwith fell asleep.
Elinour took her up
And blessed her with a cup
Of new ale in corns.
Alice found therein no thorns,
But supped it up at ones,
She found therein no bones.

(Wr. *c.*1517; pub. *c.*1545)

104 callets] lewd women, scolds
106 Without] unless
111 Gales] Galicia (north-west Spain)
112 Portingales] Portuguese
113 ywis] indeed, certainly
117 Cheap] Cheapside
119 rout] disorderly gathering
122 pose] cold in the head
123 tippet] cape, scarf
124 sippet] little sip
128 winking] dozing, nap
129 barlyhood] fit of drunkenness
135 new ale in corns] ale drawn off the malt
137 at ones] at once

from *Speak, Parrot*

9

[*The opening stanzas*]

MY name is Parrot, a bird of Paradise,
By nature devised of a wonderous kind,
Daintily dieted with divers delicate spice
Till Euphrates, that flood, driveth me into Ind;
Where men of that country by fortune me find
And send me to greate ladyes of estate:
Then Parrot must have an almond or a date.

A cage curiously carven, with silver pin,
Properly painted, to be my coverture;
A mirror of glasse, that I may toot therein: 10
These, maidens full merrily with many a divers flower,
Freshly they dress, and make sweet my bower,
With 'Speak, Parrot, I pray you!' full curtesly they say,
'Parrot is a goodly bird, a pretty popagay!'

With my beke bent, my little wanton eye,
My feathers fresh as is the emerald green,
About my neck a circulet like the rich ruby,
My little legges, my feet both feat and clean,
I am a minion to wait upon a queen.
'My proper Parrot, my little pretty fool!' 20
With ladies I learn, and go with them to school.

Hey! ha, ha! Parrot, ye can laugh prettily!'
Parrot hath not dined all this long day.
Like our puss-cat, Parrot can mewt and cry
In Latin, in Hebrew, and in Chaldy;
In Greeke tongue Parrot can both speak and say,
As Persius, that poet, doth report of me,
'*Quis expedivit psittaco suum chaire?*'

Douce French of Paris Parrot can learne,
Pronouncing my purpose after my property, 30
With '*Parlez bien,* Parrot, *ou parlez rien!*'
With Dutch, with Spanish, my tongue can agree,
In English to God Parrot can supply:

10 toot] peep, peer
14 popagay] parrot
18 feat] neat
24 mewt] mew
28 '*Quis . . . chaire?*'] 'Who taught the parrot to say his "hallo"?'
29 Douce] sweet
32 Dutch] German
33 supply] pray, supplicate

'Christ save King Henry the Eighth, our royal king,
The red rose in honour to flourish and spring!

With Katherine incomparable, our royal queen also,
That peerless pomegranate, Christ save her noble grace!'
Parrot *sabe hablar Castiliano*,
With *fidasso de cosso* in Turkey and in Thrace;
Vis consilii expers, as teacheth me Horace, 40
Mole ruit sua, whose dictates are pregnant,
Soventez foys, Parrot, *en souvenante*.

My lady mistress, Dame Philology,
Gave me a gifte, in my nest when I lay,
To learn all language, and it to speak aptely.
Now *pandez morie*, wax frantic, some men say,
Phronesis for Phrenesis may not hold her way.
An almond now for Parrot, delicately drest:
In *Salve festa dies, toto* there doth best.

Moderata juvant, but *toto* doth exceed: 50
Discretion is mother of noble virtues all.
Myden agan in Greeke tongue we read.
But reason and wit wanteth their provincial
When wilfulness is vicar general.
Haec res acu tangitur, Parrot, *par ma foy:*
Taisez-vous, Parrot, *tenez-vous coy!*

Busy, busy, busy, and business again!
Que pensez-vous, Parrot? what meaneth this business?
Vitulus in Horeb troubled Aaron's brain,
Melchisedek merciful made Moloch merciless: 60
Too wise is no virtue, too meddling, too restless.

37 pomegranate] (one of the heraldic devices of Katherine of Aragon)
38 *sabe . . . Castiliano*] can speak Castilian (Spanish)
39 *fidasso . . . cosso*] Trust in yourself (*lingua franca*?)
40 *Vis . . . sua*] Force without counsel falls under its own weight
42 *Soventez . . . souvenante*] many times in memory
46 *pandez morie*] grow mad
47 Phronesis] understanding
Phrenesis] madness, frenzy
49 In *Salve . . .* best] on holiday it is best to have everything(?)
50 *Moderata invant*] moderation pleases us
toto] everything
52 *Myden agan*] nothing in excess
53 Wanteth their provincial] lack the control of a bishop
55 *Haec . . . tangitur*] this hits the nail on the head
56 *Taisez . . . coy*] be quiet, Parrot, keep still
59 *Vitulus*] the calf

In measure is treasure, *cum sensu maturato,*
Ne tropo sanno, ne tropo mato.

Aaron was fired with Chaldee's fire called Ur,
  Jobab was brought up in the land of Hus,
The lineage of Lot took support of Assur,
  Jereboseth is Hebrew, who list the cause discuss—
  'Peace, Parrot, ye prate as ye were *ebrius*:
Hist thee, *lieber Got von Himmelsreich, ich seg!*'
In Popering grew pears when Parrot was an egg. 70

'What is this to purpose?' Over in a whinnymeg.
  Hob Lobin of Lowdeon would hae a bit a' bread;
The gibbet of Baldock was made for Jack Leg;
  An arrow unfeathered and without an head,
  A bagpipe without blowing standeth in no stead:
Some run too far before, some run too far behind,
Some be too churlish, and some be too kind.

Ich dien serveth for the ostrich feather,
  Ich dien is the language of the land of Beme;
In Afric tongue *byrsa* is a thong of leather; 80
  In Palestina there is Jerusaleme.
  Colostrum now for Parrot, white bread and sweet cream!
Our Thomasen she doth trip, our jennet she doth shale:
Parrot hath a blacke beard and a fair green tail.

'Morish mine own shelf!' the costermonger saith,
  '*Fate, fate, fate!*' ye Irish waterlag;
In flattering fables men find but little faith,
  But *moveatur terra*, let the world wag;
  Let Sir Wrig-wrag wrestle with Sir Dalyrag;
Every man after his manner of ways, 90
Paub yn ei arver, so the Welshman says.

62 *cum . . . maturato*] with mature judgement
63 *ne tropo . . . mato*] neither too sane nor too mad
65 Jobab] Job
68 *ebrius*] drunk
69 'Hist . . . *seg*] hush, dear God of Heaven, I say
71 whinnymeg] instant(?)
72 Lowdeon] Lothian
78 *Ich dien*] I serve (royal motto)
79 Beme] Bohemia
82 *Colostrum*] milk beestings
83 shale] stumble
85 'Morish . . . shelf] I, Morris, am for my own self(?) (Irish)
86 *Fate*] water (Irish)
waterlag] water-lugger, water-seller
88 *moveatur terra*] i.e. let things slide
91 *Paub yn ei arver*] every one in his manner

Such shreddes of sentence, strewed in the shop
  Of ancient Aristippus and such other mo,
I gader together and close in my crop,
  Of my wanton conceit, *unde depromo*
  Dilemmata docta in paedagogio
Sacro vatum, whereof to you I break.
I pray you, let Parrot have liberty to speak!

But 'Ware the cat, Parrot, ware the false cat!'
  With 'Who is there—a maid? Nay, nay, I trow!' 100
'Ware riot, Parrot! Ware riot, ware that!'
  'Meat, meat for Parrot, meat I say, how!'
  Thus diverse of language by learning I grow,
With 'Basse me, sweet Parrot, basse me, sweet sweet!'
To dwell among ladyes Parrot is meet.

'Parrot, Parrot, Parrot, pretty popinjay!'
  With my beak I can pick my little pretty toe;
My delight is solace, pleasure, disport, and play.
  Like a wanton, when I will, I reel to and fro.
  Parrot can say *Caesar, ave!* also. 110
But Parrot hath no favour to Esebon.
Above all other birdes, set Parrot alone.

Ulula, Esebon, for Jeremy doth weep!
  Zion is in sadness, Rachel ruly doth look;
Madionita Jethro, our Moses keepeth his sheep;
  Gideon is gone, that Zalmane undertook,
  Horeb *et* Zeb, of *Judicum* read the book.
Now Zebul, Ammon, and Abimalech—'Hark, hark!
Parrot pretendeth to be a Bible clerk!'

O Esebon, Esebon! to thee is come again 120
  Sihon, the regent *Amorraeorum*,
And Og, that fat hog of Bashan, doth retain
  The crafty *coistronus Cananaeorum*;
  And *asylum*, whilom *refugium miserorum*,
Non fanum, sed profanum, standeth in little stead.
Ulula, Esebon, for Jephthah is stark dead!

94 crop] throat
95–7 *unde . . . vatum*] whence I bring forth learned arguments in the sacred school of poets
101 riot] extravagance, dissipation
104 Basse] kiss
110 *Caesar, ave!*] Hail, Caesar!
113 *Ulula*, Esebon] Weep, Heshbon (Jeremiah 49: 3)
114 ruly] sorrowfully
122 Og] (i.e. Wolsey)
123 *coistronus Cananaeorum*] the kitchen boy of the Canaanites
124–5 *asylum . . . profanum*] asylum, formerly the refuge of wretches, is not a sanctuary but is to be made secular

Esebon, Marylebone, Whetstone next Barnet;
  A trim-tram for an horse-mill it were a nice thing!
Dainties for damoiselles, chaffer far-fet:
  Bo-ho doth bark well, but Hough-ho he ruleth the ring; 130
  From Scarpary to Tartary renown therein doth spring,
With 'He said,' and 'We said,' ich wot now what ich wot—
Quod magnus est dominus Judas Iscariot.

Ptolemy and Haly were cunning and wise
  In the volvel, in the quadrant, and in the astroloby,
To prognosticate truly the chance of Fortune's dice;
  Some treat of their tiriks, some of astrology,
  Some *pseudo-propheta* with chiromancy.
If Fortune be friendly, and grace be the guide,
Honour with renown will run on that side. 140

10 [*The conclusion*]

GALATHEA

Now, Parrot, my sweet bird, speak out yet once again,
Set aside all sophisms, and speak now true and plain.

PARROT

So many moral matters, and so little used;
  So much new making, and so mad time spent;
So much translation into English confused;
  So much noble preaching, and so little amendment;
  So much consultation, almost to none intent;
So much provision, and so little wit at need—
Since Deucalion's flood there can no clerkes rede.

So little discretion, and so much reasoning; 10
  So much hardy dardy, and so little manliness;
So prodigal expense, and so shameful reckoning;
  So gorgeous garments, and so much wretchedness;
  So much portly pride, with purses penniless;
So much spent before, and so much unpaid behind—
Since Deucalion's flood there can no clerkes find.

128 trim-tram] trifle, worthless thing
  nice] foolish
129 chaffer] merchandise
131 Scarpary] Mt. Scarperio, in Tuscany
133 *Quod . . . Iscariot*] because Judas Iscariot is a great lord
135 volvel] astronomical clock
137] tiriks] theorics

10
2 sophisms] fallacious arguments
8 provision] the practice, which Wolsey revived, of 'providing' candidates for vacant benefices, without consulting the bishops

So much forecasting, and so far an after-deal;
  So much politic prating, and so little standeth in stead;
So little secretness, and so much great counsel;
  So many bold barons, their hearts as dull as lead; 20
  So many noble bodies under one daw's head;
So royal a king as reigneth upon us all—
Since Deucalion's flood was never seen nor shall.

So many complaintes, and so smalle redress;
  So much calling on, and so small taking heed;
So much loss of merchandise, and so remediless;
  So little care for the common weal, and so much need;
  So much doubtful danger, and so little drede;
So much pride of prelates, so cruel and so keen—
Since Deucalion's flood, I trow, was never seen. 30

So many thieves hanged, and thieves never the less;
  So much prisonment for matters not worth an haw;
So much papers wearing for right a small excess;
  So much pillory-pageants under colour of good law;
  So much turning on the cuck-stool for every gee-gaw;
So much mockish making of statutes of array—
Since Deucalion's flood was never, I dare say.

So brainless calves' heads, so many sheepes tails;
  So bold a bragging butcher, and flesh sold so dear;
So many plucked partridges, and so fatte quails; 40
  So mangy a mastiff cur, the great greyhound's peer;
  So big a bulk of brow-antlers cabbaged that year;
So many swans dead, and so small revel—
Since Deucalion's flood, I trow, no man can tell.

So many truces taken, and so little perfite troth;
  So much belly-joy, and so wasteful banqueting;
So pinching and sparing, and so little profit grow'th;
  So many hugy houses building, and so small householding;
  Such statutes upon diets, such pilling and polling;
So is all thing wrought wilfully withoute reason and skill— 50
Since Deucalion's flood the world was never so ill.

17 after-deal] consequence
28 doubtful] fearful
39 butcher] Wolsey was the son of a butcher
41 greyhound] one of Henry VIII's badges was a greyhound
42 cabbaged] grown to a head like the horns of a deer
43 swans] the badge of the Duke of Buckingham, executed in 1521, was the swan

So many vagabonds, so many beggars bold;
 So much decay of monasteries and of religious places;
So hot hatred against the Church, and charity so cold;
 So much of 'my Lord's Grace,' and in him no grace is;
 So many hollow hearts, and so double faces;
So much sanctuary-breaking, and privilege barred—
Since Deucalion's flood was never seen nor lered.

So much ragged right of a rammes horn;
 So rigorous revelling in a prelate specially; 60
So bold and so bragging, and was so basely born;
 So lordly in his looks and so disdainously;
 So fat a maggot, bred of a fleshe-fly;
Was never such a filthy Gorgon, nor such an epicure,
Since Deucalion's flood, I make thee fast and sure.

So much privy watching in cold winters' nights;
 So much searching of losels, and is himself so lewd;
So much conjurations for elfish mid-day sprites;
 So many bulls of pardon published and shewed;
 So much crossing and blessing, and him all-beshrewed; 70
Such pole-axes and pillars, such mules trapt with gold—
Since Deucalion's flood in no chronicle is told.

(Wr. 1519–21; pub. *c.*1554)

ANONYMOUS

11 *The Nutbrown Maid*

First Player

BE it right or wrong, these men among
 On women do complain,
Affirming this, how that it is
 A labour spent in vain

58 lered] learned, heard of
60 rigorous] harsh
67 losels] petty criminals, worthless persons
lewd] ignorant, base, ill-bred
68 mid-day sprites] noonday devils (cf. Ps. 90: 6)
70 all-beshrewed] so depraved
71 pole-axes and pillars] Wolsey had two silver pillars and four gilt pole-axes carried before him

11

1 among] at times

To love them well, for never a dell
  They love a man again;
For let a man do what he can,
  Their favour to attain,
Yet if a new to them pursue,
  Their first true lover than 10
Laboureth for nought, and from her thought
  He is a banished man.

Second Player

I say not nay, but that all day
  It is both written and said
That woman's faith is, as who saith,
  All utterly decayed;
But nevertheless, right good witness
  In this case might be laid,
That they love true, and continue.
  Record the Nutbrown Maid, 20
Which when her love came her to prove
  To her to make his moan,
Would not depart, for in her heart
  She loved but him alone.

First Player

Then between us let us discuss,
  What was all the mannere
Between them two; we will also
  Tell all the pain, in fere,
That she was in. Now I begin,
  So that ye me answere. 30
Wherefore all ye that present be
  I pray you give an ear.
I am the knight; I come by night,
  As secret as I can,
Saying, 'Alas! thus standeth the case,
  I am a banished man.'

Puella

And I your will for to fulfil
  In this will not refuse,
Trusting to shew, in wordes few,
  That men have an ill use 40

5 dell] bit
28 in fere] together, in dialogue *Puella*] Girl

To their own shame women to blame,
  And causeless them accuse.
Therefore to you I answer now,
  All women to excuse,
'Mine own heart dear, with you what cheer?
  I pray you, tell me anon,
For in my mind of all mankind
  I love but you alone.'

Squire

'It standeth so; a deed is do,
  Whereof great harm shall grow. 50
My destiny is for to die
  A shameful death, I trow,
Or else to flee; the one must be.
  None other way I know,
But to withdraw as an outlaw,
  And take me to my bow.
Wherefore adieu, my own heart true!
  None other rede I can,
For I must to the greenwood go
  Alone, a banished man.' 60

Puella

'O Lord, what is this worldes bliss,
  That changeth as the moon?
My summer's day in lusty May
  Is darked before the noon.
I hear you say farewell. Nay, nay,
  We depart not so soon.
Why say ye so? Whither will ye go?
  Alas! what have ye done?
All my welfare to sorrow and care
  Should change, if ye were gone; 70
For in my mind of all mankind
  I love but you alone.'

Squire

'I can believe, it shall you grieve,
  And somewhat you distrain;
But afterward your paines hard
  Within a day or twain

49 do] done 58 rede] counsel 74 distrain] distress

Shall soon aslake, and ye shall take
  Comfort to you again.
Why should ye aught? for to take thought,
  Your labour were in vain. 80
And thus I do; and pray you to,
  As heartily as I can,
For I must to the greenwood go
  Alone, a banished man.'

Puella

'Now sith that ye have shewed to me
  The secret of your mind,
I shall be plain to you again,
  Like as ye shall me find.
Sith it is so, that ye will go,
  I will not leave behind; 90
Shall it never be said the Nutbrown Maid
  Was to her love unkind.
Make you ready, for so am I,
  Although it were anon,
For in my mind of all mankind
  I love but you alone.'

Squire

'Yet I you rede to take good heed
  What men will think and say;
Of young and old it shall be told
  That ye be gone away 100
Your wanton will for to fulfil,
  In greenwood you to play,
And that ye might for your delight
  No longer make delay.
Rather than ye should thus for me
  Be called an ill woman,
Yet would I to the greenwood go
  Alone, a banished man.'

Puella

'Though it be sung of old and young
  That I should be to blame, 110
Theirs be the charge that speak so large
  In hurting of my name;

79 take thought] grieve 89 sith] since 90 leave] remain

For I will prove that faithful love
  It is devoid of shame,
In your distress and heaviness
  To part with you the same.
To show all tho that do not so
  True lovers are they none;
But in my mind of all mankind
  I love but you alone.' 120

Squire

'I counsel you, remember how
  It is no maiden's law,
Nothing to doubt, but to run out
  To wood with an outlaw;
For ye must there in your hand bear
  A bow, ready to draw,
And as a thief thus must ye live,
  Ever in dread and awe,
By which to you great harm might grow.
  Yet had I liefer than, 130
That I had to the greenwood go
  Alone, a banished man.'

Puella

'I say not nay, but as ye say,
  It is no maiden's lore;
But love may make me to forsake,
  As I have said before,
To come on foot, to hunt and shoot,
  To get us meat in store.
For so that I your company
  May have, I ask no more; 140
From which to part it maketh mine heart
  As cold as any stone,
For in my mind of all mankind
  I love but you alone.'

Squire

'For an outlaw, this is the law,
  That men him take and bind,
Without pity hanged to be,
  And waver with the wind.

116 part] share   117 tho] those   135 forsake] withdraw

If I had need, as God forbid!
  What succour could ye find? 150
Forsooth, I trow, ye and your bow
  For fear would draw behind;
And no marvail, for little avail
  Were in your counsel than;
Wherefore I to the greenwood go
  Alone, a banished man.'

Puella

'Right well know ye that women be
  Full feeble for to fight;
No womanhead it is indeed
  To be bold as a knight. 160
Yet in such fear if that ye were,
  With enemies day or night
I would withstand, with bow in hand,
  To help you with my might,
And you to save, as women have
  From death many one;
For in my mind of all mankind
  I love but you alone.'

Squire

'Yet take good heed, for ever I dread
  That ye could not sustain 170
The thorny ways, the deep valleys,
  The snow, the frost, the rain,
The cold, the heat; for, dry or wet,
  We must lodge on the plain,
And, us above, none other roof
  But a brake bush or twain;
Which soon should grieve you, I believe,
  And ye would gladly than
That I had to the greenwood go
  Alone, a banished man.' 180

Puella

'Sith I have here been partener
  With you of joy and bliss,
I must also part of your woe
  Endure, as reason is.

Yet am I sure of one pleasure;
  And, shortly, it is this:
That where ye be, me seemeth, perdy,
  I could not fare amiss.
Without more speech, I you beseech
  That we were shortly gone, 190
For in my mind of all mankind
  I love but you alone.'

Squire

'If ye go thider, ye must consider,
  When ye have lust to dine,
There shall no meat be for to get,
  Neither beer, ale, ne wine;
Ne sheetes clean to lie between,
  Made of thread and twine;
None other house but leaves and boughs
  To cover your head and mine. 200
Lo, mine heart sweet, this ill diet
  Should make you pale and wan;
Wherefore I will to the greenwood go
  Alone, a banished man.'

Puella

'Among the wild deer, such an archere
  As men say that ye be,
May not fail of good vitail,
  Where is so great plenty;
And water clear of the rivere
  Shall be full sweet to me, 210
With which in heal I shall right well
  Endure, as ye shall see.
And, or we go, a bed or two
  I can provide anon,
For in my mind of all mankind
  I love but you alone.'

Squire

'Lo, yet before ye must do more,
  If ye will go with me,
As cut your hair up by your ear,
  Your kirtle by your knee, 220

187 perdy] by God
196 ne] nor
207 vitail] food
211 heal] health
213 or] before
220 kirtle] gown

With bow in hand, for to withstand
  Your enemies, if need be;
And this same night before daylight,
  To woodward will I flee.
If that ye will all this fulfil
  Do it as shortly as ye can;
Else will I to the greenwood go
  Alone, a banished man.'

Puella

'I shall as now do more for you
  Than longeth to womanhead, 230
To short my hair, a bow to bear,
  To shoot in time of need.
O my sweet mother, before all other
  For you I have most dread;
But now, adieu! I must ensue
  Where fortune doth me lead.
All this make ye. Now let us flee,
  The day cometh fast upon;
For in my mind of all mankind
  I love but you alone.' 240

Squire

'Nay, nay, not so! ye shall not go,
  And I shall tell you why.
Your appetite is to be light
  Of love, I well espy;
For, like as ye have said to me,
  In like wise, hardily,
Ye would answer whosoever it were
  In way of company.
It is said of old, soon hot, soon cold;
  And so is a woman. 250
For I must to the greenwood go,
  Alone, a banished man.'

Puella

'If ye take heed, it is no need
  Such words to say by me;
For oft ye prayed, and long assayed,
  Or I you loved, perdy.

And though that I of ancestry
   A baron's daughter be,
Yet have ye proved how I ye loved,
   A squire of low degree; 260
And ever shall, what so befall,
   To die therefore anon;
For in my mind of all mankind
   I love but you alone.'

Squire

'A baron's child to be beguiled,
   It were a cursed deed;
To be fellow with an outlaw,
   Almighty God forbid!
Yet better were the poor squier
   Alone to forest yede, 270
Than ye should say another day
   That by my cursed rede
Ye were betrayed. Wherefore, good maid;
   The best rede that I can,
Is that I to the greenwood go
   Alone, a banished man.'

Puella

'Whatever befall, I never shall
   Of this thing you upbraid;
But if ye go, and leave me so,
   Then have ye me betrayed. 280
Remember you well how that ye deal;
   For if ye be, as ye said,
Ye were unkind to leave me behind,
   Your love, the Nutbrown Maid,
Trust me truly, that I shall die
   Soon after ye be gone,
For in my mind of all mankind
   I love but you alone.'

Squire

'If that ye went, ye should repent;
   For in the forest now 290
I have purveyed me of a maid,
   Whom I love more than you.

270 yede] went   272 rede] advice

Another fairer than ever ye were,
  I dare it well avow;
And of you both each should be wroth
  With other, as I trow.
It were mine ease to live in peace,
  So will I, if I can;
Wherefore I to the wood will go
  Alone, a banished man.' 300

Puella

'Though in the wood I understood
  Ye had a paramour,
All this may nought remove my thought,
  But that I will be your;
And she shall me find soft and kind
  And courteous every hour,
Glad to fulfil all that she will
  Command me to my power.
For had ye, lo! an hundred mo,
  Yet would I be that one, 310
For in my mind of all mankind
  I love but you alone.'

Squire

'Mine own dear love, I see thee prove
  That ye be kind and true,
Of maid and wife, in all my life,
  The best that ever I knew.
Be merry and glad, be no more sad,
  The case is changed new,
For it were ruth that for your truth
  Ye should have cause to rue. 320
Be not dismayed! Whatsoever I said
  To you, when I began,
I will not to the greenwood go;
  I am no banished man.'

Puella

'These tidings be more gladder to me
  Than to be made a queen,
If I were sure they should endure;
  But it is often seen,

When men will break promise, they speak
  The wordes on the spleen. 330
Ye shape some wile me to beguile
  And steal fro me, I ween.
Then were the case worse than it was,
  And I more woebegone;
For in my mind of all mankind
  I love but you alone.'

Squire

'Ye shall not need further to dread.
  I will not disparage
You, God defend! sith you descend
  Of so great a lineage. 340
Now understand, to Westmorland,
  Which is my heritage,
I will you bring, and with a ring
  By way of marriage
I will you take, and lady make
  As shortly as I can.
Thus have ye won an earles son,
  And not a banished man.'

Both Players

Here may ye see that women be
  In love, meek, kind, and stable. 350
Let never man reprove them than,
  Or call them variable;
But rather pray God that we may
  To them be comfortable.
God sometime proveth such as he loveth,
  If they be charitable.
For sith men would that women should
  Be meek to them each one,
Much more ought they to God obey,
  And serve but him alone. 360

(Pub. 1503)

330 on the spleen] lightly
338 disparage] match unequally, degrade
354 comfortable] comforting, cheering

STEPHEN HAWES
1475?–1523?

from *The Pastime of Pleasure*

12 [*The Epitaph of Graunde Amoure*]

O Mortal folk, you may behold and see
How I lie here, sometime a mighty knight;
The end of joy and all prosperity
Is death at last, thorough his course and might;
After the day there cometh the dark night;
For though the day be never so long,
At last the bells ringeth to evensong.

And my self called La Graunde Amoure,
Seeking adventure in the worldly glory,
For to attain the riches and honour, 10
Did think full little that I should here lie,
Till death did mate me full right privily.
Lo what I am, and whereto you must!
Like as I am so shall you be all dust.

(1509)

13 [*Against Swearing*]

See
Me
Be
Kind;
Again
My pain
Retain
In mind.

My sweet blood
On the rood 10
Did thee good,
My brother.
My face right red,
Mine armes spread,
My woundes bled,
Think none other.

12
12 mate] overcome, defeat

Behold thou my side,
Wounded so right wide,
Bleeding sore that tide,
All for thine own sake. 20
Thus for thee I smarted:
Why art thou hard-hearted?
Be by me converted,
And thy swearing aslake.

Tear me now no more.
My woundes are sore.
Leave swearing therefore,
And come to my grace.
I am ready
To grant mercy 30
To thee truly
For thy trespace.

Come now near,
My friend dear,
And appear
Before me.
I so
In woe
Did go:
See, see 40
I
Cry
Hie
Thee.

(1509)

ANONYMOUS

14 *Western wind*

Western wind, when will thou blow,
The small rain down can rain?
Christ, if my love were in my arms
And I in my bed again!

(Pub. 1790)

24 aslake] lessen, abate

43 Hie] haste, speed; exert (yourself)

15 *By a bank as I lay*

BY a bank as I lay,
Musing myself alone, hey ho!
A birdes voice
Did me rejoice,
Singing before the day;
And methought in her lay
She said, winter was past, hey ho!
Then dyry come dawn, dyry come dyry, come dyry!
Come dyry, come dyry, come dawn, hey ho!

The master of music, 10
The lusty nightingale, hey ho!
Full merrily
And secretly
She singeth in the thick;
And under her breast a prick,
To keep her fro sleep, hey ho!
Then dyry come dawn, dyry come dyry, come dyry!
Come dyry, come dyry, come dawn, hey ho!

Awake therefore, young men,
All ye that lovers be, hey ho! 20
This month of May,
So fresh, so gay,
So fair be fields on fen;
Hath flourish ilka den.
Great joy it is to see, hey ho!
Then dyry come dawn, dyry come dyry, come dyry!
Come dyry, come dyry, come dawn, hey ho!

(Pub. 1889)

24 ilka den] every slope

HEATH
(First name and dates unknown)

16 *These women all*

THESE women all
Both great and small
  Are wavering to and fro,
Now here, now there,
Now everywhere;
  But I will not say so.

So they love to range,
Their minds doth change
  And make their friend their foe;
As lovers true 10
Each day they choose new;
  But I will not say so.

They laugh, they smile,
They do beguile,
  As dice that men doth throw.
Who useth them much
Shall never be rich;
  But I will not say so.

Some hot, some cold,
There is no hold 20
  But as the wind doth blow;
When all is done,
They change like the moon;
  But I will not say so.

So thus one and other
Taketh after their mother,
  As cock by kind doth crow.
My song is ended,
The best may be amended;
  But I will not say so. 30

(Pub. 1790)

ATTRIBUTED TO KING HENRY VIII

1491–1547

17 *Pastime with good company*

PASTIME with good company
I love and shall, until I die.
Grudge who list, but none deny:
So God be pleased, thus live will I.
For my pastance,
Hunt, sing and dance,
My heart is set.
All goodly sport
For my comfort
Who shall me let? 10

Youth must have some dalliance,
Of good or ill some pastance.
Company me thinks the best,
All thoughts and fancies to digest;
For idleness
Is chief mistress
Of vices all.
Then who can say
But mirth and play
Is best of all? 20

Company with honesty
Is virtue, vices to flee;
Company is good and ill,
But every man has his free will.
The best ensue,
The worst eschew!
My mind shall be,
Virtue to use,
Vice to refuse;
Thus shall I use me. 30

(Pub. 1855)

5 pastance] pastime 10 let] prevent 14 digest] disperse, dissipate

18 *Whereto should I express*

WHERETO should I express
  My inward heaviness?
No mirth can make me fain,
  Till that we meet again.

Do way, dear heart! Not so!
  Let no thought you dismay.
Though ye now part me fro,
  We shall meet when we may.

When I remember me
  Of your most gentle mind, 10
It may in no wise agree
  That I should be unkind.

The daisy delectable,
  The violet wan and blo,
Ye are not variable.
  I love you and no mo.

I make you fast and sure.
  It is to me great pain
Thus long to endure
  Till that we meet again. 20

(Pub. 1889)

19 *Green groweth the holly*

Green groweth the holly; so doth the ivy.
Though winter blastes blow never so high,
Green groweth the holly.

As the holly groweth green,
  And never changeth hue,
So I am, ever hath been
  Unto my lady true;

As the holly groweth green
  With ivy all alone,
When floweres cannot be seen 10
  And green wood leaves be gone.

3 fain] pleased   5 Do way] have done   14 blo] pale

Now unto my lady
  Promise to her I make,
From all other only
  To her I me betake.

Adieu, mine own lady,
  Adieu, my special,
Who hath my heart truly,
  Be sure, and ever shall.

(Pub. 1889)

WILLIAM CORNISH
d. 1523

20 *You and I and Amyas*

You and I and Amyas,
Amyas and you and I,
To the green wood must we go, alas!
You and I, my life, and Amyas.

THE knight knocked at the castle gate;
The lady marvelled who was thereat.

To call the porter he would not blin;
The lady said he should not come in.

The portress was a lady bright;
Strangeness that lady hight. 10

She asked him what was his name;
He said 'Desire, your man, Madame.'

She said 'Desire, what do ye here?'
He said 'Madame, as your prisoner.'

He was counselled to brief a bill,
And show my lady his own will.

'Kindness,' said she, 'would it bear,'
'And Pity,' said she, 'would be there.'

Thus how they did we cannot say;
We left them there and went our way. 20

(Pub. 1867)

15 betake] give

20
7 blin] cease
10 Strangeness] aloofness
hight] was called
15 brief a bill] draw up a petition

ANONYMOUS

21 [*The juggler and the baron's daughter*]

Draw me near, draw me near,
Draw me near, ye jolly juggler!

HERE beside dwelleth
A rich baron's daughter;
She would have no man
That for her love had sought her.
So nice she was.

She would have no man
That was made of mould,
But if he had a mouth of gold — 10
To kiss her when she would.
So dangerous she was.

Thereof heard a jolly juggler
That laid was on the green,
And at this lady's words
Ywis he had great teen.
An-angered he was.

He juggled to him a well good steed
Of an old horse-bone,
A saddle and a bridle both, — 20
And set himself thereon.
A juggler he was.

He pricked and pranced both
Before that lady's gate;
She wenned he had been an angel
Was come for her sake.
A pricker he was.

7 nice] coy; hard to please
9 mould] earth
10 But if]] unless
12 dangerous] haughty, arrogant
13 juggler] magician
16 Ywis] certainly
teen] vexation, anger
23 pricked] rode
25 wenned] thought
27 pricker] rider, horseman

He pricked and pranced
  Before that lady's bower;
She wenned he had been an angel   30
  Come from heaven tower.
    A prancer he was.

Four and twenty knights
  Led him into the hall,
And as many squires
  His horse to the stall,
    And gave him meat.

They gave him oats
  And also hay;
He was an old shrew   40
  And held his head away.
    He would not eat.

The day began to pass,
  The night began to come,
To bed was brought
  The fair gentle woman,
    And the juggler also.

The night began to pass,
  The day began to spring;
All the birds of her bower   50
  They began to sing,
    And the cuckoo also.

'Where be ye, my merry maidens,
  That ye come not me to?
The jolly windows of my bower
  Look that you undo,
    That I may see.

'For I have in mine arms
  A duke or else an earl.'
But when she looked him upon,   60
  He was a blear-eyed churl.
    'Alas!' she said.

She led him to an hill,
And hanged should he be.
He juggled himself to a meal-pock;
The dust fell in her eye.
Beguiled she was.

God and Our Lady
And sweet Saint Johan
Send every giglot of this town 70
Such another leman,
Even as he was.

(Pub. 1903)

SIR THOMAS MORE
1477 or 1478–1535

22 *A Lamentation of Queen Elizabeth*

O YE that put your trust and confidence
In worldly joy and frail prosperity,
That so live here as ye should never hence,
Remember death and look here upon me.
Ensample I think there may no better be.
Your self wot well that in this realm was I
Your queen but late, and lo now here I lie.

Was I not born of old worthy lineage?
Was not my mother queen, my father king?
Was I not a king's fere in marriage? 10
Had I not plenty of every pleasant thing?
Merciful God, this is a strange reckoning:
Riches, honour, wealth and ancestry
Hath me forsaken, and lo now here I lie.

If worship might have kept me, I had not gone.
If wit might have me saved, I needed not fear.
If money might have holp, I lacked none.
But O good God what vaileth all this gear?
When death is come, thy mighty messenger,

65 meal-pock] meal-bag
70 giglot] wench, strumpet
71 leman] lover

22
10 fere] wife
15 worship] honour
16 wit] intelligence

18 vaileth] avails
this gear] these possessions

Obey we must, there is no remedy; 20
Me hath he summoned, and lo here I lie.

Yet was I late promised otherwise,
This year to live in wealth and delice.
Lo whereto cometh thy blandishing promise,
O false astrology and divinatrice,
Of God's secrets making thy self so wise!
How true is for this year thy prophecy!
The year yet lasteth, and lo now here I lie.

O brittle wealth, aye full of bitterness,
Thy single pleasure doubled is with pain. 30
Account my sorrow first and my distress,
In sundry wise, and reckon there again
The joy that I have had, and I dare sayn,
For all my honour, endured yet have I
More woe than wealth, and lo now here I lie.

Where are our castles now, where are our towers?
Goodly Richmond, soon art thou gone from me;
At Westminster that costly work of yours,
Mine own dear lord, now shall I never see.
Almighty God vouchsafe to grant that ye 40
For you and your children well may edify.
My palace builded is, and lo now here I lie.

Adieu, mine own dear spouse, my worthy lord.
The faithful love that did us both combine
In marriage and peaceable concord
Into your hands here I clean resign
To be bestowed upon your children and mine.
Erst were you father, and now must ye supply
The mother's part also, for lo now here I lie.

Farewell, my daughter lady Margaret. 50
God wot full oft it grieved hath my mind
That ye should go where we should seldom meet.
Now am I gone, and have left you behind.
O mortal folk, that we be very blind;
That we least fear, full oft it is most nigh:
From you depart I first, and lo now here I lie.

29 wealth] well-being

Farewell, madame, my lord's worthy mother,
  Comfort your son, and be ye of good cheer.
Take all a worth, for it will be no nother.
  Farewell, my daughter Katherine late the fere 60
  To prince Arthur, mine own child so dear.
    It booteth not for me to weep or cry;
    Pray for my soul, for lo now here I lie.

Adieu, Lord Henry, my loving son, adieu.
  Our Lord increase your honour and estate.
Adieu, my daughter Mary, bright of hue.
  God make you virtuous, wise, and fortunate.
  Adieu, sweet heart, my little daughter Kate;
    Thou shalt, sweet babe, such is thy destiny,
    Thy mother never know, for lo now here I lie. 70

Lady Cecily, Anne, and Katherine,
  Farewell my well-beloved sisters three;
O Lady Bridget, other sister mine,
  Lo here the end of worldly vanity.
  Now well are ye that earthly folly flee,
    And heavenly things love and magnify.
    Farewell and pray for me, for lo now here I lie.

Adieu my lords, adieu my ladies all,
  Adieu my faithful servants every chone.
Adieu my commons whom I never shall 80
  See in this world, wherefore to Thee alone,
  Immortal God verily three and one,
    I me commend Thy infinite mercy
    Show to Thy servant, for lo now here I lie.

(Wr. 1503; pub. 1557)

59 a worth] in good part   60 fere] mate

23 *Certain metres written by master Thomas More in his youth for 'The Book of Fortune', and caused them to be printed in the beginning of that book*

The words of Fortune to the people

MINE high estate, power and auctority,
  If ye ne know, ensearch and ye shall spy
That riches, worship, wealth and dignity,
  Joy, rest, and peace, and all thing finally
  That any pleasure or profit may come by
    To man's comfort, aid and sustenance,
    Is all at my device and ordinance.

Without my favour there is nothing won.
  Many a matter have I brought at last
To good conclusion that fondly was begun, 10
  And many a purpose, bounden sure and fast
  With wise provision, I have overcast.
    Without good hap there may no wit suffice;
    Better is to be fortunate than wise.

And therefore hath there some men been or this
  My deadly foes and written many a book
To my dispraise. And other cause there nis,
  But for me list not friendly on them look.
  Thus like the fox they fare that once forsook
    The pleasant grapes and gan for to defy them, 20
    Because he leapt and yet could not come by them.

But let them write, their labour is in vain,
  For well ye wot, mirth, honour and richesse
Much better is than penury and pain.
  The needy wretch that lingereth in distress
  Without mine help is ever comfortless,
    A weary burden odious and loth
    To all the world and eke to himself both.

title '*The Book of Fortune*'] a dice game
10 fondly] foolishly
15 or this] before now
17 nis] is not
20 defy] despise, disdain
27 loth] loathsome

But he that by my favour may ascend
  To mighty power and excellent degree, 30
A common weal to govern and defend,
  O in how blest condition standeth he!
  Himself in honour and felicity,
    And over that, may further and increase
    A region whole in joyful rest and peace.

Now in this point there is no more to say:
  Each man hath of himself the governance.
Let every wight then follow his own way.
  And he that out of poverty and mischance
  List for to live, and will himself enhance 40
    In wealth and riches, come forth and wait on me,
    And he that will be a beggar, let him be.

Thomas More to them that trust in fortune

Thou that art proud of honour, shape, or kin,
  That heapest up this wretched worldes treasure,
Thy fingers shrined with gold, thy tawny skin
  With fresh apparel garnished out of measure,
  And weenest to have fortune at thy pleasure,
    Cast up thine eye and look how slipper chance
    Illudeth her men with change and variance.

Sometime she looketh as lovely fair and bright 50
  As goodly Venus mother of Cupide.
She becketh and she smileth on every wight.
  But this cheer feigned, may not long abide.
  There cometh a cloud, and farewell all our pride.
    Like any serpent she beginneth to swell,
    And looketh as fierce as any fury of hell.

Yet for all that we brotle men are fain
  (So wretched is our nature and so blind),
As soon as Fortune list to laugh again,
  With fair countenance and deceitful mind, 60
  To crouch and kneel and gape after the wind,
    Not one or twain but thousands in a rout,
    Like swarming bees come flickering her about.

39 poverty] (pronounced 'poorty')
45 shrined] enclosed
47 weenest] think
48 slipper] uncertain
49 Illudeth] deceives
52 becketh] nods
57 brotle] brittle
61 gape . . . wind] watch for a favourable moment

Then as a bait she bringeth forth her ware,
  Silver, gold, rich pearl, and precious stone,
On which the mazed people gaze and stare,
  And gape therefore, as dogs do for the bone.
  Fortune at them laugheth, and in her throne
    Amid her treasure and wavering richesse
    Proudly she hoveth as lady and empress. 70

Fast by her side doth weary Labour stand,
  Pale Fear also, and Sorrow all bewept;
Disdain and Hatred on that other hand,
  Eke restless Watch fro sleep with travail kept,
  His eyes drowsy and looking as he slept.
    Before her standeth Danger and Envy,
    Flattery, Deceit, Mischief and Tyranny.

About her cometh all the world to beg.
  He asketh land, and he to pass would bring
This toy and that, and all not worth an egg; 80
  He would in love prosper above all thing;
  He kneeleth down and would be made a king;
    He forceth not so he may money have,
    Though all the world accompt him for a knave.

Lo thus ye see divers heads, divers wits.
  Fortune alone as divers as they all
Unstable here and there among them flits,
  And at aventure down her gifts fall;
  Catch who so may she throweth great and small
    Not to all men, as cometh sun or dew, 90
    But for the most part all among a few.

And yet her brotle gifts long may not last.
  He that she gave them looketh proud and high.
She whirl'th about and pluck'th away as fast,
  And giveth them to another by and by.
  And thus from man to man continually
    She useth to give and take, and slily toss,
    One man to winning of another's loss.

66 mazed] dazed, stupefied
70 hoveth] presides
74 restless watch] sleeplessness
83 forceth] cares
88 at aventure] arbitrarily
98 One man . . . loss] gain to one man at the expense of another

And when she robbeth one, down goeth his pride.
He weepeth and waileth and curseth her full sore. 100
But he that receiveth it, on that other side,
Is glad, and bless'th her often times therefore.
But in a while she loveth him no more,
She glideth from him, and her gifts too.
And he her curseth as other fools do.

Alas the foolish people can not cease,
Ne void her train, till they the harm do feel.
About her alway, busily they preace.
But Lord, how he doth think himself full well
That may set once his hand upon her wheel. 110
He holdeth fast; but upward as he flieth,
She whippeth her wheel about, and there he lieth.

Thus fell Julius from his mighty power.
Thus fell Darius the worthy king of Perse.
Thus fell Alexander the great conquerour.
Thus many mo than I may well rehearse.
Thus double Fortune, when she list reverse
Her slipper favour fro them that in her trust,
She fleeth her way and layeth them in the dust.

She suddenly enhanceth them a loft, 120
And suddenly mischieveth all the flock.
The head that late lay easily and full soft,
In stead of pillows lieth after on the block.
And yet alas the most cruel proud mock:
The dainty mouth that ladies kissed have,
She bringeth in the case to kiss a knave.

Thus when she changeth her uncertain course,
Up start'th a knave, and down there fall'th a knight.
The beggar rich, and the rich man poor is.
Hatred is turned to love, love to despite. 130
This is her sport, thus proveth she her might.
Great boast she maketh if one be by her power
Wealthy and wretched both within an hour.

Poverty that of her gifts will nothing take
With merry cheer looketh upon the press,
And seeth how Fortune's household goeth to wrake.
Fast by her standeth the wise Socrates,
Aristippus, Pythagoras, and many a less

107 void] avoid train] snare, trap 108 preace] press

Of old philosophers. And eke against the sun
Becketh him poor Diogenes in his tun. 140

With her is Byas, whose country lacked defence,
And whilom of their foes stood so in doubt
That each man hastily gan to carry thence
And asked him why he nought carried out.
'I bear,' quod he, 'all mine with me about'.
Wisdom he meant, not Fortune's brotle fees;
For nought he counted his that he might lese.

Heraclitus eke list fellowship to keep
With glad poverty, Democritus also;
Of which the first can never cease but weep 150
To see how thick the blinded people go,
With labour great to purchase care and woe;
That other laugheth to see the foolish apes,
How earnestly they walk about their japes.

Of this poor sect it is common usage
Only to take that nature may sustain,
Banishing clean all other surplusage.
They be content, and of no thing complain.
No niggard eke is of his good so fain,
But they more pleasure have a thousand fold 160
The secret draughts of nature to behold.

Set Fortune's servants by them and ye wull,
That one is free, that other ever thrall,
That one content, that other never full,
That one in surety, that other like to fall.
Who list to advise them both, perceive he shall
As great difference between them as we see
Betwixt wretchedness and felicity.

Now have I showed you both: choose which ye list,
Stately Fortune, or humble Poverty. 170
That is to say, now lieth it in your fist
To take here bondage or free liberty.
But in this point, and ye do after me,

140 tun] tub
141 Byas] one of the seven sages of Greece
142 whilom] formerly, once
146 fees] wealth
147 lese] lose
155 this poor sect] i.e. the philosophers
159 fain] glad
161 the secret draughts of nature] the hidden designs of Nature
162 and ye wull] if you will
173 and ye do after me] if you will take my advice

Draw you to Fortune, and labour her to please,
If that ye think your self too well at ease.

And first, upon thee lovely shall she smile,
And friendly on thee cast her wandering eyes,
Embrace thee in her arms, and for a while
Put thee and keep thee in a fool's paradise;
And forthwith all what thou so list devise 180
She will thee grant it liberally perhaps:
But for all that beware of after-claps.

Reckon you never of her favour sure:
Ye may in clouds as easily trace an hare,
Or in dry land cause fishes to endure,
And make the burning fire his heat to spare,
And all this world in compass to forfare,
As her to make by craft or engine stable,
That of her nature is ever variable.

Serve her day and night as reverently 190
Upon thy knees as any servant may,
And in conclusion, that thou shalt win thereby
Shall not be worth thy service I dare say.
And look yet what she giveth thee today
With labour won she shall haply tomorrow
Pluck it again out of thine hand with sorrow.

Wherefore if thou in surety list to stand,
Take Poverty's part and let proud Fortune go;
Receive no thing that cometh from her hand.
Love manner and virtue: they be only tho 200
Which double Fortune may not take thee fro.
Then mayst thou boldly defy her turning chance:
She can thee neither hinder nor avance.

But and thou wilt needs meddle with her treasure,
Trust not therein, and spend it liberally.
Bear thee not proud, nor take not out of measure.
Build not thine house on height up in the sky.
None falleth far but he that climbeth high.
Remember Nature sent thee hither bare.
The gifts of Fortune count them borrowed ware. 210

182 after-claps] unexpected strokes (after the victim has ceased to be on his guard)
184 trace] track
187 forfare] perish
200 manner] manners (i.e. moral excellence)
tho] those
201 fro] from
202 defy] despise

Thomas More to them that seek fortune

Whoso delighteth to proven and assay
  Of wavering Fortune the uncertain lot,
If that the answer please you not alway,
  Blame ye not me; for I command you not
  Fortune to trust, and eke full well ye wot
    I have of her no bridle in my fist;
    She renneth loose, and turneth where she list.

The rolling dice in whom your luck doth stand,
  With whose unhappy chance ye be so wroth,
Ye know your self came never in mine hand. 220
  Lo in this pond be fish and frogs both.
  Cast in your net; but be you lief or loth,
    Hold you content as Fortune list assign:
    For it is your own fishing and not mine.

And though in one chance Fortune you offend,
  Grudge not thereat, but bear a merry face.
In many another she shall it amend.
  There is no man so far out of her grace
  But he sometime hath comfort and solace,
    Ne none again so far forth in her favour 230
    That is full satisfied with her behaviour.

Fortune is stately, solemn, proud, and high,
  And riches giveth to have service therefore.
The needy beggar catcheth an halfpenny;
  Some man a thousand pound, some less some more.
  But for all that she keepeth ever in store
    From every man some parcel of his will,
    That he may pray therefore and serve her still.

Some man hath good, but children hath he none,
  Some man hath both, but he can get none health. 240
Some hath all three, but up to honour's throne
  Can he not creep, by no manner of stealth.
  To some she sendeth children, riches, wealth,
    Honour, worship, and reverence all his life;
    But yet she pincheth him with a shrewd wife.

236 in store] in reserve
237 parcel] part
239 good] goods
243 wealth] well-being
245 shrewd] shrewish

Then forasmuch as it is Fortune's guise
To grant no man all thing that he will axe,
But as her self list order and devise,
Doth every man his part divide and tax,
I counsel you each one truss up your packs 250
And take no thing at all, or be content
With such reward as Fortune hath you sent.

All things in this book that ye shall read,
Do as ye list, there shall no man you bind
Them to believe as surely as your creed.
But notwithstanding certes in my mind,
I durst well swear, as true ye shall them find
In every point each answer by and by
As are the judgments of astronomy.

(Wr. c.1505; pub. 1557)

ALEXANDER BARCLAY
1475?–1552

from *Eclogues*

24 [*'The Miseries of Courtiers' by Æneas Sylvius Riccolomini*]

[*Eating in Hall*]

Cornix. BUT now hear what meat there needs eat thou must,
And then, if thou mayst, to it apply thy lust.
Thy meat in the court is neither swan nor heron,
Curlew nor crane, but coarse beef and mutton,
Fat pork or veal, and namely such as is bought
For Easter price when they be lean and nought.
Thy flesh is resty or lean, tough and old,
Or it come to board unsavoury and cold,
Sometime twice sodden, and clean without taste,
Sauced with coals and ashes all for haste. 10
When thou it eatest, it smelleth so of smoke
That every morsel is able one to choke.

247 axe] ask
253 this book] i.e. 'The Book of Fortune'
256 certes] certainly
259 astronomy] astrology

24
2 apply] bring, devote
lust] pleasure
5 namely] especially
6 Easter price] at a cheap rate
7 resty] stale
8 Or] before
9 sodden] boiled

Make hunger thy sauce be thou never so nice,
For there shalt thou find none other kind of spice.
Thy pottage is made with weeds and with ashes,
And between thy teeth ofttimes the cullis crashes.
Sometime half-sodden is both thy flesh and broth;
The water and herbs together be so wroth
That each goeth apart, they cannot well agree,
And oft be they salt as water of the sea. 20
Seldom at cheese hast thou a little lick,
And if thou aught have, within it shall be quick,
All full of maggots and like to the rainbow,
Of divers colours as red, green and yellow,
On each side gnawen with mice or with rats,
Or with vile worms, with dogs or with cats,
Unclean and scurvy, and hard as the stone,
It looketh so well thou wouldest it were gone.
If thou have butter, then shall it be as ill
Or worse than thy cheese; but hunger hath no skill, 30
And when that eggs half-hatched be almost
Then are they for thee laid in the fire to roast.
If thou have pears or apples, be thou sure
Then be they such as might no longer endure,
And if thou none eat, they be so good and fine
That after dinner they serve for the swine.
Thy oil for frying is for the lamps meet:
A man it choketh, the savour is so sweet.
A cordwainer's shop and it have equal scent,
Such pain and penance accordeth best to Lent. 40
Such is of this oil the savour perilous
That it might serpents drive out of an house.
Ofttimes it causeth thy stomach to reboke,
And oft it is ready thee suddenly to choke.
Of fish in some court thy chief and used dish
Is whiting, herring, salt-fish and stockfish.
If the day be solemn perchance thou mayst feel
The taste and the sapour of tench or eel;
Their muddy sapour shall make thy stomach ache,
And as for the eel, is cousin to a snake. 50

16 cullis] meat broth
crashes] crunches, disintegrates noisily
18 be so wroth] get on so badly
22 quick] alive
27 scurvy] scurfy, scabby; nasty
30 skill] power of discrimination
37 meet] suitable
39 cordwainer] shoemaker, worker in cordovan leather
43 reboke] regurgitate
46 stockfish] cod
47 solemn] distinguished by special ceremonies; sacred
48 sapour] flavour

But if better fish or any dishes more
Come to thy part, it nought was before,
Corrupt, ill-smelling, and five days old,
For scent thou canst not receive it if thou would.
Thy bread is black, of ill sapour and taste,
And hard as a flint because thou none should waste,
That scant be thy teeth able it to break.
Dip it in pottage if thou no shift can make,
And though white and brown be both at one price,
With brown shalt thou feed lest white might make thee nice. 60
The lords will alway that people note and see
Between them and servants some diversity,
Though it to them turn to no profit at all;
If they have pleasure, the servant shall have small.
Thy dishes be one continuing all the year:
Thou knowest what meat before thee shall appear.
This slacketh great part of lust and pleasour,
Which asketh dainties most divers of sapour.
On one dish daily needs shalt thou blow,
Till thou be all weary as dog of the bow. 70
But this might be suffered, may fortune easily,
If thou saw not sweeter meats to pass by:
For this unto courtiers most commonly doth hap,
That while they have brown bread and cheese in their lap,
On it fast gnawing as houndes ravenous,
Anon by them passeth of meat delicious,
And costly dishes a score may they tell;
Their greedy gorges are rapt with the smell,
The deinteous dishes which pass through the hall . . .

But now to the table for to return again, 80
There hast thou yet another grievous pain:
That when other talk and speak what they will,
Thou dare not whisper, but as one dumb be still.
And if thou aught by word, sign or beck,
Then Jack with the bush shall taunt thee with a check.
One reacheth thee bread with grutch and murmuring;
If thou of some other demand anything,
He hath at thy asking great scorn and disdain,
Because that thou sittest while he standeth in pain.

60 nice] fastidious, hard to please
70 weary . . . bow] as weary as a dog of hunting
82 other] others
84 beck] gesture with the hand
85 Jack with the bush] Jack in office, officious servant
check] rebuke
86 grutch] grumbling
87 if thou] even if you

Sometime the servants be blind and ignorant, 90
And spy not what thing upon the board doth want.
If they see a fault they will it not attend,
By negligent scorn disdaining it to mend.
Sometime thou wantest either bread or wine,
But nought dare thou ask, if thou should never dine.
Demand salt, trencher, spoon, or other thing,
Then art thou importune, and evermore craving:
And so shall thy name be spread to thy pain,
For at thee shall all have scorn and disdain.
Sometime art thou irked of them at the table, 100
But much more art thou of the serving rabble.
The hungry servers which at the table stand
At every morsel hath eye unto thy hand,
So much on thy morsel distract is their mind
They gape when thou gapest, oft biting the wind;
Because that thy leavings is only their part,
If thou feed thee well sore grieved is their heart.
Namely of a dish costly and dainteous,
Each piece that thou cuttest to them is tedious.
Then at the cupboard one doth another tell: 110
'See how he feedeth like the devil of hell!
Our part he eateth; nought good shall we taste.'
Then pray they to God that it be thy last.

Corydon. I had liever, Cornix, go supperless to bed
Than at such a feast to be so bested.
Better is it with cheese and bread one to fill
Than with great dainty, with anger and ill-will;
Or a small handful with rest and sure pleasance
Than twenty dishes with wrathful countenance.

Cornix. That can Amyntas record and testify; 120
But yet is in court more pain and misery.
Brought in be dishes the table for to fill,
But not one is brought in order at thy will.
That thou would have first and lovest principal
Is brought to the board ofttimes last of all.

96 trencher] plate
97 importune] importunate
104 distract] diverted
105 gape] open the mouth (to bite something)
108 Namely] especially
dainteous] dainty
109 tedious] annoying
114 I had liever] I would sooner
115 so bested] so hard put to it

With bread and rude meat when thou art satiate,
Then cometh dishes most sweet and delicate.
Then must thou either despise them utterly,
Or to thy hurt surfeit, ensuing gluttony.
Or if it fortune, as seldom doth befall, 130
That at beginning come dishes best of all,
Or thou hast tasted a morsel or twain,
Thy dish out of sight is taken soon again.
Slow be the servers in serving in alway,
But swift be they after, taking thy meat away.
A special custom is used them among,
No good dish to suffer on board to be long.
If the dish be pleasant, either flesh or fish,
Ten hands at once swarm in the dish.
And if it be flesh, ten knives shalt thou see 140
Mangling the flesh and in the platter flee.
To put there thy hands is peril without fail,
Without a gauntlet or else a glove of mail.
Among all these knives thou one of both must have,
Or else it is hard thy fingers whole to save.
Oft in such dishes, in court it is seen,
Some leave their fingers, each knife is so keen.
On one finger gnaweth some hasty glutton,
Supposing it is a piece of beef or mutton.
Beside these in court, mo pains shalt thou see: 150
At board men be set as thick as they may be.
The platters shall pass ofttimes to and fro,
And over the shoulders and head shall they go.
And oft all the broth and liquor fat
Is spilt on thy gown, thy bonnet and thy hat.
Sometime art thou thrust for little room and place,
And sometime thy fellow reboketh in thy face.
Between dish and dish is tarry tedious,
But in the mean-time, though thou have pain grievous,
Neither mayest thou rise, cough, spit or neeze, 160
Or take other easement, lest thou thy name may lese.
For such as this wise to ease them are wont,
In number of rascals courtiers them count.
Of meat is none hour nor time of certainty,
Yet from beginning absent if thou be,

126 rude] coarse, crude
130 fortune] happen
132 Or . . . tasted] before you have tasted
133 again] back
141 flee] rush
150 mo] more
154 liquor] liquid
157 reboketh] belches
158 tarry] delay
160 neeze] sneeze
161 lese] lose
164 meat] meals

Either shalt thou lose thy meat and kiss the post,
Or if by favour thy supper be not lost,
Thou shalt at the least way rebukes sour abide
For not attending and failing of thy tide.

(1514)

ANONYMOUS

from *Scottish Field*

25 [*The Battle of Flodden*]

THEN the mighty Lord Maxfield   over the mountains fleeth,
And kyred to his king   with careful tithindes,
Telleth him the truth,   and tarrieth he no longer,
Sayeth 'I am beaten back,   for all my big meinie,
And there been killed of the Scots   I know not how many.'
Then the Scottish king   full nigh his wit wanteth,
And said: 'On who was thou matched,   man, by the sooth?'
And he promised him pertly   they passed not a thousand.
'Ye been cowards,' quod the king,   'Care mote ye happen!
I will wind you to wreak,   wees, I you hete,   10
Along within that land   the length of three weeks,
And destroy all aright   that standeth me before.'
Thus he promised to the prince   that paradise wieldeth.
Then he summoned his sedges,   and set them in order.
The next way to Norham   anon then he taketh;
He umclosed that castle   clean round about,
And they defended fast,   the folk that were within.
Without some succour come soon   their sorrow is the more!
The Earl of Surrey himself   at Pomfret abideth,
And heard what unhap   all those harlots didden.   20

166 kiss the post] be shut out for arriving too late

25
Scottish Field] the battle against the Scots (i.e. Flodden, fought on 9 September, 1513, at Flodden Edge, in Northumberland, while Henry VIII was in France)
1 Maxfield] i.e. Maxwell
2 kyred] went
his king] James IV of Scotland
careful tithindes] lamentable tidings
4 meinie] body of soldiers, retainers
7 matched] opposed
8 promised] assured
pertly] promptly
9 care . . . happen] may evil come to you!
10 wind] go
wreak] avenge
wees] men
hete] assure
14 sedges] men, soldiers
15 next] most direct
16 umclosed] closed in, invested
19 Pomfret] Pontefract
20 unhap] mishap, harm
harlots] villains

He made letters boldly all the land over:
Into Lancashire belive he caused a man to ride
To the Bishop of Ely, that bode in those parts;
Courteously commanded him, in the King's name,
To summon the shire and set them in order.
He was put in more power than any prelate else.
Then the bishop full boldly bouneth forth his standard
With a captain full keen, as he was known after.
He made a wee to wind to warn his dear brother
Edward, that eager knight, that epe was of deeds, 30
A stalk of the Stanleys, stepe of him selven.
Then full radly he raiseth rinks ten thousandes;
To Skipton in Crane then he come belive.
There abideth he the banner of his dear brother,
Till a captain with it come that knowen was full wide:
Sir John Stanley, that stout knight, that stern was of deeds,
With four thousand fierce men that followed him after.
They were tenants that they took, that tenden on the bishop,
Of his household, I you hete, hope ye no other.
Every burn had on his breast, broidered with gold, 40
A foot of the fairest fowl that ever flew on wing,
With three crowns full clear, all of pure gold.
It was a seemly sight to see them together:
Fourteen thousand eagle-feet fettled in array.
Thus they coasten through the country to the new castle:
Proclamation in that place was plainly declared
That every hattel should him hie, in haste that he might,
To Bolton in Glendour, all in godly haste.
There met they at a muster men many thousand,
With knights that were keen, full well known in their country, 50
And many a lovely lord upon that land light.
Then they moved towards the mountains those meinie to seek,
Those scattle Scots, that all the scathe didden.

22 belive] with haste
23 bode] lived
27 bouneth] prepared to move
29 wee] man
wind] go
30 epe] bold
31 stalk] scion, twig
stepe] (?) scion, piece
32 radly] speedily
rinks] men, warriors
33 Crane] Craven (West Yorkshire)
36 stern] fierce
38 tenden] followed in service
40 burn] man
41 A foot . . . full clear] (the Stanley crest, an eagle's foot with three crowns)
44 fettled] prepared
array] battle order
45 coasten] travel near the coast
new castle] i.e. Newcastle-on-Tyne
47 hattel] man
hie] hasten
in haste . . . might] as hastily as possible
48 Bolton] (west of Alnwick, Northumberland)
51 light] arrived
53 scattle] harmful
scathe] damage

They would never rest, but alway raked forward
Till they had seen the sedges that they had sought after;
But they had gotten them a ground most ungracious of other
Upon the top of a high hill, I hete you for sooth.
There was no wee in this world might wind them again
But he should be killed in the cloes or he could climb the
  mountains.
When the lords had on them looked as long as them liked, 60
Every captain was commanded their company to order.
Though we were bashed of these burns, I blame us but little.
Then we tilled down our tents, that told were a thousand.
At the foot of a fine hill they settled them all night.
There they lien and lodged the length of four days,
Till every captain full keenly callen to their lords,
Bid them fettle them to fight or they would fare homeward:
Their company was clemmed, and much cold did suffer:
Water was a worthy drink, win it who might.
  Then the Lord Lieutenant looked him about, 70
And boldly unto battle busked he his meinie.
The Lord Howard, the hende knight, have should the vanward,
With 14,000 fierce men that followed him after.
The left wing to that ward was Sir Edward Howard:
He chose to him Cheshire —their chance was the worse—
Because they knew not their captain their care was the more,
For they were wont at all war to wait upon the Stanleys.
Much worship they won when they that wee served,
But now lank is their loss —Our Lord it amend!
The right wing, as I ween, was my Lord Lumley, 80
A captain full keen, with St. Cuthbert's banner.
My Lord Clifford with him came, all in clear armour;
So did Sir William Percy, that proved was of deeds;
And Sir William Bulmer, that bold hath been ever,
With many captains full keen, who so knows their names.
And if I reckon the rearward, I rest must too long,
But I shall tell you the best frekes that thereupon tenden.
The Earl of Surrey himself surely it guided;
The Lord Scroop full comely, with knights full many.
If ye would wit the wings that to that ward longed, 90

54 raked] went quickly
59 cloes] hillsides
  or he] before he
62 Though we . . . little] if we had been beaten (shamed) by these men, I would not have blamed us
63 tilled down] pitched
  told] numbered at
68 clemmed] starved
71 busked] prepared
72 hende] courteous
  vanward] main division of the army
79 lank] (?) long-standing
80 ween] think, believe
86 rearward] rear section of the army
87 frekes] men
90 wit] know

That was a bishop full bold, that born was at Lathom:
Of Ely that ilk lord, that epe was of deeds,
An egg of that bold earl that named was Stanley,
Near of nature to the Nevilles, that noble have been ever;
But now death with his dart hath driven him away.
It is a loss to the land —Our Lord have his soul!—
For his wit and his wisdom and his wale deeds,
He was a pillar of peace the people among.
His servants they may sike and sorrow for his sake;
What for pity and for pain my pen doth me fail. 100
I will meddle with this matter no more for this time,
But He that is makless of mercy have mind on his soul!
Then he sent with his company a knight that was noble:
Sir John Stanley, that stout knight, that stern was of deeds.
There was never burn born that day bare him better.
The left wing to that rearward was my Lord Mounteagle,
With many ledes of Lancashire that to him longed,
Which foughten full fiercely whiles the field lasted.
Thus the rearward in array raked ever after,
As long as the light day lasted on the ground. 110
Then the sun full soon shot under the clouds,
And it darkened full dimly, and drew toward the night.
Every rink to his rest full radly him dressed,
Beten fires full fast, and fettled them to sowp
Besides Berwick in a bank within a broad wood.
Then dayned the day, so dear God it ordained:
Clouds cast up full clearly like castles full high.
Then Phoebus full fair flourished out his beams,
With leams full light all the land over.
All was damped with dew the daisies about; 120
Flowers flourished in the fields, fair to behold;
Briddes braiden to the boughs, and boldly they songen:
It was solace to hear for any sedge living.
Then full boldly on the broad hills we busked our standards,
And on a sough us beside there seen we our enemies
Were moving over the mountains; to match us they thoughten,
As boldly as any burns that born were of mothers;
And we eagerly with ire atiled them to meet.

93 egg] offspring
97 wale] noble
99 sike] sigh
102 makless] matchless, unequalled
107 ledes] men, vassals
108 field] battle
114 beten] kindled
sowp] have supper
116 dayned] dawned
119 leams] rays
122 Briddes] birds
braiden] burst into motion
123 solace] pleasure, delight
125 sough] swampy place
128 eagerly] fiercely, impetuously
atiled] armed, equipped

Then trumpets full truly they triden together;
Many shawms in that shaw, with their shrill notes. 130
Heavenly was their melody, their mirths to hear,
How they songen with a shout all the shaws over.
There was girding forth of guns with many great stones;
Archers uttered out their arrows, and eagerly they shotten.
They proched us with spears, and put many over,
That the blood outbrast at their broken harness.
There were swingeing out of swords and swapping off heads.
We blanked them with bills through all their bright armour,
That all the dale dinned of their derf strokes.
Then betide a check that Cheshire men fledden. 140
In wing with those wees was my Lord Dacres:
He fled at the first braid, and they followed after.
When their captain was away, their comfort was gone.
They were wont at all wars to wait upon the Stanleys:
They never failed at no forward that time that they were.
Now lost is their loss —Our Lord it amend!
Many swires full swiftly were swapped to the death.
Sir John Booth of Barton was brought from his life:
A more bolder burn was never born of woman;
And of Yorkshire a young knight that epe was of deeds: 150
Sir William Warcop, as I ween, was the wee's name;
Of the same shire Sir William, that was so fierce holden;
Besides Rotherham that rink his resting-place had.
The Baron of Kinderton full keenly was killed them beside;
So was Honford, I you hete, that was a hind swyre;
Fullsewise full fell was fallen to the ground;
Christopher Savage was down cast, that kere might he never;
And of Lancashire John Lawrence —Our Lord have their souls!
These frekes would never flee for fear that might happen:
They were killed like conquerors in their King's service. 160
When the Scots and the Ketericks seen our men scatter,
They had great joy of their joining, and jollily came downward.
Then the Scots' King calleth to him a herald,
Biddeth tell him the truth, and tarry no longer:
Who were the banners of the burns that bode in the valley?

129 triden] (?) sounded
130 shaw] thicket
131 mirths] exhilarating sounds
133 girding] firing
134 uttered] shot
135 proched] pierced
136 outbrast] burst out
137 swapping] striking
138 blanked] nonplussed, disconcerted
139 derf] grievous, cruel
140 check] sudden reverse
142 braid] attack
145 forward] front line
147 swires] squires
156 Fullsewise] i.e. Robert Fouleshurst of Crewe (Cheshire)
157 kere] go, walk
161 Ketericks] Highland fighting men

'They are the standards of the Stanleys that stand by themselven.
If he be faren into France the Frenchmen to fear,
Yet is his standard in that stead with a stiff captain:
Sir Henry Keighley is called, that keen is of deeds.
Sir Thomas Gerard, that jolly knight, is joined thereunder 170
With Sir William Molyneux with a manful meinie.
These frekes will never flee for fear of no weapon,
But they will stick with their standards in their steel weeds.
Because they bashed them at Berwick, that boldeth them the more.
Lo, how he batters and beats, the bird with his wings!
We are feard of yonder fowl, so fiercely he fareth;
And yonder streamer full straight, that standeth him beside,
Is the standard of St Towder —trow ye no other—
That never beaten was in battle for burn upon live.
The third standard in that stead is my Lord's Mounteagle, 180
And of Yorkshire full epe, my young Lord Dacres,
With much puissance and power of that pure shire.'
Then the Scottish King carped these words:
'I will fight with yonder frekes that are so fierce holden:
And I beat those burns, the battle is ours!'
Then he moved toward the mountains, and manly came downward:
We met him in the midway, and matched him full even.
Then there was dealing of dents, that all the dales rung;
Many helms with heads were hewn all to pieces.
This lake lasted on the land the length of four hours. 190
Yorkshire like yorn men eagerly they foughten;
So did Derbyshire that day deyred many Scots;
Lancashire like lions laiden them about.
All had been lost, by Our Lord, had not those ledes been!
But the care of the Scots increased full sore,
For their King was down knocked and killed in their sight
Under the banner of a bishop: that was the bold Stanley.
Then they fettled them to fly as fast as they might,
But that served not, for sooth, who so truth telleth.
Our Englishmen full eagerly after them followed, 200
And killed them like caitiffs in clows all about.
There were killed of the Scots, that told were by tale,

168 stead] place
stiff] steadfast
173 bashed] disconcerted, abashed
175 batters] flaps aggressively
177 streamer] triangular elongated flag
178 St Towder] St Audrey (whose banner was the Bishop of Ely's standard)
trow] know, believe
179 burn upon live] any man alive
183 carped] spoke
185 And I beat] if I beat
190 lake] fight
191 yorn] active, fast-moving
192 deyred] injured, hurt
201 caitiffs] wretches, villains
clows] ravines, steep-sided valleys
202 told . . . tale] estimated by a careful count

That were found in the field, fifteen thousand.
Lo, what it is to be false and the fiend serve!
They have broken a book oath to their blessed king,
And the truce that was taken for the space of two years.
All the Scots that were scaped were scattered far asunder;
They removed over the moor upon the other morning,
And there stood like stakes, and stir durst no further,
For all the lords of their land were left them behind. 210
Beside Brimstone in a brink breathless they lien,
Gaping against the moon, —their ghosts were away!
Then the Earl of Surrey himself calleth to him a herald,
Bade him fare into France with these fair tithands:
'Commend me to our King these comfortable words:
Tell him I have rescued his realm, so right required.
The King of Scots is killed, with all his cursed lords.'
When the King, of his kindness, heard these words,
He saith: 'I will sing him a soul-knell with the sound of my guns.'
Such a noise, to my name, was never heard before, 220
For there was shot at a shot a thousand at once,
That all rang with the rout, rocher and other.
Now is this fierce field foughten to an end;
Many a wee wanted his horse and wandered home on foot.
All was long of the march men —a mischief them happen!
He was a gentleman, by Jesu, that this gest made,
Which said but as ye see, for sooth, and no other.
At Baguley that burn his biding-place he had;
His ancestors of old time have yerded there long,
Before William Conqueror this country inhabited. 230
Jesu, bring them to thy bliss, that brought us forth of bale,
That have hearkened me here, and heard well my tale!

(Wr. 1515(?); pub. 1856)

205 their blessed king] Henry VIII
208 other morning] next morning
211 Brimstone] (?) Branxton
brink] land by a river
220 to my name] to my knowledge
222 rout] noise
rocher] rock, rocky bank
225 march] border
226 gest] story, tale
228 Baguley] (in Cheshire)
229 yerded] dwelt
231 bale] evil

SIR THOMAS WYATT

c.1503–1542

26 *And wilt thou leave me thus?*

AND wilt thou leave me thus?
Say nay, say nay, for shame,
To save thee from the blame
Of all my grief and grame.
And wilt thou leave me thus?
  Say nay, say nay.

And wilt thou leave me thus
That hath loved thee so long
In wealth and woe among?
And is thy heart so strong 10
As for to leave me thus?
  Say nay, say nay.

And wilt thou leave me thus
That hath given thee my heart
Never for to depart,
Nother for pain nor smart?
And wilt thou leave me thus?
  Say nay, say nay.

And wilt thou leave me thus
And have no more pity 20
Of him that loveth thee?
Helas, thy cruelty!
And wilt thou leave me thus?
  Say nay, say nay!

(Pub. 1815)

27 *Madam, withouten many words*

MADAM, withouten many words,
Once, I am sure, ye will or no.
And if ye will, then leave your bourds
And use your wit and shew it so,

4 grame] sorrow
10 strong] hard, unfeeling
22 Helas] alas

27
2 Once] eventually
3 bourds] jokes

And with a beck ye shall me call.
And if of one that burneth alway
Ye have any pity at all
Answer him fair with yea or nay.

If it be yea, I shall be fain.
If it be nay, friends as before. 10
Ye shall another man obtain,
And I mine own, and yours no more.

(Pub. 1815)

28 In aeternum

In aeternum I was once determed
For to have loved, and my mind affirmed
That with my heart it should be confirmed
In aeternum.

Forthwith I found the thing that I might like
And sought with love to warm her heart alike,
For as me thought I should not see the like
In aeternum.

To trace this dance I put myself in press.
Vain hope did lead and bade I should not cease 10
To serve, to suffer, and still to hold my peace
In aeternum.

With this first rule I furthered me apace
That, as methought, my troth had taken place
With full assurance to stand in her grace
In aeternum.

It was not long ere I by proof had found
That feeble building is on feeble ground;
For in her heart this word did never sound:
In aeternum. 20

In aeternum then from my heart I cast
That I had first determined for the best.
Now in the place another thought doth rest
In aeternum.

(Pub. 1815)

5 beck] signal
9 fain] glad

28
In aeternum] for ever
1 determed] determined

29 *Whoso list to hunt*

WHOSO list to hunt, I know where is an hind,
But as for me, helas, I may no more.
The vain travail hath wearied me so sore,
I am of them that farthest cometh behind.
Yet may I by no means my wearied mind
Draw from the deer, but as she fleeth afore
Fainting I follow. I leave off therefore
Sithens in a net I seek to hold the wind.
Who list her hunt, I put him out of doubt,
As well as I may spend his time in vain. 10
And graven with diamonds in letters plain
There is written her fair neck round about:
'*Noli me tangere* for Caesar's I am,
And wild for to hold though I seem tame.'

(Pub. 1815)

30 *Farewell, Love*

FAREWELL, Love, and all thy laws forever.
Thy baited hooks shall tangle me no more.
Senec and Plato call me from thy lore
To perfect wealth my wit for to endeavour.
In blind error when I did persever,
Thy sharp repulse, that pricketh ay so sore,
Hath taught me to set in trifles no store
And scape forth since liberty is lever.
Therefore farewell. Go trouble younger hearts
And in me claim no more authority. 10
With idle youth go use thy property
And thereon spend thy many brittle darts:
For hitherto though I have lost all my time,
Me lusteth no longer rotten boughs to climb.

(Pub. 1557)

2 helas] alas
13 *Noli me tangere*] do not touch me

30
4 wealth] (spiritual) well-being
8 lever] dearer, preferable
14 Me lusteth] I want

31

Forget not yet

FORGET not yet the tried intent
Of such a truth as I have meant,
My great travail so gladly spent.
Forget not yet.

Forget not yet when first began
The weary life ye know since when,
The suit, the service none tell can.
Forget not yet.

Forget not yet the great assays,
The cruel wrong, the scornful ways, 10
The painful patience in denays.
Forget not yet.

Forget not yet, forget not this:
How long ago hath been and is
The mind that never meant amiss.
Forget not yet.

Forget not then thine own approved
The which so long hath thee so loved
Whose steadfast faith yet never moved.
Forget not this. 20

(Pub. 1815)

32

Is it possible

Is it possible
That so high debate,
So sharp, so sore, and of such rate,
Should end so soon and was begun so late?
Is it possible?

31
11 denays] denials

32
2 debate] strife

Is it possible
So cruel intent,
So hasty heat and so soon spent,
From love to hate and thence for to relent?
Is it possible? 10

Is it possible
That any may find
Within one heart so diverse mind
To change or turn as weather and wind?
Is it possible?

Is it possible
To spy it in an eye
That turns as oft as chance on die?
The truth whereof can any try?
Is it possible? 20

It is possible
For to turn so oft,
To bring that lowest that was most aloft
And to fall highest yet to light soft.
It is possible.

All is possible
Whoso list believe.
Trust therefore first and after preve,
As men wed ladies by licence and leave,
All is possible. 30

(Pub. 1815)

33 *My lute, awake!*

My lute, awake! Perform the last
Labour that thou and I shall waste,
And end that I have now begun;
For when this song is sung and past,
My lute, be still for I have done.

As to be heard where ear is none,
As lead to grave in marble stone,
My song may pierce her heart as soon.
Should we then sigh or sing or moan?
No, no, my lute, for I have done. 10

The rocks do not so cruelly
Repulse the waves continually
As she my suit and affection,
So that I am past remedy,
Whereby my lute and I have done.

Proud of the spoil that thou hast got
Of simple hearts thorough Love's shot
By whom, unkind, thou hast them won,
Think not he hath his bow forgot
Although my lute and I have done. 20

Vengeance shall fall on thy disdain
That makest but game on earnest pain.
Think not alone under the sun
Unquit to cause thy lovers plain
Although my lute and I have done.

May chance thee lie withered and old
The winter nights that are so cold,
Plaining in vain unto the moon.
Thy wishes then dare not be told.
Care then who list for I have done. 30

And then may chance thee to repent
The time that thou hast lost and spent
To cause thy lovers sigh and swoon.
Then shalt thou know beauty but lent
And wish and want as I have done.

Now cease, my lute. This is the last
Labour that thou and I shall waste,
And ended is that we begun.
Now is this song both sung and past.
My lute, be still, for I have done. 40

(Pub. 1557)

34 *They flee from me*

THEY flee from me that sometime did me seek
With naked foot stalking in my chamber.
I have seen them gentle, tame, and meek
That now are wild and do not remember
That sometime they put themself in danger
To take bread at my hand; and now they range
Busily seeking with a continual change.

Thanked be fortune it hath been otherwise
Twenty times better, but once in special,
In thin array after a pleasant guise, 10
When her loose gown from her shoulders did fall
And she me caught in her arms long and small,
Therewithal sweetly did me kiss
And softly said, 'Dear heart, how like you this?'

It was no dream: I lay broad waking.
But all is turned thorough my gentleness
Into a strange fashion of forsaking.
And I have leave to go of her goodness
And she also to use newfangleness.
But since that I so kindly am served 20
I would fain know what she hath deserved.

(Pub. 1557)

35 *With serving still*

WITH serving still
This have I won:
For my good will
To be undone.

And for redress
Of all my pain
Disdainfulness
I have again.

And for reward
Of all my smart, 10
Lo, thus unheard,
I must depart.

2 stalking] moving stealthily 5 in danger] in my power 12 small] slender

Wherefore all ye
That after shall
By fortune be,
As I am, thrall,

Example take
What I have won:
Thus for her sake
To be undone. 20

(Pub. 1913)

36 *What should I say*

WHAT should I say
Since faith is dead
And truth away
From you is fled?
Should I be led
With doubleness?
Nay, nay, mistress!

I promised you,
You promised me
To be as true 10
As I would be.
But since I see
Your double heart,
Farewell, my part!

You for to take
Is not my mind
But to forsake
Your cruel kind,
And as I find
So will I trust. 20
Farewell, unjust!

Can ye say nay
But that you said
That I alway
Should be obeyed?
And thus betrayed
Ere that I wist!
Farewell, unkissed!

(Pub. 1815)

37 *In court to serve*

IN court to serve, decked with fresh array,
Of sugared meats feeling the sweet repast,
The life in banquets and sundry kinds of play
Amid the press of lordly looks to waste,
Hath with it joined oft times such bitter taste
That whoso joys such kind of life to hold
In prison joys, fettered with chains of gold.

(Pub. 1557)

38 *Sometime I fled the fire*

SOMETIME I fled the fire that me brent
By sea, by land, by water, and by wind,
And now I follow the coals that be quent
From Dover to Calais, against my mind.
Lo, how desire is both sprung and spent!
And he may see that whilom was so blind,
And all his labour now he laugh to scorn,
Meshed in the briers that erst was all to-torn.

(Pub. 1559)

39 Quondam *was I*

Quondam was I in my lady's grace,
I think as well as now be you:
And when that you have trod the trace,
Then shall you know my words be true,
That *quondam* was I.

Quondam was I. She said, 'for ever'.
That 'ever' lasted but a short while,
A promise made not to dissever;
I thought she laughed, she did but smile.
Then *quondam* was I. 10

1 Sometime] once in the past
brent] burnt
3 quent] quenched
6 whilom] once
7 he laugh] he may laugh
8 to-torn] torn to shreds

39
1 *Quondam*] once, formerly
3 trod the trace] walked the path of hard experience
8 dissever] separate

Quondam was I that full oft lay
In her arms with kisses many a one.
It is enough that I may say,
Though 'mong the moe now I be gone,
Yet *quondam* was I.

Quondam was I: she will you tell
That since the hour she was first born
She never loved none half so well
As you. But what though she had sworn,
Sure *quondam* was I. 20

(Pub. 1961)

40 *Who list his wealth and ease retain*

WHO list his wealth and ease retain,
Himself let him unknown contain.
Press not too fast in at that gate
Where the return stands by disdain,
For sure, *circa Regna tonat.*

The high mountains are blasted oft
When the low valley is mild and soft.
Fortune with Health stands at debate.
The fall is grievous from aloft.
And sure, *circa Regna tonat.* 10

These bloody days have broken my heart.
My lust, my youth did them depart,
And blind desire of estate.
Who hastes to climb seeks to revert.
Of truth, *circa Regna tonat.*

The bell tower showed me such sight
That in my head sticks day and night.
There did I learn out of a grate,
For all favour, glory, or might,
That yet *circa Regna tonat.* 20

14 'mong the moe] among the many (no longer in her favour)

40

1 Who list] whoever wishes to
wealth] well-being
2 contain] keep
5 *circa Regna tonat*] 'he [Jupiter] thunders around thrones' (Seneca, *Phaedra*, 1140)
7 soft] free from storms, balmy
8 at debate] in conflict
12 depart] leave (me)
13 estate] high position
14 seeks to revert] is in fact bringing about the opposite
17 such sight] i.e. the execution of Anne Boleyn
18 grate] prison window

By proof, I say, there did I learn:
Wit helpeth not defence to earn,
Of innocency to plead or prate.
Bear low, therefore, give God the stern,
For sure, *circa Regna tonat.*

(Wr. probably 1536; pub. 1961)

41 *In mourning wise*

IN mourning wise since daily I increase,
Thus should I cloak the cause of all my grief:
So pensive mind with tongue to hold his peace,
My reason sayeth there can be no relief;
Wherefore, give ear, I humbly you require,
The affects to know that thus doth make me moan.
The cause is great of all my doleful cheer
For those that were and now be dead and gone.

What though to death desert be now their call
As by their faults it doth appear right plain? 10
Of force I must lament that such a fall
Should light on those so wealthily did reign,
Though some perchance will say, of cruel heart,
'A traitor's death why should we thus bemoan?'
But I, alas, set this offence apart,
Must needs bewail the death of some be gone.

As for them all I do not thus lament
But as of right my reason doth me bind.
But as the most doth all their deaths repent,
Even so do I by force of mourning mind. 20
Some say, 'Rochford, hadst thou been not so proud,
For thy great wit each man would thee bemoan.'
Since as it is so, many cry aloud,
'It is great loss that thou art dead and gone.'

Ah, Norris, Norris, my tears begin to run
To think what hap did thee so lead or guide,
Whereby thou hast both thee and thine undone,
That is bewailed in court of every side.

21 proof] experience
24 give God the stern] let God steer
41
5 require] ask, request
6 affects] feelings
9 what though . . . call] what does it matter if deserving to die is what has now called them to death
12 those . . . reign] those who held power with such well-being
19 repent] regret

In place also where thou hast never been
Both man and child doth piteously thee moan. 30
They say, 'Alas, thou art far overseen
By thine offences to be thus dead and gone.'

Ah, Weston, Weston, that pleasant was and young,
In active things who might with thee compare?
All words accept that thou didst speak with tongue,
So well esteemed with each where thou didst fare.
And we that now in court doth lead our life,
Most part in mind doth thee lament and moan.
But that thy faults we daily hear so rife,
All we should weep that thou art dead and gone. 40

Brereton, farewell, as one that least I knew.
Great was thy love with diverse, as I hear,
But common voice doth not so sore thee rue
As other twain that doth before appear.
But yet no doubt but thy friends thee lament
And other hear their piteous cry and moan.
So doth each heart for thee likewise relent
That thou giv'st cause thus to be dead and gone.

Ah, Mark, what moan should I for thee make more
Since that thy death thou hast deserved best, 50
Save only that mine eye is forced sore
With piteous plaint to moan thee with the rest?
A time thou hadst above thy poor degree,
The fall whereof thy friends may well bemoan.
A rotten twig upon so high a tree
Hath slipped thy hold and thou art dead and gone.

And thus, farewell, each one in hearty wise.
The axe is home, your heads be in the street.
The trickling tears doth fall so from my eyes,
I scarce may write, my paper is so wet. 60
But what can help when death hath played his part
Though nature's course will thus lament and moan?
Leave sobs therefore, and every Christian heart
Pray for the souls of those be dead and gone.

(Wr. after 1536, pub. 1961)

31 far overseen] much betrayed or deceived
33 pleasant] amusing, jocular
46 other hear] others hear
53 poor degree] low social rank
57 hearty] heartfelt

42 *Tagus, farewell*

TAGUS, farewell, that westward with thy streams
Turns up the grains of gold already tried,
With spur and sail for I go seek the Thames,
Gainward the sun that shew'th her wealthy pride
And, to the town which Brutus sought by dreams,
Like bended moon doth lend her lusty side.
My king, my country, alone for whom I live,
Of mighty love the wings for this me give.

(Wr. prob. 1539, pub. 1557)

43 *If waker care*

IF waker care, if sudden pale colour,
If many sighs, with little speech to plain,
Now joy, now woe, if they my cheer distain,
For hope of small, if much to fear therefore,
To haste, to slack my pace less or more
Be sign of love, then do I love again.
If thou ask whom, sure since I did refrain
Brunet that set my wealth in such a roar,
Th'unfeigned cheer of Phyllis hath the place
That Brunet had. She hath and ever shall. 10
She from myself now hath me in her grace.
She hath in hand my wit, my will, and all.
My heart alone well worthy she doth stay
Without whose help scant do I live a day.

(Wr. after 1536, pub. 1557)

44 *The pillar perished is*

THE pillar perished is whereto I leant,
The strongest stay of mine unquiet mind;
The like of it no man again can find—
From east to west still seeking though he went—
To mine unhap, for hap away hath rent
Of all my joy the very bark and rind,
And I, alas, by chance am thus assigned
Dearly to mourn till death do it relent.

4 gainward] against

43

1 waker] wakeful

2 plain] complain

3 my cheer distain] discolour my face

7 refrain] give up

8 Brunet] (?) Anne Boleyn

wealth] well-being

9 Phyllis] (?) Elizabeth Darrell

14 scant] scarcely

44

5 unhap] misfortune

But since that thus it is by destiny,
What can I more but have a woeful heart, 10
My pen in plaint, my voice in woeful cry,
My mind in woe, my body full of smart,
And I myself myself always to hate
Till dreadful death do cease my doleful state?

(Wr. *c.*1540(?), pub. 1557)

45 *Lucks, my fair falcon*

LUCKS, my fair falcon, and your fellows all,
How well pleasant it were your liberty,
Ye not forsake me that fair might ye befall.
But they that sometime liked my company
Like lice away from dead bodies they crawl.
Lo, what a proof in light adversity!
But ye, my birds, I swear by all your bells,
Ye be my friends and so be but few else.

(Wr. prob. 1540–1, pub. 1557)

46 *Sighs are my food*

SIGHS are my food, drink are my tears;
Clinking of fetters such music would crave.
Stink and close air away my life wears.
Innocency is all the hope I have.
Rain, wind, or weather I judge by mine ears.
Malice assaulteth that righteousness should save.
Sure I am, Brian, this wound shall heal again
But yet, alas, the scar shall still remain.

(Wr. prob. 1541, pub. 1557)

47 *Throughout the world, if it were sought*

THROUGHOUT the world, if it were sought,
Fair words enough a man shall find.
They be good cheap; they cost right naught;
Their substance is but only wind.
But well to say and so to mean—
That sweet accord is seldom seen.

(Pub. 1557)

2–3 How well pleasant . . . befall] however pleasing your liberty would be to you, you do not forsake me so that fair things might happen to you

6 proof] test (of friendship)

46

7 Brian] Sir Francis Brian

48 *Fortune doth frown*

FORTUNE doth frown.
What remedy?
I am down
By destiny.

(Pub. 1913)

49 [*Part of a chorus from Seneca's* Thyestes]

STAND whoso list upon the slipper top
Of court's estates, and let me here rejoice
And use me quiet without let or stop,
Unknown in court that hath such brackish joys.
In hidden place so let my days forth pass
That, when my years be done withouten noise,
I may die aged after the common trace.
For him death grip'th right hard by the crop
That is much known of other, and of himself, alas,
Doth die unknown, dazed, with dreadful face. 10

(Pub. 1557)

50 *Psalm 130*

FROM depth of sin and from a deep despair,
From depth of death, from depth of heart's sorrow,
From this deep cave, of darkness deep repair,
Thee have I called, O Lord, to be my borrow.
Thou in my voice, O Lord, perceive and hear
My heart, my hope, my plaint, my overthrow,
My will to rise, and let by grant appear
That to my voice thine ears do well intend.
No place so far that to thee is not near;

49
1 slipper] slippery
3 use me quiet] comport myself quietly
let] hindrance
7 trace] way
8 crop] throat
9 other] others
10 dazed] stunned
dreadful] full of fear

50
3 repair] dwelling-place
4 borrow] pledge, ransom
8 intend] listen sympathetically

No depth so deep that thou ne mayst extend 10
  Thine ear thereto. Hear then my woeful plaint.
  For, Lord, if thou do observe what men offend
And put thy native mercy in restraint,
  If just exaction demand recompense,
  Who may endure, O Lord? Who shall not faint
At such account? Dread and not reverence
  Should so reign large. But thou seeks rather love,
  For in thy hand is mercy's residence
By hope whereof thou dost our hearts move.
  I in thee, Lord, have set my confidence; 20
  My soul such trust doth evermore approve.
Thy holy word of eterne excellence,
  Thy mercy's promise that is alway just,
  Have been my stay, my pillar, and pretence.
My soul in God hath more desirous trust
  Than hath the watchman looking for the day
  By the relief to quench of sleep the thrust.
Let Israel trust unto the Lord alway
  For grace and favour arn his property.
  Plenteous ransom shall come with him, I say, 30
And shall redeem all our iniquity.

(Pub. 1549)

51 *Mine own John Poyntz*

MINE own John Poyntz, since ye delight to know
    The cause why that homeward I me draw
    (And flee the press of courts whereso they go
Rather than to live thrall under the awe
    Of lordly looks) wrapped within my cloak,
    To will and lust learning to set a law,
It is not because I scorn or mock
    The power of them to whom Fortune hath lent
    Charge over us, of right to strike the stroke;
But true it is that I have always meant 10
    Less to esteem them than the common sort,
    Of outward things that judge in their intent
Without regard what doth inward resort.
    I grant sometime that of glory the fire
    Doth touch my heart; me list not to report

21 approve] find to be true
27 thrust] thirst
29 arn] are
property] natural quality
51
10 meant] intended
15–16 report Blame by] cast aspersions on

Blame by honour and honour to desire.
But how may I this honour now attain
That cannot dye the colour black a liar?
My Poyntz, I cannot frame my tune to feign,
To cloak the truth for praise, without desert, 20
Of them that list all vice for to retain.
I cannot honour them that sets their part
With Venus and Bacchus all their life long,
Nor hold my peace of them although I smart.
I cannot crouch nor kneel to do such wrong
To worship them like God on earth alone
That are like wolves these silly lambs among.
I cannot with my words complain and moan
And suffer naught, nor smart without complaint,
Nor turn the word that from my mouth is gone. 30
I cannot speak with look right as a saint,
Use wiles for wit and make deceit a pleasure
And call craft counsel, for profit still to paint.
I cannot wrest the law to fill the coffer,
With innocent blood to feed myself fat,
And do most hurt where most help I offer.
I am not he that can allow the state
Of him Caesar and damn Cato to die,
That with his death did scape out of the gate
From Caesar's hands, if Livy doth not lie, 40
And would not live where liberty was lost,
So did his heart the common wealth apply.
I am not he such eloquence to boast
To make the crow singing as the swan,
Nor call 'the lion' of coward beasts the most
That cannot take a mouse as the cat can;
And he that dieth for hunger of the gold,
Call him Alexander, and say that Pan
Passeth Apollo in music many fold;
Praise Sir Thopas for a noble tale 50
And scorn the story that the knight told;
Praise him for counsel that is drunk of ale;
Grin when he laugheth that beareth all the sway,
Frown when he frowneth and groan when he is pale,

18 dye . . . liar] cf. proverbial 'black will take none other hue'. There may also be a wordplay on 'black-a-lyre', black cloth from Liere in Brabant.
22 sets their part With] commit themselves to
27 silly] simple, innocent
33 counsel] good advice
still to paint] continually to flatter
37 allow] approve
39 out of the gate] out of the way
42 apply] devote itself to
49 passeth] surpasses
52 counsel] prudence

On other's lust to hang both night and day.
    None of these points would ever frame in me.
    My wit is naught. I cannot learn the way.
And much the less of things that greater be,
    That asken help of colours of device
    To join the mean with each extremity: 60
With the nearest virtue to cloak alway the vice
    And, as to purpose likewise it shall fall,
    To press the virtue that it may not rise.
As drunkenness good fellowship to call;
    The friendly foe with his double face
    Say he is gentle and courteous therewithal;
And say that Favel hath a goodly grace
    In eloquence; and cruelty to name
    Zeal of justice and change in time and place;
And he that suffereth offence without blame 70
    Call him pitiful, and him true and plain
    That raileth reckless to every man's shame;
Say he is rude that cannot lie and feign,
    The lecher a lover, and tyranny
    To be the right of a prince's reign.
I cannot, I! No, no, it will not be!
    This is the cause that I could never yet
    Hang on their sleeves that weigh, as thou mayst see,
A chip of chance more than a pound of wit.
    This maketh me at home to hunt and to hawk 80
    And in foul weather at my book to sit;
In frost and snow then with my bow to stalk.
    No man doth mark whereso I ride or go;
    In lusty leas in liberty I walk.
And of these news I feel nor weal nor woe,
    Save that a clog doth hang yet at my heel.
    No force for that, for it is ordered so
That I may leap both hedge and dike full well.
    I am not now in France to judge the wine,
    With savoury sauce the delicates to feel; 90
Nor yet in Spain where one must him incline,
    Rather than to be, outwardly to seem.
    I meddle not with wits that be so fine.

55 lust] pleasure
56 frame in me] appeal to me
67 Favel] 'Flattery'
71 pitiful] contemptible
true and plain] truthful and plain-spoken
79 chip . . . wit] cf. proverbial 'An ounce of fortune is worth a pound of wit'
84 lusty leas] pleasant pastures

Nor Flander's cheer letteth not my sight to deem
Of black and white nor taketh my wit away
With beastliness they, beasts, do so esteem.
Nor I am not where Christ is given in prey
For money, poison, and treason at Rome—
A common practice used night and day.
But here I am in Kent and Christendom 100
Among the Muses where I read and rhyme,
Where if thou list, my Poyntz, for to come,
Thou shalt be judge how I do spend my time.

(Wr. 1536(?); pub. 1557)

52 *My mother's maids when they did sew and spin*

My mother's maids when they did sew and spin,
They sang sometime a song of the field mouse
That, for because her livelood was but thin,
Would needs go seek her townish sister's house.
She thought herself endured too much pain.
The stormy blasts her cave so sore did souse
That when the furrows swimmed with the rain
She must lie cold and wet in sorry plight.
And worse than that, bare meat there did remain
To comfort her when she her house had dight— 10
Sometime a barley corn, sometime a bean
For which she laboured hard both day and night
In harvest time whilst she might go and glean;
And when her store was 'stroyed with the flood,
Then wellaway, for she undone was clean.
Then was she fain to take instead of food
Sleep, if she might, her hunger to beguile.
'My sister,' quod she, 'hath a living good
And hence from me she dwelleth not a mile.
In cold and storm she lieth warm and dry 20
In bed of down. The dirt doth not defile
Her tender foot. She laboureth not as I.
Richly she feedeth and at the rich man's cost,
And for her meat she needs not crave nor cry.
By sea, by land, of the delicates the most
Her cater seeks and spareth for no peril.
She feedeth on boiled bacon meat and roast
And hath thereof neither charge nor travail.
And when she list, the liquor of the grape
Doth glad her heart till that her belly swell.' 30

52
6 souse] drench 10 dight] put in order 26 cater] caterer

And at this journey she maketh but a jape.
So forth she goeth, trusting of all this wealth
With her sister her part so for to shape
That, if she might keep herself in health,
To live a lady while her life doth last.
And to the door now is she come by stealth
And with her foot anon she scrapeth full fast.
Th'other for fear durst not well scarce appear,
Of every noise so was the wretch aghast.
At last she asked softly who was there. 40
And in her language as well as she could
'Peep,' quod the other, 'sister, I am here.'
'Peace,' quod the town mouse, 'why speakest thou so loud?'
And by the hand she took her fair and well.
'Welcome,' quod she, 'my sister, by the Rood.'
She feasted her, that joy it was to tell
The fare they had. They drank the wine so clear
And, as to purpose now and then it fell,
She cheered her with 'How sister, what cheer!'
Amidst this joy befell a sorry chance 50
That, wellaway, the stranger bought full dear
The fare she had. For as she looked askance,
Under a stool she spied two steaming eyes
In a round head with sharp ears. In France
Was never mouse so feared, for though th'unwise
Had not yseen such a beast before,
Yet had nature taught her after her guise
To know her foe and dread him evermore.
The towny mouse fled; she knew whither to go.
Th'other had no shift, but wondrous sore 60
Feared of her life. At home she wished her tho!
And to the door, alas, as she did skip,
Th'heaven it would, lo, and eke her chance was so,
At the threshold her silly foot did trip,
And ere she might recover it again
The traitor cat had caught her by the hip
And made her there against her will remain,
That had forgotten her poor surety and rest
For seeming wealth wherein she thought to reign.
Alas, my Poyntz, how men do seek the best 70
And find the worst by error as they stray!
And no marvel, when sight is so oppressed,

31 at this journey . . . jape] she treats this journey as a joke
48 as to purpose] as it came into the conversation
53 steaming] gleaming
61 tho] then
64 silly] poor, unfortunate

And blind the guide, anon, out of the way
  Goeth guide and all in seeking quiet life.
  O wretched minds, there is no gold that may
Grant that ye seek, no war, no peace, no strife.
  No, no, although thy head were hooped with gold,
  Sergeant with mace, halberd, sword, nor knife
Cannot repulse the care that follow should.
  Each kind of life hath with him his disease. 80
  Live in delight even as thy lust would
And thou shalt find, when lust doth most thee please,
  It irketh straight and by itself doth fade.
  A small thing it is that may thy mind appease.
None of ye all there is that is so mad
  To seek grapes upon brambles or briers,
  Nor none, I trow, that hath his wit so bad
To set his hay for conies over rivers;
  Ne ye set not a drag-net for a hare.
  And yet the thing that most is your desire 90
Ye do mis-seek with more travail and care.
  Make plain thine heart that it be not knotted
  With hope or dread, and see thy will be bare
From all affects whom vice hath ever spotted.
  Thyself content with that is thee assigned
  And use it well that is to thee allotted.
Then seek no more out of thyself to find
  The thing that thou hast sought so long before,
  For thou shalt feel it sitting in thy mind.
Mad if ye list to continue your sore! 100
  Let present pass and gape on time to come
  And deep yourself in travail more and more.
Henceforth, my Poyntz, this shall be all and sum.
  These wretched fools shall have naught else of me.
  But to the great God and to his high doom
None other pain pray I for them to be
  But, when the rage doth lead them from the right,
  That looking backward, Virtue they may see
Even as she is, so goodly fair and bright.
  And whilst they clasp their lusts in arms across, 110
  Grant them, good Lord, as thou mayst of thy might,
To fret inward for losing such a loss.

(Pub. 1557)

81 lust] pleasure, desire
88 hay] rabbit-net
89 drag-net] i.e. to catch fish
93–94 bare from all affects] free from all desires
100 Mad . . . sore] you are mad if you make your anxiety of mind continue
106 to be] to experience
110 in arms across] in an embrace

53 *A spending hand that alway poureth out*

'A SPENDING hand that alway poureth out
Had need to have a bringer-in as fast';
And 'On the stone that still doth turn about
There groweth no moss'—these proverbs yet do last.
Reason hath set them in so sure a place
That length of years their force can never waste.
When I remember this and eke the case
Wherein thou stands, I thought forthwith to write,
Brian, to thee, who knows how great a grace
In writing is to counsel man the right. 10
To thee, therefore, that trots still up and down
And never rests, but running day and night
From realm to realm, from city, street, and town,
Why dost thou wear thy body to the bones
And mightst at home sleep in thy bed of down
And drink good ale so nappy for the nonce,
Feed thyself fat and heap up pound by pound?
Likest thou not this? 'No.' Why? 'For swine so groins
In sty and chaw the turds moulded on the ground,
And drivel on pearls, the head still in the manger. 20
Then of the harp the ass do hear the sound.
So sacks of dirt be filled up in the cloister
That serves for less than do these fatted swine.
Though I seem lean and dry without moisture,
Yet will I serve my prince, my lord and thine,
And let them live to feed the paunch that list,
So I may feed to live, both me and mine.'
By God, well said, but what and if thou wist
How to bring in as fast as thou dost spend?
'That would I learn.' And it shall not be missed 30
To tell thee how. Now hark what I intend.
Thou know'st well, first, whoso can seek to please
Shall purchase friends where truth shall but offend.
Flee therefore truth: it is both wealth and ease.
For though that truth of every man hath praise,
Full near that wind goeth truth in great misease.
Use virtue as it goeth now-a-days,
In word alone to make thy language sweet,
And of the deed yet do not as thou says.

9 Brian] Sir Francis Brian, courtier and diplomat
16 nappy] heady, strong
for the nonce] for the occasion
18 groins] grunts (?); digs in the earth with the snout
moulded] turned mouldy
22 sacks of dirt] i.e. monks

Else be thou sure thou shalt be far unmeet 40
To get thy bread, each thing is now so scant.
Seek still thy profit upon thy bare feet.
Lend in no wise, for fear that thou do want,
Unless it be as to a dog a cheese;
By which return be sure to win a cant
Of half at least—it is not good to leese.
Learn at Kitson, that in a long white coat
From under the stall without lands or fees
Hath leapt into the shop; who knoweth by rote
This rule that I have told thee herebefore. 50
Sometime also rich age beginneth to dote;
See thou when there thy gain may be the more.
Stay him by the arm whereso he walk or go.
Be near alway and, if he cough too sore,
When he hath spit, tread out and please him so.
A diligent knave that picks his master's purse
May please him so that he, withouten moe,
Executor is, and what is he the worse?
But if so chance you get naught of the man,
The widow may for all thy charge deburse. 60
A rivelled skin, a stinking breath, what then?
A toothless mouth shall do thy lips no harm.
The gold is good, and though she curse or ban,
Yet where thee list thou mayst lie good and warm:
Let the old mule bite upon the bridle
Whilst there do lie a sweeter in thine arm.
In this also see you be not idle:
Thy niece, thy cousin, thy sister, or thy daughter,
If she be fair, if handsome be her middle,
If thy better hath her love besought her, 70
Advance his cause and he shall help thy need.
It is but love. Turn it to a laughter.
But ware, I say, so gold thee help and speed
That in this case thou be not so unwise
As Pandar was in such a like deed;
For he, the fool, of conscience was so nice
That he no gain would have for all his pain.
Be next thyself, for friendship bears no prize.
Laugh'st thou at me? Why? Do I speak in vain?
'No, not at thee, but at thy thrifty jest. 80
Wouldest thou I should for any loss or gain

45 cant] portion
47 Kitson] (?) wealthy Sheriff of London
60 deburse] pay
61 rivelled] wrinkled

Change that for gold that I have ta'en for best—
Next godly things, to have an honest name?
Should I leave that? Then take me for a beast!'
Nay then, farewell, and if you care for shame,
Content thee then with honest poverty,
With free tongue, what thee mislikes, to blame,
And, for thy truth, sometime adversity.
And therewithal this thing I shall thee give—
In this world now, little prosperity, 90
And coin to keep, as water in a sieve.

(Pub. 1557)

ATTRIBUTED TO SIR THOMAS WYATT

54 *I am as I am and so will I be*

I AM as I am and so will I be
But how that I am none knoweth truly.
Be it evil, be it well, be I bound, be I free,
I am as I am and so will I be.

I lead my life indifferently,
I mean no thing but honestly.
And though folks judge full diversely
I am as I am and so will I die.

I do not rejoice nor yet complain.
Both mirth and sadness I do refrain 10
And use the mean since folks will feign.
Yet I am as I am, be it pleasure or pain.

Diverse do judge as they do trow,
Some of pleasure and some of woe.
Yet for all that, nothing they know.
But I am as I am wheresoever I go.

But since that judgers do thus decay
Let every man his judgement say.
I will it take in sport and play
For I am as I am whosoever say nay. 20

Who judgeth well, well God him send.
Who judgeth evil, God them amend.
To judge the best therefore intend
For I am as I am and so will I end.

Yet some there be that take delight
To judge folks' thought for envy and spite.
But whether they judge me wrong or right
I am as I am and so do I write,

Praying you all that this do read
To trust it as you do your creed 30
And not to think I change my weed
For I am as I am however I speed.

But how that is I leave to you.
Judge as ye list, false or true.
Ye know no more than afore ye knew.
Yet I am as I am whatever ensue.

And from this mind I will not flee,
But to you all that misjudge me
I do protest, as ye may see,
That I am as I am and so will I be. 40

(Pub. 1815)

ANONYMOUS
from *The Court of Love*

55 [*The birds' matins and conclusion to the poem*]

On May-day, when the lark began to rise,
To matins went the lusty nightingale
Within a temple shapen hawthorn-wise;
He might not sleep in all the nightertale,
But '*Domine labia*', gan he cry and gale,
'My lips open, Lord of Love, I cry,
And let my mouth thy praising now bewrye.'

4 nightertale] night-time gale] sing
5 *Domine labia* (*mea aperies*)] Lord, open thou my lips (opening of Matins)
7 bewrye] disclose, utter

The eagle sang '*Venite*, bodies all,
And let us joy to love that is our health.'
And to the desk anon they gan to fall, 10
And who come late, he presseth in by stealth.
Then said the falcon, our own heartes wealth,
'*Domine, Dominus noster*, I wot,
Ye be the god that don us bren thus hot.'

'*Celi enarrant*', said the popinjay,
'Your might is told in heaven and firmament.'
And then came in the goldfinch fresh and gay,
And said this psalm with heartly glad intent,
'*Domini est terra*; this Latin intent,
The god of Love hath earth in governance.' 20
And then the wren gan skippen and to dance.

'*Jube, Domine*, Lord of Love, I pray
Command me well this lesson for to read;
This legend is of all that wolden dey
Martyrs for love; God give the soules speed!
And to thee, Venus, sing we, out of dread,
By influence of all thy virtue great,
Beseeching thee to keep us in our heat.'

The second lesson Robin redbreast sang,
'Hail to the god and goddess of our lay!' 30
And to the lectern amorously he sprang.
'Hail', quod he eke, 'O fresh season of May,
Our moneth glad that singen on the spray!
Hail to the flowers, red, and white, and blue,
Which by their virtue make our lustes new!'

13 *Domine . . . noster*] O Lord, our Lord (Ps. 8)
wot] know
14 bren] burn
15 *Celi enarrant*] The heavens declare (Ps. 19)
popinjay] parrot
19 *Domini est terra*] The earth is the Lord's (Ps. 24)
this Latin intent] this Latin means
22 *Jube, Domine* (*benedicere*)] Lord, command us to bless (versicle preceding the first lesson)
24 legend] story, history
dey] die
25 speed] good fortune, prosperity
33 Our moneth . . . singen] The glad month of us who sing

The third lesson the turtle-dove took up,
  And thereat lough the mavis as in scorn.
He said, 'O God, as mot I dine or sup,
  This foolish dove will give us all an horn!
  There been right here a thousand better born, 40
    To read this lesson, which, as well as he,
    And eke as hot, can love in all degree.'

The turtle-dove said, 'Welcome, welcome, May,
  Gladsome and light to lovers that been true!
I thank thee, Lord of Love, that doth purvey
  For me to read this lesson all of due;
  For, in good sooth, of courage I pursue
    To serve my make till death us must depart.'
    And then '*Tu autem*' sang he all apart.

'*Te deum amoris*', sang the throstle-cock: 50
  Tubal himself, the first musician,
With key of harmony could not unlock
  So sweet a tune as that the throstle can.
  'The Lord of Love we praisen', quod he than,
    'And so don all the fowles, great and lite;
    Honour we May, in false lovers' despite.'

'*Dominus regnavit*', said the peacock there,
  'The Lord of Love, that mighty prince, ywis,
He hath received here and everywhere:
  Now *Jubilate* sing.' 'What meaneth this?' 60
  Said then the linnet: 'Welcome, Lord of bliss!'
    Outstert the owl with '*Benedicite*,
    What meaneth all this merry fare?' quod he.

37 lough] laughed
mavis] thrush
38 mot I] may I
39 give . . . horn] mock, scorn
48 make] mate
depart] separate
49 *Tu autem* (*domine, miserere nobis*)] Thou, Lord, have mercy on us
50 *Te deum amoris* (*laudamus*)] we praise thee, Lord of Love
55 lite] small
57 *Dominus regnavit*] The Lord reigneth (Ps. 93)
58 ywis] indeed, certainly
60 *Jubilate*] Make a joyful noise (Ps. 100)
62 Outstert] burst out
Benedicite] Bless ye (the Lord)

'*Laudate*', sang the lark with voice full shrill;
And eke the kite, '*O admirabile*!
This choir will through mine eares pierce and thrill.
But what? Welcome this May season', quod he,
'And honour to the Lord of Love mot be,
That hath this feast so solemn and so high.'
'*Amen*', said all; and so said eke the pie. 70

And forth the cuckoo gan proceed anon,
With '*Benedictus*' thanking God in haste,
That in this May would visit them eachon,
And gladden them all while the feast shall last.
And therewithal a-laughter out he brast,
'I thank it, God, that I should end the song,
And all the service which hath been so long.'

Thus sang they all the service of the feast,
And that was done right early, to my doom;
And forth go'th all the Court, both most and least, 80
To fetch the flowers fresh, and branch and bloom;
And namely, hawthorn brought both page and groom;
With fresh garlandes, parti-blue and white,
And them rejoycen in their great delight.

Eke each at other threw the flowers bright,
The primerose, the violet, the gold;
So then, as I beheld the royal sight,
My lady gan me suddenly behold,
And with a true-love, plighted manifold,
She smote me through the very heart as blive; 90
And Venus yet I thank I am alive.

(Wr. probably *c.*1535; pub. 1561)

64 *Laudate* (*dominum*)] Praise ye (the Lord) (Ps. 148)
65 *O admirabile!*] O thou wonderful (change) (anthem)
68 mot be] let there be
70 pie] magpie
72 *Benedictus*] Blessed (be the Lord God of Israel) (part of the service)
79 doom] judgement
82 namely] especially
84 them rejoycen] exult
86 gold] marigold
89 true-love] lover's knot (i.e. a bow of ribbon)
plighted] folded
90 as blive] instantaneously

HENRY HOWARD, EARL OF SURREY
1517?–1547

56 *When raging love*

WHEN raging love with extreme pain
  Most cruelly distrains my heart,
When that my tears, as floods of rain,
  Bear witness of my woeful smart;
    When sighs have wasted so my breath
    That I lie at the point of death,

I call to mind the navy great
  That the Greeks brought to Troye town,
And how the boistous winds did beat
  Their ships, and rent their sails adown; 10
    Till Agamemnon's daughter's blood
    Appeased the gods that them withstood.

And how that in those ten years' war
  Full many a bloody deed was done,
And many a lord that came full far
  There caught his bane, alas, too soon;
    And many a good knight overrun,
    Before the Greeks had Helen won.

Then think I thus: sith such repair,
  So long time war of valiant men, 20
Was all to win a lady fair,
  Shall I not learn to suffer then,
    And think my life well spent to be,
    Serving a worthier wight than she?

(Pub. 1557)

57 *The soote season*

THE soote season, that bud and bloom forth brings,
  With green hath clad the hill and eke the vale.
The nightingale with feathers new she sings;
  The turtle to her make hath told her tale.

9 boistous] boisterous repair] coming
19 sith] since 24 wight] person

57
1 soote] sweet
4 make] mate

Summer is come, for every spray now springs.
  The hart hath hung his old head on the pale;
The buck in brake his winter coat he flings;
  The fishes float with new repaired scale;
The adder all her slough away she slings;
  The swift swallow pursueth the flies small; 10
The busy bee her honey now she mings;
  Winter is worn that was the flowers' bale.
And thus I see among these pleasant things
Each care decays; and yet my sorrow springs.

(Pub. 1557)

58 *Set me whereas the sun doth parch the green*

SET me whereas the sun doth parch the green,
  Or where his beams may not dissolve the ice,
In temperate heat, where he is felt and seen,
  With proud people, in presence sad and wise;
Set me in base, or yet in high degree,
  In the long night, or in the shortest day,
In clear weather, or where mists thickest be,
  In lusty youth, or when my hairs be gray;
Set me in earth, in heaven, or yet in hell,
  In hill, in dale, or in the foaming flood, 10
Thrall, or at large, alive whereso I dwell,
  Sick, or in health, in ill fame, or in good;
    Yours will I be, and with that only thought
    Comfort myself when that my hap is nought.

(Pub. 1557)

59 *Alas, so all things now do hold their peace*

ALAS, so all things now do hold their peace,
  Heaven and earth disturbed in no thing.
The beasts, the air, the birds their song do cease;
  The nightes chare the stars about doth bring;
Calm is the sea; the waves work less and less.
  So am not I, whom love, alas, doth wring,
Bringing before my face the great increase
  Of my desires, whereat I weep and sing,

11 mings] remembers

59
4 chare] chariot

In joy and woe, as in a doubtful ease:
  For my sweet thoughts sometime do pleasure bring; 10
But by and by, the cause of my disease
  Gives me a pang, that inwardly doth sting,
    When that I think what grief it is again,
    To live and lack the thing should rid my pain.

(Pub. 1557)

60 *O happy dames*

O HAPPY dames, that may embrace
  The fruit of your delight,
Help to bewail the woeful case
  And eke the heavy plight
Of me, that wonted to rejoice
The fortune of my pleasant choice;
Good ladies, help to fill my mourning voice.

In ship, freight with rememberance
  Of thoughts and pleasures past,
He sails that hath in governance 10
  My life while it will last;
With scalding sighs, for lack of gale,
Furthering his hope, that is his sail,
Toward me, the sweet port of his avail.

Alas, how oft in dreams I see
  Those eyes that were my food;
Which sometime so delighted me,
  That yet they do me good;
Wherewith I wake with his return,
Whose absent flame did make me burn: 20
But when I find the lack, Lord, how I mourn!

When other lovers in arms across
  Rejoice their chief delight,
Drowned in tears, to mourn my loss
  I stand the bitter night
In my window, where I may see
Before the winds how the clouds flee.
Lo, what a mariner love hath made me!

60
14 avail] disembarking
22 in arms across] embracing
23 Rejoice] enjoy

And in green waves when the salt flood
  Doth rise by rage of wind, 30
A thousand fancies in that mood
  Assail my restless mind.
Alas, now drencheth my sweet foe,
That with the spoil of my heart did go,
And left me; but, alas! why did he so?

And when the seas wax calm again
  To chase fro me annoy,
My doubtful hope doth cause me plain;
  So dread cuts off my joy.
Thus is my wealth mingled with woe, 40
And of each thought a doubt doth grow;
'Now he comes! Will he come? Alas, no, no!'

(Pub. 1557)

from *Certain Books of Virgil's 'Æneis'*

61

[*Creusa*]

And now we gan draw near unto the gate,
Right well escaped the danger, as me thought,
When that at hand a sound of feet we heard.
My father then, gazing throughout the dark,
Cried on me, 'Flee, son! They are at hand!'
With that, bright shields and sheen armours I saw.
But then I know not what unfriendly god
My troubled wit from me biraft for fear.
For while I ran by the most secret streets,
Eschewing still the common haunted track, 10
From me caitiff, alas, bereaved was
Creusa, then, my spouse, I wot not how,
Whether by fate, or missing of the way,
Or that she was by weariness retained.
But never sith these eyes might her behold,
Nor did I yet perceive that she was lost,
Ne never backward turned I my mind
Till we came to the hill whereas there stood
The old temple dedicate to Ceres.
  And when that we were there assembled all, 20
She only was away, deceiving us,

33 drencheth] drowns
61
10 haunted] frequented   11 caitiff] wretch   15 sith] since then

Her spouse, her son, and all her company.
What god or man did I not then accuse,
Near wood for ire? or what more cruel chance
Did hap to me in all Troy's overthrow?
Ascanius to my feers I then betook,
With Anchises, and eke the Troyan gods,
And left them hid within a valley deep.
And to the town I gan me hie again,
Clad in bright arms and bent for to renew 30
Adventures past, to search throughout the town,
And yield my head to perils once again.
  And first the walls and dark entry I sought
Of the same gate whereat I issued out,
Holding backward the steps where we had come
In the dark night, looking all round about.
In every place the ugsome sights I saw,
The silence self of night, aghast my sprite.
From hence again I passed unto our house,
If she by chance had been returned home. 40
The Greeks were there, and had it all beset.
The wasting fire blown up by drift of wind
Above the roofs, the blazing flame sprang up,
The sound whereof with fury pierced the skies.
To Priam's palace and the castle then
I made; and there at Juno's sanctuair
In the void porches Phoenix, Ulysses eke,
Stern guardians stood, watching of the spoil.
The riches here were set, reft from the brent
Temples of Troy; the tables of the gods, 50
The vessels eke that were of massy gold,
And vestures spoiled, were gathered all in heap.
The children orderly and mothers pale for fright
Long ranged on a row stood round about.
  So bold was I to show my voice that night,
With clepes and cries to fill the streets throughout,
With Creuse' name in sorrow, with vain tears,
And often sithes the same for to repeat.
The town restless with fury as I sought,
Th'unlucky figure of Creusa's ghost, 60
Of stature more than wont, stood fore mine eyen.
Abashed then I wox. Therewith my hair

24 wood] mad
26 feers] companions
betook] consigned
37 ugsome] fearful, horrible
48 Stern] fierce
49 brent] burnt
56 clepes] shouts
58 often sithes] oftentimes
60 unlucky] ill-omened
62 Abashed] dismayed, confounded
wox] grew

Gan start right up, my voice stuck in my throat.
When with such words she gan my hart remove:
'What helps to yield unto such furious rage,
Sweet spouse?' quod she. 'Without will of the gods
This chanced not; ne leful was for thee
To lead away Creusa hence with thee:
The king of the high heaven suff'reth it not.
A long exile thou art assigned to bear, 70
Long to furrow large space of stormy seas:
So shalt thou reach at last Hesperian land,
Where Lydian Tiber with his gentle stream
Mildly doth flow along the fruitful fields.
There mirthful wealth, there kingdom is for thee,
There a king's child prepared to be thy make.
For thy beloved Creusa stint thy tears.
For now shall I not see the proud abodes
Of Myrmidons, nor yet of Dolopes;
Ne I, a Troyan lady and the wife 80
Unto the son of Venus the goddess,
Shall go a slave to serve the Greekish dames.
Me here the gods' great mother holds.
And now farewell, and keep in father's breast
The tender love of thy young son and mine.'
This having said, she left me all in tears,
And minding much to speak; but she was gone,
And subtly fled into the weightless air.
Thrice raught I with mine arms t'accoll her neck,
Thrice did my hands' vain hold th'image escape, 90
Like nimble winds and like the flying dream.
So night spent out, return I to my feres.
And there wond'ring I find together swarmed
A new number of mates, mothers and men,
A rout exiled, a wretched multitude,
From each where flocked together, prest to pass,
With heart and goods, to whatsoever land
By sliding seas me listed them to lead.
And now rose Lucifer above the ridge
Of lusty Ide and brought the dawning light. 100
The Greeks held th'entries of the gates beset;
Of help there was no hope. Then gave I place,
Took up my sire, and hasted to the hill.

(Pub. 1557)

64 remove] move
67 leful] lawful
75 wealth] well-being, happiness
76 make] mate
77 stint] stop
89 raught] reached
accoll] embrace
92 feres] companions
96 prest] prepared

62

[*Dido in love*]

SUCH words enflamed the kindled mind with love,
Loosed all shame, and gave the doubtful hope.
And to the temples first they haste, and seek
By sacrifice for grace, with hogrels of two years
Chosen, as ought, to Ceres that gave laws,
To Phoebus, Bacchus, and to Juno chief
Which hath in care the bands of marriage.
Fair Dido held in her right hand the cup,
Which twixt the horns of a white cow she shed
In presence of the gods, passing before 10
The altars fat, which she renewed oft
With gifts that day and beasts deboweled,
Gazing for counsel on the entrails warm.
Ay me, unskilful minds of prophecy!
Temples or vows, what boot they in her rage?
A gentle flame the mary doth devour,
Whiles in the breast the silent wound keeps life.
Unhappy Dido burns, and in her rage
Throughout the town she wand'reth up and down,
Like to the stricken hind with shaft in Crete 20
Throughout the woods which chasing with his darts
Aloof, the shepherd smiteth at unwares
And leaves unwist in her the thirling head,
That through the groves and launds glides in her flight;
Amid whose side the mortal arrow sticks.
Aeneas now about the walls she leads,
The town prepared and Carthage wealth to show.
Off'ring to speak, amid her voice, she whists.
And when the day gan fail, new feasts she makes;
The Troys' travails to hear anew she lists 30
Enraged all, and stareth in his face
That tells the tale. And when they were all gone,
And the dim moon doth eft withhold the light,
And sliding stars provoked unto sleep,
Alone she mourns within her palace void,
And sets her down on her forsaken bed;
And absent him she hears, when he is gone,

4 hogrels of two years] sheep two years old
16 mary] marrow
23 unwist] unknown
thirling] piercing
24 launds] glades
28 whists] breaks off
30 Troys'] Trojans'
lists] desires
33 eft] again

And seeth eke. Oft in her lap she holds
Ascanius, trapped by his father's form,
So to beguile the love cannot be told.

(Pub. 1554)

63 [*The Happy Life*]

MARTIAL, the things for to attain
The happy life be these, I find:
The riches left, not got with pain;
The fruitful ground, the quiet mind;

The equal friend; no grudge nor strife;
No charge of rule, nor governance;
Without disease the healthful life;
The household of continuance;

The mean diet, no delicate fare;
Wisdom joined with simplicity; 10
The night discharged of all care,
Where wine may bear no sovereignty;

The chaste wife, wise, without debate;
Such sleeps as may beguile the night;
Contented with thine own estate,
Neither wish death nor fear his might.

(Pub. 1547)

64 *So cruel prison*

SO cruel prison how could betide, alas,
As proud Windsor, where I in lust and joy
With a king's son my childish years did pass
In greater feast than Priam's sons of Troy;

Where each sweet place returns a taste full sour:
The large green courts, where we were wont to hove,
With eyes cast up unto the maidens' tower,
And easy sighs, such as folk draw in love.

63
1 Martial] (the poem is a translation of Martial's epigram X.xlvii)
8 continuance] long duration
13 debate] strife

64
6 hove] linger

The stately sales, the ladies bright of hue,
  The dances short, long tales of great delight, 10
With words and looks that tigers could but rue,
  Where each of us did plead the other's right.

The palm-play where, despoiled for the game,
  With dazed eyes oft we by gleams of love
Have missed the ball and got sight of our dame,
  To bait her eyes which kept the leads above.

The gravelled ground, with sleeves tied on the helm,
  On foaming horse, with swords and friendly hearts,
With cheer as though the one should overwhelm,
  Where we have fought and chased oft with darts. 20

With silver drops the meads yet spread for ruth,
  In active games of nimbleness and strength,
Where we did strain, trailed by swarms of youth,
  Our tender limbs that yet shot up in length.

The secret groves which oft we made resound
  With pleasant plaint and of our ladies' praise,
Recording oft what grace each one had found,
  What hope of speed, what dread of long delays.

The wild forest, the clothed holts with green,
  With reins avaled, and swift ybreathed horse, 30
With cry of hounds and merry blasts between,
  Where we did chase the fearful hart a force.

The void walls eke that harboured us each night,
  Wherewith, alas, revive within my breast
The sweet accord, such sleeps as yet delight,
  The pleasant dreams, the quiet bed of rest,

The secret thoughts imparted with such trust,
  The wanton talk, the divers change of play,
The friendship sworn, each promise kept so just,
  Wherewith we passed the winter nights away. 40

And with this thought the blood forsakes my face,
  The tears berain my cheeks of deadly hue;
The which as soon as sobbing sighs, alas,
  Upsupped have, thus I my plaint renew:

9 sales] halls
13 palm-play] a game resembling tennis
16 bait] attract
19 cheer] facial expression
28 speed] success
30 avaled] lowered
ybreathed] exercised
32 chase . . . a force] (hunting term) game which is run down

'O place of bliss, renewer of my woes,
  Give me account where is my noble fere,
Whom in thy walls thou didst each night enclose,
  To other lief, but unto me most dear.'

Each stone, alas, that doth my sorrow rue,
  Returns thereto a hollow sound of plaint. 50
Thus I alone, where all my freedom grew,
  In prison pine with bondage and restraint.

And with remembrance of the greater grief,
To banish the less I find my chief relief.

(Pub. 1557)

65 *An excellent epitaph of Sir Thomas Wyatt*

WYATT resteth here, that quick could never rest,
  Whose heavenly gifts, increased by disdain
And virtue, sank the deeper in his breast,
  Such profit he by envy could obtain.

A head, where wisdom mysteries did frame,
  Whose hammers beat still in that lively brain
As on a stithy where that some work of fame
  Was daily wrought to turn to Britain's gain.

A visage stern and mild, where both did grow
  Vice to contemn, in virtue to rejoice; 10
Amid great storms whom grace assured so
  To live upright and smile at fortune's choice.

A hand that taught what might be said in rhyme,
  That reft Chaucer the glory of his wit;
A mark the which, unparfited for time,
  Some may approach but never none shall hit.

A tongue that served in foreign realms his king;
  Whose courteous talk to virtue did enflame
Each noble heart, a worthy guide to bring
  Our English youth by travail unto fame. 20

46 fere] companion
48 To other lief] to others beloved
65
1 quick] alive
4 envy] malice
5 mysteries] profound thoughts
15 unparfited for time] unfinished through want of time

An eye whose judgment none affect could blind
  Friends to allure and foes to reconcile,
Whose piercing look did represent a mind
  With virtue fraught, reposed, void of guile.

A heart where dread was never so impressed
  To hide the thought that might the truth advance;
In neither fortune loft nor yet repressed
  To swell in wealth or yield unto mischance.

A valiant corps where force and beauty met;
  Happy, alas, too happy but for foes; 30
Lived, and ran the race that Nature set,
  Of manhood's shape where she the mould did lose.

But to the heavens that simple soul is fled,
  Which left with such as covet Christ to know
Witness of faith that never shall be dead,
  Sent for our health, but not received so.

Thus for our guilt this jewel have we lost.
The earth his bones, the heavens possess his ghost.

(Pub. 1542)

66 *Th'Assyrians' king*

TH'ASSYRIANS' king, in peace with foul desire
  And filthy lust that stained his regal heart,
In war, that should set princely hearts afire,
  Vanquished did yield for want of martial art.
The dent of swords from kisses seemed strange,
  And harder than his lady's side his targe;
From glutton feasts to soldiers' fare a change,
  His helmet far above a garland's charge.
Who scarce the name of manhood did retain,
  Drenched in sloth and womanish delight, 10
Feeble of sprete, unpatient of pain,
  When he had lost his honour and his right
    (Proud time of wealth, in storms appalled with dread),
    Murdered himself to show some manful deed.

(Pub. 1557)

21 affect] passion, feeling
27 loft] proudly overweening
28 wealth] prosperity
38 ghost] spirit

66
1 Th'Assyrians' king] Sardanapalus
6 targe] shield
8 charge] weight
11 sprete] spirit
13 Proud time of wealth] proud in times of prosperity

67 [*Epitaph for Thomas Clere*]

NORFOLK sprang thee, Lambeth holds thee dead,
  Clere of the County of Cleremont though hight;
Within the womb of Ormond's race thou bred,
  And saw'st thy cousin crowned in thy sight.
Shelton for love, Surrey for lord thou chase:
  Ay me, while life did last that league was tender;
Tracing whose steps thou sawest Kelsall blaze,
  Laundersey burnt, and battered Bullen render.
At Muttrell gates, hopeless of all recure,
  Thine Earl half-dead gave in thy hand his will, 10
Which cause did thee this pining death procure,
  Ere summers four times seven thou couldst fulfil.
    Ah Clere, if love had booted, care, or cost,
    Heaven had not won, nor earth so timely lost.

(Pub. 1605)

ROBERT COPLAND
fl. 1508–1547

68 from *The High Way to the Spital House*

TO write of Sol in his exaltation,
  Of his solstice or declination,
Or in what sign, planet, or degree,
  As he in course is used for to be;
Scorpio, Pisces, or Sagittary;
  Or when the moon her way doth contrary,
Or her eclipse, her wane, or yet her full,
  It were but lost for blockish braines dull;

1 sprang thee] gave thee birth
2 hight] called
3 bred] wert reared
5 chase] chose
7 Kelsall] town in Scotland
8 Laundersey] Landrecy
Bullen] Boulogne
render] surrender
9 Muttrell] Montreuil
recure] recovery, cure
13 had booted] had availed
14 timely] untimely

68
Spital House] hospital, charitable foundation for the poor and diseased
1 Sol] the sun
exaltation] position of greatest (astrological) influence
2 solstice] one of two times in the year, when the sun is furthest from the equator
declination] the angular distance of the sun from the celestial equator
3 sign] sign of the zodiac
5 Sagittary] Sagittarius
6 contrary] opposite

But plainly to say, even as the time was,
  About a fortnight after Hallowmass, 10
I chanced to come by a certain spital,
  Where I thought best to tarry a little,
And under the porch for to take succour,
  To abide the passing of a stormy shower;
For it had snowen, and frozen very strong,
  With great icicles on the eaves long.
The sharp north wind hurled bitterly,
  And with black clouds darked was the sky,
Like as, in winter, some days be natural
  With frost, and rain, and storms over all. 20
So still I stood; as chanced to be,
  The porter of the house stood also by me,
With whom I reasoned of many divers things
  Touching the course of all such weatherings.
And as we talked, there gathered at the gate
  People, as me thought, of very poor estate,
With bag and staff, both crooked, lame and blind,
  Scabby and scurvy, pock-eaten flesh and rind,
Lousy and scald, and pilled like as apes,
  With scantly a rag for to cover their shapes, 30
Breechless, bare-footed, all stinking with dirt,
  With a thousand of tatters drabbling to the skirt,
Boyes, girles, and luskish strong knaves,
  Diddering and daddering, leaning on their staves,
Saying 'Good master, for your mother's blessing,
  Give us a halfpenny toward our lodging.'
The porter said: 'What need you to crave,
  That in the spital shall your lodging have?
Ye shall be entreated as ye ought to be:
  For I am charged that daily to see. 40
The sisters shall do their observance
  As of the house is the due ordinance.'

Copland. 'Porter', said I, 'God's blessing and Our Lady,
  Have ye for speaking so courteously

10 Hallowmass] 1 November
17 hurled] howled, blustered
23 reasoned] talked
24 weatherings] weather conditions
28 rind] skin
  scald] scabby, with sores
  pilled] deprived of hair, bald
32 drabbling] trailing wet
33 luskish] lazy, idle
34 Diddering and daddering] trembling and quaking
37 crave] beg
39 entreated] treated
41 sisters] nuns

To these poor folk, and God his soul pardon
  That for their sake made this foundation.
But sir, I pray you, do ye lodge them all
  That do ask lodging in this hospital?'

Porter. Forsooth, yea, we do all such folk in take,
  That do ask lodging for Our Lord's sake; 50
And indeed it is our custom and use
  Sometime to take in and some to refuse.

Copland. Then is it common to every wight,
  How they live all day, to lie here at night?
As losels, mighty beggars and vagabonds,
  And truants that walk over the londs,
Michers, hedge-creepers, fillocks, and lusks,
  That all the summer keep ditches and busks,
Loitering, and wandering fro place to place,
  And will not work but the by-paths trace, 60
And live with haws, and hunt the blackberry,
  And with hedge-breaking make themself merry;
But in winter they draw to the town,
  And will do nothing but go up and down,
And all for lodging that they have here by night?
  Methink that therein ye do no right,
Nor all such places of hospitality
  To comfort people of such iniquity.
But, sir, I pray you of your goodness and favour,
  Tell me which ye leave, and which ye do succour? 70
For I have seen at sundry hospitals
  That many have lain dead without the walls,
And for lack of succour have died wretchedly.
  Unto your foundation I think contrary,
Much people resort here, and have lodging.
  But yet I marvel greatly of one thing,
That in the night so many lodge without:
  For in the watches when that we go about,

45 his soul] i.e. the founder, Rahere
53 wight] person
54 How they live] in whatever way they live
55 losels] profligates, rakes
mighty] able-bodied
56 truants] idle rogues
londs] strips of a cornfield
57 Michers] petty thieves, mischievous wanderers
hedge-creepers] men who skulk under hedges for dishonest purposes
fillocks] loose-living young girls
lusks] lay-abouts
58 keep] stay in, camp in
busks] bushes
62 hedge-breaking] disturbing the peace
78 watches] watches of the night, i.e. night-time

Under the stalls, in porches, and in doors,
  I wot not whether they be thieves or whores. 80
But surely, every night there is found
  One or other lying by the pound,
In the sheep cotes, or in the hay loft,
  And at Saint Bartholmew's church door full oft,
And even here alway by this brick wall
  We do them find that do both chide and brawl,
And like as beasts together they be throng,
  Both lame, and sick, and whole them among,
And in many corners where that we go,
  Whereof I wonder greatly why they do so; 90
But oft-times when that they us see,
  They do run a great deal faster than we.

Porter. Such folks be they that we do abject:
  We are not bound to have to them aspect.
Those be michers that live in truandise:
  Hospitality doth them alway despise.

Copland. Sir, I pray you, who hath of your relief?

Porter. Forsooth they that be at such mischief
That for their living can do no labour,
  And have no friends to do them succour; 100
As old people, sick and impotent;
  Poor women in childbed have here easement;
Weak men sore wounded by great violence,
  And sore men eaten with pox and pestilence,
And honest folk fallen in great poverty
  By mischance or other infirmity;
Wayfaring men and maimed soldiers
  Have their relief in this poor house of ours;
And all other, which we seem good and plain,
  Have here lodging for a night or twain; 110
Bedrid folk, and such as can not crave,
  In these places most relief they have;
And if they hap within our place to die,
  Then are they buried well and honestly;

79 stalls] outside tables, booths
82 pound] enclosure, fold
86 chide] wrangle
87 throng] pressed together, crowded
88 whole] healthy
89 where] wherever
93 abject] exclude, reject
94 aspect] consideration, regard
98 mischief] misfortune, distress
101 impotent] helpless
106 other] others
109 seem] think, deem
plain] honest
111 Bedrid] bedridden
crave] beg

But not every unsick stubborn knave,
For then we should over-many have.

Copland. How say you by these common beggars that cry
Daily on the world, and in the highways lie
At Westminster and at Saint Paul's,
And in all streets they sit as desolate souls? 120
Methink it a very well done deed
With devotion such people to feed.

Porter. Where any giveth almesse with good intent
The reward can not be nowise mis-spent.

Copland. Yea, but sir, I will not lie, by my soul,
As I walked to the church of Saint Paul,
There sat beggars, on each side the way two,
As is seen daily they be wont to do.
Sir, one there was, a mighty stubborn slave,
That for the other began to beg and crave: 130
'Now, master, in the way of your good speed,
To us all four behold where it is need;
And make this farthing worth a halfpenny,
For the five joys of Our Blessed Lady!
Now turn again for Saint Erasmus' sake!
And on my bare knees here a vow I make,
Our Lady's psalter three times even now
To turn again, as God shall turn to you!
Now, master, do that no man did this day,
On yon poor wretch that rotteth in the way, 140
Now, master, for Him that died on tree,
Let us not die for lack of charity!'
Thus he prated, as he full well can,
Till at last an honest serving man
Came by the way, and by compassion
Of his words did his devotion.
When he was gone a little fro thence
I saw the beggar pull out eleven pence,
Saying to his fellows: 'See, what here is:
Many a knave have I called master for this. 150

120 desolate] destitute
123 almesse] alms, charitable relief
129 mighty] strong, able-bodied
130 the other] the others
131 speed] success, fortune
135 Saint Erasmus] (invoked against epidemic diseases, especially the plague)
135, 138 turn again] turn back
143 can] knows how to

Let us go dine, this is a simple day,
  My master therewith shall I scantly pay.'
Come these folks hither, good master porter?

Porter. No, in sooth; this house is of no such supporter.
They have houses, and keep full ill jesting,
  And to them resort all the whole offspring
In the Barbican and in Turnbull street,
  In Houndsditch and behind the Fleet;
And in twenty places mo than there,
  Where they make revel and gaudy cheer, 160
With 'Fill the pot full' and 'Go fill me the can',
  'Here is my penny' and 'I am a gentleman'.
And there they bid and fill as doth a gull;
  And when that they have their heads full,
Then they fall out, and make reviling,
  And in this wise make the drunken reckoning:
'Thou beggarly knave, bag nor staff hast thou none,
  But as I am fain daily to lend thee one;
Thou gettest it no more, though it lie and rot,
  Nor my long cloak, nor my new patched coat.' 170
This rule make they every day and night,
  Till like as swine they lie sleeping upright.
Some beggarly churls, to whom they resort,
  Be the maintainers of a great sort
Of mighty lubbers, and have them in service,
  Some journeymen, and some to their prentice,
And they walk to each market and fair,
  And to all places where folk do repair,
By day on stilts, or stooping on crouches,
  And so dissimule as false loitering slouches, 180
With bloody clouts all about their leg,
  And plasters on their skin, when they go beg;
Some counterfeit lepry, and other some
  Put soap in their mouth to make it scum,
And fall down as Saint Cornelis' evil.
  These deceits they use worse than any devil;

151 simple] unprofitable
159 mo] more
160 gaudy] luxurious
163 bid] give orders
  gull] extravagant fool
168 fain] obliged, necessitated
172 upright] flat on their backs
174 sort] band, company
175 mighty lubbers] able-bodied wastrels
176 journeymen] trained craftsmen
179 crouches] crutches
180 dissimule] dissemble
  slouches] louts, lazy fellows
181 clouts] cloths
183 lepry] leprosy
184 scum] foam
185 Saint Cornelis' evil] epilepsy

And when they be in their own company,
They be as whole as either you or I.
But at the last, when sickness cometh indeed,
Then to the spital house must they come need. 190

(Wr. after 1535; first printing without date)

JOHN HARINGTON
d. 1582

69 *To his mother*

THERE was a battle fought of late,
Yet was the slaughter small;
The strife was, whether I should write,
Or send nothing at all.
Of one side were the captains' names
Short Time and Little Skill;
One fought alone against them both,
Whose name was Great Good-will.
Short Time enforced me in a strait,
And bade me hold my hand; 10
Small Skill also withstood desire
My writing to withstand.
But Great Good-will, in show though small,
To write encouraged me,
And to the battle held on still,
No common thing to see.
Thus gan these busy warriors three
Between themselves to fight
As valiantly as though they had
Been of much greater might. 20
Till Fortune, that unconstant dame,
Which rules such things alway,
Did cause the weaker part in fight
To bear the greater sway.
And then the victor caused me,
However was my skill,
To write these verses unto you
To show my great good-will.

(Wr. 1540; pub. 1769)

9 strait] narrow place or way; dilemma
10 hold my hand] refrain, forbear
24 sway] force, pressure

70 [*Husband to wife*]

IF duty, wife, lead thee to deem
  That trade most fit I hold most dear,
First, God regard, next me esteem,
  Our children then respect thou near.

Our house both sweet and cleanly see,
  Order our fare, thy maids keep short;
Thy mirth with mean well mixed be;
  Thy courteous parts in chaste wise sort.

In sober weed thee cleanly dress;
  When joys me raise, thy cares down cast; 10
When griefs me seize, thy solace cease;
  Whoso me friend, friend them as fast.

In peace give place, whatso I say;
  Apart complain, if cause thou find;
Let liberal lips no trust bewray,
  Nor jealous humour pain thy mind.

If I thee wrong, thy griefs unfold;
  If thou me vex, thine error graunt;
To seek strange toils be not too bold;
  The strifeless bed no jars may haunt. 20

Small sleep and early prayer intend;
  The idle life, as poison, hate;
No credit light nor much speech spend;
  In open place no cause debate.

No thwarts, no frowns, no grudge, no strife;
  Eschew the bad, embrace the best;
To troth of word join honest life,
  And in my bosom build thy nest.

(Wr. 1564; pub 1775)

2 trade] manner of life
6 fare] food
  keep short] keep in short supply (i.e. don't over-indulge them in food or household provisions)
11 solace] pleasures, amusements
15 liberal] unrestrained, licentious
  bewray] betray
16 humour] fancy, impulse
18 graunt] grant, concede
19 toils] disputes, verbal contentions
24 debate] contest, fight for

71 [*Wife to husband*]

HUSBAND, if you will be my dear,
  Your other self you must me make;
So, next to God, you shall be near;
  So, of our babes, care will I take.

An wholesome house and strong-built, give;
  See needful things be never scace;
Provide your men unidly live;
  Use courteous speech, show friendly face.

T'observe your times, if time I choose,
  To know my time you must take pain; 10
And how your friends you would I use,
  So, look my friends you entertain.

Your storms for stubborn servants stay,
  And gently warn me in mine ear;
As you may at your pleasure play,
  So, when I sport, be not severe.

That I you please, doth not alone
  In all respects myself suffice;
For good, of moe, I would you known,
  And long from home stayed in no wise. 20

If no suspicions rise, you read
  Suspicious cause, t'eschew were best;
Whatever care the day doth breed,
  Agree, the night yield pleasant rest.

Whatso, a wooer, you me behight,
  Now, husband good, perform as due;
Penelop's path if I hold right,
  Ulysses' steps see you tread true.

(Wr. 1564; pub. 1769)

6 scace] scarce
19 moe] more
21 read] make out
25 behight] promised

72 *A sonnet written upon my Lord Admiral Seymour*

OF person rare, strong limbs, and manly shape;
Of nature framed to serve on sea or land;
Of friendship firm in good state and ill hap;
In peace, head wise; in war, skill great, bold hand;
On horse, on foot, in peril or in play,
None could excel, though many did assay;
A subject true to king, and servant great;
Friend to God's truth, en'my to Rome's deceit;
Sumptuous abroad, for honour of the land;
Temp'rate at home, yet kept great state with stay 10
And noble house, and gave moe mouths more meat
Than some advanced on higher steps to stand.
Yet against Nature, Reason, and just Laws,
His blood was spilt, guiltless, without just cause.

(Wr. 1549–69, pub. 1769)

ANONYMOUS

73 *[How to obtain her]*

THE more ye desire her, the sooner ye miss;
The more ye require her, the stranger she is.
The more ye pursue her, the faster she flyeth;
The more ye eschew her, the sooner she plyeth.
But if ye refrain her, and use not to crave her,
So shall ye obtain her if ever ye have her.

(Pub. 1960)

9 Sumptuous] spending lavishly, magnificent in style of life
10 stay] staying power
11 moe] more
12 advanced] raised

73
2 require] beg, beseech
4 plyeth] will yield, submit
5 refrain] avoid, abstain from

ANNE ASKEW
1521–1546

74 *The Ballad which Anne Askew made and sang when she was in Newgate*

LIKE as the armed knight
  Appointed to the field,
With this world will I fight,
  And faith shall be my shield.

Faith is that weapon strong
  Which will not fail at need;
My foes therefore among
  Therewith will I proceed.

As it is had in strength
  And force of Christes way, 10
It will prevail at length
  Though all the devils say nay.

Faith in the Fathers old
  Obtained righteousness,
Which makes me very bold
  To fear no world's distress.

I now rejoice in heart,
  And hope bid me do so,
For Christ will take my part
  And ease me of my woe. 20

Thou say'st, Lord, who so knock
  To them wilt thou attend.
Undo therefore the lock
  And thy strong power send.

More en'mies now I have
  Than hairs upon my head.
Let them not me deprave,
  But fight thou in my stead.

2 appointed] equipped field] battle-field 27 deprave] corrupt

On thee my care I cast,
  For all their cruel spite; 30
I set not by their haste,
  For thou art my delight.

I am not she that list
  My anchor to let fall,
For every drizzling mist
  My ship substantial.

Not oft use I to write
  In prose nor yet in rhyme,
Yet will I show one sight
  That I saw in my time. 40

I saw a royal throne
  Where Justice should have sit,
But in her stead was one
  Of moody cruel wit.

Absorbed was righteousness
  As of the raging flood;
Satan in his excess
  Sucked up the guiltless blood.

Then thought I, Jesus Lord,
  When thou shalt judge us all, 50
Hard is it to record
  On these men what will fall.

Yet, Lord, I thee desire
  For that they do to me,
Let them not taste the hire
  Of their iniquity.

(1546)

31 set not by] have no regard for
haste] rashness, blind precipitancy
44 moody] angry, wilful
wit] mind
45 Absorbed] swallowed up, sucked in
46 As of] as by
51 Hard] painful
record] call to mind, ponder
54 For that] i.e. for what
55 hire] payment, reward

SIR THOMAS SEYMOUR (Baron Seymour of Sudeley)
1508?–1549

75 *Forgetting God*

FORGETTING God
  To love a king
Hath been my rod
  Or else nothing,

In this frail life,
  Being a blast
Of care and strife
  Till it be past.

Yet God did call
  Me in my pride, 10
Lest I should fall
  And from Him slide.

For whom loves He
  And not correct,
That they may be
  Of His elect?

Then, Death, haste thee.
  Thou shalt me gain
Immortally
  With Him to reign; 20

Who send the king
  Like years as Noye
In governing
  His realm in joy;

And after this
  Frail life, such grace
As in His bliss
  I may have place.

(Wr. 1549; pub. 1769)

22 Noye] Noah

JOHN HEYWOOD

*c.*1497–*c.*1580

76 *[A quiet neighbour]*

ACCOUNTED our commodities,
Few more commodious reason sees
Than is this one commodity,
Quietly neighboured to be.
Which neighbourhood in thee appears.
For we two having ten whole years
Dwelt wall to wall, so joiningly,
That whispering soundeth through well-nigh,
I never heard thy servants brawl
More than thou hadst had none at all. 10
Nor I can no way make avaunt
That ever I heard thee give them taunt.
Thou art to them and they to thee
More mild than mute—mum ye be.
I hear no noise mine ease to break,
Thy butt'ry door I hear not creak.
Thy kitchen cumbreth not by heat,
Thy cooks chop neither herbs nor meat.
I never heard thy fire once spark,
I never heard thy dog once bark. 20
I never heard once in thy house
So much as one peep of one mouse.
I never heard thy cat once mew.
These praises are not small nor few.
I bear all water of thy soil,
Whereof I feel no filthy foil,
Save water which doth wash thy hands,
Wherein there none annoyance stands.
Of all thy guests set at thy board
I never heard one speak one word. 30
I never heard them cough nor hem.
I think thence to Jerusalem,

1 Accounted] counted, reckoned
commodities] advantages, benefits
11 avaunt] boast
16 butt'ry] room for storing liquor or provisions
17 cumbreth] causes trouble or inconvenience
22 peep] squeak
25 soil] sewage, drainage
26 foil] dirt, dung
31 hem] clear the throat

For this neighbourly quietness
Thou art the neighbour neighbourless.
For ere thou wouldst neighbour annoy
These kinds of quiet to destroy,
Thou rather wouldst to help that matter
At home alone fast bread and water.

(1556)

NICHOLAS GRIMALD
1519?–1562?

77

Description of Virtue

WHAT one art thou, thus in torn weed yclad?
'Virtue, in price whom ancient sages had.'
Why poorly rayed? 'For fading goods past care.'
Why double-faced? 'I mark each fortune's fare.'
This bridle, what? 'Mind's rages to restrain.'
Tools why bear you? 'I love to take great pain.'
Why wings? 'I teach above the stars to fly.'
Why tread you death? 'I only cannot die.'

(1557)

THOMAS, LORD VAUX
1510–1556

78

The Aged Lover Renounceth Love

I LOATHE that I did love,
  In youth that I thought sweet;
As time requires, for my behove,
  Methinks they are not meet.

My lusts they do me leave,
  My fancies all be fled,
And tract of time begins to weave
  Grey hairs upon my head.

34 neighbourless] (1) solitary; (2) unsurpassed
38 fast] fast on, live on nothing but

77
1 weed] garment
3 rayed] arrayed, dressed

78
3 behove] use, benefit
4 meet] suitable

For age with stealing steps
  Hath clawed me with his clutch, 10
And lusty life away she leaps,
  As there had been none such.

My Muse doth not delight
  Me as she did before;
My hand and pen are not in plight,
  As they have been of yore.

For reason me denies
  This youthly idle rhyme;
And day by day to me she cries,
  'Leave off these toys in time.' 20

The wrinkles in my brow,
  The furrows in my face,
Say, limping age will lodge him now
  Where youth must give him place.

The harbinger of death,
  To me I see him ride;
The cough, the cold, the gasping breath
  Doth bid me to provide

A pickaxe and a spade,
  And eke a shrouding sheet, 30
A house of clay for to be made
  For such a guest most meet.

Methinks I hear the clerk
  That knolls the careful knell,
And bids me leave my woeful work,
  Ere nature me compel.

My keepers knit the knot
  That youth did laugh to scorn,
Of me that clean shall be forgot,
  As I had not been born. 40

Thus must I youth give up,
  Whose badge I long did wear;
To them I yield the wanton cup,
  That better may it bear.

10 clawed] seized, gripped
15 plight] good condition
20 toys] trifles
33 clerk] sexton
34 careful] sad, sorrowful

Lo, here the bared skull,
By whose bald sign I know
That stooping age away shall pull
That youthful years did sow.

For beauty with her band
These crooked cares hath wrought, 50
And shipped me into the land
From whence I first was brought.

And ye that bide behind,
Have ye none other trust;
As ye of clay were cast by kind,
So shall ye waste to dust.

(1557)

79 [*The Pleasures of Thinking*]

WHEN all is done and said, in the end thus shall you find,
He most of all doth bathe in bliss that hath a quiet mind;
And, clear from worldly cares, to deem can be content
The sweetest time in all his life in thinking to be spent.

The body subject is to fickle Fortune's power,
And to a million of mishaps is casual every hour,
And death in time doth change it to a clod of clay;
Whenas the mind, which is divine, runs never to decay.

Companion none is like unto the mind alone,
For many have been harmed by speech; through thinking, few or none. 10
Fear oftentimes restraineth words, but makes not thoughts to cease;
And he speaks best that hath the skill when for to hold his peace.

Our wealth leaves us at death, our kinsmen at the grave;
But virtue of the mind unto the heavens with us we have.
Wherefore, for virtue's sake, I can be well content
The sweetest time of all my life to deem in thinking spent.

(Pub. 1576)

55 cast] formed kind] nature

79 6 casual] liable 8 Whenas] whereas

80 [*Death in Life*]

HOW can the tree but waste and wither away
  That hath not sometime comfort of the sun?
How can that flower but fade and soon decay
  That always is with dark clouds over-run?
Is this a life? Nay, death you may it call,
That feels each pain and knows no joy at all.

What foodless beast can live long in good plight?
  Or is it life where senses there be none?
Or what availeth eyes without their light?
  Or else a tongue to him that is alone? 10
Is this a life? Nay, death you may it call,
That feels each pain and knows no joy at all.

Whereto serve ears if that there be no sound?
  Or such a head where no device doth grow
But all of plaints, since sorrow is the ground
  Whereby the heart doth pine in deadly woe?
Is this a life? Nay, death you may it call,
That feels each pain and knows no joy at all.

(Pub. 1576)

81 [*Age looks back at Youth*]

WHEN I look back and in myself behold
The wand'ring ways that youth could not descry,
And see the fearful course that youth did hold,
And mete in mind each step I strayed awry,
My knees I bow and from my heart I call,
O Lord forget youth's faults and follies all.

For now I see how scant youth was of skill;
I find by proof his pleasures all be pain;
I feel the sour that sweetness then did still;
I taste the gall hid under sugared train; 10
And with a mind repentant of all crimes
Pardon I ask for youth ten thousand times.

81
4 mete] measure
8 proof] experience
9 still] allay, assuage

The humble heart hath daunted the proud mind;
Knowledge hath given to ignorance the fall;
Wisdom hath taught that folly could not find,
And age hath youth his subject and his thrall;
Wherefore I pray, O Lord of life and truth,
Cancel the crimes committed in my youth.

Thou that didst grant the wise king his request,
Thou that in whale thy prophet didst preserve, 20
Thou that forgav'st the wounding of thy breast,
Thou that didst save the thief in state to starve,
Thou only good and giver of all grace,
Forgive the guilts that grew in youth's green race.

Thou that by power to life didst raise the dead,
Thou that of grace restored'st the blind to sight,
Thou that for love thy life and blood out-bled,
Thou that of favour mad'st the lame go right,
Thou that canst heal and help in all assays,
Wipe out of mind the wants of youth's vain ways. 30

And now since hope by grace with doubtless mind
Doth press to Thee by prayer t'appease Thine ire,
And since with trust to speed I seek to find
And wait through faith t'attain this just desire,
Lord, mind no more youth's error nor unskill,
But able age to do Thy holy will.

(Pub. 1576)

GEORGE CAVENDISH

1499?–1561?

82 *An Epitaph of our late Queen Mary*

DESCEND from heaven O muse Melpomene,
Thou mournful goddess with thy sisters all.
Pass in your plaints the woeful Niobe;
Turn music to moan with tears eternal.
Black be your habits, dim and funeral,

15 taught that] taught what
28 go right] walk properly
29 in all assays] in every crisis or time of need
36 able] enable
82
Epitaph] elegy

For death hath bereft, to our great dolour,
Mary our mistress, our queen of honour.

Our queen of honour, compared aptly
To *veritas victrix*, daughter of time,
By God assisted, amassed in army, 10
When she a virgin clear, without crime,
By right, without might, did happily climb
To the stage royal, just inheritor,
Proclaimed Mary our queen of honour.

And as a victrix valorous endued
With justice, prudence, high mercy and force,
Dreadless of danger with sword subdued
Her vassals rebellious, yet having remorse
With loss of few she saved the corse:
Such was thy mercy surmounting rigour, 20
O Mary mistress, O queen of honour!

To a virgin's life which liked thee best
Professed was thine heart, when moved with zeal
And tears of subjects expressing request,
For no lust but love for the common weal,
Virginity's vow thou diddest repeal,
Knit with a king co-equal in valour,
Thine estate to conserve as queen of honour.

The Rose and Pomegranate joined in one,
England and Spain by spousal allied. 30
Yet of these branches blossoms came none
Whereby their kingdoms might be supplied;
For this conjunction a comet envied,
Influence casting of mortal vapour
On Mary the Rose, our queen of honour.

Then faded the flower that whilom was fresh,
For Boreas' blasts did wither away
The spirit and life from the tender flesh
Of that imp royal, that prime rose gay,
Equal in odour to Flora in May; 40

9 *veritas victrix*] victorious truth. *Veritas filia temporis* (truth is the Daughter of Time) was Mary Tudor's personal motto.
13 stage] seat, position
16 force] strength
18 remorse] mercy
19 corse] body (i.e. the 'body politic')
29 Rose and Pomegranate] emblems of England and Spain
33 envied] was hostile to
36 whilom] once
37 Boreas] the north wind
39 imp] scion, shoot of a tree
prime] spring

The virtue vanished   with vital vigour
From our fair Mary,   our queen of honour.

Though virtue vital   did vanish away,
Her virtues inward   remain immortal,
Eterne and exempt   from death and decay,
As fountains flowing   with course continual,
As Ver in verdure   and green perpetual,
Or lamps ever light   supplied with liquor,
Enduring endless   to Mary's honour.

Add then to virtue   blood and parentage,   50
In all Europa   no princes equal,
So noble of birth,   descent and lineage,
As no man can number   the joints legal
Of emperors old   and houses regal.
No herald huked   in king's coat-armour
Sufficeth to blaze   our Mary's honour.

Lament, ye lords   and ladies of estate,
You puissant princes   and dukes of degree!
Let never nobles   appear so ingrate
As to forget   the great gratuity   60
Of graces granted   and benefits free,
Given and restored   only by favour
Of noble Mary,   your queen of honour.

High Priest of Rome,   O Paul Apostolic,
And College Conscript   of Cardinals all,
And ye that confess   the faith Catholic
Of Christ's Church   in earth universal,
O clerks and religious,   to you I call:
Pray for your patron,   your friend and founder,
Mary our mistress,   our queen of honour.   70

Which late restored   the right religion
And faith of fathers   observed of old,
Subdued sects   and all division,
Reducing the flock   to the former fold;
A pillar most firm   the Church to uphold.
Lo where she lieth   true faith's defender,
Mary our mistress,   our queen of honour!

47 Ver] spring
48 liquor] oil
53 joints] inter-relationships
55 huked] cloaked
56 blaze] blazon, proclaim
60 gratuity] graciousness
64 Paul] Pope Paul IV
74 Reducing] leading back

When sacred altars were all defaced,
  Images of saints with outrage burned;
Instead of priests apostatas placed; 80
  Holy sacraments with spite down spurned;
  When spoil and ravin had all overturned,
    This chaos confuse, this heap of horror
    Dissolveth Mary, as queen of honour.

Elizabeth excellent of God elect,
  With sceptre to sit in seat imperial,
In throne triumphant where thou art erect,
  Have death alway in thy memorial.
  Death is th'end of flesh universal.
    The world is but vain: make for your mirror 90
    Mary thy sister, late queen of honour.

So shall the Almighty stablish thy throne
  In quiet concord and due obeisance,
And send thee a prince to appease our moan,
  With happy reign of long continuance,
  This thing reposed in deep remembrance.
    Say and pray all: 'O Christ, O Saviour,
    Have mercy on Mary our queen of honour!'

O Virgin Mary, mother of Jesu,
  O spouse unspotted and queen eternal, 100
As our Queen Mary was handmaid true
  To thee, O Lady, in this life mortal,
  So of thy grace and bounty special
    To the King on high be intercessor
    In heaven to crown her a queen of honour.

(Wr. 1558–9; pub. 1825)

80 apostatas] apostates
82 ravin] robbery, plunder
83 confuse] confused
84 Dissolveth] causes to vanish, does away with
88 memorial] memory

THOMAS PHAER

1510?–1560

from *The nine first books of the Eneidos*

83 [*Euryalus and Nisus meet their deaths*]

A WOOD with bushes broad there was, begrown with bigtree boughs,
Whom thick entangling thorns and briery brambles filled with brows:
No trade but trattling paths, some here, some there that secret strays.
Euryalus the branches dark of trees, and heavy preys,
Done let; he clean contrary runs, beguiled by wandering ways.
Nisus went on, and en'mies all unwares had scaped quite,
And passed that place which afterwards Albanus mountain hight
Of Alba's name; King Latin there great pastures did maintain;
When first he stood, and for his absent friend did look in vain.
'Euryalus, poor lad, what country now shall I thee seek? 10
What path should I pursue?' Straight back again from creek to creek
Through that deceitful wood unwinding ways perplexed he sought,
Still tracking marking steps through thickets silent straggling blind.
He hears their horse, he hears their rustling noise and en'mies' wind.
Not long between there was when to his ears the cry came hot,
And first Euryalus he seeth, whom all men's hands had got,
Through fraud of night and place, of troublous tumult wareless
  trapped,
Vain-struggling, working much, but round about him all they wrapped.
What should he do? What strength? How could he shift or dare dispose
To rescue thus this lad? Should he run rashly mids his foes? 20
Enforcing fair to death with comely wounds his life to lose?
He swiftly shook his dart, and high beholding bright the moon
He whirling bent his arm, and thus he fervent made his boon:
'Thou goddess, thou this time, thou in our labours lend relief,
Thou beauteous queen of stars, in forests virgin keeper chief.
If ever gift for me sir Hyrtacus my father gave
Unto thine offerings' seats, if ever I increased have
Thy sacred altars' fees, with hunting daily through my coasts,
Or decked thy church with spoils, or hanged about thy holy posts,

2 brows] prickles
3 trade] way, path
trattling paths] sheep tracks (from 'trattle', sheep-droppings)
4 preys] spoils, plunder
5 Done let] impede
7 hight] was called
8 Latin] i.e. Latinus
10 country] region
11 creek] crevice, hidden corner
14 wind] sounds of tracking
20 mids] midst
23 boon] prayer
28 fees] tributes
my coasts] my usual resorts, places

Give me to break this plump, and through the skies now guide my dart.' 30
He spake; and straining total strength his tool with hand and heart
Cast forth. It whirling flew, and through the shade of shimm'ring night
It passed, and into Sulmon's back with noise did sharply light;
In pieces there it brake, and to the heart-strings pierced the wood.
He tumbling (cold) outspewed all hot from breast his reeking flood,
Far-fetching, vexing slow; his guts upgathering smites his sides.
Each man about them look. Lo, yet again a smarter glides,
Which he with force outflang and level cast direct from ear.
Whiles all they troubled stood, to Tagus whistling ran that spear;
Athwart his head it came, and thirled him quite through temples twain 40
With noise, where fixed fast it stack warm-waxing through his brain.
Duke Volscens storming frets, nor him that did that weapon fling
He one where could behold, nor whither fervent mad to spring.
'But thou this while,' quoth he, 'these two men's death shalt surely rue,
If any hot blood in thy heart there be'—and straight outdrew
Against Euryalus his sword. Then verily indeed dismayed
Did Nisus loudly shriek, nor more to lurk in darkness stayed.
Such torments then him took, he cried amain with voice afraid:
''Tis I, 'tis I, here, here I am that did, turn all at me,
O Rutils, with your tools! My only craft here 'tis, not he! 50
He neither durst nor could, this heaven, these stars I witness take.
Only for too much love his wretched friend he nould forsake.'
Such words he gave, but deep with dint the sword enforced first
Had ransacked through his ribs, and sweet white breast at once had burst.
Down falls Euryalus in death; his limbs, his fair fine flesh
All runs on blood; his neck down-fainting nods on shoulders nesh
Well like the purple flower that, cut with plough, letfalling lops
In languish withering dies, or like weak necks of poppies' crops
Down peising heavy heads, when rain doth lading grieve their tops.
But Nisus to his en'mies fiercely ran, and through their mids 60
Duke Volscens out he seeks, he only Volscens battle bids;
Whom Rutils clustering close on each side shoves and stout withstands.
Yet ne'ertheless his sword like lightning bright with both his hands
He swingeing stirred, and as Duke Volscens cried he smote him so
That through his throat in went, and even in death he killed his foe.

30 break this plump] destroy this troop
36 Far-fetching] lashing out convulsively
vexing] writhing
40 thirled] pierced
41 warm-waxing] growing warm
48 amain] with full force
49 that did] that did it
50 Rutils] i.e. Rutilians
52 nould] would not
54 ransacked] searched deeply
56 nesh] soft, tender
58 crops] heads, tops
59 peising] weighing, pressing
60 mids] midst
64 swingeing] brandishing

Then weary, digged with wounds, on his dead friend himself he kest
Expiring life at last, and took his death for pleasant rest.
O fortunate both twain! And if my verse may get good luck,
Shall never day nor time from mindful age your praises pluck,
While Prince Aeneas' house, while Capitol most stately stone 70
Unmovable shall stand, while Roman rules this world in one.

(Pub. 1562)

BARNABY GOOGE
1540–1594

84

To Doctor Bale

GOOD aged Bale,
That with thy hoary hairs
Dost yet persist
To turn the painful book,
O happy man
That hast obtained such years,
And leav'st not yet
On papers pale to look,
Give over now
To beat thy wearied brain, 10
And rest thy pen
That long hath laboured sore;
For aged men
Unfit sure is such pain,
And thee beseems
To labour now no more.
But thou, I think,
Don Plato's part will play,
With book in hand
To have thy dying day. 20

(1563)

66 digged] pierced
kest] cast

84
1 Bale] (John Bale, 1495–1563, antiquary, controversialist, and dramatist)

85 *Of Money*

GIVE money me, take
Friendship whoso list,
For friends are gone
Come once adversity,
When money yet
Remaineth safe in chest,
That quickly can
Thee bring from misery.
Fair face show friends
When riches do abound: 10
Come time of proof,
Farewell, they must away.
Believe me well,
They are not to be found,
If God but send
Thee once a lowering day.
Gold never starts
Aside, but in distress
Finds ways enough
To ease thine heaviness. 20

(1563)

86 *Coming homeward out of Spain*

O RAGING seas
And mighty Neptune's reign,
In monstrous hills
That throwest thyself so high,
That with thy floods
Dost beat the shores of Spain,
And break the cliffs
That dare thy force envy:
Cease now thy rage
And lay thine ire aside. 10
And thou that hast
The governance of all,
O mighty God,
Grant weather, wind and tide,
Till on my coun-
try coast our anchor fall.

(1563)

THOMAS SACKVILLE, EARL OF DORSET
1536–1608

from *The Mirror for Magistrates*

87

The Induction

THE wrathful winter, 'proaching on apace,
With blustering blasts had all ybared the treen,
And old Saturnus, with his frosty face,
With chilling cold had pierced the tender green;
The mantles rent, wherein enwrapped been
The gladsome groves that now lay overthrown,
The tapets torn, and every bloom down blown.

The soil, that erst so seemly was to seen,
Was all despoiled of her beauty's hue;
And soote fresh flowers, wherewith the summer's queen 10
Had clad the earth, now Boreas' blasts down blew;
And small fowls flocking in their song did rue
The winter's wrath, wherewith each thing defaced
In woeful wise bewailed the summer past.

Hawthorn had lost his motley livery,
The naked twigs were shivering all for cold,
And dropping down the tears abundantly;
Each thing, methought, with weeping eye me told
The cruel season, bidding me withhold
Myself within; for I was gotten out 20
Into the fields, whereas I walked about.

When lo, the night with misty mantles spread
Gan dark the day and dim the azure skies;
And Venus in her message Hermes sped
To bloody Mars, to will him not to rise,
While she herself approached in speedy wise;
And Virgo, hiding her disdainful breast,
With Thetis now had laid her down to rest.

7 tapets] tapestries 10 soote] sweet

Whiles Scorpio, dreading Sagittarius' dart,
  Whose bow prest bent in fight the string had slipped, 30
Down slid into the ocean flood apart;
  The Bear, that in the Irish seas had dipped
  His grisly feet, with speed from thence he whipped;
    For Thetis, hasting from the Virgin's bed,
    Pursued the Bear, that ere she came was fled.

And Phaethon now, near reaching to his race
  With glistering beams, gold streaming where they bent,
Was prest to enter in his resting place.
  Erythius, that in the cart first went,
  Had even now attained his journey's stent; 40
    And, fast declining, hid away his head,
    While Titan couched him in his purple bed.

And pale Cynthia, with her borrowed light,
  Beginning to supply her brother's place,
Was past the noonstead six degrees in sight,
  When sparkling stars amid the heaven's face
  With twinkling light shone on the earth apace,
    That, while they brought about the nightes chare,
    The dark had dimmed the day ere I was ware.

And sorrowing I to see the summer flowers, 50
  The lively green, the lusty leas, forlorn,
The sturdy trees so shattered with the showers,
  The fields so fade that flourished so beforn,
  It taught me well all earthly things be born
    To die the death, for nought long time may last;
    The summer's beauty yields to winter's blast.

Then looking upward to the heaven's leams,
  With nightes stars thick powdered everywhere,
Which erst so glistened with the golden streams
  That cheerful Phoebus spread down from his sphere, 60
  Beholding dark oppressing day so near;
    The sudden sight reduced to my mind
    The sundry changes that in earth we find:

33 grisly] fear-arousing
38 prest] ready
40 stent] destination, end
48 chare] chariot
51 lusty] pleasant, delightful
leas] meadows
57 leams] lights
62 reduced] brought back

That musing on this worldly wealth in thought,
  Which comes and goes more faster than we see
The flickering flame that with the fire is wrought,
  My busy mind presented unto me
  Such fall of peers as in this realm had be,
    That oft I wished some would their woes descrive,
    To warn the rest whom Fortune left alive. 70

And straight forth stalking with redoubled pace,
  For that I saw the night drew on so fast,
In black all clad there fell before my face
  A piteous wight, whom woe had all forwaste;
  Forth from her eyne the crystal tears outbrast,
    And sighing sore, her hands she wrung and fold,
    Tare all her hair, that ruth was to behold.

Her body small, forwithered and forspent,
  As is the stalk that summer's drought oppressed;
Her welked face with woeful tears besprent, 80
  Her colour pale, and, as it seemed her best,
  In woe and plaint reposed was her rest;
    And as the stone that drops of water wears,
    So dented were her cheeks with fall of tears:

Her eyes swollen with flowing streams afloat;
  Wherewith, her looks thrown up full piteously,
Her forceless hands together oft she smote,
  With doleful shrieks that echoed in the sky;
  Whose plaint such sighs did straight accompany,
    That, in my doom, was never man did see 90
    A wight but half so woebegone as she.

I stood aghast, beholding all her plight,
  'Tween dread and dolour so distrained in heart
That, while my hairs upstarted with the sight,
  The tears outstreamed for sorrow of her smart;
  But when I saw no end that could apart
    The deadly dule which she so sore did make,
    With doleful voice then thus to her I spake:

64 wealth] well-being
69 descrive] describe
80 welked] wrinkled
90 doom] opinion
96 apart] put away
97 dule] woe

'Unwrap thy woes, whatever wight thou be,
And stint betime to spill thyself with plaint; 100
Tell what thou art, and whence, for well I see
Thou canst not dure, with sorrow thus attaint.'
And with that word of sorrow, all forfaint
She looked up, and prostrate as she lay,
With piteous sound, lo, thus she gan to say:

'Alas, I wretch whom thus thou seest distrained
With wasting woes that never shall aslake,
Sorrow I am, in endless torments pained
Among the Furies in the infernal lake,
Where Pluto, god of Hell, so grisly black, 110
Doth hold his throne, and Lethe's deadly taste
Doth reave remembrance of each thing forepast.

'Whence come I am, the dreary destiny
And luckless lot for to bemoan of those
Whom Fortune, in this maze of misery,
Of wretched chance most woeful mirrors chose;
That when thou seest how lightly they did lose
Their pomp, their power, and that they thought most sure,
Thou mayst soon deem no earthly joy may dure.'

Whose rueful voice no sooner had out brayed 120
Those woeful words wherewith she sorrowed so,
But out, alas, she shright and never stayed,
Fell down, and all to-dashed herself for woe:
The cold pale dread my limbs gan overgo,
And I so sorrowed at her sorrows eft
That, what with grief and fear, my wits were reft.

I stretched myself and straight my heart revives,
That dread and dolour erst did so appale;
Like him that with the fervent fever strives,
When sickness seeks his castle health to scale, 130
With gathered spirits so forced I fear to avale;
And rearing her with anguish all fordone,
My spirits returned and then I thus begun:

100 stint] cease
spill] destroy
112 reave] rob
120 out brayed] ejaculated
122 shright] shrieked
125 eft] again
131 avale] sink

'O Sorrow, alas, sith Sorrow is thy name,
And that to thee this drear doth well pertain,
In vain it were to seek to cease the same;
But as a man himself with sorrow slain,
So I, alas, do comfort thee in pain,
That here in sorrow art forsunk so deep
That at thy sight I can but sigh and weep.' 140

I had no sooner spoken of a sike,
But that the storm so rumbled in her breast,
As Aeolus could never roar the like;
And showers down rained from her eyne so fast
That all bedrent the place, till at the last
Well eased they the dolour of her mind,
As rage of rain doth swage the stormy wind.

For forth she paced in her fearful tale:
'Come, come,' quod she, 'and see what I shall show;
Come hear the plaining and the bitter bale 150
Of worthy men by Fortune's overthrow;
Come thou and see them ruing all in row.
They were but shades that erst in mind thou rolled;
Come, come with me, thine eyes shall them behold.'

What could these words but make me more aghast,
To hear her tell whereon I mused while ere?
So was I mazed therewith, till at the last,
Musing upon her words, and what they were,
All suddenly well lessoned was my fear;
For to my mind returned how she telled 160
Both what she was and where her wone she held.

Whereby I knew that she a goddess was,
And therewithal resorted to my mind
My thought, that late presented me the glass
Of brittle state, of cares that here we find,
Of thousand woes to silly men assigned;
And how she now bid me come and behold,
To see with eye that erst in thought I rolled.

141 sike] sigh
145 bedrent] drenched
150 bale] torment, woe
161 wone] abode
166 silly] helpless, weak
168 that erst] what before

Flat down I fell, and with all reverence
Adored her, perceiving now that she, 170
A goddess sent by godly providence,
In earthly shape thus showed herself to me,
To wail and rue this world's uncertainty;
And while I honoured thus her godhead's might,
With plaining voice these words to me she shright:

'I shall thee guide first to the grisly lake
And thence unto the blissful place of rest,
Where thou shalt see and hear the plaint they make
That whilom here bare swing among the best;
This shalt thou see, but great is the unrest 180
That thou must bide before thou canst attain
Unto the dreadful place where these remain.'

And with these words, as I upraised stood,
And gan to follow her that straight forth paced,
Ere I was ware, into a desert wood
We now were come, where, hand in hand embraced,
She led the way and through the thick so traced
As, but I had been guided by her might,
It was no way for any mortal wight.

But lo, while thus amid the desert dark 190
We passed on with steps and pace unmeet,
A rumbling roar, confused with howl and bark
Of dogs, shook all the ground under our feet,
And struck the din within our ears so deep
As, half distraught, unto the ground I fell,
Besought return, and not to visit hell.

But she, forthwith uplifting me apace,
Removed my dread, and with a steadfast mind
Bade me come on, for here was now the place,
The place where we our travail end should find; 200
Wherewith I arose, and to the place assigned
Astoined I stalk, when straight we approached near
The dreadful place that you will dread to hear.

179 bare swing] enjoyed power
187 thick] thicket
191 unmeet] unevenly matched
202 Astoined] astonished

An hideous hole all vast, withouten shape,
Of endless depth, o'erwhelmed with ragged stone,
With ugly mouth and grisly jaws doth gape,
And to our sight confounds itself in one.
Here entered we, and yeding forth, anon
An horrible loathly lake we might discern,
As black as pitch, that cleped is Averne: 210

A deadly gulf where nought but rubbish grows,
With foul black swelth in thickened lumps that lies,
Which up in the air such stinking vapours throws
That over there may fly no fowl but dies,
Choked with the pestilent savours that arise;
Hither we come, whence forth we still did pace,
In dreadful fear amid the dreadful place.

And first, within the porch and jaws of hell,
Sat deep Remorse of Conscience, all besprent
With tears, and to herself oft would she tell 220
Her wretchedness, and cursing never stent
To sob and sigh, but ever thus lament
With thoughtful care as she that, all in vain,
Would wear and waste continually in pain.

Her eyes unsteadfast, rolling here and there,
Whirled on each place, as place that vengeance brought,
So was her mind continually in fear,
Tossed and tormented with the tedious thought
Of those detested crimes which she had wrought;
With dreadful cheer and looks thrown to the sky, 230
Wishing for death, and yet she could not die.

Next saw we Dread, all trembling how he shook,
With foot uncertain proffered here and there,
Benumbed of speech, and with a ghastly look
Searched every place, all pale and dead for fear,
His cap borne up with staring of his hair,
'Stoined and amazed at his own shade for dread,
And fearing greater dangers than was need.

208 yeding] going
210 cleped] called
212 swelth] filth
219 besprent] sprinkled
221 stent] stopped
236 staring] standing up

And next, within the entry of this lake,
Sat fell Revenge, gnashing her teeth for ire, 240
Devising means how she may vengeance take,
Never in rest till she have her desire;
But frets within so far forth with the fire
Of wreaking flames, that now determines she
To die by death, or venged by death to be.

When fell Revenge, with bloody foul pretence,
Had showed herself as next in order set,
With trembling limbs we softly parted thence,
Till in our eyes another sight we met,
When from my heart a sigh forthwith I fet, 250
Ruing, alas, upon the woeful plight
Of Misery, that next appeared in sight.

His face was lean and somedeal pined away,
And eke his hands consumed to the bone,
But what his body was I cannot say,
For on his carcass raiment had he none,
Save clouts and patches, pieced one by one;
With staff in hand and scrip on shoulders cast,
His chief defence against the winter's blast.

His food, for most, was wild fruits of the tree, 260
Unless sometimes some crumbs fell to his share,
Which in his wallet long, God wot, kept he,
As on the which full daintily would he fare;
His drink, the running stream; his cup, the bare
Of his palm closed; his bed, the hard cold ground;
To this poor life was Misery ybound.

Whose wretched state when we had well beheld,
With tender ruth on him and on his fears,
In thoughtful cares forth then our pace we held;
And by and by another shape appears, 270
Of greedy Care, still brushing up the breres,
His knuckles knobbed, his flesh deep dented in,
With tawed hands and hard ytanned skin.

250 fet] fetched
253 somedeal] somewhat
271 brushing up the breres] brushing off the briars
273 tawed] hardened

The morrow gray no sooner hath begun
  To spread his light, even peeping in our eyes,
When he is up and to his work yrun;
  But let the night's black misty mantles rise,
  And with foul dark never so much disguise
    The fair bright day, yet ceaseth he no while,
    But hath his candles to prolong his toil. 280

By him lay heavy Sleep, the cousin of Death,
  Flat on the ground and still as any stone,
A very corpse, save yielding forth a breath.
  Small keep took he whom Fortune frowned on
  Or whom she lifted up into the throne
    Of high renown; but as a living death,
    So, dead alive, of life he drew the breath.

The body's rest, the quiet of the heart,
  The travail's ease, the still night's fere was he,
And of our life in earth the better part; 290
  Reaver of sight, and yet in whom we see
  Things oft that tide, and oft that never be;
    Without respect esteeming equally
    King Croesus' pomp and Irus' poverty.

And next in order sad Old Age we found,
  His beard all hoar, his eyes hollow and blind,
With drooping cheer still poring on the ground,
  As on the place where nature him assigned
  To rest, when that the sisters had untwined
    His vital thread and ended with their knife 300
    The fleeting course of fast declining life.

There heard we him with broken and hollow plaint
  Rue with himself his end approaching fast,
And all for nought his wretched mind torment
  With sweet remembrance of his pleasures past,
  And fresh delights of lusty youth forewaste;
    Recounting which, how would he sob and shriek,
    And to be young again of Jove beseek!

289 fere] companion
291 Reaver] depriver
292 tide] happen
308 beseek] beseech

But and the cruel fates so fixed be
  That time forepast cannot return again, 310
This one request of Jove yet prayed he,
  That in such withered plight and wretched pain
  As eld, accompanied with his loathsome train,
    Had brought on him, all were it woe and grief,
    He might a while yet linger forth his life:

And not so soon descend into the pit
  Where Death, when he the mortal corpse hath slain,
With reckless hand in grave doth cover it,
  Thereafter never to enjoy again
  The gladsome light, but in the ground ylain, 320
    In depth of darkness waste and wear to nought,
    As he had never into the world been brought.

But who had seen him sobbing, how he stood
  Unto himself and how he would bemoan
His youth forepast, as though it wrought him good
  To talk of youth, all were his youth foregone,
  He would have mused and marvelled much, whereon
    This wretched Age should life desire so fain,
    And knows full well life doth but length his pain.

Crookbacked he was, tooth-shaken, and blear-eyed, 330
  Went on three feet, and sometime crept on four,
With old lame bones that rattled by his side,
  His scalp all pilled and he with eld forlore;
  His withered fist still knocking at Death's door,
    Fumbling and drivelling as he draws his breath;
    For brief, the shape and messenger of Death.

And fast by him pale Malady was placed,
  Sore sick in bed, her colour all foregone,
Bereft of stomach, savour, and of taste,
  Ne could she brook no meat, but broths alone; 340
  Her breath corrupt, her keepers every one
    Abhorring her, her sickness past recure,
    Detesting physic and all physic's cure.

309 But and] But if   333 pilled] bald

But oh, the doleful sight that then we see!
We turned our look and on the other side
A grisly shape of Famine mought we see,
With greedy looks and gaping mouth that cried
And roared for meat, as she should there have died;
Her body thin and bare as any bone,
Whereto was left nought but the case alone. 350

And that, alas, was gnawn on everywhere,
All full of holes, that I ne mought refrain
From tears to see how she her arms could tear,
And with her teeth gnash on the bones in vain,
When all for nought she fain would so sustain
Her starven corpse, that rather seemed a shade
Than any substance of a creature made.

Great was her force, whom stone wall could not stay,
Her tearing nails snatching at all she saw;
With gaping jaws that by no means ymay 360
Be satisfied from hunger of her maw,
But eats herself as she that hath no law;
Gnawing, alas, her carcass all in vain,
Where you may count each sinew, bone, and vein.

On her while we thus firmly fixed our eyes,
That bled for ruth of such a dreary sight,
Lo, suddenly she shright in so huge wise,
As made hell gates to shiver with the might;
Wherewith a dart we saw, how it did light
Right on her breast, and therewithal pale Death 370
Enthrilling it, to reave her of her breath.

And by and by a dumb dead corpse we saw,
Heavy and cold, the shape of Death aright,
That daunts all earthly creatures to his law;
Against whose force in vain it is to fight;
Ne peers, ne princes, nor no mortal wight,
No towns, ne realms, cities, ne strongest tower,
But all, perforce, must yield unto his power.

344 see] saw 346 mought] might 371 enthrilling] piercing

His dart, anon, out of the corpse he took,
And in his hand, a dreadful sight to see, 380
With great triumph eftsoons the same he shook,
That most of all my fears affrayed me;
His body dight with nought but bones, perdy,
The naked shape of man there saw I plain,
All save the flesh, the sinew, and the vein.

Lastly, stood War, in glittering arms yclad,
With visage grim, stern looks, and blackly hued;
In his right hand a naked sword he had,
That to the hilts was all with blood imbrued;
And in his left, that kings and kingdoms rued, 390
Famine and fire he held, and therewithal
He razed towns and threw down towers and all.

Cities he sacked, and realms, that whilom flowered
In honour, glory, and rule above the best,
He overwhelmed and all their fame devoured,
Consumed, destroyed, wasted, and never ceased,
Till he their wealth, their name, and all oppressed;
His face forhewed with wounds, and by his side
There hung his targe, with gashes deep and wide.

In midst of which, depainted there, we found 400
Deadly Debate, all full of snaky hair,
That with a bloody fillet was ybound,
Out-breathing nought but discord everywhere.
And round about were portrayed, here and there,
The hugy hosts, Darius and his power,
His kings, princes, his peers, and all his flower:

Whom great Macedo vanquished there in fight
With deep slaughter, despoiling all his pride,
Pierced through his realms and daunted all his might.
Duke Hannibal beheld I there beside, 410
In Canna's field victor how he did ride,
And woeful Romans that in vain withstood,
And consul Paulus covered all in blood.

381 eftsoons] afterwards
383 perdy] 'by God'; indeed
387 stern] fierce
401 Debate] Conflict
407 Macedo] Alexander

Yet saw I more: the fight at Thrasimene,
And Trebery field, and eke when Hannibal
And worthy Scipio last in arms were seen
Before Carthago gate, to try for all
The world's empire, to whom it should befall;
There saw I Pompey and Caesar clad in arms,
Their hosts allied and all their civil harms: 420

With conquerors' hands, forbathed in their own blood,
And Caesar weeping over Pompey's head.
Yet saw I Sulla and Marius where they stood,
Their great cruelty and the deep bloodshed
Of friends; Cyrus I saw and his host dead,
And how the queen with great despite hath flung
His head in blood of them she overcome.

Xerxes, the Persian king, yet saw I there
With his huge host that drank the rivers dry,
Dismounted hills, and made the vales uprear, 430
His host and all yet saw I plain, perdy!
Thebes I saw, all razed how it did lie
In heaps of stones, and Tyrus put to spoil,
With walls and towers flat evened with the soil.

But Troy, alas, methought, above them all,
It made mine eyes in very tears consume,
When I beheld the woeful weird befall,
That by the wrathful will of gods was come;
And Jove's unmoved sentence and foredoom
On Priam king and on his town so bent, 440
I could not lin, but I must there lament.

And that the more, sith destiny was so stern
As, force perforce, there might no force avail,
But she must fall, and by her fall we learn
That cities, towers, wealth, world, and all shall quail.
No manhood, might, nor nothing mought prevail;
All were there prest full many a prince and peer,
And many a knight that sold his death full dear:

437 weird] destiny, fate 441 lin] cease 447 prest] ready, eager

Not worthy Hector, worthiest of them all,
Her hope, her joy: his force is now for nought. 450
O Troy, Troy, there is no boot but bale;
The hugy horse within thy walls is brought;
Thy turrets fall, thy knights, that whilom fought
In arms amid the field, are slain in bed,
Thy gods defiled and all thy honour dead.

The flames upspring and cruelly they creep
From wall to roof till all to cinders waste;
Some fire the houses where the wretches sleep,
Some rush in here, some run in there as fast;
In every where or sword or fire they taste; 460
The walls are torn, the towers whirled to the ground;
There is no mischief but may there be found.

Cassandra yet there saw I how they haled
From Pallas' house, with sparkled tress undone,
Her wrists fast bound, and with Greeks' rout empaled;
And Priam eke, in vain how did he run
To arms, whom Pyrrhus with despite hath done
To cruel death, and bathed him in the baign
Of his son's blood, before the altar slain.

But how can I describe the doleful sight 470
That in the shield so livelike fair did shine?
Sith in this world I think was never wight
Could have set forth the half, not half so fine.
I can no more but tell how there is seen
Fair Ilium fall in burning red gledes down,
And from the soil great Troy, Neptunus' town.

Herefrom when scarce I could mine eyes withdraw,
That filled with tears as doth the springing well,
We passed on so far forth till we saw
Rude Acheron, a loathsome lake to tell, 480
That boils and bubs up swelth as black as hell;
Where grisly Charon, at their fixed tide,
Still ferries ghosts unto the farther side.

451 no boot but bale] no remedy but woe
464 sparkled] dishevelled
465 rout] troop
empaled] hemmed in
468 baign] bath
475 gledes] glowing ashes
481 bubs] bubbles

The aged god no sooner Sorrow spied,
But hasting straight unto the bank apace,
With hollow call unto the rout he cried
To swerve apart and give the goddess place.
Straight it was done, when to the shore we pace,
Where, hand in hand as we then linked fast,
Within the boat we are together placed. 490

And forth we launch full fraughted to the brink,
When with the unwonted weight the rusty keel
Began to crack as if the same should sink.
We hoise up mast and sail, that in a while
We fet the shore, where scarcely we had while
For to arrive, but that we heard anon
A three-sound bark confounded all in one.

We had not long forth passed but that we saw
Black Cerberus, the hideous hound of hell,
With bristles reared and with a three-mouthed jaw 500
Fordinning the air with his horrible yell,
Out of the deep dark cave where he did dwell.
The goddess straight he knew, and by and by,
He 'peased and couched while that we passed by.

Thence come we to the horror and the hell,
The large great kingdoms and the dreadful reign
Of Pluto in his throne where he did dwell,
The wide waste places and the hugy plain,
The wailings, shrieks, and sundry sorts of pain,
The sighs, the sobs, the deep and deadly groan, 510
Earth, air, and all, resounding plaint and moan.

Here puled the babes, and here the maids unwed
With folded hands their sorry chance bewailed,
Here wept the guiltless slain, and lovers dead,
That slew themselves when nothing else availed;
A thousand sorts of sorrow here, that wailed
With sighs and tears, sobs, shrieks, and all yfere,
That oh, alas, it was a hell to hear.

517 yfere] together

We stayed us straight, and with a rueful fear
  Beheld this heavy sight, while from mine eyes 520
The vapoured tears downstilled here and there,
  And Sorrow eke, in far more woeful wise,
  Took on with plaint, upheaving to the skies
    Her wretched hands, that with her cry the rout
    Gan all in heaps to swarm us round about.

Lo here, quod Sorrow, princes of renown,
  That whilom sat on top of Fortune's wheel,
Now laid full low, like wretches whirled down,
  Even with one frown, that stayed but with a smile;
  And now behold the thing that thou, erewhile, 530
    Saw only in thought, and what thou now shalt hear,
    Recount the same to kesar, king, and peer.

(1563)

ANONYMOUS

88 *A Dialogue between Death and Youth*

Death. Come on, good fellow, make an end,
    For you and I must talk.
  You may no longer sojourn here,
    But hence you must go walk.

Youth. What woeful words, alas,
    Be these that I do hear!
  Alas, and shall I now forthwith
    Forsake my life so dear?

Death. Come on, come on, and linger not;
    Ye trifle but the time. 10
  Ye make too much of that, ywis,
    Which is but dirt and slime.

Youth. O cursed death, what dost thou mean,
    So cruel for to be,
  To him that never thought thee harm
    Nor once offended thee!

11 ywis] indeed, certainly

O death, behold: I am but young
  And of a pleasant age.
Take thou some old and crooked wight,
  And spare me in thy rage. 20

Behold, my limbs be lively now,
  My mind and courage strong,
And by the verdict of all men
  Like to continue long;

My beauty like the rose so red,
  My hair like glist'ring gold;
And canst thou now of pity then
  Transform me into mould?

O gentle death, be not extreme;
  Thy mercy here I crave. 30
It is not for thine honour now
  To fetch me to my grave.

But rather let me live a while,
  Till youth consumed be.
When crooked age doth me oppress,
  Then welcome death to me.

Death. O foolish man, what dost thou mean
  To strive against the stream?
Nothing there is that can thee now
  Out of my hands redeem. 40

Thy time is past, thy days are gone,
  Thy race is fully run.
Thou must of force now make an end,
  As thou hadst once begun.

O fool, why dost thou beg and boast
  Of these thy youthful days?—
Which passeth fast and fadeth swift
  As flowers fresh decays.

Both youth and age to me be one—
  I care not whom I strike: 50
The child, the man, the father old
  Do I reward alike.

19 wight] person 28 mould] earth 30 crave] beg

The proudest of them all, ywis,
  Can not escape my dart:
The lady fair, the lazar foul
  Shall both possess a part.

Thou art not now the first, I say,
  That I have eared up;
Ne yet shalt be the last, pardie,
  That drinketh of my cup. 60

For he that doth us now behold—
  Perusing this our talk—
He knoweth not yet how soon, God wot,
  With thee and me to walk!

Dispatch, therefore, and make an end,
  For needs you must obey;
And as thou camest into this world,
  So shalt thou now away.

Youth. And must I pass out of this world
  Indeed, and shall I so? 70
May no man me restrain a while,
  But needs now must I go?

Why, then, farewell my life and lands,
  Adieu my pleasures all!
Lo dreadful death doth us depart,
  And me away doth call.

My cheerful days be worn away,
  My pleasant time is past;
My youthful years are spent and gone,
  My life it may not last. 80

And I (for lack of life and breath)
  Whose like hath not been seen,
Shall straight consumed be to dust,
  As I had never been.

But though I yield as now to thee,
  When nothing can me save,
Yet I am sure that I shall live
  When thou thy death shalt have.

(Pub. 1564)

55 lazar] leper, poor and diseased person
58 eared] ploughed
59 pardie] i.e. by God; certainly, indeed
62 Perusing] surveying
71 restrain] save, keep free
75 depart] separate

EDWARD DE VERE, EARL OF OXFORD
1550–1604

89 *The lively lark stretched forth her wing*

THE lively lark stretched forth her wing,
The messenger of morning bright,
And with her cheerful voice did sing
The day's approach, discharging night,
  When that Aurora, blushing red,
  Descried the guilt of Thetis' bed.

I went abroad to take the air,
And in the meads I met a knight,
Clad in carnation colour fair.
I did salute this gentle wight; 10
  Of him I did his name enquire.
  He sighed, and said 'I am Desire'.

Desire I did desire to stay;
A while with him I craved to talk.
The courteous knight said me no nay,
But hand in hand with me did walk.
  Then of Desire I asked again
  What thing did please, and what did pain.

He smiled, and thus he answered then:
'Desire can have no greater pain 20
Than for to see another man
That he desireth, to obtain;
  Nor greater joy can be than this,
  Than to enjoy that others miss'.

(Wr. before 1566; pub. 1576)

90 *If women could be fair and yet not fond*

IF women could be fair, and yet not fond,
Or that their loves were firm, not fickle still,
I would not wonder that they make men bond,
By service long to purchase their good will.
  But when I see how frail these creatures are,
  I laugh that men forget themselves so far.

22 That he desireth] i.e. what he desires 24 that others] i.e. what others

To mark the choice they make, and how they change,
How oft from Phoebus they do cleave to Pan,
Unsettled still, like haggards wild they range,
These gentle birds, that fly from man to man. 10
 Who would not scorn, and shake them from the fist,
 And let me go, fair fools, which way they list?

Yet, for disport, we fawn and flatter both,
To pass the time when nothing else can please,
And train them to our lure with subtle oath,
Till, weary of our wills, ourselves we ease.
 And then we say, when we their fancies try,
 To play with fools, O what a dolt was I!

(Pub. 1588)

91 *The labouring man, that tills the fertile soil*

THE labouring man, that tills the fertile soil,
 And reaps the harvest fruit, hath not in deed
The gain, but pain; and if for all his toil
 He gets the straw, the lord will have the seed.

The manchet fine falls not unto his share;
 On coarsest cheat his hungry stomach feeds.
The landlord doth possess the finest fare;
 He pulls the flowers, the other plucks but weeds.

The mason poor, that builds the lordly halls,
 Dwells not in them; they are for high degree. 10
His cottage is compact in paper walls,
 And not with brick or stone, as others be.

The idle drone, that labours not at all,
 Sucks up the sweet of honey from the bee.
Who worketh most, to their share least doth fall;
 With due desert reward will never be.

9 haggards] untamed hawks
91
5 manchet] finest kind of wheaten bread
6 cheat] wheaten bread of poorer quality
10 degree] rank
11 compact in] made of

The swiftest hare unto the mastiff slow
  Oft times doth fall to him as for a prey;
The greyhound thereby doth miss his game, we know,
  For which he made such speedy haste away. 20

So he that takes the pain to pen the book
  Reaps not the gifts of goodly golden Muse;
But those gain that who on the work shall look,
  And from the sour the sweet by skill doth choose.
For he that beats the bush the bird not gets,
But who sits still and holdeth fast the nets.

(Pub. 1573)

92 *Sitting alone upon my thought*

SITTING alone upon my thought, in melancholy mood,
In sight of sea, and at my back an ancient hoary wood,
I saw a fair young lady come, her secret tears to wail,
Clad all in colour of a vow, and covered with a veil.
Yet, for the day was clear and calm, I might discern her face,
As one might see a damask rose, though hid with crystal glass.
Three times with her soft hand full hard on her left side she knocks,
And sighed so sore as might have moved some mercy in the rocks.
From sighs, and shedding amber tears, into sweet song she brake,
And thus the echo answered her to every word she spake. 10

'O heavens', quoth she, 'who was the first that bred in me this
  fever?' *Echo*: Vere.
'Who was the first that gave the wound, whose scar I wear for
  ever?' *Echo*: Vere.
'What tyrant Cupid to my harms usurps the golden quiver?' *Echo*: Vere.
'What wight first caught this heart, and can from bondage it
  deliver?' *Echo*: Vere.
'Yet who doth most adore this wight, O hollow caves, tell true?' *Echo*:
  You.
'What nymph deserves his liking best, yet doth in sorrow rue?' *Echo*:
  You.
'What makes him not regard good will with some remorse or
  ruth?' *Echo*: Youth.
'What makes him show, besides his birth, such pride and such
  untruth?' *Echo*: Youth.

19 greyhound] (pronounced with one syllable, i.e. 'greund')
21 the book] (this poem was prefaced to Thomas Bedingfield's book *Cardanus Comforte* (1573))

'May I his beauty match with love, if he my love will try?' *Echo*; Aye.
'May I requite his birth with faith? Then faithful will I die.'
Echo: Aye. 20

And I that knew this lady well,
Said: 'Lord, how great a miracle,
To hear the echo tell the truth,
As 'twere Apollo's oracle'.

(Probably wr. 1580–1; pub. 1872)

93 [*A Court Lady addresses her Lover*]

THOUGH I be strange, sweet friend, be thou not so;
Do not annoy thyself with sullen will.
My heart hath vowed, although my tongue say no,
To rest thine own, in friendly liking still.

Thou seest we live amongst the lynx's eyes,
That pries and spies each privy thought of mind;
Thou knowest right well what sorrows may arise
If once they chance my settled looks to find.

Content thyself that once I made an oath
To shield myself in shroud of honest shame; 10
And when thou list, make trial of my troth,
So that thou save the honour of my name.

And let me seem, although I be not coy,
To cloak my sad conceits with smiling cheer;
Let not my gestures show wherein I joy,
Nor by my looks let not my love appear.

We silly dames, that false suspect do fear,
And live within the mouth of envy's lake,
Must in our hearts a secret meaning bear,
Far from the show that outwardly we make. 20

So where I like, I list not vaunt my love;
Where I desire, there must I feign debate.
One hath my hand, another hath my glove,
But he my heart whom most I seem to hate.

93
1 strange] unfriendly, distant
18 envy] malice
22 debate] quarrel, strife

Thus farewell, friend: I will continue strange;
  Thou shalt not hear by word or writing aught.
Let it suffice, my vow shall never change;
  As for the rest, I leave it to thy thought.

(Prob. wr. 1580–1; pub. here for the first time (?))

94 *When wert thou born, Desire?*

WHEN wert thou born, Desire?
  In pomp and prime of May.
By whom, sweet boy, wert thou begot?
  By good conceit, men say.

Tell me who was thy nurse?
  Fresh youth in sugared joy.
What was thy meat and dainty food?
  Sad sighs, with great annoy.

What hadst thou then to drink?
  Unfeigned lovers' tears. 10
What cradle wert thou rocked in?
  In hope devoid of fears.

What brought thee then asleep?
  Sweet speech that liked me best.
And where is now thy dwelling place?
  In gentle hearts I rest.

Doth company displease?
  It doth in many a one.
Where would Desire then choose to be?
  He likes to muse alone. 20

What feedeth most thy sight?
  To gaze on favour still.
What findst thou most to be thy foe?
  Disdain of my good will.

Will ever age or death
  Bring thee unto decay?
No, no, Desire both lives and dies
  Ten thousand times a day.

(Pub. 1591)

94
4 good conceit] favourable opinion

95 *What cunning can express*

WHAT cunning can express
The favour of her face,
To whom in this distress
I do appeal for grace.
A thousand Cupids fly
About her gentle eye.

From whence each throws a dart
That kindleth soft sweet fire
Within my sighing heart,
Possessed by desire. 10
No sweeter life I try
Than in her love to die.

The lily in the field,
That glories in his white,
For pureness now must yield
And render up his right.
Heaven pictured in her face
Doth promise joy and grace.

Fair Cynthia's silver light,
That beats on running streams, 20
Compares not with her white,
Whose hairs are all sunbeams.
Her virtues so do shine,
As day unto mine eyne.

With this there is a red
Exceeds the damask rose,
Which in her cheeks is spread,
Whence every favour grows.
In sky there is no star,
That she surmounts not far. 30

When Phoebus from the bed
Of Thetis doth arise,
The morning blushing red
In fair carnation wise,
He shows it in her face,
As queen of every grace.

24 eyne] eyes

This pleasant lily white,
This taint of roseate red,
This Cynthia's silver light,
This sweet fair Dea spread 40
These sunbeams in mine eye;
These beauties make me die.

(Pub. 1593)

ATTRIBUTED TO EDWARD DE VERE, EARL OF OXFORD

96 *When I was fair and young*

WHEN I was fair and young, then favour graced me,
Of many was I sought, their mistress for to be;
But I did scorn them all, and answered them therefore,
Go, go, go, seek some other where, importune me no more!

How many weeping eyes I made to pine in woe,
How many sighing hearts, I have not skill to show,
But I the prouder grew, and still this spake therefore,
Go, go, go, seek some other where, importune me no more.

Then spake fair Venus' son, that brave, victorious boy,
Saying: 'You dainty dame, for that you be so coy, 10
I will so pull your plumes, as you shall say no more,
Go, go, go, seek some other where, importune me no more.'

As soon as he had said, such change grew in my breast
That neither night nor day I could take any rest;
Wherefore I did repent that I had said before,
Go, go, go, seek some other where, importune me no more.

(Pub. 1801)

38 taint] tint, colour
40 Dea] goddess

96
1 When I was fair and young] (the speaker is supposed to be Queen Elizabeth I)

ANONYMOUS

97 *The lover compareth himself to the painful falconer*

THE soaring hawk from fist that flies
  Her falconer doth constrain
Sometime to range the ground unknown
  To find her out again:
And if by sight or sound of bell
  His falcon he may see,
'Wo ho' he cries, with cheerful voice,
  The gladdest man is he.

By lure then, in finest sort,
  He seeks to bring her in; 10
But if that she full gorged be,
  He cannot her so win.
Although her becks and bending eyes
  She many proffers makes,
'Wo ho ho' he cries, away she flies,
  And so her leave she takes.

This woeful man with weary limbs
  Runs wandering round about;
At length by noise of chattering pies
  His hawk again found out. 20
His heart was glad his eyes had seen
  His falcon swift of flight:
'Wo ho ho' he cries: she empty-gorged
  Upon his lure doth light.

How glad was then the falconer there
  Nor pen nor tongue can tell.
He swam in bliss that lately felt
  Like pains of cruel hell.
His hand sometime upon her train,
  Sometime upon her breast: 30
'Wo ho ho' he cries with cheerful voice;
  His heart was now at rest.

3 range] search
7 Wo ho] (the cry of the falconer calling the falcon back to the lure)
13 bending eyes] eyes looking towards him, as if consenting to return
19 pies] magpies
29 train] raw meat used as a lure
29–30] His hand sometime . . . Sometime . . . breast] i.e. 'His hand one moment holding the meat, the next touching her breast'

My dear, likewise behold thy love,
  What pains he doth endure:
And now at length let pity move
  To stoop into his lure.
A hood of silk and silver bells,
  New gifts I promise thee:
'Wo ho ho' I cry; 'I come', then say:
  Make me as glad as he. 40

(Wr. by 1566; pub. 1584 and probably in lost edition of 1566)

ARTHUR GOLDING

*c.*1536–1605

from *Ovid's 'Metamorphoses'*

98 [*Ceyx and Alcyone*]

In this meantime the Trachin king sore vexed in his thought
With signs that both before and since his brother's death were wrought,
For counsel at the sacred spells, which are but toys to food
Fond fancies, and not counsellors in peril to do good,
Did make him ready to the God of Claros for to go.
For heathenish Phorbas and the folk of Phlegia had as tho
The way to Delphos stopped, that none could travel to or fro.
And ere he on his journey went, he made his faithful make,
Alcyone, privy to the thing. Immediately there struck
A chillness to her very bones, and pale was all her face 10
Like box, and down her heavy cheeks the tears did gush apace.
Three times about to speak, three times she washed her face with tears,
And stinting oft with sobs she thus complained in his ears:
'What fault of mine, O husband dear, hath turned thy heart fro me?
Where is that care of me that erst was wont to be in thee?
And canst thou, having left thy dear Alcyone, merry be?
Do journeys long delight thee now? Doth now mine absence please
Thee better than my presence doth? Think I that thou at ease
Shalt go by land? Shall I have cause but only for to mourn
And not to be afraid? And shall my care of thy return 20
Be void of fear? No, no. The sea me sore afraid doth make.
To think upon the sea doth cause my flesh for fear to quake.

36 stoop] descend swiftly, swoop
98
1 Trachin king] King of Trachis
3 food] (fode) give encouragement to
6 tho] then
8 make] mate
11 box] boxwood
13 stinting] stopping

I saw the broken ribs of ships alate upon the shore;
And oft on tombs I read their names whose bodies long before
The sea had swallowed. Let not fond vain hope seduce thy mind,
That Aeolus is thy father-in-law who holds the boistous wind
In prison, and can calm the seas at pleasure. When the winds
Are once let loose upon the sea, no order then them binds.
Then neither land hath privilege, nor sea exemption finds.
Yea, even the clouds of heaven they vex, and with their meeting stout 30
Enforce the fire with hideous noise to burst in flashes out.
The more that I do know them, for right well I know their power,
And saw them oft, a little wench, within my father's bower,
So much the more I think them to be feared. But if thy will
By no entreatance may be turned at home to tarry still,
But that thou needs wilt go, then me, dear husband, with thee take.
So shall the sea us equally together toss and shake.
So worser than I feel I shall be certain not to fear.
So shall we whatsoever haps together jointly bear.
So shall we on the broad main sea, together jointly sail.' 40
These words and tears wherewith the imp of Aeolus did assail
Her husband born of heavenly race, did make his heart relent,
For he loved her no less than she loved him. But fully bent
He seemed, neither for to leave the journey which he meant
To take by sea, nor yet to give Alcyone leave as tho
Companion of his parlous course by water for to go.
He many words of comfort spake her fear away to chase,
But nought he could persuade therein to make her like the case.
This last assuagement of her grief he added in the end,
Which was the only thing that made her loving heart to bend: 50
'All tarriance will assuredly seem overlong to me;
And by my father's blazing beams I make my vow to thee
That at the furthest ere the time, if God thereto agree,
The moon do fill her circle twice, again I will here be.'
When in some hope of his return this promise had her set,
He willed a ship immediately from harbour to be fet,
And throughly rigged for to be that neither mast nor sail
Nor tackling, no, nor other thing should appertaining fail.
Which when Alcyone did behold, as one whose heart misgave
The haps at hand, she quaked again, and tears outgushing drave. 60
And straining Ceyx in her arms with pale and piteous look,
Poor wretched soul, her last farewell at length she sadly took,
And swounded flat upon the ground. Anon the watermen,
As Ceyx sought delays and was in doubt to turn again,
Set hand to oars, of which there were two rows on either side,
And all at once with equal stroke the swelling sea divide.

23 alate] lately 26 boistous] violent 41 imp] offspring

She lifting up her watery eyes beheld her husband stand
Upon the hatches, making signs by beckoning with his hand,
And she made signs to him again. And after that the land
Was far removed from the ship, and that the sight began 70
To be unable to discern the face of any man,
As long as e'er she could she looked upon the rowing keel,
And when she could no longer time for distance ken it well,
She looked still upon the sails that flasked with the wind
Upon the mast. And when she could the sails no longer find
She gat her to her empty bed with sad and sorry heart,
And laid her down. The chamber did renew afresh her smart,
And of her bed did bring to mind the dear departed part.
  From harbour now they quite were gone, and now a pleasant gale
  Did blow. The master made his men their oars aside to hale 80
And hoisted up the topsail on the highest of the mast,
And clapped on all his other sails because no wind should waste.
Scarce full th' one half, or sure not much above, the ship had run
Upon the sea, and every way the land did far them shun,
When toward night the wallowing waves began to waxen white,
And eke the heady eastern wind did blow with greater might.
Anon the master cried, 'Strike the topsail, let the main
Sheet fly and fardel it to the yard.' Thus spake he, but in vain;
For why, so hideous was the storm upon the sudden braid
That not a man was able there to hear what other said. 90
And loud the sea with meeting waves extremely raging roars.
Yet fell they to it of themselves. Some haled aside the oars,
Some fenced in the galley's sides, some down the sailcloths rend,
Some pump the water out, and sea to sea again do send.
Another hales the sailyards down. And while they did each thing
Disorderly, the storm increased, and from each quarter fling
The winds with deadly feud, and bounce the raging waves together.
The pilot being sore dismayed saith plain he knows not whither
To wend himself, nor what to do or bid, nor in what state
Things stood; so huge the mischief was, and did so overmate 100
All art. For why, of rattling ropes, of crying men and boys,
Of flushing waves and thundering air, confused was the noise.
The surges mounting up aloft did seem to mate the sky,
And with their sprinkling for to wet the clouds that hang on high.
One while the sea, when from the brink it raised the yellow sand,
Was like in colour to the same. Another while did stand
A colour on it blacker than the Lake of Styx. Anon
It lieth plain and loometh white with seething froth thereon;

73 ken] see
74 flasked] flapped
85 wallowing] rolling
88 fardel] furl
89 braid] onset
100 overmate] overcome
103 mate] equal

And with the sea the Trachin ship aye alteration took.
One while as from a mountain's top it seemed down to look 110
To valleys and the depth of hell. Another while beset
With swelling surges round about which ne'er above it met,
It looked from the bottom of the whirlpool up aloft
As if it were from hell to heaven. A hideous flushing oft
The waves did make in beating full against the galley's side.
The galley being stricken gave as great a sound that tide,
As did sometime the battle-ram of steel, or now the gun
In making battery to a tower. And as fierce lions run
Full breast with all their force against the armed men that stand
In order bent to keep them off with weapons in their hand; 120
Even so as often as the waves by force of wind did rave,
So oft upon the netting of the ship they mainly drave,
And mounted far above the same. Anon off fell the hoops,
And having washed the pitch away, the sea made open loops
To let the deadly water in. Behold, the clouds did melt,
And showers large came pouring down. The seamen that them felt
Might think that all the heaven had fallen upon them that same time,
And that the swelling sea likewise above the heaven would climb.
The sails were throughly wet with showers, and with the heavenly rain
Was mixed the waters of the sea. No lights at all remain 130
Of sun or moon or stars in heaven. The darkness of the night
Augmented with the dreadful storm takes double power and might.
Howbeit, the flashing lightnings oft do put the same to flight,
And with their glancing now and then do give a sudden light.
The lightning sets the waves on fire. Above the netting skip
The waves, and with a violent force do light within the ship;
And as a soldier stouter than the rest of all his band
That oft assails a city's walls defended well by hand,
At length attains his hope, and for to purchase praise withal
Alone among a thousand men gets up upon the wall; 140
So when the lofty waves had long the galley's sides assayed,
At length the tenth wave rising up with huger force and braid,
Did never cease assaulting of the weary ship till that
Upon the hatches like a foe victoriously it gat.
A part thereof did still as yet assault the ship without,
And part had gotten in. The men all trembling ran about,
As in a city comes to pass when of the enemies some
Did down the walls without, and some already in are come.
All art and cunning was to seek. Their hearts and stomachs fail;
And look how many surges came their vessel to assail, 150
So many deaths did seem to charge and break upon them all.
One weeps, another stands amazed, the third them blessed doth call

116 tide] time 122 mainly] with force

Whom burial doth remain. To God another makes his vow,
And holding up his hands to heaven the which he sees not now
Doth pray in vain for help. The thought of this man is upon
His brother and his parents whom he clearly hath forgone.
Another calls his house and wife and children unto mind,
And every man in general the things he left behind.
Alcyone moveth Ceyx' heart. In Ceyx' mouth is none
But only one Alcyone. And though she were alone 160
The wight that he desired most, yet was he very glad
She was not there. To Trachinward to look desire he had
And homeward fain he would have turned his eyes which nevermore
Should see the land, but then he knew not which way was the shore
Nor where he was. The raging sea did roll about so fast,
And all the heaven with clouds as black as pitch was overcast
That never night was half so dark. There came a flaw at last
That with his violence broke the mast and struck the stern away.
A billow proudly pranking up as vaunting of his prey
By conquest gotten, walloweth whole and breaketh not asunder, 170
Beholding with a lofty look the water working under.
And look, as if a man should from the places where they grow
Rend down the mountains Athe and Pind, and whole them overthrow
Into the open sea, so soft the billow tumbling down
With weight and violent stroke did sink and in the bottom drown
The galley. And the most of them that were within the same
Went down therewith, and never up to open aier came,
But died strangled in the gulf. Another sort again
Caught pieces of the broken ship. The king himself was fain
A shiver of the sunken ship in that same hand to hold 180
In which he erst a royal mace had held of yellow gold.
His father and his father-in-law he calls upon (alas,
In vain!), but chiefly in his mouth his wife Alcyone was,
In heart was she, in tongue was she. He wished that his corse
To land where she might take it up the surges might enforce,
And that by her most loving hands he might be laid in grave.
In swimming still, as often as the surges leave him gave
To ope his lips, he harped still upon Alcyone's name,
And when he drowned in the waves he muttered still the same.
Behold, even full upon the wave a flake of water black 190
Did break, and underneath the sea the head of Ceyx strack.
That night the lightsome Lucifer for sorrow was so dim
As scarcely could a man discern or think it to be him.
And forasmuch as out of heaven he might not step aside,
With thick and darksome clouds that night his countenance he did
hide.

153 remain] await
169 pranking] swaggering
179 fain] glad
190 flake] mass, heavy weight

Alcyone of so great mischance not knowing aught as yet
Did keep a reckoning of the nights that in the while did flit,
And hasted garments both for him and for herself likewise
To wear at his homecoming which she vainly did surmise.
To all the gods devoutly she did offer frankincense, 200
But most above them all the church of Juno she did cense.
And for her husband, who as then was none, she kneeled before
The altar, wishing health and soon arrival at the shore,
And that none other woman might before her be preferred.
Of all her prayers this one piece effectually was heard.
For Juno could not find in heart entreated for to be
For him that was already dead; but to th'intent that she
From dame Alcyone's deadly hands might keep her altars free,
She said: 'Most faithful messenger of my commandments, O
Thou rainbow, to the sluggish House of Slumber swiftly go, 210
And bid him send a dream in shape of Ceyx to his wife
Alcyone, for to show her plain the losing of his life.'
Dame Iris takes her pall wherein a thousand colours were,
And bowing like a stringed bow upon a cloudy sphere
Immediately descended to the drowsy House of Sleep
Whose courts the clouds continually do closely overdreep.
Among the dark Cimmerians is a hollow mountain found
And in the hill a cave that far doth run within the ground,
The chamber and the dwelling-place where slothful Sleep doth couch.
The light of Phoebus' golden beams this place can never touch. 220
A foggy mist with dimness mixed streams upward from the ground,
And glimmering twilight evermore within the same is found.
No watchful bird with barbed bill and combed crown doth call
The morning forth with crowing out. There is no noise at all
Of waking dog, nor gaggling goose more waker than the hound,
To hinder sleep; of beast nor wild nor tame there is no sound.
No boughs are stirred with blasts of wind, no noise of tattling tongue
Of man or woman ever yet within that bower rung.
Dumb quiet dwelleth there. Yet from the roche's foot doth go
The river of forgetfulness, which runneth trickling so 230
Upon the little pebble stones which in the channel lie
That unto sleep a great deal more it doth provoke thereby.
Before the entry of the cave there grows of poppy store,
With seeded heads, and other weeds innumerable more,
Out of the milky juice of which the night doth gather sleeps,
And over all the shadowed earth with dankish dew them dreeps.
Because the creaking hinges of the door no noise should make,
There is no door in all the house, nor porter at the gate.

213 pall] mantle
216 overdreep] droop over
229 roche] rock
236 dreeps] drips

Amid the cave, of ebony a bedstead standeth high,
And on the same a bed of down with coverings black doth lie, 240
In which the drowsy God of Sleep his lither limbs doth rest.
About him, forging sundry shapes as many dreams lie prest
As ears of corn do stand in fields in harvest time, or leaves
Do grow on trees, or sea to shore of sandy cinder heaves.
As soon as Iris came within this house, and with her hand
Had put aside the dazzling dreams that in her way did stand,
The brightness of her robe through all the sacred house did shine.
The God of Sleep scarce able for to raise his heavy eyen,
A three or four times at the least did fall again to rest,
And with his nodding head did knock his chin against his breast. 250
At length he shaking of himself upon his elbow leant,
And though he knew for what she came, he asked her what she meant.
'O Sleep,' quoth she, 'the rest of things, O gentlest of the gods,
Sweet Sleep, the peace of mind, with whom crook'd care is aye at odds,
Which cherishest men's weary limbs appalled with toiling sore,
And makest them as fresh to work and lusty as before,
Command a dream that in their kinds can everything express,
To Trachin, Hercles' town, himself this instant to address,
And let him lively counterfeit to Queen Alcyone
The image of her husband who is drowned in the sea 260
By shipwreck. Juno willeth so.' Her message being told,
Dame Iris went her way; she could her eyes no longer hold
From sleep, but when she felt it come she fled that instant time,
And by the bow that brought her down to heaven again did climb.
  Among a thousand sons and more that father Slumber had,
  He called up Morph, the feigner of man's shape, a crafty lad.
None other could so cunningly express man's very face,
His gesture and his sound of voice, and manner of his pace,
Together with his wonted weed, and wonted phrase of talk;
But this same Morphy only in the shape of man doth walk. 270
There is another who the shapes of beast or bird doth take,
Or else appeareth unto men in likeness of a snake;
The gods do call him Icilos, and mortal folk him name
Phobetor. There is also yet a third who from these same
Works diversely, and Phantasos he highteth; into streams
This turns himself, and into stones, and earth, and timber beams,
And into every other thing that wanteth life. These three
Great kings and captains in the night are wonted for to see;
The meaner and inferior sort of others haunted be.
Sir Slumber overpassed the rest, and of the brothers all 280
To do dame Iris' message he did only Morphy call;

241 lither] lazy

266 Morph, Morphy] Morpheus, god of dreams

Which done, he waxing luskish straight laid down his drowsy head
And softly shrunk his lazy limbs within his sluggish bed.
Away flew Morphy through the air—no flickering made his wings—
And came anon to Trachin; there his feathers off he flings
And in the shape of Ceyx stands before Alcyone's bed,
Pale, wan, stark naked, and like a man that was but lately dead.
His beard seemed wet, and of his head the hair was dropping dry,
And leaning on her bed, with tears he seemed thus to cry:
'Most wretched woman, knowest thou thy loving Ceyx now? 290
Or is my face by death disformed? Behold me well, and thou
Shalt know me; for thy husband thou thy husband's ghost shalt see.
No good thy prayers and thy vows have done at all to me,
For I am dead; in vain of my return no reckoning make.
The cloudy south amid the sea our ship did tardy take
And tossing it with violent blasts asunder did it shake,
And floods have filled my mouth which called in vain upon thy name.
No person whom thou may'st misdeem brings tidings of the same,
Thou hearest not thereof by false report of flying fame,
But I myself: I presently my shipwreck to thee show. 300
Arise therefore, and woeful tears upon thy spouse bestow,
Put mourning raiment on, and let me not to Limbo go
Unmourned for.' In showing of this shipwreck Morphy so
Did feign the voice of Ceyx that she could none other deem
But that it should be his indeed. Moreover, he did seem
To weep in earnest, and his hands the very gesture had
Of Ceyx. Queen Alcyone did groan, and being sad
Did stir her arms, and thrust them forth his body to embrace.
Instead whereof she caught but air. The tears ran down her face.
She cried, 'Tarry. Whither fliest? Together let us go.' 310
And all this while she was asleep. Both with her crying so
And flaighted with the image of her husband's ghastly sprite,
She started up, and sought about if find him there she might
(For why, her grooms awaking with the shriek had brought a light).
And when she nowhere could him find, she gan her face to smite,
And tore her nightclothes from her breast, and struck it fiercely, and
Not passing to untie her hair she rent it with her hand.
And when her nurse of this her grief desired to understand
The cause: 'Alcyone is undone, undone and cast away
With Ceyx her dear spouse,' she said. 'Leave comforting, I pray. 320
By shipwreck he is perished, I have seen him, and I knew
His hands. When in departing I to hold him did pursue,
I caught a ghost, but such a ghost as well discern I might
To be my husband's. Natheless he had not to my sight
His wonted countenance, neither did his visage shine so bright

282 luskish] lazy, sluggish 312 flaighted] frightened 317 passing] caring
298 misdeem] mistake

As heretofore it had been wont. I saw him, wretched wight,
Stark naked, pale, and with his hair still wet; even very here
I saw him stand.' With that she looks if any print appear
Of footing whereas he did stand upon the floor behind.
'This, this is it that I did fear in far forecasting mind 330
When fleeing me I thee desired thou shouldst not trust the wind,
But sith thou wentest to thy death, I would that I had gone
With thee. Ah, meet, it meet had been thou shouldst not go alone
Without me! So it should have come to pass that neither I
Had overlived thee, nor yet been forced twice to die.
Already absent in the waves now tossed have I be,
Already have I perished. And yet the sea hath thee
Without me. But the cruelness were greater far of me
Than of the sea if after thy decease I still would strive
In sorrow and in anguish still to pine away alive. 340
But neither will I strive in care to lengthen still my life,
Nor, wretched wight, abandon thee, but like a faithful wife
At leastwise now will come as thy companion. And the hearse
Shall join us though not in the selfsame coffin, yet in verse.
Although in tomb the bones of us together may not couch,
Yet in a graven epitaph my name thy name shall touch.'
Her sorrow would not suffer her to utter any more;
She sobbed and sighed at every word until her heart was sore.
  The morning came, and out she went right pensive to the shore
  To that same place in which she took her leave of him before. 350
While there she musing stood, and said: 'He kissed me even here,
Here weighed he his anchors up, here loosed he from the pier,'
And while she called to mind the things there marked with her eyes,
In looking on the open sea, a great way off she spies
A certain thing much like a corse come hovering on the wave.
At first she doubted what it was. As tide it nearer drave,
Although it were a good way off, yet did it plainly show
To be a corse. And though that whose it was she did not know,
Yet for because it seemed a wreck, her heart thereat did rise;
And as it had some stranger been, with water in her eyes 360
She said: 'Alas, poor wretch, whoe'er thou art, alas for her
That is thy wife, if any be!' And as the waves did stir,
The body floated nearer land, the which the more that she
Beheld, the less began in her of stayed wit to be.
Anon it did arrive to shore. Then plainly she did see
And know it that it was her fere. She shrieked, 'It is he.'
And therewithal her face, her hair, and garments she did tear,
And unto Ceyx stretching out her trembling hands with fear,

333 meet] fitting 359 wreck] shipwrecked person 366 fere] mate

Said: 'Com'st thou home in such a plight to me, O husband dear?
Return'st in such a wretched plight?' There was a certain pier 370
That builded was by hand, of waves the first assaults to break,
And at the haven's mouth to cause the tide to enter weak.
She leapt thereon—a wonder sure it was she could do so—
She flew, and with her new-grown wings did beat the air as tho.
And on the waves a wretched bird she whisked to and fro,
And with her croaking neb then grown to slender bill and round,
Like one that wailed and mourned still she made a moaning sound.
Howbeit, as soon as she did touch his dumb and bloodless flesh,
And had embraced his loved limbs with wings made new and fresh,
And with her hardened neb had kissed him coldly though in vain, 380
Folk doubt if Ceyx feeling it to raise his head did strain,
Or whether that the waves did lift it up. But surely he
It felt: and through compassion of the gods both he and she
Were turned to birds. The love of them eke subject to their fate
Continued after; neither did the faithful bond abate
Of wedlock in them being birds, but stands in steadfast state.
They tread, and lay, and bring forth young, and now the halcyon sits
In wintertime upon her nest, which on the water flits
A sevennight, during all which time the sea is calm and still,
And every man may to and fro sail safely at his will, 390
For Aeolus for his offspring's sake the winds at home doth keep,
And will not let them go abroad for troubling of the deep.

(1567)

JOHN PIKERYNG

*c.*1567

from *The History of Horestes*

99

[*Haltersick's Song*]

FAREWELL, adieu, that courtly life,
To war we tend to go;
It is good sport to see the strife
Of soldiers on a row.
How merrily they forward march,
These enemies to slay;
With hey, trim, and tricksy too
Their banners they display.

374 tho] then
376 neb] bill

99
2 tend] intend
7 hey, trim and tricksy] hey, neat and finely dressed

Now shall we have the golden cheats,
  When others want the same; 10
And soldiers have full many feats
  Their enemies to tame.
    With cucking here and booming there,
      They break their foe's array;
    And lusty lads amid the fields
      Their ensigns do display.

The drum and flute play lustily,
  The trumpet blows amain,
And venturous knights courageously
  Do march before their train. 20
    With spear in rest so lively dressed,
      In armour bright and gay,
    With hey, trim, and tricksy too
      Their banners they display.

100 [*Song sung by Egistus and Clytemnestra*]

Egistus. And was it not a worthy sight
  Of Venus' child, King Priam's son,
To steal from Greece a lady bright,
  For whom the wars of Troy begun?
    Naught fearing danger that might fall,
      Lady, lady, 30
    From Greece to Troy he went withal,
      My dear lady.

Clytemnestra. When Paris first arrived there,
  Whereas Dame Venus' worship is,
And blustering Fame abroad did bear
  His lively fame, she did not miss
    To Helena for to repair,
      Her for to tell
    Of praise and shape so trim and fair
      That did excel. 40

Egistus. Her beauty caused Paris pain,
  And bare chief sway within his mind.
No thing was able to restrain
  His will, some way forth for to find

9 cheats] booty   13 cucking] fighting   booming] charging

Whereby he might have his desire,
Lady, lady,
So great in him was Cupid's fire,
My dear lady.

Clytemnestra. And eke as Paris did desire
Fair Helena for to possess, 50
Her heart, enflamed with like fire,
Of Paris' love desired no less,
And found occasion him to meet
In Cytheron,
Where each of them the other did greet
The feast upon.

Egistus. If that in Paris Cupid's shaft,
O Clytemnestra, took such place,
That time ne way he never left
Till he had got her comely grace, 60
I think my chance not ill to be,
Lady, lady,
That ventured life to purchase ye,
My dear lady.

Clytemnestra. King Priam's son loved not so sore
The Grecian dame (thy brother's wife)
But she esteemed his person more,
Not for his sake, saving her life,
Which caused her people to be slain
With him to fly 70
And he requite her love again
Most faithfully.

Egistus. And as he recompense again
The fair Queen Helen for the same,
So while I live I will take pain
My will always to yours to frame,
Sith that you have vouchsafed to be,
Lady, lady,
A queen and lady unto me,
My dear lady. 80

Clytemnestra. And as she loved him best while life
Did last, so tend I you to do,
If that devoid of war and strife
The gods shall please to grant us two;

63 purchase] possess 77 sith] since

Sith you vouchsaf'st me for to take,
O my good knight,
And me thy lady for to make,
My heart's delight.

101 [*The Vice's Song*]

STAND back, ye sleeping jacks at home,
And let me go. 90
You lie, sir knave, am I a mome?
Why say you so?
Tut, tut, you dare not come in field
For fear you should the ghost up yield.
With blows he goes, the gun shot fly—
It fears, it sears, and there doth lie!

A hundred in a moment be
Destroyed quite.
Sir sauce, in faith, if you should see
The gun shot light, 100
To quake for fear you would not stint,
Whenas by force of gunshot's dint
The ranks in ray are took away,
As pleaseth Fortune oft to play.

But in this stour who bears the fame
But only I?
Revenge, Revenge, will have the name,
Or he will die.
I spare no wight, I fear none ill,
But with this blade I will them kill, 110
For when mine ire is set on fire
I rap them, I snap them—that is my desire!

Farewell! Adieu! To wars I must
In all the haste.
My cousin Cutpurse will, I trust,
Your purse well taste.
But to it, man, and fear for nought,
Me say to thee, it is well fraught
With ruddocks red. Be at a beck!
Beware the arse! Break not thy neck! 120

(1567–8)

91 mome] fool
99 Sir sauce] i.e. one of the audience
103 in ray] in order
105 stour] combat
112 snap] stab
119 ruddocks red] gold coins

ANONYMOUS

102 *Fain would I have a pretty thing*

Fain would I have a pretty thing
To give unto my lady:
I name no thing, nor I mean no thing,
But as pretty a thing as may be.

Twenty journeys would I make
And twenty ways would hie me,
To make adventure for her sake
To set some matter by me: But fain, etc.

Some do long for pretty knacks,
And some for strange devices: 10
God send me that my lady lacks,
I care not what the price is, Thus fain, etc.

Some go here, and some go there,
Where gazes be not geason;
And I go gaping everywhere
But still come out of season. Yet fain, etc.

I walk the town, and tread the street,
In every corner seeking
The pretty thing I cannot meet
That's for my lady's liking. Fain, etc. 20

The mercers pull me going by;
The silk wives say 'What lack ye?'
'The thing you have not', then say I,
'Ye foolish fools, go pack ye'. But fain, etc.

It is not all the silk in Cheap,
Nor all the golden treasure,
Nor twenty bushels on a heap
Can do my lady pleasure. But fain, etc.

6 hie me] hurry
14 geason] rare, uncommon
25 Cheap] Cheapside

The gravers of the golden shows
With jewels do beset me; 30
The shemsters in the shops that sews,
They do nothing but let me. But fain, etc.

But were it in the wit of man
By any means to make it,
I could for money buy it than,
And say, 'Fair lady, take it'. Thus, fain, etc.

O lady, what a luck is this,
That my good willing misseth:
To find what pretty thing it is
That my good lady wisheth. 40

Thus fain would I have had this pretty thing
To give unto my lady:
I said no harm, nor I meant no harm,
But as pretty a thing as may be.

(Pub. 1566)

GEORGE TURBERVILLE

*c.*1548–*c.*1597

103 *A poor Ploughman to a Gentleman for whom he had taken a little pains*

YOUR coulter cuts the soil that erst was sown;
Your harvest was fore-reaped long ago;
Your sickle shears the meadow that was mown
Ere you the toil of tillman's trade did know.
Good faith, you are beholding to the man
That so for you your husbandry began.

He craves of you no silver for his seed,
Ne doth demand a penny for his grain;
But if you stand at any time in need
(Good master), be as bold with him again. 10
You can not do a greater pleasure than
To choose you such a one to be your man.

(1567)

29 gravers] engravers 32 let] obstruct, hinder 35 than] then
31 shemsters] sempstresses

104 *To his friend P. of courting, travelling, dicing, and tennis*

To live in court among the crew is care;
  Is nothing there but daily diligence;
Nor cap nor knee nor money must thou spare:
  The prince's hall is place of great expense.

In rotten ribbed bark to pass the seas,
  The foreign lands and strangy seas to see,
Doth danger dwell; the passage breeds unease,
  Nor safe the soil; the men unfriendly be.

Admit thou see the strangest things of all:
  When eye is turned the pleasant sight is gone. 10
The treasure then of travel is but small.
  Wherefore (friend P.) let all such toys alone.

To shake the bones, and cog the crafty dice,
  To card in care of sudden loss of pence,
Unseemly is, and taken for a vice:
  Unlawful play can have no good pretence.

To band the ball doth cause the coin to waste:
  It melts as butter doth against the sun.
Naught save thy pain, when play doth cease, thou hast.
  To study, then, is best when all is done; 20
    For study stays and brings a pleasant gain,
    When play doth pass as glare with gushing rain.

(1567)

105 [*Epigram* from Plato]

My girl, thou gazest much
  Upon the golden skies.
Would I were heaven: I would behold
  Thee then with all mine eyes.

(1567)

13 cog] practise tricks, cheat (in playing dice)
14 card] play cards
17 band] bandy, throw or strike a ball (as in tennis)

105
1 My girl] (Plato addresses a boy)

106 [*A Letter from Russia*]

To Spencer

If I should now forget, or not remember thee,
Thou, Spencer, might'st a foul rebuke and shame impute to me.
For I to open show did love thee passing well,
And thou wert he at parture whom I loathed to bid farewell.
And as I went thy friend, so I continue still:
No better proof thou canst than this desire of true good will.
I do remember well when needs I should away,
And that the post would license us no longer time to stay:
Thou wrungst me by the fist, and holding fast my hand,
Didst crave of me to send thee news, and how I liked the land. 10
It is a sandy soil, no very fruitful vein,
More waste and woody grounds there are than closes fit for grain.
Yet grain there growing is, which they untimely take,
And cut or e'er the corn be ripe; they mow it on a stake.
And laying sheaf by sheaf, their harvest so they dry,
They make the greater haste, for fear the frost the corn destroy.
For in the winter time, so glary is the ground,
As neither grass, nor other grain, in pastures may be found.
In comes the cattle then, the sheep, the colt, the cow,
Fast by his bed the mousik then a lodging doth allow, 20
Whom he with fodder feeds, and holds as dear as life:
And thus they wear the winter with the mousik and his wife.
Seven months the winter dures, the glare it is so great,
As it is May before he turn his ground to sow his wheat.
The bodies eke that die unburied lie they then,
Laid up in coffins made of fir, as well the poorest men
As those of greater state: the cause is lightly found,
For that in winter time they cannot come to break the ground.
And wood so plenteous is, quite throughout all the land,
As rich, and poor, at time of death assured of coffins stand. 30
Perhaps thou musest much, how this may stand with reason,
That bodies dead can uncorrupt abide so long a season.
Take this for certain troth, as soon as heat is gone,
The force of cold the body binds as hard as any stone,

2 rebuke] disgrace
impute] assign
4 parture] departure
8 post] man in charge of the post-horses
license] permit
9 wrungst] clasped
11 vein] stretch of soil
12 closes] enclosed places
13 untimely] not fully ripened, immature
14 or e'er] before
on a stake] in a pile
17 glary] icy, frozen
19 cattle] livestock
20 mousik] moujik, i.e. Russian peasant
23 dures] lasts
glare] icy condition
27 lightly] readily, easily
33 troth] truth

Without offence at all to any living thing:
As so they lie in perfect state, till next return of spring.
Their beasts be like to ours, as far as I can see,
For shape, and show, but somewhat less of bulk and bone they be.
Of wat'rish taste, the flesh not firm, like English beef,
And yet it serves them very well, and is a good relief. 40
Their sheep are very small, sharp-cingled, handful long;
Great store of fowl on sea and land, the moorish reeds among.
The greatness of the store doth make the prices less;
Besides in all the land they know not how good meat to dress.
They use neither broach nor spit, but when the stove they heat
They put their victuals in a pan, and so they bake their meat.
No pewter to be had, no dishes but of wood,
No use of trenchers, cups cut out of birch are very good.
They use but wooden spoons, which hanging in a case
Each mousik at his girdle ties, and thinks it no disgrace. 50
With whittles two or three, the better man the moe,
The chiefest Russies in the land with spoon and knives do go.
Their houses are not huge of building, but they say,
They plant them in the loftiest ground, to shift the snow away,
Which in the winter time each where full thick doth lie:
Which makes them have the more desire, to set their houses high.
No stone work is in use, their roofs of rafters be,
One linked in another fast, their walls are all of tree.
Of masts both long, and large, with moss put in between,
To keep the force of weather out—I never erst have seen 60
A gross device so good—and on the roof they lay
The burthen bark, to rid the rain and sudden showers away.
In every room a stove, to serve the winter turn;
Of wood they have sufficient store, as much as they can burn.
They have no English glass; of slices of a rock
Hight sluda they their windows make, that English glass doth mock.
They cut it very thin, and sew it with a thread
In pretty order like to panes, to serve their present need.
No other glass, good faith, doth give a better light:
And sure the rock is nothing rich, the cost is very slight. 70
The chiefest place is that, where hangs the god by it,
The owner of the house himself doth never sit,

40 relief] sustenance
41 sharp-cingled] tapering in the waist
43 store] abundance
48 trenchers] plates, platters
51 whittles] knives
moe] more
52 Russies] Russians
58 tree] wood, timber
59 large] broad
61 gross] rough, clumsy
66 hight] called
sluda] Russian mica (in thin transparent plates)
mock] imitate, counterfeit

Unless his better come, to whom he yields the seat:
The stranger bending to the god, the ground with brow must beat.
And in that very place which they most sacred deem,
The stranger lies: a token that his guest he doth esteem.
Where he is wont to have a bear's skin for his bed,
And must, instead of pillow, clap his saddle to his head.
In Russia other shift there is not to be had,
For where the bedding is not good, the bolsters are but bad. 80
I mused very much, what made them so to lie,
Sith in their country down is rife, and feathers out of cry:
Unless it be because the country is so hard,
They fear by niceness of a bed their bodies would be marred.
I wished thee oft with us, save that I stood in fear
Thou wouldst have loathed to have laid thy limbs upon a bear,
As I and Stafford did, that was my mate in bed:
And yet (we thank the God of heaven) we both right well have sped.
Lo thus I make an end: none other news to thee,
But that the country is too cold, the people beastly be. 90
I write not all I know, I touch but here and there,
For if I should, my pen would pinch, and eke offend I fear.
Whoso shall read this verse, conjecture of the rest,
And think by reason of our trade, that I do think the best.
But if no traffic were, then could I boldly pen
The hardness of the soil, and eke the manners of the men.
They say the lion's paw gives judgment of the beast:
And so may you deem of the great, by reading of the least.

(Wr. 1568; pub. 1587)

QUEEN ELIZABETH I
1533–1603

107 *The doubt of future foes*

THE doubt of future foes exiles my present joy,
And wit me warns to shun such snares as threaten mine annoy;
For falsehood now doth flow, and subjects' faith doth ebb,
Which should not be if reason ruled or wisdom weaved the web.

74 god] image of God, icon
82 rife] abundant
out of cry] beyond measure, to excess
84 niceness] luxury, effeminacy
88 sped] flourished, prospered
92 pinch] carp, find fault
95 traffic] trade, commerce
98 the great] the socially powerful, the ruling classes
the least] the lower orders

107
1 doubt] fear, apprehension
2 wit] reason, mind

But clouds of joys untried do cloak aspiring minds,
Which turn to rain of late repent by changed course of winds.
The top of hope supposed the root upreared shall be,
And fruitless all their grafted guile, as shortly ye shall see.
The dazzled eyes with pride, which great ambition blinds,
Shall be unseeled by worthy wights whose foresight falsehood finds. 10
The daughter of debate that discord aye doth sow
Shall reap no gain where former rule still peace hath taught to know.
No foreign banished wight shall anchor in this port;
Our realm brooks not seditious sects, let them elsewhere resort.
My rusty sword through rest shall first his edge employ
To poll their tops that seek such change or gape for future joy.

(Wr. 1568–9; pub. 1589)

from Boethius' *The Consolation of Philosophy*

108 *All human kind on earth*

ALL human kind on earth
From like beginning comes:
One father is of all,
One only all doth guide.
He gave to sun the beams
And horns on moon bestowed;
He men to earth did give
And signs to heaven.
He closed in limbs our souls
Fetched from highest seat. 10
A noble seed therefore
Brought forth all mortal folk.
What crake you of your stock
Or forefathers old?
If your first spring and author
God you view,
No man bastard be,
Unless with vice the worst he feed
And leaveth so his birth.

(Wr. 1593; pub. 1899)

5 cloak] conceal
10 unseeled] opened, unclosed
wights] men, persons
11 daughter of debate] i.e. Mary Queen of Scots
debate] strife, conflict

108
13 crake] boast

109 *Ah, silly pug, wert thou so sore afraid?*

Ah, silly pug, wert thou so sore afraid?
Mourn not, my Wat, nor be thou so dismayed.
It passeth fickle fortune's power and skill
To force my heart to think thee any ill.

No fortune base, thou sayest, shall alter thee,
And may so blind a witch so conquer me?
No, no, my pug, though fortune were not blind,
Assure thyself she could not rule my mind.

Fortune, I know, sometime doth conquer kings,
And rules and reigns on earth and earthly things; 10
But never think fortune can bear the sway,
If virtue watch and will her not obey.

Ne chose I thee by fickle fortune's rede,
Ne she shall force me alter with such speed;
But if to try this mistress jest with thee,

. . .

Pull up thy heart, suppress thy brackish tears,
Torment thee not, but put away thy fears.

Dead to all joys and living unto woe,
Slain quite by her that ne'er gave wise man blow,
Revive again and live without all dread; 20
The less afraid, the better thou shalt speed.

(Pub. 1960)

ANONYMOUS

110 *Christ was the Word that spake it*

Christ was the Word that spake it;
He took the bread and brake it;
And what the Word did make it,
That I believe and take it.

(Pub. 1960)

1 pug] (term of endearment)
2 Wat] i.e. Walter (Sir Walter Ralegh)
3 passeth] surpasses, exceeds
13 rede] counsel
14 speed] success

THOMAS TUSSER

1524?–1580

from *Five Hundred Points of Good Husbandry*

111 [*December's Husbandry*]

O dirty December *Forgotten month past,*
For Christmas remember. *Do now at the last.*

WHEN frost will not suffer to dike and to hedge,
then get thee a heat with thy beetle and wedge:
Once Hallowmas come, and a fire in the hall,
such slivers do well for to lie by the wall.

Get grindstone and whetstone, for tool that is dull,
or often be letted and fret belly-ful.
A wheel-barrow also be ready to have
at hand of thy servant, thy compass to save. 10

Give cattle their fodder in plot dry and warm
and count them for miring or other like harm.
Young colts with thy wennels together go serve,
lest lurched by others they happen to starve.

The rack is commended for saving of dung,
so set as the old cannot mischief the young;
In tempest (the wind being northly or east)
warm barth under hedge is a succour to beast.

The housing of cattle while winter doth hold,
is good for all such as are feeble and old: 20
It saveth much compass, and many a sleep,
and spareth the pasture for walk of thy sheep.

For charges so little much quiet is won,
if strongly and handsomely all thing be done:
But use to untackle them once in a day,
to rub and to lick them, to drink and to play.

3 dike] dig, make a ditch
4 get thee a heat] get warm
beetle] heavy mallet
8 letted] hindered
10 compass] compost, manure
11 cattle] livestock (incl. horses)
13 wennels] calves just weaned
14 lurched] robbed of their food
18 barth] sheltered place
25 untackle] unharness

Get trusty to tend them, not lubberly squire,
that all the day long hath his nose at the fire.
Nor trust unto children poor cattle to feed,
but such as be able to help at a need. 30

Serve ryestraw out first, then wheatstraw and pease,
then oatstraw and barley, then hay if ye please:
But serve them with hay while the straw stover last,
then love they no straw, they had rather to fast.

Yokes, forks, and such other, let bailie spy out,
and gather the same as he walketh about.
And after at leisure let this be his hire,
to beath them and trim them at home by the fire.

As well at the full of the moon as the change,
sea rages in winter be suddenly strange. 40
Then look to thy marshes, if doubt be to fray,
for fear of (*ne forte*) have cattle away.

Both saltfish and lingfish (if any ye have)
through shifting and drying from rotting go save:
Lest winter with moistness do make it relent,
and put it in hazard before it be spent.

Broom faggot is best to dry haberdine on,
lay board upon ladder if faggots be gone.
For breaking (in turning) have very good eye,
and blame not the wind, so the weather be dry. 50

Good fruit and good plenty doth well in the loft,
then make thee an orchard and cherish it oft:
For plant or for stock lay aforehand to cast,
but set or remove it ere Christmas be past.

Set one fro other full forty foot wide,
to stand as he stood is a part of his pride.
More fair, more worthy, of cost to remove,
more steady ye set it, more likely to prove.

27 lubberly squire] a lazy servant
33 stover] winter food for cattle
35 bailie] steward, bailiff, land-agent
38 beath] heat (unseasoned wood so as to straighten it)
41 fray] trouble
42 *ne forte*] lest perhaps
45 relent] become soft
47 haberdine] salt cod

To teach and unteach in a school is unmeet,
  to do and undo to the purse is unsweet. 60
Then orchard or hopyard, so trimmed with cost,
  should not through folly be spoiled and lost.

Ere Christmas be passed let horse be let blood,
  for many a purpose it doth them much good.
The day of St. Stephen old fathers did use:
  if that do mislike thee some other day choose.

Look well to thy horses in stable thou must,
  that hay be not foisty, nor chaff full of dust:
Nor stone in their provender, feather, nor clots,
  nor fed with green peason, for breeding of bots. 70

Some horsekeeper lasheth out provender so,
  some Gillian spend-all so often doth go.
For hogs' meat and hens' meat, for that and for this,
  that corn-loft is emptied ere chapman hath his.

Some countries are pinched of meadow for hay,
  yet ease it with fitches as well as they may.
Which inned and threshed and husbandly dight,
  keeps labouring cattle in very good plight.

In threshing out fitches one point I will show,
  first thresh out for seed of the fitches a few: 80
Thresh few fro thy plough-horse, thresh clean for the cow,
  this order in Norfolk good husbands allow.

If frost do continue, take this for a law,
  the strawberries look to be covered with straw.
Laid overly trim upon crutches and boughs,
  and after uncovered as weather allows.

The gillyflower also, the skilful do know,
  do look to be covered, in frost and in snow.
The knot, and the border, and rosemary gay,
  do crave the like succour for dying away. 90

59 unmeet] unfit
68 foisty] musty
70 peason] pease
bots] disease (worms) troublesome to horses
74 chapman] purchaser, tradesman
75 countries] rural districts
76 fitches] vetches
77 inned] saved, taken indoors
husbandly dight] prepared in a way favoured by a skilled farmer
82 husbands] farmers
allow] approve
89 knot] intricately designed flower-bed
90 for] to prevent

Go look to thy bees, if the hive be too light,
set water and honey, with rosemary dight.
Which set in a dish full of sticks in the hive,
from danger of famine ye save them alive.

In meadow or pasture (to grow the more fine)
let campers be camping in any of thine:
Which if ye do suffer when low is the spring,
you gain to your self a commodious thing.

Thus endeth December's husbandry.

(1571)

112 [*Advice to Housewives*]

AFTERNOON WORKS

Make company break,
Go cherish the weak.

WHEN dinner is ended, set servants to work,
and follow such fellows as loveth to lurk.

To servant in sicknesse see nothing ye grutch,
a thing of a trifle shall comfort him much.

Who many do feed,
Save much they had need.

Put chippings in dippings, use parings to save,
fat capons or chickens that lookest to have. 10

Save droppings and skimmings, how ever ye do,
for medicine for cattle, for cart and for shoe.

Lean capon unmeet,
Dear fed is unsweet.

Such offcorn as cometh give wife to her fee,
feed willingly such as do help to feed thee.

Though fat fed is dainty, yet this I thee warn,
be cunning in fatting for robbing thy barn.

96 campers] football players
camping] playing football
97 spring] young growth, shoots

112
4 lurk] be idle
9 chippings] fragments of bread
dippings] dripping, grease
15 offcorn] waste or offal corn

Piece hole to defend.
Things timely amend. 20

Good sempsters be sewing of fine pretty knacks,
good huswifes be mending and piecing their sacks.

Though making and mending be huswifely ways,
yet mending in time is the huswife to praise.

Buy new as is meet,
Mark blanket and sheet.

Though ladies may rend and buy new ery day,
good huswifes must mend and buy new as they may.

Call quarterly servants to court and to leet,
write every coverlet, blanket, and sheet. 30

Shift slovenly elf,
Be gaoler thy self.

Though shifting too oft be a thief in a house,
yet shift slut and sloven for fear of a louse.

Grant doubtful no key of his chamber in purse,
lest chamber door locked be to thievery a nurse.

Save feathers for guest,
These other rob chest.

Save wing for a thresher, when gander doth die,
save feather of all thing, the softer to lie. 40

Much spice is a thief, so is candle and fire,
sweet sauce is as crafty as ever was friar.

Wife make thine owne candle,
Spare penny to handle.

Provide for thy tallow, ere frost cometh in,
and make thine owne candle, ere winter begin.

If penny for all thing be suffered to trudge,
trust long, not to penny, to have him thy drudge.

EVENING WORKS

Time drawing to night,
See all things go right. 50

WHEN hens go to roost go in hand to dress meat,
serve hogs and to milking and some to serve neat.

Where twain be enow, be not served with three,
more knaves in a company worser they be.

27 ery] every 29 leet] manor court 31 elf] young person 52 neat] cattle

Make lackey to trudge,
Make servant thy drudge.

For every trifle leave jaunting thy nag,
but rather make lackey of Jack boy thy wag.

Make servant at night lug in wood or a log,
let none come in empty but slut and thy dog. 60

False knave ready prest,
All safe is the best.

Where pullen use nightly to perch in the yard,
there two legged foxes keep watches and ward.

See cattle well served, without and within,
and all thing at quiet ere supper begin.

Take heed it is needful,
True pity is meedful.

No clothes in garden, no trinkets without,
no door leave unbolted, for fear of a doubt. 70

Thou woman whom pity becometh the best,
grant all that hath laboured time to take rest.

SUPPER MATTERS

Use mirth and good word,
At bed and at board.

PROVIDE for thy husband, to make him good cheer,
make merry together, while time ye be here.

At bed and at board, howsoever befall,
what ever God sendeth be merry withal.

No brawling make,
No jealousy take. 80

No taunts before servants, for hindering of fame,
no jarring too loud for avoiding of shame.

As frenzy and heresy roveth together,
so jealousy leadeth a fool ye wot whither.

Tend such as ye have,
Stop talkative knave.

Young children and chickens would ever be eating,
good servants look duly for gentle entreating.

61 prest] ready
63 pullen] poultry
68 meedful] meritorious
70 doubt] risk, danger
84 wot] know
88 entreating] treatment, handling

No servant at table use saucely to talk,
lest tongue set at large out of measure do walk. 90

No snatching at all,
Sirs, hearken now all.

No lurching, no snatching, no striving at all,
lest one go without and another have all.

Declare after supper, take heed thereunto,
what work in the morning each servant shall do.

(1573)

ISABELLA WHITNEY

fl. 1567–1573

from *The Manner of her Will and What she left to London and to All Those in it, at her Departing*

113

I WHOLE in body and in mind,
But very weak in purse,
Do make and write my testament
For fear it will be worse.
And first I wholly do commend
My soul and body eke
To God the Father and the Son,
So long as I can speak.
And after speech, my soul to Him
And body to the grave, 10
Till time that all shall rise again
Their Judgment for to have.
And then I hope they both shall meet
To dwell for aye in joy,
Whereas I trust to see my friends
Released from all annoy.
Thus have you heard touching my soul
And body what I mean;

89 saucely] saucily, rudely
93 lurching] eating rapidly so as to have more than one's share

113
6 eke] also
15 Whereas] when

I trust you all will witness bear
  I have a steadfast brain. 20
And now let me dispose such things
  As I shall leave behind,
That those which shall receive the same
  May know my willing mind.
I first of all to London leave,
  Because I there was bred,
Brave buildings rare, of churches store,
  And Pauls unto the head.
Between the same, fair streets there be
  And people goodly store; 30
Because their keeping craveth cost
  I yet will leave them more.
First for their food, I butchers leave,
  That every day shall kill;
By Thames you shall have brewers store
  And bakers at your will.
And such as others do observe
  And eat fish thrice a week,
I leave two streets full fraught therewith—
  They need not far to seek. 40
Watling Street and Canwick Street
  I full of woollen leave,
And linen store in Friday Street,
  If they me not deceive.
And those which are of calling such
  That costlier they require,
I mercers leave, with silk so rich
  As any would desire.
In Cheap, of them they store shall find,
  And likewise in that street 50
I goldsmiths leave with jewels such
  As are for ladies meet.
And plate to furnish cupboards with
  Full brave there shall you find,
With purl of silver and of gold
  To satisfy your mind;

28 Pauls . . . head] St Paul's Cathedral to its top
30 store] abundance
41 Canwick Street] (also known as Candlewick Street, known for sellers of woollen cloth)
47 mercers] dealers in silks, velvets, and other costly materials
49 Cheap] Cheapside
52 meet] suitable
54 brave] fine, splendid
55 purl] thread

With hoods, bongraces, hats or caps
  Such store are in that street
As if on tone side you should miss
  The tother serves you feat. 60
For nets of every kind of sort
  I leave within the pawn,
French ruffs, high purls, gorgets and sleeves
  Of any kind of lawn.
For purse or knives, for combs or glass,
  Or any needful knack,
I by the stocks have left a boy
  Will ask you what you lack.
I hose do leave in Birchin Lane
  Of any kind of size, 70
For women stitched, for men both trunks
  And those of Gascoyns' guise.
Boots, shoes, or pantables good store
  St. Martin's hath for you;
In Cornwall, there I leave you beds
  And all that longs thereto.
For women shall you tailors have,
  By Bow the chiefest dwell:
In every lane you some shall find
  Can do indifferent well. 80
And for the men few streets or lanes
  But body-makers be,
And such as make the sweeping cloaks
  With guards beneath the knee.
Artillery at Temple Bar
  And dags at Tower Hill;
Swords and bucklers of the best
  Are aye the Fleet until.

57 bongraces] shades worn on the front of women's bonnets or caps
59–60 tone . . . the tother] the one . . . the other
60 feat] nicely, elegantly
61 nets] hairnets
62 pawn] gallery or upper walk of the Royal Exchange
63 purls] pleats or folds of a ruff or band
gorgets] wimples, coverings for neck or breast
64 lawn] a kind of fine linen
69 hose] stockings, leggings, breeches
71 stitched] embroidered
trunks] trunk-hose (full bag-like breeches)
72 Gascoyns'] i.e. gaskins, wide breeches
73 pantables] (i.e. pantofles) slippers, over-shoes
75 Cornwall] 'Cornwallish ground', in Vintry Ward
78 Bow] St Mary Bow
80 indifferent well] fairly well
82 body-makers] tailors
84 guards] ornamental borders or trimmings
85 Artillery] weaponry, arms
86 dags] heavy pistols
88 until] unto

Now when thy folk are fed and clad
 With such as I have named, 90
For dainty mouths and stomachs weak
 Some junkets must be framed.
Wherefore I potecaries leave,
 With banquets in their shop;
Physicians also for the sick,
 Diseases for to stop.
Some roisters still must bide in thee
 And such as cut it out,
That with the guiltless quarrel will
 To let their blood about. 100
For them I cunning surgeons leave,
 Some plasters to apply,
That ruffians may not still be hanged
 Nor quiet persons die.
For salt, oatmeal, candles, soap,
 Or what you else do want,
In many places shops are full—
 I left you nothing scant.
If they that keep what I you leave
 Ask money when they sell it, 110
At Mint there is such store it is
 Unpossible to tell it.
At Steelyard, store of wines there be
 Your dulled minds to glad,
And handsome men that must not wed
 Except they leave their trade.
They oft shall seek for proper girls
 And some perhaps shall find
That need compels or lucre lures
 To satisfy their mind. 120
And near the same I houses leave
 For people to repair
To bathe themselves, so to prevent
 Infection of the air.

92 junkets] sweetmeats
93 potecaries] (i.e. apothecaries) those who sold spices, drugs, comfits, preserves, etc.
94 banquets] dainty dishes, sweetmeats
97 roisters] riotous fellows
98 cut it out] flaunt, make a show
101 cunning] skilful
108 scant] scarce
111 Mint] where money was officially coined
112 tell] count
113 Steelyard] i.e. the place of business of the Hanseatic merchants, who were noted for their Rhenish wines
116 Except . . . trade] (apprentices were not allowed to get married)
122 repair] resort, go to in large numbers

On Saturdays I wish that those
Which all the week do drug
Shall thither trudge to trim them up
On Sundays to look smug.
If any other thing be lacked
In thee, I wish them look; 130
For there it is: I little brought
But nothing from thee took.

(1573)

GEORGE GASCOIGNE
1534–1577

114 *Gascoigne's Woodmanship*

My worthy Lord, I pray you wonder not
To see your woodman shoot so oft awry,
Nor that he stands amazed like a sot,
And lets the harmless deer (unhurt) go by.
Or if he strike a doe which is but carren
Laugh not good Lord, but favour such a fault,
Take will in worth, he would fain hit the barren,
But though his heart be good, his hap is naught.
And therefore now I crave your Lordship's leave,
To tell you plain what is the cause of this. 10
First if it please your honour to perceive,
What makes your woodman shoot so oft amiss,
Believe me, Lord, the case is nothing strange,
He shoots awry almost at every mark,
His eyes have been so used for to range,
That now God knows they be both dim and dark.
For proof he bears the note of folly now,
Who shot sometimes to hit Philosophy,
And ask you why? forsooth I make avow,
Because his wanton wits went all awry. 20

126 drug] drudge
128 smug] neat, smart
114
2 woodman] one who hunts game in a forest or wood, huntsman
3 amazed] confused, completely 'lost'
5 carren] carrion, i.e. unfit for eating because with young
7 Take will in worth] take the intention in good part
8 hap is naught] luck is bad
17 note] mark, sign
19 make avow] explain

Next that, he shot to be a man in law,
And spent some time with learned Littleton,
Yet in the end, he proved but a daw,
For law was dark and he had quickly done.
Then could he wish Fitzherbert such a brain
As Tully had, to write the law by art,
So that with pleasure, or with little pain,
He might perhaps, have caught a truant's part.
But all too late, he most misliked the thing,
Which most might help to guide his arrow straight; 30
He winked wrong, and so let slip the string,
Which cast him wide, for all his quaint conceit.
From thence he shot to catch a courtly grace,
And thought even there to wield the world at will,
But out, alas, he much mistook the place,
And shot awry at every rover still.
The blazing baits which draw the gazing eye
Unfeathered there his first affection;
No wonder then although he shot awry,
Wanting the feathers of discretion. 40
Yet more than them, the marks of dignity
He much mistook, and shot the wronger way,
Thinking the purse of prodigality
Had been best mean to purchase such a prey.
He thought the flattering face which fleereth still,
Had been full fraught with all fidelity,
And that such words as courtiers use at will,
Could not have varied from the verity.
But when his bonnet buttoned with gold,
His comely cap beguarded all with gay, 50
His bumbast hose, with linings manifold,
His knit silk stocks and all his quaint array,
Had picked his purse of all the Peter-pence,
Which might have paid for his promotion,
Then (all too late) he found that light expense
Had quite quenched out the court's devotion.

22 Littleton] author of a standard textbook for law students
23 daw] fool
24 dark] hard to understand
25 Fitzherbert] legal author
31 winked] aimed (keeping one eye closed)
32 quaint conceit] cunning judgement
36 rover] a mark selected at will or at random by an archer
37 blazing] resplendent
38 unfeathered] stripped bare
affection] inclination
45 fleereth] smiles subserviently
50 beguarded] ornamented
gay] bright trimmings
51 bumbast] stuffed
52 stocks] stockings
53 Peter-pence] i.e. money (orig. tax paid to the Papacy)

So that since then the taste of misery
Hath been always full bitter in his bit,
And why? forsooth because he shot awry,
Mistaking still the marks which others hit. 60
But now behold what mark the man doth find:
He shoots to be a soldier in his age;
Mistrusting all the virtues of the mind,
He trusts the power of his personage.
As though long limbs led by a lusty heart,
Might yet suffice to make him rich again,
But Flushing frays have taught him such a part
That now he thinks the wars yield no such gain.
And sure I fear, unless your lordship deign
To train him yet into some better trade, 70
It will be long before he hit the vein
Whereby he may a richer man be made.
He cannot climb as other catchers can,
To lead a charge before himself be led.
He cannot spoil the simple sakeless man,
Which is content to feed him with his bread.
He cannot pinch the painful soldier's pay,
And shear him out his share in ragged sheets,
He cannot stoop to take a greedy prey
Upon his fellows grovelling in the streets. 80
He cannot pull the spoil from such as pill,
And seem full angry at such foul offence,
Although the gain content his greedy will,
Under the cloak of contrary pretence:
And nowadays, the man that shoots not so,
May shoot amiss, even as your woodman doth:
But then you marvel why I let them go,
And never shoot, but say farewell forsooth:
Alas my Lord, while I do muse hereon,
And call to mind my youthful years mis-spent, 90
They give me such a bone to gnaw upon,
That all my senses are in silence pent.
My mind is rapt in contemplation,
Wherein my dazzled eyes only behold
The black hour of my constellation
Which framed me so luckless on the mould.

58 bit] eating (i.e. in his mouth)
67 Flushing frays] i.e. fighting in the Low Countries
75 simple] poor, humble
sakeless] innocent
77 pinch] stint, give in short measure
81 pill] pillage
96 mould] workman's or builder's pattern

Yet therewithal I cannot but confess,
That vain presumption makes my heart to swell,
For thus I think, not all the world (I guess,)
Shoots bet than I, nay some shoots not so well. 100
In Aristotle somewhat did I learn,
To guide my manners all by comeliness,
And Tully taught me somewhat to discern,
Between sweet speech and barbarous rudeness.
Old Parkins, Rastell, and Dan Bracton's books,
Did lend me somewhat of the lawless law;
The crafty courtiers with their guileful looks,
Must needs put some experience in my maw:
Yet cannot these with many mast'ries moe
Make me shoot straight at any gainful prick,
Where some that never handled such a bow
Can hit the white or touch it near the quick,
Who can nor speak nor write in pleasant wise,
Nor lead their life by Aristotle's rule,
Nor argue well on questions that arise,
Nor plead a case more than my Lord Mayor's mule,
Yet can they hit the marks that I do miss,
And win the mean which may the man maintain.
Now when my mind doth mumble upon this,
No wonder then although I pine for pain: 120
And whiles mine eyes behold this mirror thus,
The herd goeth by, and farewell gentle does:
So that your lordship quickly may discuss
What blinds mine eyes so oft (as I suppose).
But since my Muse can to my Lord rehearse
What makes me miss, and why I do not shoot,
Let me imagine in this worthless verse,
If right before me, at my standing's foot
There stood a doe, and I should strike her dead,
And then she prove a carrion carcase too, 130
What figure might I find within my head,
To scuse the rage which ruled me so to do?
Some might interpret with plain paraphrase,
That lack of skill or fortune led the chance,
But I must otherwise expound the case;
I say Jehovah did this doe advance,

105 Parkins, Rastell . . . Bracton] authors of law-books
109 mast'ries] skills, arts
moe] more
110 prick] spot in the centre of the target, bull's eye
112 white] target
123 discuss] declare, pronounce
128 standing] a hunter's station or stand

And made her bold to stand before me so,
Till I had thrust mine arrow to her heart,
That by the sudden of her overthrow
I might endeavour to amend my part 140
And turn mine eyes that they no more behold
Such guileful marks as seem more than they be:
And though they glister outwardly like gold,
Are inwardly like brass, as men may see:
And when I see the milk hang in her teat,
Methinks it saith, old babe now learn to suck,
Who in thy youth couldst never learn the feat
To hit the whites which live with all good luck.
Thus have I told my Lord (God grant in season),
A tedious tale in rhyme, but little reason. 150

(1573)

115 Magnum vectigal parsimonia

THE common speech is, spend and God will send.
But what sends he? a bottle and a bag,
A staff, a wallet and a woeful end,
For such as list in bravery so to brag.
Then if thou covet coin enough to spend,
Learn first to spare thy budget at the brink,
So shall the bottom be the faster bound:
But he that list with lavish hand to link
(In like expense) a penny with a pound,
May chance at last to sit aside and shrink 10
His harebrained head without Dame Dainty's door.
Hick, Hob and Dick, with clouts upon their knee,
Have many times more gunhole groats in store
And change of crowns more quick at call than he,
Which let their lease and took their rent before.
For he that raps a royal on his cap,
Before he put one penny in his purse,
Had need turn quick and broach a better tap,
Or else his drink may chance go down the worse.
I not deny but some men have good hap, 20

139 sudden] suddenness

115

Magnum ... parsimonia thrift makes a good income

2–3 a bottle ... end] attributes of a beggar

4 bravery] fine clothing

10 shrink] turn aside (in shame)

11 without] outside

12 clouts] patches

13 gunhole] meaning not known

16 raps] throws away

royal] i.e. a gold coin worth 15 shillings

To climb aloft by scales of courtly grace,
And win the world with liberality:
Yet he that yerks old angels out apace,
And hath no new to purchase dignity,
When orders fall, may chance to lack his grace.
For haggard hawks mislike an empty hand:
So stiffly some stick to the mercer's stall,
Till suits of silk have sweat out all their land.
So oft thy neighbours banquet in thy hall,
Till Davie Debit in thy parlour stand, 30
And bids thee welcome to thine own decay.
I like a lion's looks not worth a leek
When every fox beguiles him of his prey:
What sauce but sorrow serveth him a week,
Which all his cates consumeth in one day?
First use thy stomach to a stound of ale,
Before thy malmsey come in merchants' books,
And rather wear (for shift) thy shirt of mail,
Than tear thy silken sleeves with tenterhooks.
Put feathers in thy pillows great and small, 40
Let them be prinked with plumes that gape for plums,
Heap up both gold and silver safe in hooches,
Catch, snatch, and scratch for scrapings and for crumbs,
Before thou deck thy hat (on high) with brooches.
Let first thine one hand hold fast all that comes,
Before that other learn his letting fly:
Remember still that soft fire makes sweet malt,
No haste but good (who means to multiply):
Bought wit is dear, and dressed with sour salt,
Repentance comes too late, and then say I, 50
Who spares the first and keeps the last unspent,
Shall find that sparing yields a goodly rent.

(1573)

23 yerks] flings
angels] gold coins worth from 6 to 10 shillings
25 When orders fall] when offices fall vacant (?)
26 haggard] untamed
36 stound] drinking measure
39 tenterhooks] i.e. for stretching cloth
41 prinked] decked
42 hooches] (hutches) coffers, chests
47 soft] gentle, slow

116 *Gascoigne's Lullaby*

SING lullaby, as women do,
  Wherewith they bring their babes to rest,
And lullaby can I sing too
  As womanly as can the best.
    With lullaby they still the child,
    And if I be not much beguiled,
    Full many wanton babes have I
    Which must be stilled with lullaby.

First lullaby my youthful years.
  It is now time to go to bed, 10
For crooked age and hoary hairs
  Have won the haven within my head.
    With lullaby then youth be still,
    With lullaby content thy will,
    Since courage quails and comes behind,
    Go sleep, and so beguile thy mind.

Next lullaby my gazing eyes,
  Which wonted were to glance apace:
For every glass may now suffice
  To show the furrows in my face. 20
    With lullaby then wink a while,
    With lullaby your looks beguile.
    Let no fair face nor beauty bright
    Entice you eft with vain delight.

And lullaby my wanton will.
  Let reason's rule now reign thy thought,
Since all too late I find by skill
  How dear I have thy fancies bought.
    With lullaby now take thine ease,
    With lullaby thy doubts appease: 30
    For trust to this, if thou be still,
    My body shall obey thy will.

Eke lullaby my loving boy:
  My little Robin take thy rest;
Since age is cold and nothing coy,
  Keep close thy coin, for so is best.

15 courage] vigour, lustiness 21 wink] close the eyes 24 eft] again 27 skill] experience 34 Robin] i.e. male organ

With lullaby be thou content,
With lullaby thy lusts relent;
Let others pay which have more pence,
Thou art too poor for such expense. 40

Then lullaby my youth, mine eyes,
My will, my ware, and all that was.
I can no more delays devise,
But welcome pain, let pleasure pass.
With lullaby now take your leave,
With lullaby your dreams deceive;
And when you rise with waking eye,
Remember Gascoigne's Lullaby.

(1573)

117 *Gascoigne's Good Morrow*

YOU that have spent the silent night
In sleep and quiet rest,
And joy to see the cheerful light
That riseth in the east:
Now clear your voice, now cheer your heart,
Come help me now to sing:
Each willing wight come bear a part,
To praise the heavenly King.

And you whom care in prison keeps,
Or sickness doth suppress, 10
Or secret sorrow breaks your sleeps,
Or dolours do distress:
Yet bear a part in doleful wise,
Yea think it good accord,
And acceptable sacrifice,
Each sprite to praise the Lord.

The dreadful night with darksomeness
Had overspread the light,
And sluggish sleep with drowsiness
Had overpressed our might: 20

117
15 acceptable] (pronounced with stress on first and third syllables)
16 sprite] spirit

A glass wherein we may behold
Each storm that stays our breath,
Our bed the grave, our clothes like mould,
And sleep like dreadful death.

Yet as this deadly night did last,
But for a little space,
And heavenly day, now night is past,
Doth show his pleasant face:
So must we hope to see God's face,
At last in heaven on high, 30
When we have changed this mortal place
For immortality.

And of such haps and heavenly joys,
As then we hope to hold,
All earthly sights, all worldly toys,
Are tokens to behold:
The day is like the day of doom,
The sun, the Son of Man,
The skies the heavens, the earth the tomb
Wherein we rest till than. 40

The rainbow bending in the sky,
Bedecked with sundry hues,
Is like the seat of God on high,
And seems to tell these news:
That as thereby he promised
To drown the world no more,
So by the blood which Christ hath shed,
He will our health restore.

The misty clouds that fall sometime,
And overcast the skies, 50
Are like to troubles of our time,
Which do but dim our eyes:
But as such dews are dried up quite,
When Phoebus shows his face,
So are such fancies put to flight,
Where God doth guide by grace.

40 than] then

The carrion crow, that loathsome beast,
Which cries against the rain,
Both for her hue and for the rest,
The Devil resembleth plain: 60
And as with guns we kill the crow,
For spoiling our relief,
The Devil so must we overthrow,
With gunshot of belief.

The little birds which sing so sweet,
Are like the angel's voice,
Which render God his praises meet,
And teach us to rejoice:
And as they more esteem that mirth,
Then dread the night's annoy, 70
So must we deem our days on earth.
But hell to heavenly joy.

Unto which joys for to attain,
God grant us all his grace,
And send us after worldly pain,
In heaven to have a place.
Where we may still enjoy that light,
Which never shall decay:
Lord for thy mercy lend us might
To see that joyful day. 80

(1573)

118 *Gascoigne's Goodnight*

When thou hast spent the lingering day in pleasure and delight,
Or after toil and weary way, dost seek to rest at night:
Unto thy pains or pleasures past, add this one labour yet,
Ere sleep close up thine eye too fast, do not thy God forget,
But search within thy secret thoughts what deeds did thee befall:
And if thou find amiss in aught, to God for mercy call.
Yea though thou find nothing amiss, which thou canst call to mind,
Yet evermore remember this, there is the more behind:
And think how well soever it be that thou hast spent the day,
It came of God, and not of thee, so to direct thy way. 10
Thus if thou try thy daily deeds, and pleasure in this pain,
Thy life shall cleanse thy corn from weeds, and thine shall be the gain.
But if thy sinful sluggish eye will venture for to wink,
Before thy wading will may try how far thy soul may sink,

Beware and wake, for else thy bed, which soft and smooth is made,
May heap more harm upon thy head, than blows of en'my's blade.
Thus if this pain procure thine ease, in bed as thou dost lie,
Perhaps it shall not God displease, to sing thus soberly:
'I see that sleep is lent me here, to ease my weary bones,
As death at last shall eke appear, to ease my grievous groans. 20
My daily sports, my paunch full fed, have caused my drowsy eye,
As careless life in quiet led, might cause my soul to die;
The stretching arms, the yawning breath, which I to bedward use,
Are patterns of the pangs of death, when life will me refuse;
And of my bed each sundry part in shadows doth resemble
The sundry shapes of death, whose dart shall make my flesh to tremble.
My bed itself is like the grave, my sheets the winding sheet,
My clothes the mould which I must have to cover me most meet;
The hungry fleas which frisk so fresh, to worms I can compare
Which greedily shall gnaw my flesh, and leave the bones full bare; 30
The waking cock that early crows to wear the night away,
Puts in my mind the trump that blows before the latter day.
And as I rise up lustily, when sluggish sleep is past,
So hope I to rise joyfully, to Judgement at the last.'
Thus will I wake, thus will I sleep, thus will I hope to rise,
Thus will I neither wail nor weep, but sing in godly wise.
My bones shall in this bed remain, my soul in God shall trust,
By whom I hope to rise again from death and earthly dust.

(1573)

119 [*No haste but good*]

In haste post haste when first my wandering mind
Beheld the glistering court with gazing eye,
Such deep delights I seemed therein to find
As might beguile a graver guest than I.
The stately pomp of princes and their peers
Did seem to swim in floods of beaten gold;
The wanton world of young delightful years
Was not unlike a heaven to behold.
Wherein did swarm (for every saint) a dame,
So fair of hue, so fresh of their attire 10
As might excel Dame Cynthia for fame,
Or conquer Cupid with his own desire.
These and such like were baits that blazed still
Before mine eye to feed my greedy will.

119
1 In haste post haste] with all possible speed

Before mine eye to feed my greedy will
  Gan muster eke mine old acquainted mates,
Who helped the dish (of vain delight) to fill
  My empty mouth with dainty delicates;
And foolish boldness took the whip in hand
  To lash my life into this trustless trace, 20
Till all in haste I leapt aloof from land,
  And hoist up sail to catch a courtly grace.
Each lingering day did seem a world of woe,
  Till in that hapless haven my head was brought;
Waves of wanhope so tossed me to and fro,
  In deep despair to drown my dreadful thought.
    Each hour a day, each day a year did seem,
    And every year a world my will did deem.

And every year a world my will did deem,
  Till, lo, at last, to court now am I come, 30
A seemly swain, that might the place beseem,
  A gladsome guest embraced of all and some.
Not there content with common dignity,
  My wandering eye in haste (yea post post haste)
Beheld the blazing badge of bravery,
  For want whereof I thought myself disgraced.
Then peevish pride puffed up my swelling heart
  To further forth so hot an enterprise;
And comely cost began to play his part
  In praising patterns of mine own device. 40
    Thus all was good that might be got in haste,
    To prink me up, and make me higher placed.

To prink me up and make me higher placed,
  All came too late that tarried any time.
Piles of provision pleased not my taste,
  They made my heels too heavy for to climb.
Me thought it best that boughs of boist'rous oak
  Should first be shred to make my feathers gay,
Till at the last a deadly dinting stroke
  Brought down the bulk with edgetools of decay. 50

20 trace] track, way
21 aloof] with a clear space intervening, a distance from
25 wanhope] despair, dejection
26 dreadful] full of dread
31 seemly swain] handsome young man
35 bravery] fine clothing, display
37 peevish] perverse, obstinate, self-willed
42 prink . . . up] dress up, show off
47 boist'rous] massive, bulky
48 feathers gay] i.e. fine attire
50 decay] destruction

Of every farm I then let fly a lease,
  To feed the purse that paid for peevishness,
Till rent and all were fallen in such disease
  As scarce could serve to maintain cleanliness.
    The bough, the body, fine, farm, lease and land,
    All were too little for the merchant's hand.

All were too little for the merchant's hand,
  And yet my bravery bigger than his book.
But when this hot account was coldly scanned,
  I thought high time about me for to look. 60
With heavy cheer I cast my head aback
  To see the fountain of my furious race,
Compared my loss, my living, and my lack,
  In equal balance with my jolly grace,
And saw expenses grating on the ground
  Like lumps of lead to press my purse full oft,
When light reward and recompense were found
  Fleeting like feathers in the wind aloft.
    These thus compared, I left the court at large.
    For why? The gains doth seldom quit the charge. 70

For why? The gains doth seldom quit the charge.
  And so say I, by proof too dearly bought.
My haste made waste, my brave and brainsick barge
  Did float too fast, to catch a thing of nought.
With leisure, measure, mean, and many moe,
  I might have kept a chair of quiet state,
But hasty heads cannot be settled so
  Till crooked Fortune give a crabbed mate.
As busy brains must beat on tickle toys,
  As rash invention breeds a raw device, 80
So sudden falls do hinder hasty joys;
  And as swift baits do fleetest fish entice,
    So haste makes waste, and therefore now I say,
    No haste but good, where wisdom makes the way.

52 peevishness] foolish perversity
58 bravery] (1) showy clothes; (2) bravado
62 fountain] source
69 at large] altogether
70 quit] requite, repay
72 proof] experience
73 brave] finely bedecked, splendid
75 moe] more
78 crabbed] ill-natured; sour
   mate] checkmate
79 beat on] be preoccupied with
   tickle toys] fickle, changing fancies
80 raw] unripe, immature, crude

No haste but good, where wisdom makes the way.
For proof whereof behold the silly snail
(Who sees the soldier's carcass cast away,
With hot assault the castle to assail)
By line and leisure climbs the lofty wall,
And wins the turret's top more cunningly 90
Than doughty Dick, who lost his life and all
With hoisting up his head too hastily.
The swiftest bitch brings forth the blindest whelps,
The hottest fevers coldest cramps ensue,
The naked'st need hath ever latest helps.
With Nevile then I find this proverb true,
That *haste makes waste*, and therefore still I say,
No haste but good, where wisdom makes the way.

(1573)

120 *The Green Knight's Farewell to Fancy*

FANCY (quoth he) farewell, whose badge I long did bear,
And in my hat full harebrainedly thy flowers did I wear.
Too late I find (at last) thy fruits are nothing worth,
Thy blossoms fall and fade full fast, though bravery bring them forth.
By thee I hoped always in deep delights to dwell,
But since I find thy fickleness, *Fancy* (quoth he) *farewell.*

Thou mad'st me live in love, which wisdom bids me hate.
Thou bleared'st mine eyes and mad'st me think that faith was mine by fate.
By thee those bitter sweets did please my taste alway;
By thee I thought that love was light, and pain was but a play. 10
I thought that beauty's blaze was meet to bear the bell,
And since I find myself deceived, *Fancy* (quoth he) *farewell.*

The gloss of gorgeous courts, by thee did please mine eye:
A stately sight methought it was to see the brave go by,
To see their feathers flaunt, to make their strange device,
To lie along in ladies' laps, to lisp and make it nice.

86 silly] mindless, helpless
89 By line and leisure] in a slow but sure way
90 cunningly] wisely, cleverly
96 Nevile] (Alexander Nevile, Gascoigne's friend)

120
Green Knight] (a nickname of Gascoigne's)
1 Fancy] fantasy, daydreams
4 bravery] appearing in fine clothes
11 bear the bell] take first place
14 brave] showily dressed
16 along] at full length
lisp] speak in an affected way
make it nice] display reluctance

To fawn and flatter both, I liked sometime well,
But since I see how vain it is, *Fancy* (quoth he) *farewell.*

When court had cast me off, I toiled at the plough.
My fancy stood in strange conceits, to thrive I wot not how: 20
By mills, by making malt, by sheep and eke by swine,
By duck and drake, by pig and goose, by calves and keeping kine,
By feeding bullocks fat, when price at markets fell—
But since my swains eat up the gains, *Fancy* (quoth he) *farewell.*

In hunting of the deer my fancy took delight.
All forests knew my folly still; the moonshine was my light.
In frosts I felt no cold, a sunburnt hue was best;
I sweat and was in temper still; my watching seemed rest.
What dangers deep I passed, it folly were to tell,
And since I sigh to think thereon, *Fancy* (quoth he) *farewell.* 30

A fancy fed me once to write in verse and rhyme,
To wray my grief, to crave reward, to cover still my crime;
To frame a long discourse on stirring of a straw,
To rumble rhyme in raff and ruff, yet all not worth a haw;
To hear it said 'There goeth the man that writes so well'.
But since I see what poets be, *Fancy* (quoth he) *farewell.*

At music's sacred sound my fancies eft begun,
In concords, discords, notes and clefs, in tunes of unison;
In hierarchies and strains, in rests, in rule and space,
In monochords and moving modes, in burdens underbass. 40
In descants and in chants I strained many a yell,
But since musicians be so mad, *Fancy* (quoth he) *farewell.*

To plant strange country fruits, to sow such seeds likewise,
To dig and delve for new-found roots, where old might well suffice;
To prune the water-boughs, to pick the mossy trees
(Oh how it pleased my fancy once), to kneel upon my knees

20 wot] know
22 kine] cattle
24 swains] servants, labourers
28 sweat] sweated
in temper] in good condition
watching] staying awake
32 wray] betray, display
34 raff and ruff] alliterative verse
37 eft] then, afterwards
39 strains] portions of a musical movement
40 monochords] harmonious combinations of sound
modes] particular schemes or systems of sounds
burdens underbass] accompaniments or 'undersongs' in the bass part
45 water-boughs] undergrowth

To griff a pippin stock, when sap begins to swell:
But since the gains scarce quit the cost, *Fancy* (quoth he) *farewell.*

Fancy (quoth he) *farewell*, which made me follow drums,
Where powdered bullets serves for sauce, to every dish that comes; 50
Where treason lurks in trust, where hope all hearts beguiles,
Where mischief lieth still in wait, where fortune friendly smiles;
Where one day's prison proves that all such heavens are hell:
And such I feel the fruits thereof—*Fancy* (quoth he) *farewell.*

If reason rule my thoughts, and God vouchsafe me grace,
Then comfort of philosophy shall make me change my race;
And fond I shall it find, that Fancy sets to show,
For weakly stands that building still, which lacketh grace below.
But since I must accept my fortunes as they fell,
I say God send me better speed, and *Fancy now farewell.* 60

(1575)

BEWE

*c.*1576

121 *I would I were Actaeon*

I WOULD I were Actaeon, whom Diana did disguise,
To walk the woods unknown whereas my lady lies;
A hart of pleasant hue I wish that I were so,
So that my lady knew alone me and no mo;

To follow thick and plain, by hill and dale alow,
To drink the water fain, and feed me with the sloe.
I would not fear the frost, to lie upon the ground,
Delight should quite the cost, what pain so that I found.

The shaling nuts and mast that falleth from the tree
Should serve for my repast, might I my lady see; 10
Sometime that I might say when I saw her alone,
'Behold thy slave, all day that walks these woods unknown!'

(1578)

47 griff] graft
48 quit] requite, repay
56 race] course of life
57 fond] foolish
that Fancy] i.e. what Fancy

121
5 thick] thicket
8 quite] requite
9 shaling] falling from the husk

THOMAS PROCTOR

*c.*1578

122 Respice Finem

Lo, here the state of every mortal wight,
See here the fine of all their gallant joys;
Behold their pomp, their beauty, and delight,
Whereof they vaunt as safe from all annoys.

To earth the stout, the proud, the rich shall yield,
The weak, the meek, the poor shall shrouded lie
In dampish mould; the stout with spear and shield
Cannot defend himself when he shall die.

The proudest wight, for all his lively shows,
Shall leave his pomp, cut off by dreadful death; 10
The rich, whose hutch with golden ruddocks flows,
At length shall rest uncoined in dampish earth.

By Nature's law we all are born to die,
But where or when, the best uncertain be;
No time prefixed, no goods our life shall buy,
Of dreadful death no friends shall set us free.

We subject be a thousand ways to death;
Small sickness moves the valianst heart to fear;
A little push bereaves your breathing breath
Of brave delights, whereto you subject are. 20

Your world is vain; no trust in earth you find;
Your valianst prime is but a brittle glass;
Your pleasures vade, your thoughts a puff of wind;
Your ancient years are but a withered grass.

(1578)

Respice Finem] regard your end
11 golden ruddocks] gold coins
18 valianst] most valiant
23 vade] fade

THOMAS CHURCHYARD

1520?–1604

123 *A Tale of a Friar and a Shoemaker's Wife*

In Wales there is a borough town,
  Carmarthen hight the same,
Where dwelt sometimes a lusty friar;
  I need not show his name.
This friar was fat and full of flesh,
  A jolly merry knave,
Who with the gossips of the town
  Himself could well behave.
Those wealthy wives and thrifty dames
  Could never make good cheer, 10
Nor well dispute of Peter's keys
  If absent were this frere.
He said his matins in their ears
  And gospel at their bed,
And spared no service for the quick,
  Nor cared for the dead.
With abbot's ease and faring well
  This friar so wanton was,
That neither maid nor married wife
  His dortour door might pass 20
Without some stop: such stales he laid
  To make them stumble in,
That by his life men guessed he thought
  That lechery was no sin.
A loving friar, good-fellow-like
  In those days was he held;
In every corner of the town
  Good comp'ny out he smelled;
And as ye know, in haunting long
  All sorts of people there, 30
He must find out some baiting place,
  A mistress foul or fair,
A dainty morsel for his tooth:
  These friars loved well to fare,
Though some were pleased with cheeses still,
  Some found a better share,

2 hight] was called
12 frere] friar
20 dortour] monastic bed-chamber
21 stales] snares, traps
31 baiting place] place where a halt is made for refreshment

As did this honest brother in Christ,
By gossiping about;
Who, when he would a hackney ride,
Had found a palfrey out: 40
A nag much of a woman's height,
That used for to bear
More sacks perchance unto the mill
Than corn was grinded there.
I not declare what trim conceits
He gave her all the while,
Ere he obtained the thing he sought;
How he his tongue could file,
To talk and mince the matter well,
The better to digest; 50
And how full oft at morrow mass
His mistress he could feast,
And after noon to gardens walk
And gathered posies gay,
And wore them closely in his cowl
As he did service say;
Nor cannot show you half the feats
He wrought to please his trull;
But those most fit for you to read
Here put in rime I wull. 60
A shoemaker, that held a shop
Far from his dwelling-place,
A fair wife had, a good brown wench,
And come of no ill race.
Some say of wagtails, pretty fools,
A kindred great and good,
That knows what shears will serve the turn
When shrews will shape a hood.
The chief of this great lineage leads
Their lives like holy nuns, 70
That for relief in gadding time
About the cloister runs;
A caterwauling once a week,
In breath to keep them well,
Lest virgins should from surfeit take
When they lead apes in hell.

39 hackney] horse let out for hire
40 palfrey] saddle-horse (often for ladies)
45 conceits] fancy presents
48 file] smooth, polish
51 morrow mass] the first mass of the day
63 brown] dark complexioned
76 lead apes in hell] i.e. the (fancied) consequence to a woman of dying unmarried

This woman went not out of kind,
  And, sure, for Simon's sake
She used great deeds of charity
  And much ado did make. 80
Saint Simon was a godly man,
  The friar might so be called;
I touch no further lest he kick,
  For, sure, his back is galled.
Alive the man was many years,
  Since abbeys were suppressed,
And dwelt not far from Cardiff town
  When written was this jest.
But to my tale let me return:
  This woman seldom failed 90
The morrow mass at four o'clock,
  To see how Christ was nailed
Unto the cross: to whom she kneeled,
  With book and beads in fist;
And for devotion many times
  This gentle friar she kissed,
At every pater noster while,
  Which was a precious thing,
And Jesus! how it did her good
  To hear her lubber sing. 100
And when he turned about his face
  And looked through the quere,
She scrat her head, she sat on pricks,
  And crept the altar near.
This custom kept she many days,
  The friar thereof full glad;
Yet still referred his other sport
  Till better time were had.
You must conceive, this merry man
  In jests and light conceits 110
His head was set, and for the same
  Full oft he laid his baits.
To laugh and pass the time away
  Such toys he would devise,
That few men, for the mirth thereof,
  The matter could despise.
Upon a day appointed was
  This wife, as was her use,

77 went . . . out of kind] lose the character of her family, become degenerate
100 lubber] idle monk or friar; lazy fellow
102 quere] choir
103 scrat] scratched
107 referred] deferred, postponed

Should early come to morrow mass;
  There might be made no scuse. 120
She kept her hour, and hard she kneeled
  Without the dortour door;
The friar came forth and haled her in,
  And flang her on the floor.
'Fie, fie, sir friar!' she cried apace,
  But what should more be said?
She was content to take her ease
  And leap into the bed.
And, as mine author doth declare,
  The sounder for to sleep, 130
She had no more upon her tho
  Than hath a shoren sheep.
Sir Simkin had no points to loose:
  In, cowl and all, he skips.
God send my friar well forth again:
  The moon was in the 'clipse.
How long he lay, or what he did,
  In sooth, I cannot tell;
But at the length the sexton went,
  And rang the service bell. 140
The friar wished rope about his neck;
  The matins was begun
That he that morn would sing or say,
  And all the lessons done.
Yet up he must for fear of check,
  His course was come to rise;
The night before he took his rest,
  To heal his bleared eyes.
A law there was within that house,
  Who slept the service out, 150
In fratry should be hoist full high
  And whipped like breechless lout.
Wherefore to tinder box he stepped,
  And light a fire in haste;
And as he girded knotted cord
  Full hard about his waist,
'Lie still', said he unto his guest,
  'I must go take some pain,
And sing a psalm within the quere;
  But I will come again.' 160

131 tho] then
133 points] tagged laces or cords for attaching the hose to the doublet
151 fratry] monastic common-room
160 again] back

Out goes he then: that liked her not,
  She durst not lie alone,
For fear of bugs. Thus leave I now
  Abed this good wife Joan,
And tell you how in quere full loud
  This shaven cock he crows,
And drowned his fellows everichon,
  He sang so in the nose.
But as he turned the plainsong book,
  Full smoothly could he smile, 170
Yet none of all the covent could
  Perceive him all the while.
To mend his mirth and make him laugh,
  A fancy fell in thought:
He saw the owner of the beast
  That he had rid for naught,
The husband of the wife, indeed,
  That he in bed had laft,
Who walked within the church beneath,
  All careless of this craft. 180
'By God', thought he, 'I will go prove
  This man if he do know
His wife by measuring her foot
  Or mark upon her toe;
For if I do deceive the fool
  And make the wife afeard,
He nor his wife is ne'er the worse,
  A hair not of his beard,
And I shall much the better be
  And laughing have at will. 190
Thus every way, and be my luck,
  I shall have sport my fill.'
Down went this good religious man
  Where hornsby husband walked,
And curchie made, and ducked full low;
  And as he with him talked,
'I have', quod he, 'known thee right long,
  And still, the truth to say,
I have thee found a faithful friend
  In every kind of way. 200
A customer thou hast of me;
  My money I bestow
On thee, before all other men
  That dwells within thy row.

163 bugs] bugbears, bogies 194 hornsby] i.e. cuckold 195 curchie] curtsey, bow

And to be plain, I love thee well,
  And plainer now I am;
Then, give good ear, I shall declare
  Wherefore to thee I cam.
But wise and warely use my words,
  And keep my counsel both: 210
Thy promise is sufficient band;
  I will no further oath.'
This man full well he knew his good,
  Who curchied to the ground:
'Sweet sir', quod he, 'tell on your mind,
  I am your beadman bound.'
'Thou know'st, my neighbour, men must live,
  And have a wench sometime,
And we, poor friars, must keep it close,
  For fear of open crime. 220
It were a spot unto our house,
  A slander to our name,
When we have sport, if all the world
  Should understand the same.
For God Himself doth give us leave,
  As thou hast heard ere now,
Although the world we do deceive
  In keeping of our vow.
I am too long in preaching thus,
  And time I do abuse: 230
I have a wench for whom thou must
  Go make a pair of shoes.
Let them be good; when I thee pay,
  A penny more to boot
I shall thee give.' 'I lack', quod he,
  'The measure of her foot.'
Then boldly spake this bare-foot frere:
  'By God, that shalt thou have,
If thou keep close and follow me,
  Else call the friar a knave.' 240
The straight plain path to dortour, then,
  They took the way full right,
The friar before; but you must note,
  It was not full day-light.
Wherefore the man came far behind,
  The friar went in apace,
And caused his wench, the other's wife,
  Right close to hide her face.

216 beadman] 'humble servant'
221 spot] stain, discredit
234 to boot] in addition, into the bargain
246 apace] quickly

When entered was this honest man,
  'Put forth thy foot', quod he 250
—The friar I mean, which at that time
  The bolder man might be.
She thrust her leg out of the bed,
  But head fast under clothes
She kept; and cursed the saucy friar
  A hundred times, God knows.
The workman took his measure well,
  And had no further care;
The friar well laughed within his sleeve:
  Thus pleased both they are. 260
But how the wife contented was
  Let wives be judge herein,
That from their husband's bed sometime
  In suchlike case hath bin.
Yet let me show how she did quake
  And tremble all the while,
And wished the roperipe hanged full high
  That did her thus beguile;
And how for fear her body was
  On water every part. 270
Hereafter shall you know likewise
  What hate was in her heart,
Which for the time she covered well
  And ne'er a word she spake.
Her husband hasted to his shop,
  And so his leave did take.
'I have a pair of shoes', quod he,
  'Which I shall bring anon
All ready made; for my wife's foot
  And hers I think both one.' 280
'Ye say the truth, good moon', thought she,
  'The friar hath played the knave.
Make for your wife what shoes ye list,
  The measure twice you have.'
The friar runs forth, the man went home,
  The woman lay a space,
As she had bin in swadling clouts,
  And durst not show her face.
When she had found herself alone,
  She rose, and speed did make 290
To be at home ere her good man
  His breakfast came to take.

267 roperipe] one ripe for the gallows   281 moon] man

As in her house she did arrive,
  She barred the door full fast,
And burst a-weeping like a babe,
  And this she said at last:
'O, shameless knave! not pleased to spoil
  Me of my wifely fame,
But at my faults thy frantic head
  Must make thereof a game. 300
Could not my breach of wedlock's band
  Content thee, but in spite
Thou must devise so lewd a fact
  My faith with fraud to quite?
How didst thou know I durst not stir
  That touched was so near?
I might have scaped my husband's wrath,
  But thou hadst bought it dear.
If I had spoke, as once I thought
  To do, my fear was such, 310
Thy folly had been ten times more,
  Though mine were very much.
He might have took his wife again,
  And knocked full well thy pate,
And shaved thy crown another sort
  Than falls for thine estate;
Or else he might have shamed us both,
  And so refused his wife.
I could have lived, but where wouldst thou
  Have led a friar's life? 320
O beastly wretch! that of thy self
  Hast had so small regard.
As for the knavery showed to me,
  I will it well reward;
Not for the malice due therefore
  But that I mind to leave
Example to thy fellows all
  How they their friends deceive.
Did I procure thee to this deed?
  Did not thy gospels sweet, 330
And mumbling oft, make me believe
  A devil was no sprete?
Didst thou not seek me every hour
  To show me thy good will?
And brought me grapes and goodly fruits
  Among my gossips still?

297 spoil] rob   303 fact] crime, action   332 sprete] spirit

Thou car'st not if ten couple of hounds
  Did follow me full fast,
And I a fox were in the field,
  Since now thy gear is past. 340
Did not thy fleering face full oft
  Frame me thus to thy fist?
Then wast thou hot, now art thou cold,
  Or warms thee where thou list.
A warming place within the town
  Hereafter mayst thou lack,
And miss perchance so meet a seat
  To drink a cup of sack.
Thou keepst not such a diet still,
  Nor art not so precise, 350
But as the thirst doth come again
  Thy appetite will rise.
I pray to God it be my lot
  To see thee at that stay!'
So thus the woman held her peace,
  And out she went her way
Unto the market, for to seek
  Such things as housewives do:
You know, that have more skill than I,
  What doth belong thereto. 360
The poor man brought the friar his shoes,
  And thought no harm therein,
And to his labour did return,
  His living for to win.
His wife and he, as they were wont,
  Full quiet days did lead:
He ne'er perceived by her shoe
  Where she awry did tread.
She went as upright in the street
  And with as good a grace, 370
And set upon her follies past
  Indeed as bold a face,
As she that never made offence;
  For custom breeds a law,
And makes them keep their count'nance trim
  That once have broke a straw.

340 gear] affair, business
341 fleering] grinning, grimacing
342 frame] train, prepare to thy fist (i.e. like a falcon)
350 precise] strict, punctilious
354 stay] set-back, obstacle, pause

Well, all the winter passed forth
  This couple at their will;
The wife her counsel kept full close,
  The poor man meant none ill. 380
But as the spring came on apace
  The friar waxed wanton too,
And fain would nag; but credit lost,
  He knew not where to woo;
And so bethought him of the prank
  He played in way of sport,
And sought to salve the sore again
  With words and medicines short.
So he devised amends to make,
  And turn it to a jest, 390
And thought to laugh the matter out,
  As it was meet and best.
And as by chance he met this wife,
  'God speed, sweet heart', quod he,
'I marvel why these many days
  You are so strange to me.'
The fowler's merry whistle now
  Must needs betray the bird;
The wily wife now shaped her tongue
  To give the friar a gird. 400
'Not strange', quod she, 'but that in faith
  I did unkindly take
The part ye played; and yet I thought
  It was for favour's sake,
Or for some mirth; for if of spite
  It had been wrought, I know
I should have had some shame ere this,
  But sure I find not so.'
'I swear by good saint Francis, dame,
  The truth thou say'st indeed; 410
Wherefore let pass such follies old
  That may new quarrels breed,
And be my friend; thou hast good wit,
  Thou know'st now what I mean:
Let all old jests, long gone and past,
  Be now forgotten clean.'
The wife, thus finding fortune good
  To compass that she would,
A gentle limetwig gan she make
  To take the friar in hold; 420

383 nag] gnaw, nibble   387 salve] heal   400 gird] hit, dig

Yet shaped to save them harmless both
  From blot and worldly shame,
And quit the knack, so she might laugh,
  And have thereat some game.
'Well, sir', quoth she, 'I know at full
  The meaning of your mind,
And would to God some honest way
  For you now I might find.
My husband haply may me miss
  If I should come to you; 430
Then, our old fatches will not serve,
  We must devise anew.
A colour must the painter cast
  On posts and painted walls;
Who takes away a stumbling stock
  Shall freely scape from falls.
A jealous toy is taken soon,
  A trifle breeds mistrust;
Great danger follows foul delights,
  As slander follows lust. 440
If will be won with worldly shame,
  The pleasures turns to pain;
Wherefore we need a double clock
  To keep us from the rain.
When that my husband is in shop,
  If you the pains will take
To come unto my house betimes,
  There we will merry make.
But come as soon, and if you may,
  As any day appears; 450
The way ye know unto my house,
  It standeth by the Freres.'
'I will', said he, and sighed therewith,
  So wrung her by the hand;
But little of the matter yet
  The fool did understand.
As beetle-brains are brought in briars
  Before they see the snare,
So this wise woodcock in a net
  Was caught ere he was ware. 460

423 quit the knack] repay (him) for his trick or prank
431 fatches] tricks, dodges
437 toy] fancy, whim
447 betimes] early
457 beetle-brains] fools
459 woodcock] fool, dolt

The time came on, the friar was there,
  And up the stairs he went.
'A cup of malmsey', quod the wife,
  'Now would us both content.
The little boy that is beneath
  Shall soon go fetch the same.'
'Take money with thee', quod the frere.
  So thus goes down the dame
Unto the boy, and bade him run
  Unto the shop above, 470
And bid his master come in haste
  If he his wife did love,
For sick she was. 'But, boy', quod she,
  'Then, trudge thou for the drink.
Oh boy, I fear that I shall sound
  Before thou come, I think.'
Out flings the lad, up goes the wife,
  And at a window pried,
Until at length far off full well
  Her husband had she spied. 480
'Alas, go hide thee quickly, frere',
  Said she, 'if that thou can,
For here at hand, I do not feign,
  There cometh my good man.
Here is no corner to get out,
  Full woe is me therefore!
Now shall we buy our pastime dear
  And pay for pleasures sore.
Now all the mischief will be mine,
  Because I have thee here; 490
Now shall my honest name be brought
  In question by a frere.
Well, now there is no nother shift
  But here the brunt to bide—
Except that in this little chest
  Thyself now canst thou hide.
Now choose thou whether open blame
  Or secret prison sweet
In these extremes and haste is most
  For present mischief meet.' 500
The friar to find some ready help
  Was pleased and well apaid,
So in the chest this great wise man
  Is crept full sore afraid.

475 sound] swoon, faint   502 apaid] pleased, satisfied

She locked the same, and clapped the keys
  Close under bolster sure,
So laid her down upon the bed
  And did sore fits endure,
Or feigned to feel, about her breast;
  Such gripes she said she felt, 510
The groaning of the same did make
  Her husband's heart to melt.
'How now, dear wife! what aileth thee?'
  The simple soul said then.
'Fie, wife! pluck up a woman's heart.'
  'Yea, husband, God knows when',
Quoth she, 'if aquavitae now
  I drink not out of hand,
I have a stitch so sore, God wot,
  I can nor sit nor stand.' 520
'Thou hast a bottle in the house,
  I dare well say', quod he.
'Of aquavitae lately bought
  There may no better be
Within thy chest. Where are thy keys?'
  'I know not, by my life',
Said she, 'you set more by a lock
  Than you do by your wife.
Ye wus and ye were sick, I should
  The lock right soon up break.' 530
'That shall be done', quod he, 'you need
  Thereof no more to speak.'
A hatchet took he in his hand,
  And struck it such a blow
The chamber shaked, the friar he quaked,
  And stank for fear and woe.
The chest with iron bars was bound,
  Which made the goodman sweat:
The friar, like doctor Dolt, lay still,
  In dread and danger great. 540
And durst not stir for all the world,
  His courage quite was gone.
The poor man had a pig in poke,
  Had he looked well thereon.
The lock was good, that knew the wife,
  Who bade her husband strike:
He laid on load, the friar within
  That sport did little like.

529 Ye wus] certainly, truly
and ye] if you

543 had a pig in poke] didn't know the value of what he had

At length the bands began to loose:
  The wife had eye thereto; 550
She feared if he did strike again
  The lock would sure undo.
Then thought she on a woman's wile,
  Which never fails at need:
If friar were seen, then was she shamed;
  No, no, she took more heed.
'Oh hold your hand! You kill my head',
  Quod she, 'to hear you knock.
Now am I eased; great harm it were
  To spill so good a lock. 560
My stitch is gone, then let me sleep
  And rest myself a while.'
The goodman went unto his shop;
  The wife began to smile.
When she had sent away the boy
  All things in quiet were;
She rose and went to ease the friar
  That lay half dead for fear:
Which resurrection who had seen
  Must needs have laughed at least— 570
First how he lay, then how he looked
  And trembled like a beast.
'Now am I quit', quod she, 'sir frere,
  And yet you are not shamed,
And through a woman who you scorned
  Your folly now is tamed.'
This tale so ends, and by the same
  You see what friars have bin;
And how their outward holy lives
  Was but a cloak for sin. 580
Here may you see how plain poor men
  That labours for their food
Are soon deceived with subtle snakes
  Of wicked serpents' brood.
Here, under cloud of matter light,
  Some words of weight may pass,
To make the lewd abhor foul life
  And see themselves in glass.
Here is no terms to stir up vice;
  The writer meant not so, 590
For by the foil that folly takes
  The wise may blotless go.

560 spill] destroy   591 foil] defeat, repulse

The more we see the wicked plagued
  And painted plain to sight,
The more we pace the path of grace
  And seek to walk upright.

(1575)

TIMOTHY KENDALL

fl. 1577

from *Flowers of Epigrams*

[*Epigrams by Sir Thomas More*]

124 *The difference between a King and a Tyrant*

BETWEEN a Tyrant and a King
  Would you the difference have?
The King each subject counts his child,
  The Tyrant each his slave.

125 *A Tyrant in sleep, naught differeth from a common man*

DOST therefore swell and pout with pride
  And rear thy snout on high,
Because the crowd doth crouch and couch
  Whereso thou comest by?
Because the people bonnetless
  Before thee still do stand? 10
Because the life and death doth lie
  Of divers in thy hand?
But when that drowsy sleep of thee
  Hath every part possessed,
Tell then where is thy pomp and pride,
  Thy port, and all the rest?
Then snorting lozel as thou art,
  Then liest thou like a block,
Or as a carrion corpse late dead,
  And senseless as a stock. 20
And if it were not that thou wert
  Closed up in walls of stone,
And fenced round, thy life would be
  In hands of every one.

125
12 divers] any number (of people)
16 port] state, stately bearing
17 lozel] worthless person, scoundrel

126

Of a good prince and an evil

A GOOD prince, what? The dog that keeps
His flock aye safe in rest,
And hunts the wolf away. An ill?
Himself the ravening beast.

127

Desire of Dominion

AMONGEST many kings,
Skant one king shall you see 30
Content with kingdom one alone,
Skant one, if one there be.
Amongest many kings,
Skant one king shall you see
That rules one only kingdom right,
Skant one, if one there be.

128

[Epigram by Jovianus Pontanus]

Upon the grave of a beggar

WHILE as I lived no house I had,
Now dead I have a grave.
In life I lived in loathsome lack,
Now dead I nothing crave. 40
In life I lived an exile poor,
Now death brings rest to me.
In life poor naked soul unclad,
Now clad in clods ye see.

(1577)

26 aye] always, ever

127
30 skant] scarcely

NICHOLAS BRETON
*c.*1555–1626

129 [*Service is no Heritage*]

THAT I would not persuaded be
  In my young reckless youth,
By plain experience I see
  That now it proved truth:
    It is Tom's song, my lady's page,
    That service is no heritage.

I heard him sing this other night,
  As he lay all alone:
Was never boy in such a plight—
  Where should he make his moan? 10
    'O Lord', quoth he, 'to be a page:
    This service is none heritage.

'Mine uncle told me tother day
  That I must take great pain;
And I must cast all sloth away
  If I seek aught to gain.
    "For sure", quoth he, "a painful page
    Will make service an heritage."

'Yea sure, a great commodity,
  If once Madam he do displease: 20
A cuff on the ear, two or three
  He shall have, smally for his ease.
    I would for me he were a page
    For to possess his heritage.

'I rub and brush almost all day,
  I make clean many a coat;
I seek all honest means I may
  How to come by a groat.
    I think I am a painful page,
    Yet I can make no heritage. 30

6 service is no heritage] (proverbial: i.e. a servant will make little or nothing to leave to his heirs)
7 this other] the other
17 painful] painstaking, hardworking

'Why, I to get have much ado
 A kirtle now and than,
For making clean of many a shoe
 For Alice or Mistress Anne.
  My lady's maids will wipe the page
  Always of such an heritage.

'The wenches they get coifs and cauls,
 French hoods and partlets eke;
And I get naught but checks and brawls,
 A thousand in a week. 40
  These are rewards meet for a page,
  Surely a goodly heritage.

'My lady's maids too must I please,
 But chiefly Mistress Anne:
For else, by the Mass, she will disease
 Me vilely, now and than.
  "Faith", she will say, "you whoreson page,
  I'll purchase you an heritage."

'And if she say so, by the rood,
 'Tis Cock I warrant it. 50
But God he knows, I were as good
 To be withouten it.
  For all the gains I get, poor page,
  Is but a slender heritage.

'I have so many folks to please,
 And creep and kneel unto,
That I shall never live at ease
 Whatever so I do:
  I'll therefore be no more a page,
  But seek some other heritage. 60

'But was there ever such a patch
 To speak so loud as I?
Knowing what hold the maids will catch
 At every fault they spy;
  And all for spite at me, poor page,
  To purchase me an heritage.

32 kirtle] short jacket or tunic
35 wipe] rob, get dishonestly
37 coifs and cauls] close-fitting caps, often ornamented
38 French hoods] fashionable women's head-dress
 partlets] neckerchiefs, collars, ruffs
45 disease] make uncomfortable
49 rood] cross
50 Cock] (vague oath, from 'God')
61 patch] fool

'And if that they may hear of this,
I were as good as hanged:
"My lady shall know it, by Gis"—
And I shall sure be banged. 70
I shall be used like a page,
I shall not lose mine heritage.

'Well, yet I hope the time to see
When I may run as fast
For wands for them, as they for me,
Ere many days be past:
For when I am no longer page,
I'll give them up mine heritage.

'Well, I a while must stand content,
Till better hap do fall, 80
With such poor state as God hath sent,
And give Him thanks for all:
Who will, I hope, send me, poor page,
Than this, some better heritage.'

With this, with hands and eyes
Lift up to heaven on high,
He sighed twice or thrice,
And wept too, piteously.
Which when I saw, I wished the page,
In faith, some better heritage. 90

And weeping thus: 'Good God', quoth he,
'Have mercy on my soul,
That ready I may be for thee
When that the bell doth knoll—
To make me free of this bondage
And partner of Thine heritage.

'Lord, grant me grace so Thee to serve,
That at the latter day,
Although I can no good deserve,
Yet Thou to me mayest say: 100
"Be thou now free, that wert a page,
And here in heaven have heritage".'

(1577)

69 by Gis] by Jesus (a genteel oath) 76 wands] rods (for beating)

130 *In the merry month of May*

IN the merry month of May,
In a morn by break of day,
Forth I walked by the wood side,
Whereas May was in his pride.
There I spied all alone
Phyllida and Corydon.
Much ado there was, God wot,
He would love and she would not.
She said, never man was true;
He said, none was false to you. 10
He said, he had loved her long;
She said, love should have no wrong.
Corydon would kiss her then;
She said, maids must kiss no men,
Till they did for good and all.
Then she made the shepherd call
All the heavens to witness truth,
Never loved a truer youth.
Thus with many a pretty oath,
Yea and nay, and faith and troth, 20
Such as silly shepherds use,
When they will not love abuse,
Love, which had been long deluded,
Was with kisses sweet concluded:
And Phyllida with garlands gay
Was made the Lady of the May.

(1591)

131 *The Chess Play*

A SECRET many years unseen,
In play at chess, who knows the game:
First of the King, and then the Queen,
Knight, Bishop, Rook, and so by name
Of every Pawn I will descry
The nature with the quality.

131
5 descry] describe

The King

The King himself is haughty care,
  Which overlooketh all his men,
And when he seeth how they fare,
  He steps among them now and then; 10
    Whom, when his foe presumes to check,
    His servants stand to give the neck.

The Queen

The Queen is quaint and quick conceit,
  Which makes her walk which way she list,
And roots them up that lie in wait
  To work her treason, ere she wist;
    Her force is such against her foes
    That whom she meets she overthrows.

The Knight

The Knight is knowledge how to fight
  Against his Prince's enemies. 20
He never makes his walk outright,
  But leaps and skips, in wily wise,
    To take by sleight a trait'rous foe
    Might slily seek their overthrow.

The Bishop

The Bishop he is witty brain
  That chooseth crossest paths to pace,
And evermore he pries with pain
  To see who seeks him most disgrace.
    Such stragglers when he finds astray,
    He takes them up, and throws away. 30

The Rooks

The Rooks are reason on both sides,
  Which keep the corner-houses still,
And warily stand to watch their tides,
  By secret art to work their will,
    To take sometime a thief unseen
    Might mischief mean to King or Queen.

12 give the neck] move to cover a 'check'
13 quaint] clever, cunning
16 wist] knows
25 witty] clever, cunning

The Pawns

The Pawn before the King is peace,
Which he desires to keep at home;
Practice, the Queen's, which doth not cease
Amid the world abroad to roam, 40
To find and fall upon each foe
Whereas his mistress means to go.

Before the Knight is peril placed,
Which he, by skipping, overgoes,
And yet that Pawn can work a cast
To overthrow his greatest foes;
The Bishop's, prudence, prying still
Which way to work his master's will.

The Rooks' poor Pawns are silly swains,
Which seldom serve, except by hap, 50
And yet those Pawns can lay their trains
To catch a great man in a trap:
So that I see sometime a groom
May not be spared from his room.

The nature of the Chess men

The King is stately, looking high;
The Queen doth bear like majesty;
The Knight is hardy, valiant, wise;
The Bishop, prudent and precise;
The Rooks, no rangers out of ray;
The Pawns, the pages in the play. 60

Lenvoy

Then rule with care and quick conceit,
And fight with knowledge, as with force;
So bear a brain to dash deceit,
And work with reason and remorse;
Forgive a fault when young men play,
So give a mate, and go your way.

39 Practice] plotting, contrivance
49 silly] simple
59 rangers out of ray] i.e. they don't straggle out of ranks (array, order)

And when you play, beware of Check;
Know how to save, and give, a neck;
And with a Check, beware of Mate;
But chief, ware 'had I wist' too late. 70
Lose not the Queen, for ten to one,
If she be lost, the game is gone.

(1593)

132

A Report Song

SHALL we go dance the hay, the hay?
Never pipe could ever play
Better shepherd's roundelay.

Shall we go sing the song, the song?
Never Love did ever wrong.
Fair maids, hold hands all along.

Shall we go learn to woo, to woo?
Never thought came ever to,
Better deed could better do.

Shall we go learn to kiss, to kiss? 10
Never heart could ever miss
Comfort, where true meaning is.

Thus at base they run, they run,
When the sport was scarce begun.
But I waked, and all was done.

(Pub. 1600)

133

Who can live in heart so glad

WHO can live in heart so glad
As the merry country lad?
Who upon a fair green balk
May at pleasure sit and walk.
And amid the azure skies
See the morning sun arise;
While he hears in every spring
How the birds do chirp and sing;

70 'had I wist'] had I known (proverbial)

132 *Report Song*] echo song

133 3 balk] ridge, mound

Or before the hounds in cry
See the hare go stealing by; 10
Or along the shallow brook
Angling with a baited hook,
See the fishes leap and play
In a blessed sunny day;
Or to hear the partridge call
Till she have her covey all;
Or to see the subtle fox,
How the villain plies the box,
After feeding on his prey
How he closely sneaks away 20
Through the hedge and down the furrow,
Till he gets into his burrow;
Then the bee to gather honey,
And the little black-haired coney
On a bank for sunny place
With her forefeet wash her face:
Are not these, with thousands mo
Than the courts of kings do know,
The true pleasing spirit's sights
That may breed true love's delights? 30
But with all this happiness
To behold that shepherdess
To whose eyes all shepherds yield,
All the fairest of the field,
Fair Aglaia, in whose face
Lives the shepherds' highest grace,
In whose worthy-wonder praise
See what her true shepherd says.
She is neither proud nor fine,
But in spirit more divine; 40
She can neither lour nor leer,
But a sweeter smiling cheer;
She had never painted face,
But a sweeter smiling grace;
She can never love dissemble,
Truth doth so her thoughts assemble
That, where wisdom guides her will,
She is kind and constant still.
All in sum, she is that creature
Of that truest comfort's nature, 50
That doth show (but in exceedings)
How their praises had their breedings.

18 plies the box] dodges 24 coney] rabbit

Let, then, poets feign their pleasure
In their fictions of love's treasure,
Proud high spirits seek their graces
In their idol-painted faces;
My love's spirit's lowliness,
In affection's humbleness,
Under heaven no happiness
Seeks but in this shepherdess. 60
For whose sake I say and swear
By the passions that I bear,
Had I got a kingly grace,
I would leave my kingly place
And in heart be truly glad
To become a country lad,
Hard to lie, and go full bare,
And to feed on hungry fare;
So I might but live to be
Where I might but sit to see 70
Once a day, or all day long,
The sweet subject of my song;
In Aglaia's only eyes
All my worldly paradise.

(Pub. 1604)

134

In time of yore

In time of yore when shepherds dwelt
 Upon the mountain rocks;
And simple people never felt
 The pain of lovers' mocks:
But little birds would carry tales
 'Twixt Susan and her sweeting,
And all the dainty nightingales
 Did sing at lovers' meeting:
Then might you see what looks did pass
 Where shepherds did assemble, 10
And where the life of true love was
 When hearts could not dissemble.

Then 'yea' and 'nay' was thought an oath
 That was not to be doubted,
And when it came to 'faith' and 'troth',
 We were not to be flouted.

Then did they talk of curds and cream,
  Of butter, cheese, and milk;
There was no speech of sunny beam
  Nor of the golden silk. 20
Then for a gift a row of pins,
  A purse, a pair of knives,
Was all the way that love begins;
  And so the shepherd wives.

But now we have so much ado,
  And are so sore aggrieved,
That when we go about to woo
  We cannot be believed;
Such choice of jewels, rings, and chains,
  That may but favour move, 30
And such intolerable pains
  Ere one can hit on love;
That if I still shall bide this life
  'Twixt love and deadly hate,
I will go learn the country life
  Or leave the lover's state.

(Pub. 1879)

EDMUND SPENSER

*c.*1552–1599

135 *To the right worshipful my singular good friend, Master Gabriel Harvey, Doctor of the Laws*

HARVEY, the happy above happiest men,
I read; that, sitting like a looker-on
Of this world's stage, dost note, with critic pen,
The sharp dislikes of each condition;
And, as one careless of suspicion,
Ne fawnest for the favour of the great,
Ne fearest foolish reprehension
Of faulty men, which danger to thee threat:

135
2 read] consider
4 dislikes] (1) annoyances, pains; (2) repugnant features
6–7] Ne . . . ne] neither . . . nor

But freely dost of what thee list entreat,
Like a great lord of peerless liberty; 10
Lifting the good up to high Honour's seat,
And the evil damning evermore to die;
For Life, and Death, is in thy doomful writing!
So thy renowm lives ever by indicting.

(Wr. 1586; pub. 1592)

from *Mother Hubbard's Tale*

136 [*The Fox and the Ape go to Court*]

THEN gan this crafty couple to devise,
How for the court themselves they might aguise;
For thither they themselves meant to address,
In hope to find there happier success.
So well they shifted, that the ape anon
Himself had clothed like a gentleman,
And the sly fox, as like to be his groom,
That to the court in seemly sort they come;
Where the fond ape, himself uprearing high
Upon his tiptoes, stalketh stately by, 10
As if he were some great magnifico,
And boldly doth amongst the boldest go;
And his man Reynold, with fine counterfeasance,
Supports his credit and his countenance.
Then gan the courtiers gaze on every side,
And stare on him, with big looks basin-wide,
Wond'ring what mister wight he was, and whence:
For he was clad in strange accoutrements,
Fashioned with quaint devices, never seen
In court before, yet there all fashions bene; 20
Yet he them in newfangleness did pass.
But his behaviour altogether was
Alla Turchesca, much the more admired;
And his looks lofty, as if he aspired
To dignity and sdeigned the low degree;
That all which did such strangeness in him see

9 entreat] treat
10 peerless] (1) matchless; (2) without lords
13 doomful] fateful; concerned with judgement
14 indicting] (1) accusing, bringing charges; (2) literary composition (enditing)

136
2 aguise] adorn
13 counterfeasance] counterfeiting
17 what mister wight] what sort of creature
23 *Alla Turchesca*] in the Turkish manner
admired] wondered at
25 sdeigned] disdained

By secret means gan of his state inquire,
And privily his servant thereto hire:
Who, throughly armed against such coverture,
Reported unto all, that he was sure 30
A noble gentleman of high regard,
Which through the world had with long travel fared,
And seen the manners of all beasts on ground;
Now here arrived, to see if like he found.
  Thus did the ape at first him credit gain,
Which afterwards he wisely did maintain
With gallant show, and daily more augment
Through his fine feats and courtly complement;
For he could play, and dance, and vault, and spring,
And all that else pertains to revelling, 40
Only through kindly aptness of his joints.
Besides, he could do many other points,
The which in court him served to good stead;
For he 'mongst ladies could their fortunes read
Out of their hands, and merry leasings tell,
And juggle finely, that became him well.
But he so light was at legerdemain,
That what he touched came not to light again;
Yet would he laugh it out, and proudly look,
And tell them that they greatly him mistook. 50
So would he scoff them out with mockery,
For he therein had great felicity;
And with sharp quips joyed others to deface,
Thinking that their disgracing did him grace:
So whilst that other like vain wits he pleased,
And made to laugh, his heart was greatly eased.
But the right gentle mind would bite his lip
To hear the javel so good men to nip;
For, though the vulgar yield an open ear,
And common courtiers love to gibe and fleer 60
At everything which they hear spoken ill,
And the best speeches with ill meaning spill,
Yet the brave courtier, in whose beauteous thought
Regard of honour harbours more than aught,
Doth loathe such base condition, to backbite
Any's good name for envy or despite:
He stands on terms of honourable mind,
Ne will be carried with the common wind

30 sure] assuredly
45 leasings] falsehoods
53 other] others
58 javel] worthless wretch
60 fleer] sneer, gibe

Of court's inconstant mutability,
Ne after every tattling fable fly; 70
But hears and sees the follies of the rest,
And thereof gathers for himself the best.
He will not creep, nor crouch with feigned face,
But walks upright with comely steadfast pace,
And unto all doth yield due courtesy;
But not with kissed hand below the knee,
As that same apish crew is wont to do:
For he disdains himself t'embase thereto.
He hates foul leasings, and vile flattery,
Two filthy blots in noble gentry; 80
And loathful idleness he doth detest,
The canker-worm of every gentle breast;
The which to banish with fair exercise
Of knightly feats, he daily doth devise;
Now managing the mouths of stubborn steeds,
Now practising the proof of warlike deeds,
Now his bright arms assaying, now his spear,
Now the nigh aimed ring away to bear.
At other times he casts to sue the chase
Of swift wild beasts, or run on foot a race, 90
T'enlarge his breath (large breath in arms most needful),
Or else by wrestling to wex strong and heedful,
Or his stiff arms to stretch with yewen bow,
And manly legs, still passing to and fro,
Without a gowned beast him fast beside,
A vain ensample of the Persian pride;
Who, after he had won th'Assyrian foe,
Did ever after scorn on foot to go.
  Thus when this courtly gentleman with toil
Himself hath wearied, he doth recoil 100
Unto his rest, and there with sweet delight
Of music's skill revives his toiled sprite;
Or else with loves, and ladies' gentle sports,
The joy of youth, himself he recomforts;
Or lastly, when the body list to pause,
His mind unto the Muses he withdraws:
Sweet lady Muses, ladies of delight,
Delights of life, and ornaments of light!
With whom he close confers with wise discourse,
Of Nature's works, of heaven's continual course, 110

88 ring] i.e. for spear practice. A ring was hung up, and the horseman, charging at it, attempted to carry it off on his spear.

92 wex] wax, grow

Of foreign lands, of people different,
Of kingdoms' change, of divers government,
Of dreadful battles of renowned knights;
With which he kindleth his ambitious sprites
To like desire and praise of noble fame,
The only upshot whereto he doth aim:
For all his mind on honour fixed is,
To which he levels all his purposes.
And in his prince's service spends his days,
Not so much for to gain, or for to raise — 120
Himself to high degree, as for His Grace,
And in his liking to win worthy place,
Through due deserts and comely carriage,
In whatso please employ his personage,
That may be matter meet to gain him praise:
For he is fit to use in all assays,
Whether for arms and warlike amenance,
Or else for wise and civil governance.
For he is practised well in policy,
And thereto doth his courting most apply: — 130
To learn the interdeal of princes strange,
To mark th'intent of counsels, and the change
Of states, and eke of private men somewhile,
Supplanted by fine falsehood and fair guile;
Of all the which he gathereth what is fit
T'enrich the storehouse of his powerful wit,
Which through wise speeches and grave conference
He daily ekes, and brings to excellence.
Such is the rightful courtier in his kind,
But unto such the ape lent not his mind: — 140
Such were for him no fit companions,
Such would descry his lewd conditions;
But the young lusty gallants he did choose
To follow, meet to whom he might disclose
His witless pleasance, and ill-pleasing vein.
A thousand ways he them could entertain,
With all the thriftless games that may be found;
With mumming and with masking all around,
With dice, with cards, with billiards far unfit,
With shuttlecocks, misseeming manly wit, — 150
With courtesans, and costly riotise,
Whereof still somewhat to his share did rise:

118 levels] aims
127 amenance] behaviour, carriage
131 interdeal] negotiations
137 conference] conversation
144 meet] suitable
145 pleasance] pleasantries

Nor, them to pleasure, would he sometimes scorn
A pander's coat (so basely was he born).
Thereto he could fine loving verses frame,
And play the poet oft. But ah! for shame,
Let not sweet poets' praise, whose only pride
Is virtue to advance, and vice deride,
Be with the work of losels' wit defamed,
Ne let such verses poetry be named: 160
Yet he the name on him would rashly take,
Maugre the sacred Muses, and it make
A servant to the vile affection
Of such, as he depended most upon;
And with the sugary sweet thereof allure
Chaste ladies' ears to fantasies impure.
To such delights the noble wits he led
Which him relieved, and their vain humours fed
With fruitless follies and unsound delights.
But if perhaps into their noble sprites 170
Desire of honour or brave thought of arms
Did ever creep, then with his wicked charms
And strong conceits he would it drive away,
Ne suffer it to house there half a day.
And whenso love of letters did inspire
Their gentle wits, and kindle wise desire,
That chiefly doth each noble mind adorn,
Then he would scoff at learning, and eke scorn
The sectaries thereof, as people base
And simple men, which never came in place 180
Of world's affairs, but, in dark corners mewed,
Mutt'red of matters as their books them shewed,
Ne other knowledge ever did attain,
But with their gowns their gravity maintain.
From them he would his impudent lewd speech
Against God's holy ministers oft reach,
And mock divines and their profession.
What else then did he by progression,
But mock high God himself, whom they profess?
But what cared he for God, or godliness? 190
All his care was himself how to advance,
And to uphold his courtly countenance
By all the cunning means he could devise:
Were it by honest ways, or otherwise,

159 losels] worthless profligates
162 Maugre] despite
163 affection] (sexual) passion
179 sectaries] members of a sect, disciples
180 simple] low-born

He made small choice; yet sure his honesty
Got him small gains, but shameless flattery,
And filthy brocage, and unseemly shifts,
And borrow base, and some good ladies' gifts:
But the best help, which chiefly him sustained,
Was his man Reynold's purchase which he gained. 200
For he was schooled by kind in all the skill
Of close conveyance, and each practice ill
Of cozenage and cleanly knavery,
Which oft maintained his master's bravery.
Besides, he used another slippery sleight,
In taking on himself, in common sight,
False personages fit for every stead,
With which he thousands cleanly cozened:
Now like a merchant, merchants to deceive,
With whom his credit he did often leave 210
In gage, for his gay master's hopeless debt;
Now like a lawyer, when he land would let,
Or sell fee-simples in his master's name,
Which he had never, nor aught like the same.
Then would he be a broker, and draw in
Both wares and money, by exchange to win;
Then would he seem a farmer, that would sell
Bargains of woods, which he did lately fell,
Or corn, or cattle, or such other ware,
Thereby to cozen men not well aware: 220
Of all the which there came a secret fee,
To th'ape, that he his countenance might be.
  Besides all this, he used oft to beguile
Poor suitors, that in court did haunt somewhile;
For he would learn their business secretly,
And then inform his master hastily,
That he by means might cast them to prevent,
And beg the suit the which the other meant.
Or otherwise false Reynold would abuse
The simple suitor, and wish him to choose 230
His master, being one of great regard
In court, to compass any suit not hard,
In case his pains were recompensed with reason.
So would he work the silly man by treason

197 brocage] pimping
198 borrow] pledge, ransom
200 purchase] acquisition, loot
204 bravery] fine clothes
215 broker] second-hand dealer, middle-man
222 countenance] patron and protector
227 cast] scheme, contrive
228 prevent] forestall, baffle
232 hard] heard

To buy his master's frivolous good will,
That had not power to do him good or ill.
So pitiful a thing is suitor's state!
Most miserable man, whom wicked fate
Hath brought to court, to sue for had-I-wist,
That few have found, and many one hath missed! 240
Full little knowest thou, that hast not tried,
What hell it is in suing long to bide:
To lose good days, that might be better spent;
To waste long nights in pensive discontent;
To speed today, to be put back tomorrow;
To feed on hope, to pine with fear and sorrow;
To have thy prince's grace, yet want her peers';
To have thy asking, yet wait many years;
To fret thy soul with crosses and with cares;
To eat thy heart through comfortless despairs; 250
To fawn, to crouch, to wait, to ride, to run,
To spend, to give, to want, to be undone.
Unhappy wight, born to disastrous end,
That doth his life in so long 'tendance spend!
  Whoever leaves sweet home, where mean estate
In safe assurance, without strife or hate,
Finds all things needful for contentment meek,
And will to court for shadows vain to seek,
Or hope to gain, himself will a daw try:
That curse God send unto mine enemy! 260
For none but such as this bold ape, unblest,
Can ever thrive in that unlucky quest;
Or such as hath a Reynold to his man,
That by his shifts his master furnish can.
But yet this fox could not so closely hide
His crafty feats, but that they were descried
At length by such as sate in justice seat,
Who for the same him foully did entreat;
And having worthily him punished,
Out of the court for ever banished. 270

(Pub. 1591)

239 had-I-wist] had I known, i.e. a vain pursuit
241 tried] experienced it
259 daw] fool
try] prove
268 entreat] treat

from *The Faerie Queene*

137 [*Guyon's Voyage to the Bower of Bliss*]

TWO dayes now in that sea he sayled has,
Ne euer land beheld, ne liuing wight,
Ne ought saue perill, still as he did pas:
Tho when appeared the third *Morrow* bright,
Vpon the waues to spred her trembling light,
An hideous roaring farre away they heard,
That all their senses filled with affright,
And streight they saw the raging surges reard
Vp to the skyes, that them of drowning made affeard.

Said then the Boteman, Palmer stere aright, 10
And keepe an euen course; for yonder way
We needes must passe (God do vs well acquight,)
That is the *Gulfe of Greedinesse*, they say,
That deepe engorgeth all this worldes pray:
Which hauing swallowd vp excessiuely,
He soone in vomit vp againe doth lay,
And belcheth forth his superfluity,
That all the seas for feare do seeme away to fly.

On th'other side an hideous Rocke is pight,
Of mightie *Magnes* stone, whose craggie clift 20
Depending from on high, dreadfull to sight,
Ouer the waues his rugged armes doth lift,
And threatneth downe to throw his ragged rift
On who so commeth nigh; yet nigh it drawes
All passengers, that none from it can shift:
For whiles they fly that Gulfes deuouring iawes,
They on this rock are rent, and sunck in helplesse wawes.

Forward they passe, and strongly he them rowes,
Vntill they nigh vnto that Gulfe arriue,
Where streame more violent and greedy growes: 30
Then he with all his puissance doth striue
To strike his oares, and mightily doth driue
The hollow vessell through the threatfull waue,
Which gaping wide, to swallow them aliue,
In th'huge abysse of his engulfing graue,
Doth rore at them in vaine, and with great terror raue.

4 Tho] then
11 euen] steady
19 pight] placed
25 passengers] passers-by
27 helpless] affording no help
wawes] waves

They passing by, that griesly mouth did see,
Sucking the seas into his entralles deepe,
That seem'd more horrible then hell to bee,
Or that darke dreadfull hole of *Tartare* steepe, 40
Through which the damned ghosts doen often creepe
Backe to the world, bad liuers to torment:
But nought that falles into this direfull deepe,
Ne that approcheth nigh the wide descent,
May backe returne, but is condemned to be drent.

On th'other side, they saw that perilous Rocke,
Threatning it selfe on them to ruinate,
On whose sharpe clifts the ribs of vessels broke,
And shiuered ships, which had bene wrecked late,
Yet stuck, with carkasses exanimate 50
Of such, as hauing all their substance spent
In wanton ioyes, and lustes intemperate,
Did afterwards make shipwracke violent,
Both of their life, and fame for euer fowly blent.

For thy, this hight *The Rocke of* vile *Reproch*,
A daungerous and detestable place,
To which nor fish nor fowle did once approch,
But yelling Meawes, with Seagulles hoarse and bace,
And Cormoyrants, with birds of rauenous race,
Which still sate waiting on that wastfull clift, 60
For spoyle of wretches, whose vnhappie cace,
After lost credite and consumed thrift,
At last them driuen hath to this despairefull drift.

The Palmer seeing them in safetie past,
Thus said; Behold th'ensamples in our sights,
Of lustfull luxurie and thriftlesse wast:
What now is left of miserable wights,
Which spent their looser daies in lewd delights,
But shame and sad reproch, here to be red,
By these rent reliques, speaking their ill plights? 70
Let all that liue, hereby be counselled,
To shunne *Rocke of Reproch*, and it as death to dred.

45 drent] drowned
50 exanimate] lifeless
54 blent] stained
55 For thy] therefore
hight] is called
69 red] made known

So forth they rowed, and that *Ferryman*
With his stiffe oares did brush the sea so strong,
That the hoare waters from his frigot ran,
And the light bubbles daunced all along,
Whiles the salt brine out of the billowes sprong.
At last farre off they many Islands spy,
On euery side floting the floods emong:
Then said the knight, Loe I the land descry, 80
Therefore old Syre thy course do thereunto apply.

That may not be, said then the *Ferryman*
Least we vnweeting hap to be fordonne:
For those same Islands, seeming now and than,
Are not firme lande, nor any certein wonne,
But straggling plots, which to and fro do ronne
In the wide waters: therefore are they hight
The *wandring Islands*. Therefore doe them shonne;
For they haue oft drawne many a wandring wight
Into most deadly daunger and distressed plight. 90

Yet well they seeme to him, that farre doth vew,
Both faire and fruitfull, and the ground dispred
With grassie greene of delectable hew,
And the tall trees with leaues apparelled,
Are deckt with blossomes dyde in white and red,
That mote the passengers thereto allure;
But whosoeuer once hath fastened
His foot thereon, may neuer it recure,
But wandreth euer more vncertein and vnsure.

As th'Isle of *Delos* whylome men report 100
Amid th' *Aegæan* sea long time did stray,
Ne made for shipping any certaine port,
Till that *Latona* traueiling that way,
Flying from *Iunoes* wrath and hard assay,
Of her faire twins was there deliuered,
Which afterwards did rule the night and day;
Thenceforth it firmely was established,
And for *Apolloes* honor highly herried.

83 fordonne] killed
85 wonne] dwelling-place
100 whylome] formerly
108 herried] honoured, praised

They to him hearken, as beseemeth meete,
And passe on forward: so their way does ly, 110
That one of those same Islands, which doe fleet
In the wide sea, they needes must passen by,
Which seemd so sweet and pleasant to the eye,
That it would tempt a man to touchen there:
Vpon the banck they sitting did espy
A daintie damzell, dressing of her heare,
By whom a little skippet floting did appeare.

She them espying, loud to them can call,
Bidding them nigher draw vnto the shore;
For she had cause to busie them withall; 120
And therewith loudly laught: But nathemore
Would they once turne, but kept on as afore:
Which when she saw, she left her lockes vndight,
And running to her boat withouten ore
From the departing land it launched light,
And after them did driue with all her power and might.

Whom ouertaking, she in merry sort
Them gan to bord, and purpose diuersly,
Now faining dalliance and wanton sport,
Now throwing forth lewd words immodestly; 130
Till that the Palmer gan full bitterly
Her to rebuke, for being loose and light:
Which not abiding, but more scornefully
Scoffing at him, that did her iustly wite,
She turnd her bote about, and from them rowed quite.

That was the wanton *Phædria*, which late
Did ferry him ouer the *Idle lake*:
Whom nought regarding, they kept on their gate,
And all her vaine allurements did forsake,
When them the wary Boateman thus bespake; 140
Here now behoueth vs well to auyse,
And of our safetie good heede to take;
For here before a perlous passage lyes,
Where many Mermayds haunt, making false melodies.

114 touchen] land, call
117 skippet] small boat
118 can call] called
128 bord] address
purpose diuersly] talk of various matters
134 wite] blame, reprimand
138 gate] way

But by the way, there is a great Quicksand,
And a whirlepoole of hidden ieopardy,
Therefore, Sir Palmer, keepe an euen hand;
For twixt them both the narrow way doth ly.
Scarse had he said, when hard at hand they spy
That quicksand nigh with water couered; 150
But by the checked waue they did descry
It plaine, and by the sea discoloured:
It called was the quicksand of *Vnthriftyhed*.

They passing by, a goodly Ship did see,
Laden from far with precious merchandize,
And brauely furnished, as ship might bee,
Which through great disauenture, or mesprize,
Her selfe had runne into that hazardize;
Whose mariners and merchants with much toyle,
Labour'd in vaine, to haue recur'd their prize, 160
And the rich wares to saue from pitteous spoyle,
But neither toyle nor trauell might her backe recoyle.

On th'other side they see that perilous Poole,
That called was the *Whirlepoole of decay*,
In which full many had with haplesse doole
Beene suncke, of whom no memorie did stay:
Whose circled waters rapt with whirling sway,
Like to a restlesse wheele, still running round,
Did couet, as they passed by that way,
To draw their boate within the vtmost bound 170
Of his wide *Labyrinth*, and then to haue them dround.

But th'heedfull Boateman strongly forth did stretch
His brawnie armes, and all his body straine,
That th'vtmost sandy breach they shortly fetch,
Whiles the dred daunger does behind remaine.
Suddeine they see from midst of all the Maine,
The surging waters like a mountaine rise,
And the great sea puft vp with proud disdaine,
To swell aboue the measure of his guise,
As threatning to deuoure all, that his powre despise. 180

145 by the way] on the way
151 checked] chequered
156 brauely] finely
157 mesprize] misfortune
160 recur'd] recovered
162 recoyle] retrieve
165 doole] grief
174 fetch] reach
179 guise] usual appearance

The waues come rolling, and the billowes rore
  Outragiously, as they enraged were,
  Or wrathfull *Neptune* did them driue before
  His whirling charet, for exceeding feare:
  For not one puffe of wind there did appeare,
  That all the three thereat woxe much afrayd,
  Vnweeting, what such horrour straunge did reare.
  Eftsoones they saw an hideous hoast arrayd,
Of huge Sea monsters, such as liuing sence dismayd.

Most vgly shapes, and horrible aspects, 190
  Such as Dame Nature selfe mote feare to see,
  Or shame, that euer should so fowle defects
  From her most cunning hand escaped bee;
  All dreadfull pourtraicts of deformitee:
  Spring-headed *Hydraes*, and sea-shouldring Whales,
  Great whirlpooles, which all fishes make to flee,
  Bright Scolopendraes, arm'd with siluer scales,
Mighty *Monoceroses*, with immeasured tayles.

The dreadfull Fish, that hath deseru'd the name
  Of Death, and like him lookes in dreadfull hew, 200
  The griesly Wasserman, that makes his game
  The flying ships with swiftnesse to pursew,
  The horrible Sea-satyre, that doth shew
  His fearefull face in time of greatest storme,
  Huge *Ziffius*, whom Mariners eschew
  No lesse, then rockes, (as trauellers informe,)
And greedy *Rosmarines* with visages deforme.

All these, and thousand thousands many more,
  And more deformed Monsters thousand fold,
  With dreadfull noise, and hollow rombling rore, 210
  Came rushing in the fomy waues enrold,
  Which seem'd to fly for feare, them to behold:
  Ne wonder, if these did thc knight appall;
  For all that here on earth we dreadfull hold,
  Be but as bugs to fearen babes withall,
Compared to the creatures in the seas entrall.

188 Eftsoons] at once, then
195 *Hydraes*] seven-headed serpents
197 Scolopendraes] centipede-like fish
198 *Monoceroses*] sea-unicorns
immeasured] immense
199 The dreadfull Fish] i.e. the morse or walrus
201 Wasserman] merman
203 Sea-satyre] i.e. part-man, part-fish (?)
205 *Ziffius*] sword-fish
207 *Rosmarines*] walruses
215 bugs] bugbears
216 entrall] entrails

Feare nought, (then said the Palmer well auiz'd;)
For these same Monsters are not these in deed,
But are into these fearefull shapes disguiz'd
By that same wicked witch, to worke vs dreed, 220
And draw from on this iourney to proceede.
Tho lifting vp his vertuous staffe on hye,
He smote the sea, which calmed was with speed,
And all that dreadfull Armie fast gan flye
Into great *Tethys* bosome, where they hidden lye.

Quit from that daunger, forth their course they kept,
And as they went, they heard a ruefull cry
Of one, that wayld and pittifully wept,
That through the sea the resounding plaints did fly:
At last they in an Island did espy 230
A seemely Maiden, sitting by the shore,
That with great sorrow and sad agony,
Seemed some great misfortune to deplore,
And lowd to them for succour called euermore.

Which *Guyon* hearing, streight his Palmer bad,
To stere the boate towards that dolefull Mayd,
That he might know, and ease her sorrow sad:
Who him auizing better, to him sayd;
Faire Sir, be not displeasd, if disobayd:
For ill it were to hearken to her cry; 240
For she is inly nothing ill apayd,
But onely womanish fine forgery,
Your stubborne hart t'affect with fraile infirmity.

To which when she your courage hath inclind
Through foolish pitty, then her guilefull bayt
She will embosome deeper in your mind,
And for your ruine at the last awayt.
The knight was ruled, and the Boateman strayt
Held on his course with stayed stedfastnesse,
Ne euer shruncke, ne euer sought to bayt 250
His tyred armes for toylesome wearinesse,
But with his oares did sweepe the watry wildernesse.

220 wicked witch] Acrasia 241 ill apayd] distressed 250 bayt] abate
222 Tho] then

And now they nigh approched to the sted,
Where as those Mermayds dwelt: it was a still
And calmy bay, on th'one side sheltered
With the brode shadow of an hoarie hill,
On th'other side an high rocke toured still,
That twixt them both a pleasaunt port they made,
And did like an halfe Theatre fulfill:
There those fiue sisters had continuall trade, 260
And vsd to bath themselues in that deceiptfull shade.

They were faire Ladies, till they fondly striu'd
With th'*Heliconian* maides for maistery;
Of whom they ouer-comen, were depriu'd
Of their proud beautie, and th'one moyity
Transform'd to fish, for their bold surquedry,
But th'vpper halfe their hew retained still,
And their sweet skill in wonted melody;
Which euer after they abusd to ill,
T'allure weake trauellers, whom gotten they did kill. 270

So now to *Guyon*, as he passed by,
Their pleasaunt tunes they sweetly thus applide;
O thou faire sonne of gentle Faery,
That art in mighty armes most magnifide
Aboue all knights, that euer battell tride,
O turne thy rudder hither-ward a while:
Here may thy storme-bet vessell safely ride;
This is the Port of rest from troublous toyle,
The worlds sweet In, from paine and wearisome turmoyle.

With that the rolling sea resounding soft, 280
In his big base them fitly answered,
And on the rocke the waues breaking aloft,
A solemne Meane vnto them measured,
The whiles sweet *Zephirus* lowd whisteled
His treble, a straunge kinde of harmony;
Which *Guyons* senses softly tickeled,
That he the boateman bad row easily,
And let him heare some part of their rare melody.

257 toured] towered
260 trade] occupation
265 moyity] half
266 surquedry] pride
283 Meane] middle part of the harmony

But him the Palmer from that vanity,
With temperate aduice discounselled, 290
That they it past, and shortly gan descry
The land, to which their course they leueled;
When suddeinly a grosse fog ouer spred
With his dull vapour all that desert has,
And heauens chearefull face enueloped,
That all things one, and one as nothing was,
And this great Vniuerse seemd one confused mas.

Thereat they greatly were dismayd, ne wist
How to direct their way in darkenesse wide,
But feard to wander in that wastfull mist, 300
For tombling into mischiefe vnespide.
Worse is the daunger hidden, then descride.
Suddeinly an innumerable flight
Of harmefull fowles about them fluttering, cride,
And with their wicked wings them oft did smight,
And sore annoyed, groping in that griesly night.

Euen all the nation of vnfortunate
And fatall birds about them flocked were,
Such as by nature men abhorre and hate,
The ill-faste Owle, deaths dreadfull messengere, 310
The hoars Night-rauen, trump of dolefull drere,
The lether-winged Bat, dayes enimy,
The ruefull Strich, still waiting on the bere,
The Whistler shrill, that who so heares, doth dy,
The hellish Harpies, prophets of sad destiny.

All those, and all that else does horrour breed,
About them flew, and fild their sayles with feare:
Yet stayd they not, but forward did proceed,
Whiles th'one did row, and th'other stifly steare;
Till that at last the weather gan to cleare, 320
And the faire land it selfe did plainly show.
Said then the Palmer, Lo where does appeare
The sacred soile, where all our perils grow;
Therefore, Sir knight, your ready armes about you throw.

308 fatall] of ill portent
310 ill-faste] ugly
311 drere] sadness, gloom
313 Strich] screech-owl
bere] bier, i.e. sepulchre
314 Whistler] plover
323 sacred] accursed

He hearkned, and his armes about him tooke,
The whiles the nimble boate so well her sped,
That with her crooked keele the land she strooke,
Then forth the noble *Guyon* sallied,
And his sage Palmer, that him gouerned;
But th'other by his boate behind did stay. 330
They marched fairly forth, of nought ydred,
Both firmely armd for euery hard assay,
With constancy and care, gainst daunger and dismay.

(1590)

138 [*The House of Busyrane*]

WITH huge impatience, he inly swelt,
More for great sorrow, that he could not pas,
Then for the burning torment, which he felt,
That with fell woodnesse he effierced was,
And wilfully him throwing on the gras,
Did beat and bounse his head and brest full sore;
The whiles the Championesse now entred has
The vtmost rowme, and past the formest dore,
The vtmost rowme, abounding with all precious store.

For round about, the wals yclothed were 10
With goodly arras of great maiesty,
Wouen with gold and silke so close and nere,
That the rich metall lurked priuily,
As faining to be hid from enuious eye;
Yet here, and there, and euery where vnwares
It shewd it selfe, and shone vnwillingly;
Like a discolourd Snake, whose hidden snares
Through the greene gras his long bright burnisht backe declares.

And in those Tapets weren fashioned
Many faire pourtraicts, and many a faire feate, 20
And all of loue, and all of lusty-hed,
As seemed by their semblaunt did entreat;
And eke all *Cupids* warres they did repeate,
And cruell battels, which he whilome fought
Gainst all the Gods, to make his empire great;
Besides the huge massacres, which he wrought
On mighty kings and kesars, into thraldome brought.

1 he inly swelt] i.e. Scudamour
4 fell woodnesse] fierce madness
effierced] angered, frenzied
7 Championesse] Britomart
8 vtmost] outermost
formest] foremost
11 arras] tapestry

Therein was writ, how often thundring *Ioue*
Had felt the point of his hart-percing dart,
And leauing heauens kingdome, here did roue 30
In straunge disguize, to slake his scalding smart;
Now like a Ram, faire *Helle* to peruart,
Now like a Bull, *Europa* to withdraw:
Ah, how the fearefull Ladies tender hart
Did liuely seeme to tremble, when she saw
The huge seas vnder her t'obay her seruaunts law.

Soone after that into a golden showre
Him selfe he chaung'd faire *Danaë* to vew,
And through the roofe of her strong brasen towre
Did raine into her lap an hony dew, 40
The whiles her foolish garde, that little knew
Of such deceipt, kept th'yron dore fast bard,
And watcht, that none should enter nor issew;
Vaine was the watch, and bootlesse all the ward,
Whenas the God to golden hew him selfe transfard.

Then was he turnd into a snowy Swan,
To win faire *Leda* to his louely trade:
O wondrous skill, and sweet wit of the man,
That her in daffadillies sleeping made,
From scorching heat her daintie limbes to shade: 50
Whiles the proud Bird ruffing his fethers wyde,
And brushing his faire brest, did her inuade;
She slept, yet twixt her eyelids closely spyde,
How towards her he rusht, and smiled at his pryde.

Then shewd it, how the *Thebane Sémelee*
Deceiu'd of gealous *Iuno*, did require
To see him in his soueraigne maiestee,
Armd with his thunderbolts and lightning fire,
Whence dearely she with death bought her desire.
But faire *Alcmena* better match did make, 60
Ioying his loue in likenesse more entire;
Three nights in one, they say, that for her sake
He then did put, her pleasures lenger to partake.

Twise was he seene in soaring Eagles shape,
And with wide wings to beat the buxome ayre,
Once, when he with *Asterie* did scape,
Againe, when as the *Troiane* boy so faire
He snatcht from *Ida* hill, and with him bare:
Wondrous delight it was, there to behould,
How the rude Shepheards after him did stare, 70
Trembling through feare, least down he fallen should,
And often to him calling, to take surer hould.

In *Satyres* shape *Antiopa* he snatcht:
And like a fire, when he *Aegin'* assayd:
A shepheard, when *Mnemosyne* he catcht:
And like a Serpent to the *Thracian* mayd.
Whiles thus on earth great *Ioue* these pageaunts playd,
The winged boy did thrust into his throne,
And scoffing, thus vnto his mother sayd,
Lo now the heauens obey to me alone, 80
And take me for their *Ioue*, whiles *Ioue* to earth is gone.

And thou, faire *Phœbus*, in thy colours bright
Wast there enwouen, and the sad distresse,
In which that boy thee plonged, for despight,
That thou bewray'dst his mothers wantonnesse,
When she with *Mars* was meynt in ioyfulnesse:
For thy he thrild thee with a leaden dart,
To loue faire *Daphne*, which thee loued lesse:
Lesse she thee lou'd, then was thy iust desart,
Yet was thy loue her death, and her death was thy smart. 90

So louedst thou the lusty *Hyacinct*,
So louedst thou the faire *Coronis* deare:
Yet both are of thy haplesse hand extinct,
Yet both in flowres do liue, and loue thee beare,
The one a Paunce, the other a sweet breare:
For griefe whereof, ye mote haue liuely seene
The God himselfe rending his golden heare,
And breaking quite his gyrlond euer greene,
With other signes of sorrow and impatient teene.

65 buxome] yielding
86 meynt] mingled
87 For thy] for which
thrild] pierced
95 Paunce] pansy
99 teene] woe

Both for those two, and for his owne deare sonne, 100
The sonne of *Climene* he did repent,
Who bold to guide the charet of the Sunne,
Himselfe in thousand peeces fondly rent,
And all the world with flashing fier brent;
So like, that all the walles did seeme to flame.
Yet cruell *Cupid*, not herewith content,
Forst him eftsoones to follow other game,
And loue a Shepheards daughter for his dearest Dame.

He loued *Isse* for his dearest Dame,
And for her sake her cattell fed a while, 110
And for her sake a cowheard vile became,
The seruant of *Admetus* cowheard vile,
Whiles that from heauen he suffered exile.
Long were to tell each other louely fit,
Now like a Lyon, hunting after spoile,
Now like a Stag, now like a faulcon flit:
All which in that faire arras was most liuely writ.

Next vnto him was *Neptune* pictured,
In his diuine resemblance wondrous lyke:
His face was rugged, and his hoarie hed 120
Dropped with brackish deaw; his three-forkt Pyke
He stearnly shooke, and therewith fierce did stryke
The raging billowes, that on euery syde
They trembling stood, and made a long broad dyke,
That his swift charet might haue passage wyde,
Which foure great *Hippodames* did draw in temewise tyde.

His sea-horses did seeme to snort amayne,
And from their nosethrilles blow the brynie streame,
That made the sparckling waues to smoke agayne,
And flame with gold, but the white fomy creame, 130
Did shine with siluer, and shoot forth his beame.
The God himselfe did pensiue seeme and sad,
And hong adowne his head, as he did dreame:
For priuy loue his brest empierced had,
Ne ought but deare *Bisaltis* ay could make him glad.

114 louely] amorous
116 flit] swift
122 stearnly] fiercely
126 *Hippodames*] sea-horses
127 amayne] violently

He loued eke *Iphimedia* deare,
And *Aeolus* fair daughter *Arne* hight,
For whom he turnd him selfe into a Steare,
And fed on fodder, to beguile her sight.
Also to win *Deucalions* daughter bright, 140
He turnd him selfe into a Dolphin fayre;
And like a winged horse he tooke his flight,
To snaky-locke *Medusa* to repayre,
On whom he got faire *Pegasus*, that flitteth in the ayre.

Next *Saturne* was, (but who would euer weene,
That sullein *Saturne* euer weend to loue?
Yet loue is sullein, and *Saturnlike* seene,
As he did for *Erigone* it proue,)
That to a *Centaure* did him selfe transmoue.
So proou'd it eke that gracious God of wine, 150
When for to compasse *Philliras* hard loue,
He turnd himselfe into a fruitfull vine,
And into her faire bosome made his grapes decline.

Long were to tell the amorous assayes,
And gentle pangues, with which he maked meeke
The mighty *Mars*, to learne his wanton playes:
How oft for *Venus*, and how often eek
For many other Nymphes he sore did shreek,
With womanish teares, and with vnwarlike smarts,
Priuily moystening his horrid cheek. 160
There was he painted full of burning darts,
And many wide woundes launched through his inner parts.

Ne did he spare (so cruell was the Elfe)
His owne deare mother, (ah why should he so?)
Ne did he spare sometime to pricke himselfe,
That he might tast the sweet consuming woe,
Which he had wrought to many others moe.
But to declare the mournfull Tragedyes,
And spoiles, wherewith he all the ground did strow,
More eath to number, with how many eyes 170
High heauen beholds sad louers nightly theeueryes.

145 weene] think 162 launched] lanced 170 eath] easy
160 horrid] bristly

Kings Queenes, Lords Ladies, Knights and Damzels gent
Were heap'd together with the vulgar sort,
And mingled with the raskall rablement,
Without respect of person or of port,
To shew Dan *Cupids* powre and great effort:
And round about a border was entrayld,
Of broken bowes and arrowes shiuered short,
And a long bloudy riuer through them rayld,
So liuely and so like, that liuing sence it fayld. 180

And at the vpper end of that faire rowme,
There was an Altar built of pretious stone,
Of passing valew, and of great renowme,
On which there stood an Image all alone,
Of massy gold, which with his owne light shone;
And wings it had with sundry colours dight,
More sundry colours, then the proud *Pauone*
Beares in his boasted fan, or *Iris* bright,
When her discoloured bow she spreds through heauen bright.

Blindfold he was, and in his cruell fist 190
A mortall bow and arrowes keene did hold,
With which he shot at randon, when him list,
Some headed with sad lead, some with pure gold;
(Ah man beware, how thou those darts behold)
A wounded Dragon vnder him did ly,
Whose hideous tayle his left foot did enfold,
And with a shaft was shot through either eye,
That no man forth might draw, ne no man remedye.

And vnderneath his feet was written thus,
Vnto the Victor of the Gods this bee: 200
And all the people in that ample hous
Did to that image bow their humble knee,
And oft committed fowle Idolatree.
That wondrous sight faire *Britomart* amazed,
Ne seeing could her wonder satisfie,
But euermore and more vpon it gazed,
The whiles the passing brightnes her fraile sences dazed.

179 rayld] flowed
180 fayld] deceived
187 *Pauone*] peacock
189 discoloured] variously coloured
207 passing] surpassing

Tho as she backward cast her busie eye,
To search each secret of that goodly sted,
Ouer the dore thus written she did spye 210
Be bold: she oft and oft it ouer-red,
Yet could not find what sence it figured:
But what so were therein or writ or ment,
She was no whit thereby discouraged
From prosecuting of her first intent,
But forward with bold steps into the next roome went.

Much fairer, then the former, was that roome,
And richlier by many partes arayd:
For not with arras made in painefull loome,
But with pure gold it all was ouerlayd, 220
Wrought with wilde Antickes, which their follies playd,
In the rich metall, as they liuing were:
A thousand monstrous formes therein were made,
Such as false loue doth oft vpon him weare,
For loue in thousand monstrous formes doth oft appeare.

And all about, the glistring walles were hong
With warlike spoiles, and with victorious prayes,
Of mighty Conquerours and Captaines strong,
Which were whilome captiued in their dayes
To cruell loue, and wrought their owne decayes: 230
Their swerds and speres were broke, and hauberques rent;
And their proud girlonds of tryumphant bayes
Troden in dust with fury insolent,
To shew the victors might and mercilesse intent.

The warlike Mayde beholding earnestly
The goodly ordinance of this rich place,
Did greatly wonder, ne could satisfie
Her greedy eyes with gazing a long space,
But more she meruaild that no footings trace,
Nor wight appear'd, but wastefull emptinesse, 240
And solemne silence ouer all that place:
Straunge thing it seem'd, that none was to possesse
So rich purueyance, ne them keepe with carefulnesse.

218 by many partes] by much
221 wilde Antickes] grotesque carvings
231 hauberques] coats of mail

And as she lookt about, she did behold,
How ouer that same dore was likewise writ,
Be bold, be bold, and euery where *Be bold*,
That much she muz'd, yet could not construe it
By any ridling skill, or commune wit.
At last she spyde at that roomes vpper end,
Another yron dore, on which was writ, 250
Be not too bold; whereto though she did bend
Her earnest mind, yet wist not what it might intend.

Thus she there waited vntill euentyde,
Yet liuing creature none she saw appeare:
And now sad shadowes gan the world to hyde,
From mortall vew, and wrap in darkenesse dreare;
Yet nould she d'off her weary armes, for feare
Of secret daunger, ne let sleepe oppresse
Her heauy eyes with natures burdein deare,
But drew her selfe aside in sickernesse, 260
And her welpointed weapons did about her dresse.

(1590)

139 [*The Vision of the Graces*]

ONE day as he did raunge the fields abroad,
Whilest his faire *Pastorella* was elsewhere,
He chaunst to come, far from all peoples troad,
Vnto a place, whose pleasaunce did appere
To passe all others, on the earth which were:
For all that euer was by natures skill
Deuized to worke delight, was gathered there,
And there by her were poured forth at fill,
As if this to adorne, she all the rest did pill.

It was an hill plaste in an open plaine, 10
That round about was bordered with a wood
Of matchlesse hight, that seem'd th'earth to disdaine,
In which all trees of honour stately stood,
And did all winter as in sommer bud,
Spredding pauilions for the birds to bowre,
Which in their lower braunches sung aloud;
And in their tops the soring hauke did towre,
Sitting like King of fowles in maiesty and powre.

260 sickernesse] safety, security
261 welpointed] well-appointed
dresse] set in position, prepare

139
1 he] Sir Calidore
3 troad] tread
9 pill] pillage, rob

And at the foote thereof, a gentle flud
  His siluer waues did softly tumble downe, 20
  Vnmard with ragged mosse or filthy mud,
  Ne mote wylde beastes, ne mote the ruder clowne
  Thereto approch, ne filth mote therein drowne:
  But Nymphes and Faeries by the bancks did sit,
  In the woods shade, which did the waters crowne,
  Keeping all noysome things away from it,
And to the waters fall tuning their accents fit.

And on the top thereof a spacious plaine
  Did spred it selfe, to serue to all delight,
  Either to daunce, when they to daunce would faine, 30
  Or else to course about their bases light;
  Ne ought there wanted, which for pleasure might
  Desired be, or thence to banish bale:
  So pleasauntly the hill with equall hight,
  Did seeme to ouerlooke the lowly vale;
Therefore it rightly cleeped was mount *Acidale*.

They say that *Venus*, when she did dispose
  Her selfe to pleasaunce, vsed to resort
  Vnto this place, and therein to repose
  And rest her selfe, as in a gladsome port, 40
  Or with the Graces there to play and sport;
  That euen her owne Cytheron, though in it
  She vsed most to keepe her royall court,
  And in her soueraine Maiesty to sit,
She in regard hereof refusde and thought vnfit.

Vnto this place when as the Elfin Knight
  Approcht, him seemed that the merry sound
  Of a shrill pipe he playing heard on hight,
  And many feete fast thumping th'hollow ground,
  That through the woods their Eccho did rebound. 50
  He nigher drew, to weete what mote it be;
  There he a troupe of Ladies dauncing found
  Full merrily, and making gladfull glee,
And in the midst a Shepheard piping he did see.

51 weete] know

He durst not enter into th'open greene,
For dread of them vnwares to be descryde,
For breaking of their daunce, if he were seene;
But in the couert of the wood did byde,
Beholding all, yet of them vnespyde.
There he did see, that pleased much his sight, 60
That euen he him selfe his eyes enuyde,
An hundred naked maidens lilly white,
All raunged in a ring, and dauncing in delight.

All they without were raunged in a ring,
And daunced round; but in the midst of them
Three other Ladies did both daunce and sing,
The whilest the rest them round about did hemme,
And like a girlond did in compasse stemme:
And in the middest of those same three, was placed
Another Damzell, as a precious gemme, 70
Amidst a ring most richly well enchaced,
That with her goodly presence all the rest much graced.

Looke how the Crowne, which *Ariadne* wore
Vpon her yuory forehead that same day,
That *Theseus* her vnto his bridale bore,
When the bold *Centaures* made that bloudy fray,
With the fierce *Lapithes*, which did them dismay;
Being now placed in the firmament,
Through the bright heauen doth her beams display,
And is vnto the starres an ornament, 80
Which round about her moue in order excellent.

Such was the beauty of this goodly band,
Whose sundry parts were here too long to tell:
But she that in the midst of them did stand,
Seem'd all the rest in beauty to excell,
Crownd with a rosie girlond, that right well
Did her beseeme. And euer, as the crew
About her daunst, sweet flowres, that far did smell,
And fragrant odours they vppon her threw;
But most of all, those three did her with gifts endew. 90

60 that pleased] what pleased
68 did in compasse stemme] encircled
73 Looke how] like

Those were the Graces, daughters of delight,
  Handmaides of *Venus*, which are wont to haunt
  Vppon this hill, and daunce there day and night:
  Those three to men all gifts of grace do graunt,
  And all, that *Venus* in her selfe doth vaunt,
  Is borrowed of them. But that faire one,
  That in the midst was placed parauaunt,
  Was she to whom that shepheard pypt alone,
That made him pipe so merrily, as neuer none.

She was to weete that iolly Shepheards lasse, 100
  Which piped there vnto that merry rout,
  That iolly shepheard, which there piped, was
  Poore *Colin Clout* (who knowes not *Colin Clout?*)
  He pypt apace, whilest they him daunst about.
  Pype iolly shepheard, pype thou now apace
  Vnto thy loue, that made thee low to lout:
  Thy loue is present there with thee in place,
Thy loue is there aduaunst to be another Grace.

Much wondred *Calidore* at this straunge sight,
  Whose like before his eye had neuer seene, 110
  And standing long astonished in spright,
  And rapt with pleasaunce, wist not what to weene;
  Whether it were the traine of beauties Queene,
  Or Nymphes, or Faeries, or enchaunted show,
  With which his eyes mote haue deluded beene.
  Therefore resoluing, what it was, to know,
Out of the wood he rose, and toward them did go.

But soone as he appeared to their vew,
  They vanisht all away out of his sight,
  And cleane were gone, which way he neuer knew; 120
  All saue the shepheard, who for fell despight
  Of that displeasure, broke his bag-pipe quight,
  And made great mone for that vnhappy turne.
  But *Calidore*, though no lesse sory wight,
  For that mishap, yet seeing him to mourne,
Drew neare, that he the truth of all by him mote learne.

97 parauaunt] pre-eminently 106 lout] bow

And first him greeting, thus vnto him spake,
Haile iolly shepheard, which thy ioyous dayes
Here leadest in this goodly merry make,
Frequented of these gentle Nymphes alwayes, 130
Which to thee flocke, to heare thy louely layes;
Tell me, what mote these dainty Damzels be,
Which here with thee doe make their pleasant playes?
Right happy thou, that mayst them freely see:
But why when I them saw, fled they away from me?

Not I so happy, answerd then that swaine,
As thou vnhappy, which them thence didst chace,
Whom by no meanes thou canst recall againe,
For being gone, none can them bring in place,
But whom they of them selues list so to grace. 140
Right sory I, (said then Sir *Calidore*,)
That my ill fortune did them hence displace.
But since things passed none may now restore,
Tell me, what were they all, whose lacke thee grieues so sore.

Tho gan that shepheard thus for to dilate;
Then wote thou shepheard, whatsoeuer thou bee,
That all those Ladies, which thou sawest late,
Are *Venus* Damzels, all within her fee,
But differing in honour and degree:
They all are Graces, which on her depend, 150
Besides a thousand more, which ready bee
Her to adorne, when so she forth doth wend:
But those three in the midst, doe chiefe on her attend.

They are the daughters of sky-ruling Ioue,
By him begot of faire *Eurynome*,
The Oceans daughter, in this pleasant groue,
As he this way comming from feastfull glee,
Of *Thetis* wedding with *Æacidee*,
In sommers shade him selfe here rested weary.
The first of them hight mylde *Euphrosyne*, 160
Next faire *Aglaia*, last *Thalia* merry:
Sweete Goddesses all three which me in mirth do cherry.

145 Tho] then
146 wote] know
148 fee] service
162 cherry] cheer

These three on men all gracious gifts bestow,
  Which decke the body or adorne the mynde,
  To make them louely or well fauoured show,
  As comely carriage, entertainement kynde,
  Sweete semblaunt, friendly offices that bynde,
  And all the complements of curtesie:
  They teach vs, how to each degree and kynde
  We should our selues demeane, to low, to hie; 170
To friends, to foes, which skill men call Ciuility.

Therefore they alwaies smoothly seeme to smile,
  That we likewise should mylde and gentle be,
  And also naked are, that without guile
  Or false dissemblaunce all them plaine may see,
  Simple and true from couert malice free:
  And eeke them selues so in their daunce they bore,
  That two of them still froward seem'd to bee,
  But one still towards shew'd her selfe afore;
That good should from vs goe, then come in greater store. 180

Such were those Goddesses, which ye did see;
  But that fourth Mayd, which there amidst them traced,
  Who can aread, what creature mote she bee,
  Whether a creature, or a goddesse graced
  With heauenly gifts from heuen first enraced?
  But what so sure she was, she worthy was,
  To be the fourth with those three other placed:
  Yet was she certes but a countrey lasse,
Yet she all other countrey lasses farre did passe.

So farre as doth the daughter of the day, 190
  All other lesser lights in light excell,
  So farre doth she in beautyfull array,
  Aboue all other lasses beare the bell,
  Ne lesse in vertue that beseemes her well,
  Doth she exceede the rest of all her race,
  For which the Graces that here wont to dwell,
  Haue for more honor brought her to this place,
And graced her so much to be another Grace.

178 froward] i.e. 'fromward', turned away
179 afore] facing
182 traced] danced
183 aread] tell
185 enraced] implanted
193 beare the bell] win the prize

Another Grace she well deserues to be,
In whom so many Graces gathered are, 200
Excelling much the meane of her degree;
Diuine resemblaunce, beauty soueraine rare,
Firme Chastity, that spight ne blemish dare;
All which she with such courtesie doth grace,
That all her peres cannot with her compare,
But quite are dimmed, when she is in place.
She made me often pipe and now to pipe apace.

Sunne of the world, great glory of the sky,
That all the earth doest lighten with thy rayes,
Great *Gloriana*, greatest Maiesty, 210
Pardon thy shepheard, mongst so many layes,
As he hath sung of thee in all his dayes,
To make one minime of thy poore handmayd,
And vnderneath thy feete to place her prayse,
That when thy glory shall be farre displayd
To future age of her this mention may be made.

(1596)

140 [*Mutability claims to rule the world*]

THIS great Grandmother of all creatures bred
Great *Nature*, euer young yet full of eld,
Still moouing, yet vnmoued from her sted;
Vnseene of any, yet of all beheld;
Thus sitting in her throne as I haue teld,
Before her came dame *Mutabilitie*;
And being lowe before her presence feld,
With meek obaysance and humilitie,
Thus gan her plaintif Plea, with words to amplifie;

To thee O greatest goddesse, onely great, 10
An humble suppliant loe, I lowely fly
Seeking for Right, which I of thee entreat;
Who Right to all dost deale indifferently,
Damning all Wrong and tortious Iniurie,
Which any of thy creatures doe to other
(Oppressing them with power, vnequally)
Sith of them all thou art the equall mother,
And knittest each to each, as brother vnto brother.

201 meane] norm
degree] rank
213 minime] brief musical note

140
2 eld] age
7 feld] fallen, bowed
13 indifferently] impartially
14 tortious] wrongful

To thee therefore of this same *Ioue* I plaine,
And of his fellow gods that faine to be, 20
That challenge to themselues the whole worlds raign;
Of which, the greatest part is due to me,
And heauen it selfe by heritage in Fee:
For, heauen and earth I both alike do deeme,
Sith heauen and earth are both alike to thee;
And, gods no more then men thou doest esteeme:
For, euen the gods to thee, as men to gods do seeme.

Then weigh, O soueraigne goddesse, by what right
These gods do claime the worlds whole souerainty;
And that is onely dew vnto thy might 30
Arrogate to themselues ambitiously:
As for the gods owne principality,
Which *Ioue* vsurpes vniustly; that to be
My heritage, *Ioue's* self cannot deny,
From my great Grandsire *Titan*, vnto mee,
Deriv'd by dew descent; as is well knowen to thee.

Yet mauger *Ioue*, and all his gods beside,
I doe possesse the worlds most regiment;
As, if ye please it into parts diuide,
And euery parts inholders to conuent, 40
Shall to your eyes appeare incontinent.
And first, the Earth (great mother of vs all)
That only seems vnmov'd and permanent,
And vnto *Mutability* not thrall;
Yet is she chang'd in part, and eeke in generall.

For, all that from her springs, and is ybredde,
How-euer fayre it flourish for a time,
Yet see we soone decay; and, being dead,
To turne again vnto their earthly slime:
Yet, out of their decay and mortall crime, 50
We daily see new creatures to arize;
And of their Winter spring another Prime,
Vnlike in forme, and chang'd by strange disguise:
So turne they still about, and change in restlesse wise.

21 challenge] claim
23 in Fee] in right possession
30 that is] what is
37 mauger] despite
38 most regiment] greatest power
40 inholders] tenants
conuent] assemble
41 incontinent] at once
50 mortall crime] corruption
52 Prime] spring

As for her tenants; that is, man and beasts,
The beasts we daily see massacred dy,
As thralls and vassals vnto mens beheasts:
And men themselues doe change continually,
From youth to eld, from wealth to pouerty,
From good to bad, from bad to worst of all. 60
Ne doe their bodies only flit and fly:
But eeke their minds (which they immortall call)
Still change and vary thoughts, as new occasions fall.

Ne is the water in more constant case;
Whether those same on high, or these belowe.
For, th'Ocean moueth stil, from place to place;
And euery Riuer still doth ebbe and flowe:
Ne any Lake, that seems most still and slowe,
Ne Poole so small, that can his smoothnesse holde,
When any winde doth vnder heauen blowe; 70
With which, the clouds are also tost and roll'd;
Now like great Hills; and, streight, like sluces, them vnfold.

So likewise are all watry liuing wights
Still tost, and turned, with continuall change,
Neuer abyding in their stedfast plights.
The fish, still floting, doe at randon range,
And neuer rest; but euermore exchange
Their dwelling places, as the streames them carrie:
Ne haue the watry foules a certaine grange,
Wherein to rest, ne in one stead do tarry; 80
But flitting still doe flie, and still their places vary.

Next is the Ayre: which who feeles not by sense
(For, of all sense it is the middle meane)
To flit still? and, with subtill influence
Of his thin spirit, all creatures to maintaine,
In state of life? O weake life! that does leane
On thing so tickle as th'vnsteady ayre;
Which euery howre is chang'd, and altred cleane
With euery blast that bloweth fowle or faire:
The faire doth it prolong; the fowle doth it impaire. 90

72 them vnfold] open themselves
79 grange] lodging
83 middle meane] intermediary
87 tickle] unstable, insubstantial

Therein the changes infinite beholde,
Which to her creatures euery minute chaunce;
Now, boyling hot: streight, friezing deadly cold:
Now, faire sun-shine, that makes all skip and daunce:
Streight, bitter storms and balefull countenance,
That makes them all to shiuer and to shake:
Rayne, hayle, and snowe do pay them sad penance,
And dreadfull thunder-claps (that make them quake)
With flames and flashing lights that thousand changes make.

Last is the fire: which, though it liue for euer, 100
Ne can be quenched quite; yet, euery day,
Wee see his parts, so soone as they do seuer,
To lose their heat, and shortly to decay;
So, makes himself his owne consuming pray.
Ne any liuing creatures doth he breed:
But all, that are of others bredd, doth slay;
And, with their death, his cruell life dooth feed;
Nought leauing, but their barren ashes, without seede.

Thus, all these fower (the which the ground-work bee
Of all the world, and of all liuing wights) 110
To thousand sorts of *Change* we subiect see:
Yet are they chang'd (by other wondrous slights)
Into themselues, and lose their natiue mights;
The Fire to Aire, and th'Ayre to Water sheere,
And Water into Earth: yet Water fights
With Fire, and Aire with Earth approaching neere:
Yet all are in one body, and as one appeare.

So, in them all raignes *Mutabilitie*;
How-euer these, that Gods themselues do call,
Of them doe claime the rule and souerainty: 120
As, *Vesta*, of the fire æthereall;
Vulcan, of this, with vs so vsuall;
Ops, of the earth; and *Iuno* of the Ayre;
Neptune, of Seas; and Nymphes, of Riuers all.
For, all those Riuers to me subiect are:
And all the rest, which they vsurp, be all my share.

113 natiue mights] natural powers   114 sheere] clear

Which to approuen true, as I haue told,
Vouchsafe, O goddesse, to thy presence call
The rest which doe the world in being hold:
As, times and seasons of the yeare that fall: 130
Of all the which, demand in generall,
Or iudge thy selfe, by verdit of thine eye,
Whether to me they are not subiect all.
Nature did yeeld thereto; and by-and-by,
Bade *Order* call them all, before her Maiesty.

So, forth issew'd the Seasons of the yeare;
First, lusty *Spring*, all dight in leaues of flowres
That freshly budded and new bloosmes did beare
(In which a thousand birds had built their bowres
That sweetly sung, to call forth Paramours): 140
And in his hand a iauelin he did beare,
And on his head (as fit for warlike stoures)
A guilt engrauen morion he did weare;
That as some did him loue, so others did him feare.

Then came the iolly *Sommer*, being dight
In a thin silken cassock coloured greene,
That was vnlyned all, to be more light:
And on his head a girlond well beseene
He wore, from which as he had chauffed been
The sweat did drop; and in his hand he bore 150
A boawe and shaftes, as he in forrest greene
Had hunted late the Libbard or the Bore,
And now would bathe his limbes, with labor heated sore.

Then came the *Autumne* all in yellow clad,
As though he ioyed in his plentious store,
Laden with fruits that made him laugh, full glad
That he had banisht hunger, which to-fore
Had by the belly oft him pinched sore.
Vpon his head a wreath that was enrold
With eares of corne, of euery sort he bore: 160
And in his hand a sickle he did holde,
To reape the ripened fruits the which the earth had yold.

136 issew'd] came forth
137 dight] decked
142 stoures] storms
143 morion] helmet
149 chauffed] heated
152 Libbard] leopard

Lastly, came *Winter* cloathed all in frize,
Chattering his teeth for cold that did him chill,
Whil'st on his hoary beard his breath did freese;
And the dull drops that from his purpled bill
As from a limbeck did adown distill.
In his right hand a tipped staffe he held,
With which his feeble steps he stayed still:
For, he was faint with cold, and weak with eld; 170
That scarse his loosed limbes he hable was to weld.

These, marching softly, thus in order went,
And after them, the Monthes all riding came;
First, sturdy *March* with brows full sternly bent,
And armed strongly, rode vpon a Ram,
The same which ouer *Hellespontus* swam:
Yet in his hand a spade he also hent,
And in a bag all sorts of seeds ysame,
Which on the earth he strowed as he went,
And fild her womb with fruitfull hope of nourishment. 180

Next came fresh *Aprill* full of lustyhed,
And wanton as a Kid whose horne new buds:
Vpon a Bull he rode, the same which led
Europa floting through th'*Argolick* fluds:
His hornes were gilden all with golden studs
And garnished with garlonds goodly dight
Of all the fairest flowres and freshest buds
Which th'earth brings forth, and wet he seem'd in sight
With waues, through which he waded for his loues delight.

Then came faire *May*, the fayrest mayd on ground, 190
Deckt all with dainties of her seasons pryde,
And throwing flowres out of her lap around:
Vpon two brethrens shoulders she did ride,
The twinnes of *Leda*; which on eyther side
Supported her like to their soueraine Queene.
Lord! how all creatures laught, when her they spide,
And leapt and daunc't as they had rauisht beene!
And *Cupid* selfe about her fluttred all in greene.

163 frize] coarse wool
167 limbeck] alembic, retort
171 loosed] weakened, made unstable
172 softly] slowly
174 sternly] fiercely
177 hent] held

And after her, came iolly *Iune*, arrayd
All in greene leaues, as he a Player were; 200
Yet in his time, he wrought as well as playd,
That by his plough-yrons mote right well appeare:
Vpon a Crab he rode, that him did beare
With crooked crawling steps an vncouth pase,
And backward yode, as Bargemen wont to fare
Bending their force contrary to their face,
Like that vngracious crew which faines demurest grace.

Then came hot *Iuly* boyling like to fire,
That all his garments he had cast away:
Vpon a Lyon raging yet with ire 210
He boldly rode and made him to obay:
It was the beast that whylome did forray
The Nemæan forrest, till th'*Amphytrionide*
Him slew, and with his hide did him array;
Behinde his back a sithe, and by his side
Vnder his belt he bore a sickle circling wide.

The sixt was *August*, being rich arrayd
In garment all of gold downe to the ground:
Yet rode he not, but led a louely Mayd
Forth by the lilly hand, the which was cround 220
With eares of corne, and full her hand was found;
That was the righteous Virgin, which of old
Liv'd here on earth, and plenty made abound;
But, after Wrong was lov'd and Iustice solde,
She left th'vnrighteous world and was to heauen extold.

Next him, *September* marched eeke on foote;
Yet was he heauy laden with the spoyle
Of haruests riches, which he made his boot,
And him enricht with bounty of the soyle:
In his one hand, as fit for haruests toyle, 230
He held a knife-hook; and in th'other hand
A paire of waights, with which he did assoyle
Both more and lesse, where it in doubt did stand,
And equall gaue to each as Iustice duly scann'd.

201 wrought] worked
205 yode] went
213 th'*Amphytrionide*] Hercules
225 extold] raised
232 assoyle] determine

Then came *October* full of merry glee:
For, yet his noule was totty of the must,
Which he was treading in the wine-fats see,
And of the ioyous oyle, whose gentle gust
Made him so frollick and so full of lust:
Vpon a dreadfull Scorpion he did ride, 240
The same which by *Dianaes* doom vniust
Slew great *Orion*: and eeke by his side
He had his ploughing share, and coulter ready tyde.

Next was *Nouember*, he full grosse and fat,
As fed with lard, and that right well might seeme;
For, he had been a fatting hogs of late,
That yet his browes with sweat, did reek and steem,
And yet the season was full sharp and breem;
In planting eeke he took no small delight:
Whereon he rode, not easie was to deeme: 250
For it a dreadfull *Centaure* was in sight,
The seed of *Saturne*, and faire *Nais*, *Chiron* hight.

And after him, came next the chill *December*:
Yet he through merry feasting which he made,
And great bonfires, did not the cold remember;
His Sauiours birth his mind so much did glad:
Vpon a shaggy-bearded Goat he rode,
The same wherewith *Dan Ioue* in tender yeares,
They say, was nourisht by th'*Idæan* mayd;
And in his hand a broad deepe boawle he beares; 260
Of which, he freely drinks an health to all his peeres.

Then came old *Ianuary*, wrapped well
In many weeds to keep the cold away;
Yet did he quake and quiuer like to quell,
And blowe his nayles to warme them if he may:
For, they were numbd with holding all the day
An hatchet keene, with which he felled wood,
And from the trees did lop the needlesse spray:
Vpon an huge great Earth-pot steane he stood;
From whose wide mouth, there flowed forth the Romane floud. 270

236 noule] head
totty of the must] dizzy from the new wine
237 wine-fats see] sea of the wine-vats
238 gust] taste
248 breem] cold
264 like to quell] as if he were dying
269 steane] urn

And lastly, came cold *February*, sitting
In an old wagon, for he could not ride;
Drawne of two fishes for the season fitting,
Which through the flood before did softly slyde
And swim away: yet had he by his side
His plough and harnesse fit to till the ground,
And tooles to prune the trees, before the pride
Of hasting Prime did make them burgein round:
So past the twelue Months forth, and their dew places found.

And after these, there came the *Day*, and *Night*, 280
Riding together both with equall pase,
Th'one on a Palfrey blacke, the other white;
But *Night* had couered her vncomely face
With a blacke veile, and held in hand a mace,
On top whereof the moon and stars were pight,
And sleep and darknesse round about did trace:
But *Day* did beare, vpon his scepters hight,
The goodly Sun, encompast all with beames bright.

Then came the *Howres*, faire daughters of high *Ioue*,
And timely *Night*, the which were all endewed 290
With wondrous beauty fit to kindle loue;
But they were Virgins all, and loue eschewed,
That might forslack the charge to them fore-shewed
By mighty *Ioue*; who did them Porters make
Of heauens gate (whence all the gods issued)
Which they did dayly watch, and nightly wake
By euen turnes, ne euer did their charge forsake.

And after all came *Life*, and lastly *Death*;
Death with most grim and griesly visage seene,
Yet is he nought but parting of the breath; 300
Ne ought to see, but like a shade to weene,
Vnbodied, vnsoul'd, vnheard, vnseene.
But *Life* was like a faire young lusty boy,
Such as they faine *Dan Cupid* to haue beene,
Full of delightfull health and liuely ioy,
Deckt all with flowres, and wings of gold fit to employ.

278 Prime] spring
281 with equall pase] abreast
286 trace] dance
293 forslack] neglect

When these were past, thus gan the *Titanesse*;
Lo, mighty mother, now be iudge and say,
Whether in all thy creatures more or lesse
CHANGE doth not raign and beare the greatest sway: 310
For, who sees not, that *Time* on all doth pray?
But *Times* do change and moue continually.
So nothing here long standeth in one stay:
Wherefore, this lower world who can deny
But to be subiect still to *Mutabilitie*?

(Pub. 1609)

141 [A *Faerie Queene* Miscellany]

(i)

He making speedy way through spersed ayre,
And through the world of waters wide and deepe,
To *Morpheus* house doth hastily repaire.
Amid the bowels of the earth full steepe,
And low, where dawning day doth neuer peepe,
His dwelling is; there *Tethys* his wet bed
Doth euer wash, and *Cynthia* still doth steepe
In siluer deaw his euer-drouping hed,
Whiles sad Night ouer him her mantle black doth spred.

Whose double gates he findeth locked fast, 10
The one faire fram'd of burnisht Yuory,
The other all with siluer ouercast;
And wakefull dogges before them farre do lye,
Watching to banish Care their enimy,
Who oft is wont to trouble gentle Sleepe.
By them the Sprite doth passe in quietly,
And vnto *Morpheus* comes, whom drowned deepe
In drowsie fit he findes: of nothing he takes keepe.

And more, to lulle him in his slumber soft,
A trickling streame from high rocke tumbling downe 20
And euer-drizling raine vpon the loft,
Mixt with a murmuring winde, much like the sowne
Of swarming Bees, did cast him in a swowne:
No other noyse, nor peoples troublous cryes,
As still are wont t'annoy the walled towne,
Might there be heard: but carelesse Quiet lyes,
Wrapt in eternall silence farre from enemyes.

(1590)

1 spersed] dispersed, thin 18 keepe] heed

(ii)

By this the Northerne wagoner had set
His seuenfold teme behind the stedfast starre,
That was in Ocean waues yet neuer wet, 30
But firme is fixt, and sendeth light from farre
To all, that in the wide deepe wandring arre:
And chearefull Chaunticlere with his note shrill
Had warned once, that *Phœbus* fiery carre
In hast was climbing vp the Easterne hill,
Full enuious that night so long his roome did fill.

(1590)

(iii)

The noble hart, that harbours vertuous thought,
And is with child of glorious great intent,
Can neuer rest, vntill it forth haue brought
Th'eternall brood of glorie excellent: 40
Such restlesse passion did all night torment
The flaming corage of that Faery knight,
Deuizing, how that doughtie turnament
With greatest honour he atchieuen might;
Still did he wake, and still did watch for dawning light.

At last the golden Orientall gate
Of greatest heauen gan to open faire,
And *Phœbus* fresh, as bridegrome to his mate,
Came dauncing forth, shaking his deawie haire:
And hurld his glistring beames through gloomy aire. 50
Which when the wakeful Elfe perceiu'd, streight way
He started vp, and did him selfe prepaire,
In sun-bright armes, and battailous array:
For with that Pagan proud he combat will that day.

(1590)

(iv)

Right well I wote most mighty Soueraine,
That all this famous antique history,
Of some th'aboundance of an idle braine
Will iudged be, and painted forgery,
Rather then matter of iust memory,
Sith none, that breatheth liuing aire, does know, 60
Where is that happy land of Faery,
Which I so much do vaunt, yet no where show,
But vouch antiquities, which no body can know.

But let that man with better sence aduize,
  That of the world least part to vs is red:
  And dayly how through hardy enterprize,
  Many great Regions are discouered,
  Which to late age were neuer mentioned.
  Who euer heard of th'Indian *Peru*?
  Or who in venturous vessell measured   70
  The *Amazons* huge riuer now found trew?
Or fruitfullest *Virginia* who did euer vew?

Yet all these were, when no man did them know;
  Yet haue from wisest ages hidden beene:
  And later times things more vnknowne shall show.
  Why then should witlesse man so much misweene
  That nothing is, but that which he hath seene?
  What if within the Moones faire shining spheare?
  What if in euery other starre vnseene
  Of other worldes he happily should heare?   80
He wonder would much more: yet such to some appeare.

Of Faerie lond yet if he more inquire,
  By certaine signes here set in sundry place
  He may it find; ne let him then admire,
  But yield his sence to be too blunt and bace,
  That no'te without an hound fine footing trace.
  And thou, O fairest Princesse vnder sky,
  In this faire mirrhour maist behold thy face,
  And thine owne realmes in lond of Faery,
And in this antique Image thy great auncestry.   90

(1590)

(v)

And is there care in heauen? and is there loue
  In heauenly spirits to these creatures bace,
  That may compassion of their euils moue?
  There is: else much more wretched were the cace
  Of men, then beasts. But O th'exceeding grace
  Of highest God, that loues his creatures so,
  And all his workes with mercy doth embrace,
  That blessed Angels, he sends to and fro,
To serue to wicked man, to serue his wicked foe.

65 red] made known   76 misweene] misbelieve   84 admire] wonder

How oft do they, their siluer bowers leaue, 100
To come to succour vs, that succour want?
How oft do they with golden pineons, cleaue
The flitting skyes, like flying Pursuiuant,
Against foule feends to aide vs millitant?
They for vs fight, they watch and dewly ward,
And their bright Squadrons round about vs plant,
And all for loue, and nothing for reward:
O why should heauenly God to men haue such regard?

(1590)

(vi)

Nought vnder heauen so strongly doth allure
The sence of man, and all his minde possesse, 110
As beauties louely baite, that doth procure
Great warriours oft their rigour to represse,
And mighty hands forget their manlinesse;
Drawne with the powre of an heart-robbing eye,
And wrapt in fetters of a golden tresse,
That can with melting pleasaunce mollifye
Their hardned hearts, enur'd to bloud and cruelty.

So whylome learnd that mighty Iewish swaine,
Each of whose lockes did match a man in might,
To lay his spoiles before his lemans traine: 120
So also did that great Oetean Knight
For his loues sake his Lions skin vndight:
And so did warlike *Antony* neglect
The worlds whole rule for *Cleopatras* sight.
Such wondrous powre hath wemens faire aspect,
To captiue men, and make them all the world reiect.

(1596)

(vii)

When I bethinke me on that speech whyleare,
Of *Mutability*, and well it way:
Me seemes, that though she all vnworthy were
Of the Heav'ns Rule; yet very sooth to say, 130
In all things else she beares the greatest sway.
Which makes me loath this state of life so tickle,
And loue of things so vaine to cast away;
Whose flowring pride, so fading and so fickle,
Short *Time* shall soon cut down with his consuming sickle.

118 whylome] formerly
120 lemans] beloved's, mistress's
127 whyleare] just past
132 tickle] insecure

Then gin I thinke on that which Nature sayd,
Of that same time when no more *Change* shall be,
But stedfast rest of all things firmely stayd
Vpon the pillours of Eternity,
That is contrayr to *Mutabilitie*: 140
For, all that moueth, doth in *Change* delight:
But thence-forth all shall rest eternally
With Him that is the God of Sabbaoth hight:
O that great Sabbaoth God, graunt me that Sabaoths sight.

(1609)

from *Amoretti*

142 NEW year, forth looking out of Janus' gate,
Doth seem to promise hope of new delight:
And, bidding th'old adieu, his passed date
Bids all old thoughts to die in dumpish sprite;
And, calling forth out of sad winter's night
Fresh Love, that long hath slept in cheerless bower,
Wills him awake, and soon about him dight
His wanton wings and darts of deadly power.
For lusty spring now in his timely hour
Is ready to come forth, him to receive; 10
And warns the earth with divers-coloured flower
To deck herself, and her fair mantle weave.
Then you, fair flower, in whom fresh youth doth reign,
Prepare yourself new love to entertain.

143 MOST glorious Lord of life, that on this day
Didst make thy triumph over death and sin;
And, having harrowed hell, didst bring away
Captivity thence captive, us to win:
This joyous day, dear Lord, with joy begin;
And grant that we, for whom thou diddest die,
Being with thy dear blood clean washed from sin,
May live for ever in felicity.
And that thy love we weighing worthily,
May likewise love thee for the same again; 10
And for thy sake, that all like dear didst buy,
With love may one another entertain.

143 hight] called

142
3 date] end of a period of time

7 dight] fit, furnish

So let us love, dear love, like as we ought:
Love is the lesson which the Lord us taught.

144

ONE day I wrote her name upon the strand,
But came the waves, and washed it away.
Again I wrote it with a second hand;
But came the tide, and made my pains his prey.
'Vain man', said she, 'that dost in vain assay
A mortal thing so to immortalise;
For I myself shall like to this decay,
And eke my name be wiped out likewise.'
'Not so', quoth I; 'let baser things devise
To die in dust, but you shall live by fame: 10
My verse your virtues rare shall eternise,
And in the heavens write your glorious name,
Where, whenas death shall all the world subdue,
Our love shall live, and later life renew.'

145

LACKING my love, I go from place to place,
Like a young fawn that late hath lost the hind;
And seek eachwhere, where last I saw her face,
Whose image yet I carry fresh in mind.
I seek the fields with her late footing signed;
I seek her bower with her late presence decked;
Yet nor in field nor bower I her can find,
Yet field and bower are full of her aspect.
But when mine eyes I thereunto direct,
They idly back return to me again; 10
And when I hope to see their true object,
I find myself but fed with fancies vain.
Cease then, mine eyes, to seek herself to see;
And let my thoughts behold herself in me.

(1595)

146

Epithalamion

YE learned sisters which have oftentimes
Been to me aiding, others to adorn,
Whom ye thought worthy of your graceful rimes,
That even the greatest did not greatly scorn
To hear their names sung in your simple lays,
But joyed in their praise;

And when ye list your own mishaps to mourn,
Which death, or love, or fortune's wreck did raise,
Your string could soon to sadder tenour turn,
And teach the woods and waters to lament 10
Your doleful dreariment;
Now lay those sorrowful complaints aside,
And having all your heads with garland crowned,
Help me mine own love's praises to resound,
Ne let the same of any be envied:
So Orpheus did for his own bride,
So I unto myself alone with sing;
The woods shall to me answer and my echo ring.

Early before the world's light-giving lamp
His golden beam upon the hills doth spread, 20
Having dispersed the night's uncheerful damp,
Do ye awake, and with fresh lustihead
Go to the bower of my beloved love,
My truest turtle dove;
Bid her awake; for Hymen is awake,
And long since ready forth his mask to move,
With his bright tead that flames with many a flake,
And many a bachelor to wait on him,
In their fresh garments trim.
Bid her awake therefore and soon her dight, 30
For lo, the wished day is come at last,
That shall for all the pains and sorrows past
Pay to her usury of long delight;
And whilst she doth her dight,
Do ye to her of joy and solace sing,
That all the woods may answer and your echo ring.

Bring with you all the nymphs that you can hear
Both of the rivers and the forests green,
And of the sea that neighbours to her near,
All with gay garlands goodly well beseen. 40
And let them also with them bring in hand
Another gay garland
For my fair love of lilies and of roses,
Bound true-love-wise with a blue silk ribband.
And let them make great store of bridal posies,
And let them eke bring store of other flowers
To deck the bridal bowers.

27 tead] torch 30 dight] dress

And let the ground whereas her foot shall tread,
For fear the stones her tender foot should wrong,
Be strewed with fragrant flowers all along, 50
And diapered like the discoloured mead.
Which done, do at her chamber door await,
For she will waken straight,
The whiles do ye this song unto her sing;
The woods shall to you answer and your echo ring.

Ye nymphs of Mulla, which with careful heed
The silver scaly trouts do tend full well,
And greedy pikes which use therein to feed,
(Those trouts and pikes all others do excel)
And ye likewise which keep the rushy lake, 60
Where none do fishes take,
Bind up the locks the which hang scattered light,
And in his waters, which your mirror make,
Behold your faces as the crystal bright,
That when you come whereas my love doth lie,
No blemish she may spy.
And eke ye lightfoot maids which keep the deer,
That on the hoary mountain use to tower,
And the wild wolves, which seek them to devour,
With your steel darts do chase from coming near, 70
Be also present here,
To help to deck her and to help to sing,
That all the woods may answer and your echo ring.

Wake now, my love, awake; for it is time.
The rosy morn long since left Tithones bed,
All ready to her silver coach to climb,
And Phoebus gins to shew his glorious head.
Hark how the cheerful birds do chant their lays
And carol of love's praise.
The merry lark her matins sings aloft, 80
The thrush replies, the mavis descant plays,
The ouzel shrills, the ruddock warbles soft,
So goodly all agree with sweet consent,
To this day's merriment.
Ah, my dear love, why do ye sleep thus long,
When meeter were that ye should now awake,
T' await the coming of your joyous make,
And harken to the birds' love-learned song,
The dewy leaves among?

81 mavis] song-thrush ruddock] robin 87 make] mate
82 ouzel] blackbird

For they of joy and pleasance to you sing, 90
That all the woods them answer and their echo ring.

My love is now awake out of her dream,
And her fair eyes, like stars that dimmed were
With darksome cloud, now shew their goodly beams
More bright than Hesperus his head doth rear.
Come now ye damsels, daughters of delight,
Help quickly her to dight,
But first come ye, fair hours, which were begot
In Jove's sweet paradise, of day and night,
Which do the seasons of the year allot, 100
And all that ever in this world is fair
Do make and still repair.
And ye three handmaids of the Cyprian queen,
The which do still adorn her beauty's pride,
Help to adorn my beautifullest bride;
And as ye her array, still throw between
Some graces to be seen,
And as ye use to Venus, to her sing,
The whiles the woods shall answer and your echo ring.

Now is my love all ready forth to come; 110
Let all the virgins therefore well await,
And ye fresh boys that tend upon her groom
Prepare yourselves; for he is coming straight.
Set all your things in seemly good array
Fit for so joyful day,
The joyful'st day that ever sun did see.
Fair sun, shew forth thy favourable ray,
And let thy lifeful heat not fervent be,
For fear of burning her sunshiny face,
Her beauty to disgrace. 120
O fairest Phoebus, father of the Muse,
If ever I did honour thee aright,
Or sing the thing, that mote thy mind delight,
Do not thy servant's simple boon refuse,
But let this day, let this one day, be mine,
Let all the rest be thine.
Then I thy sovereign praises loud will sing,
That all the woods shall answer and their echo ring.

Hark how the minstrels gin to shrill aloud
Their merry music that resounds from far, 130
The pipe, the tabor, and the trembling crowd,
That well agree withouten breach or jar.

120 disgrace] mar the grace of 131 crowd] fiddle

But most of all the damsels do delight,
When they their timbrels smite,
And thereunto do dance and carol sweet,
That all the senses they do ravish quite,
The whiles the boys run up and down the street,
Crying aloud with strong confused noise,
As if it were one voice.
Hymen, Io Hymen, Hymen, they do shout, 140
That even to the heavens their shouting shrill
Doth reach, and all the firmament doth fill,
To which the people standing all about,
As in approvance do thereto applaud
And loud advance her laud,
And evermore they Hymen, Hymen, sing,
That all the woods them answer and their echo ring.

Lo, where she comes along with portly pace
Like Phoebe from her chamber of the east,
Arising forth to run her mighty race, 150
Clad all in white, that seems a virgin best.
So well it her beseems that ye would ween
Some angel she had been.
Her long loose yellow locks like golden wire,
Sprinkled with pearl, and pearling flowers a-tween,
Do like a golden mantle her attire,
And being crowned with a garland green,
Seem like some maiden queen.
Her modest eyes abashed to behold
So many gazers, as on her do stare, 160
Upon the lowly ground affixed are.
Ne dare lift up her countenance too bold,
But blush to hear her praises sung so loud,
So far from being proud.
Nathless do ye still loud her praises sing,
That all the woods may answer and your echo ring.

Tell me, ye merchants' daughters, did ye see
So fair a creature in your town before,
So sweet, so lovely, and so mild as she,
Adorned with beauty's grace and virtue's store? 170
Her goodly eyes like sapphires shining bright,
Her forehead ivory white,
Her cheeks like apples which the sun hath rudded,
Her lips like cherries charming men to bite,
Her breast like to a bowl of cream uncrudded,
Her paps like lilies budded,

Her snowy neck like to a marble tower,
And all her body like a palace fair,
Ascending up with many a stately stair,
To honour's seat and chastity's sweet bower. 180
Why stand ye still, ye virgins, in amaze,
Upon her so to gaze,
Whiles ye forget your former lay to sing,
To which the woods did answer and your echo ring?

But if ye saw that which no eyes can see,
The inward beauty of her lively spright,
Garnished with heavenly gifts of high degree,
Much more then would ye wonder at that sight,
And stand astonished like to those which read
Medusa's mazeful head. 190
There dwells sweet love and constant chastity,
Unspotted faith and comely womanhead,
Regard of honour and mild modesty;
There virtue reigns as queen in royal throne,
And giveth laws alone,
The which the base affections do obey,
And yield their services unto her will;
Ne thought of thing uncomely ever may
Thereto approach to tempt her mind to ill.
Had ye once seen these her celestial treasures, 200
And unrevealed pleasures,
Then would ye wonder and her praises sing,
That all the woods should answer and your echo ring.

Open the temple gates unto my love,
Open them wide that she may enter in,
And all the posts adorn as doth behove,
And all the pillars deck with garlands trim,
For to receive this saint with honour due,
That cometh in to you.
With trembling steps and humble reverence, 210
She cometh in, before th'Almighty's view.
Of her ye virgins learn obedience,
When so ye come into those holy places,
To humble your proud faces:
Bring her up to th'high altar, that she may
The sacred ceremonies there partake,
The which do endless matrimony make,
And let the roaring organs loudly play

189 read] see 190 mazeful] stupefying

The praises of the Lord in lively notes,
The whiles with hollow throats   220
The choristers the joyous anthem sing,
That all the woods may answer and their echo ring.

Behold whiles she before the altar stands,
Hearing the holy priest that to her speaks
And blesseth her with his two happy hands,
How the red roses flush up in her cheeks,
And the pure snow with goodly vermeil stain,
Like crimson dyed in grain,
That even th'angels which continually
About the sacred altar do remain,   230
Forget their service and about her fly,
Oft peeping in her face that seems more fair,
The more they on it stare.
But her sad eyes, still fastened on the ground,
Are governed with goodly modesty,
That suffers not one look to glance awry,
Which may let in a little thought unsound.
Why blush ye, love, to give to me your hand,
The pledge of all our band?
Sing, ye sweet angels, Alleluia sing,   240
That all the woods may answer and your echo ring.

Now all is done; bring home the bride again,
Bring home the triumph of our victory,
Bring home with you the glory of her gain,
With joyance bring her and with jollity.
Never had man more joyful day than this,
Whom heaven would heap with bliss.
Make feast therefore now all this live-long day,
This day for ever to me holy is;
Pour out the wine without restraint or stay,   250
Pour not by cups, but by the bellyful,
Pour out to all that wull,
And sprinkle all the posts and walls with wine,
That they may sweat, and drunken be withal.
Crown ye god Bacchus with a coronal,
And Hymen also crown with wreaths of vine,
And let the Graces dance unto the rest;
For they can do it best:
The whiles the maidens do their carol sing,
To which the woods shall answer and their echo ring.   260

234 sad] steadfast   239 band] bond

Ring ye the bells, ye young men of the town,
And leave your wonted labours for this day:
This day is holy; do ye write it down,
That ye for ever it remember may.
This day the sun is in his chiefest height,
With Barnaby the bright,
From whence declining daily by degrees,
He somewhat loseth of his heat and light,
When once the Crab behind his back he sees.
But for this time it ill ordained was, 270
To choose the longest day in all the year,
And shortest night, when longest fitter were:
Yet never day so long, but late would pass.
Ring ye the bells, to make it wear away,
And bonfires make all day,
And dance about them, and about them sing,
That all the woods may answer, and your echo ring.

Ah, when will this long weary day have end,
And lend me leave to come unto my love?
How slowly do the hours their numbers spend! 280
How slowly does sad Time his feathers move!
Haste thee, O fairest planet, to thy home
Within the western foam;
Thy tired steeds long since have need of rest.
Long though it be, at last I see it gloom,
And the bright evening star with golden crest
Appear out of the east.
Fair child of beauty, glorious lamp of love,
That all the host of heaven in ranks dost lead,
And guidest lovers through the nightes dread, 290
How cheerfully thou lookest from above,
And seem'st to laugh atween thy twinkling light,
As joying in the sight
Of these glad many which for joy do sing,
That all the woods them answer and their echo ring.

Now cease, ye damsels, your delights forepast;
Enough is it, that all the day was yours.
Now day is done, and night is nighing fast;
Now bring the bride into the bridal bowers.
Now night is come, now soon her disarray, 300
And in her bed her lay;
Lay her in lilies and in violets,
And silken curtains over her display,
And odoured sheets, and Arras coverlets.

Behold how goodly my fair love does lie
In proud humility;
Like unto Maia, when as Jove her took,
In Tempe, lying on the flowery grass,
'Twixt sleep and wake, after she weary was,
With bathing in the Acidalian brook. 310
Now it is night, ye damsels may be gone,
And leave my love alone,
And leave likewise your former lay to sing;
The woods no more shall answer, nor your echo ring.

Now welcome, night, thou night so long expected,
That long day's labour dost at last defray,
And all my cares, which cruel love collected,
Hast summed in one, and cancelled for aye:
Spread thy broad wing over my love and me,
That no man may us see, 320
And in thy sable mantle us enwrap,
From fear of peril and foul horror free.
Let no false treason seek us to entrap,
Nor any dread disquiet once annoy
The safety of our joy:
But let the night be calm and quietsome,
Without tempestuous storms or sad affray;
Like as when Jove with fair Alcmena lay,
When he begot the great Tirynthian groom;
Or like as when he with thyself did lie, 330
And begot majesty.
And let the maids and young men cease to sing;
Ne let the woods them answer, nor their echo ring.

Let no lamenting cries, nor doleful tears,
Be heard all night within nor yet without;
Ne let false whispers, breeding hidden fears,
Break gentle sleep with misconceived doubt.
Let no deluding dreams nor dreadful sights
Make sudden sad affrights;
Ne let housefires, nor lightning's helpless harms, 340
Ne let the Puck, nor other evil sprights,
Ne let mischievous witches with their charms,
Ne let hobgoblins, names whose sense we see not,
Fray us with things that be not.
Let not the screech owl, nor the stork be heard;
Nor the night raven that still deadly yells,
Nor damned ghosts called up with mighty spells,

Nor grisly vultures make us once affeared:
Ne let th'unpleasant quire of frogs still croaking
Make us to wish they're choking. 350
Let none of these their dreary accents sing;
Ne let the woods them answer, nor their echo ring.

But let still silence true night watches keep,
That sacred peace may in assurance reign,
And timely sleep, when it is time to sleep,
May pour his limbs forth on your pleasant plain,
The whiles an hundred little winged loves,
Like divers feathered doves,
Shall fly and flutter round about your bed,
And in the secret dark, that none reproves, 360
Their pretty stealths shall work, and snares shall spread
To filch away sweet snatches of delight,
Concealed through covert night.
Ye sons of Venus, play your sports at will,
For greedy pleasure, careless of your toys,
Thinks more upon her paradise of joys,
Than what ye do, albeit good or ill.
All night therefore attend your merry play,
For it will soon be day:
Now none doth hinder you, that say or sing; 370
Ne will the woods now answer, nor your echo ring.

Who is the same, which at my window peeps,
Or whose is that fair face, that shines so bright?
Is it not Cynthia, she that never sleeps,
But walks about high heaven all the night?
O fairest goddess, do thou not envy
My love with me to spy;
For thou likewise didst love, though now unthought,
And for a fleece of wool, which privily
The Latmian shepherd once unto thee brought, 380
His pleasures with thee wrought.
Therefore to us be favourable now;
And sith of women's labours thou hast charge,
And generation goodly dost enlarge,
Incline thy will t'effect our wishful vow,
And the chaste womb inform with timely seed,
That may our comfort breed:
Till which we cease our hopeful hap to sing;
Ne let the woods us answer, nor our echo ring.

And thou great Juno, which with awful might 390
The laws of wedlock still dost patronize,
And the religion of the faith first plight
With sacred rites hast taught to solemnize,
And eke for comfort often called art
Of women in their smart,
Eternally bind thou this lovely band,
And all thy blessings unto us impart.
And thou glad Genius, in whose gentle hand
The bridal bower and genial bed remain,
Without blemish or stain, 400
And the sweet pleasures of their love's delight
With secret aid dost succour and supply,
Till they bring forth the fruitful progeny,
Send us the timely fruit of this same night.
And thou fair Hebe, and thou Hymen free,
Grant that it may so be.
Till which we cease your further praise to sing;
Ne any woods shall answer, nor your echo ring.

And ye high heavens, the temple of the gods,
In which a thousand torches flaming bright 410
Do burn, that to us wretched earthly clods
In dreadful darkness lend desired light;
And all ye powers which in the same remain,
More than we men can feign,
Pour out your blessing on us plenteously,
And happy influence upon us rain,
That we may raise a large posterity,
Which from the earth, which they may long possess,
With lasting happiness,
Up to your haughty palaces may mount, 420
And for the guerdon of their glorious merit
May heavenly tabernacles there inherit,
Of blessed saints for to increase the count.
So let us rest, sweet love, in hope of this,
And cease till then our timely joys to sing;
The woods no more us answer, nor our echo ring.

Song, made in lieu of many ornaments,
With which my love should duly have been decked,
Which cutting off through hasty accidents,
Ye would not stay your due time to expect, 430
But promised both to recompense,
Be unto her a goodly ornament,
And for short time an endless monument.

(1595)

147 *Prothalamion*

CALM was the day, and through the trembling air
Sweet breathing Zephyrus did softly play,
A gentle spirit, that lightly did delay
Hot Titan's beams, which then did glister fair;
When I whose sullen care,
Through discontent of my long fruitless stay
In prince's court, and expectation vain
Of idle hopes, which still do fly away
Like empty shadows, did afflict my brain,
Walked forth to ease my pain 10
Along the shore of silver streaming Thames,
Whose rutty bank, the which his river hems,
Was painted all with variable flowers,
And all the meads adorned with dainty gems,
Fit to deck maidens' bowers,
And crown their paramours,
Against the bridal day, which is not long:
Sweet Thames, run softly, till I end my song.

There, in a meadow, by the river's side,
A flock of nymphs I chanced to espy, 20
All lovely daughters of the flood thereby,
With goodly greenish locks all loose untied,
As each had been a bride;
And each one had a little wicker basket,
Made of fine twigs entrailed curiously,
In which they gathered flowers to fill their flasket,
And with fine fingers cropped full featously
The tender stalks on high.
Of every sort, which in that meadow grew,
They gathered some; the violet pallid blue, 30
The little daisy, that at evening closes,
The virgin lily, and the primrose true,
With store of vermeil roses,
To deck their bridegrooms' posies,
Against the bridal day, which was not long:
Sweet Thames, run softly, till I end my song.

With that, I saw two swans of goodly hue
Come softly swimming down along the Lee;
Two fairer birds I yet did never see.

3 delay] allay, mitigate
26 flasket] long shallow basket
27 featously] nimbly, dexterously

The snow, which doth the top of Pindus strew, 40
Did never whiter shew,
Nor Jove himself, when he a swan would be
For love of Leda, whiter did appear:
Yet Leda was they say as white as he,
Yet not so white as these, nor nothing near.
So purely white they were,
That even the gentle stream, the which them bare,
Seemed foul to them, and bade his billows spare
To wet their silken feathers, lest they might
Soil their fair plumes with water not so fair, 50
And mar their beauties bright,
That shone as heaven's light,
Against their bridal day, which was not long:
  Sweet Thames, run softly, till I end my song.

Eftsoons the nymphs, which now had flowers their fill,
Ran all in haste, to see that silver brood,
As they came floating on the crystal flood.
Whom when they saw, they stood amazed still,
Their wondering eyes to fill.
Them seemed they never saw a sight so fair, 60
Of fowls so lovely, that they sure did deem
Them heavenly born, or to be that same pair
Which through the sky draw Venus' silver team;
For sure they did not seem
To be begot of any earthly seed,
But rather angels or of angels' breed:
Yet were they bred of Somers-heat they say,
In sweetest season, when each flower and weed
The earth did fresh array,
So fresh they seemed as day, 70
Even as their bridal day, which was not long:
  Sweet Thames, run softly, till I end my song.

Then forth they all out of their baskets drew
Great store of flowers, the honour of the field,
That to the sense did fragrant odours yield,
All which upon those goodly birds they threw,
And all the waves did strew,
That like old Peneus' waters they did seem,
When down along by pleasant Tempe's shore,
Scattered with flowers, through Thessaly they stream, 80
That they appear through lilies' plenteous store,
Like a bride's chamber floor.

54 Eftsoons] soon afterwards

67 Somers-heat] summer's heat = Somerset

Two of those nymphs meanwhile, two garlands bound,
Of freshest flowers which in that mead they found,
The which presenting all in trim array,
Their snowy foreheads therewithal they crowned,
Whilst one did sing this lay,
Prepared against that day,
Against their bridal day, which was not long:
  Sweet Thames, run softly, till I end my song. 90

'Ye gentle birds, the world's fair ornament,
And heaven's glory, whom this happy hour
Doth lead unto your lovers' blissful bower,
Joy may you have and gentle heart's content
Of your love's couplement:
And let fair Venus, that is queen of love,
With her heart-quelling son upon you smile,
Whose smile, they say, hath virtue to remove
All love's dislike, and friendship's faulty guile
For ever to assoil. 100
Let endless peace your steadfast hearts accord,
And blessed plenty wait upon your board,
And let your bed with pleasures chaste abound,
That fruitful issue may to you afford,
Which may your foes confound,
And make your joys redound,
Upon your bridal day, which is not long:
  Sweet Thames, run softly, till I end my song.'

So ended she; and all the rest around
To her redoubled that her undersong, 110
Which said, their bridal day should not be long.
And gentle echo from the neighbour ground
Their accents did resound.
So forth those joyous birds did pass along,
Adown the Lee, that to them murmured low,
As he would speak, but that he lacked a tongue,
Yet did by signs his glad affection show,
Making his stream run slow.
And all the fowl which in his flood did dwell
Gan flock about these twain, that did excel 120
The rest so far as Cynthia doth shend
The lesser stars. So they, enranged well,
Did on those two attend,
And their best service lend,
Against their wedding day, which was not long:
  Sweet Thames, run softly, till I end my song.

100 assoil] purge 121 shend] put to shame

At length they all to merry London came,
To merry London, my most kindly nurse,
That to me gave this life's first native source;
Though from another place I take my name, 130
An house of ancient fame.
There when they came, whereas those bricky towers,
The which on Thames' broad aged back do ride,
Where now the studious lawyers have their bowers
There whilom wont the Templar Knights to bide,
Till they decayed through pride:
Next whereunto there stands a stately place,
Where oft I gained gifts and goodly grace
Of that great lord, which therein wont to dwell,
Whose want too well now feels my friendless case. 140
But ah, here fits not well
Old woes but joys to tell
Against the bridal day, which is not long:
Sweet Thames, run softly, till I end my song.

Yet therein now doth lodge a noble peer,
Great England's glory and the world's wide wonder,
Whose dreadful name late through all Spain did thunder,
And Hercules' two pillars standing near
Did make to quake and fear.
Fair branch of honour, flower of chivalry, 150
That fillest England with thy triumph's fame,
Joy have thou of thy noble victory,
And endless happiness of thine own name
That promiseth the same:
That through thy prowess and victorious arms,
Thy country may be freed from foreign harms;
And great Elisa's glorious name may ring
Through all the world, filled with thy wide alarms,
Which some brave Muse may sing
To ages following, 160
Upon the bridal day, which is not long:
Sweet Thames, run softly, till I end my song.

From those high towers this noble lord issuing,
Like radiant Hesper when his golden hair
In th'Ocean billows he hath bathed fair,
Descended to the river's open viewing,
With a great train ensuing.

135 whilom] formerly
139 great lord] Earl of Leicester
145 noble peer] Earl of Essex

Above the rest were goodly to be seen
Two gentle knights of lovely face and feature
Beseeming well the bower of any queen, 170
With gifts of wit and ornaments of nature,
Fit for so goodly stature;
That like the twins of Jove they seemed in sight,
Which deck the baldric of the heavens bright.
They two forth pacing to the river's side,
Received those two fair birds, their love's delight,
Which at th'appointed tide
Each one did make his bride,
Against their bridal day, which is not long:
Sweet Thames, run softly, till I end my song. 180

(1596)

SIR PHILIP SIDNEY
1554–1586

from *The Countess of Pembroke's Arcadia*

148

My sheep are thoughts, which I both guide and serve;
Their pasture is fair hills of fruitless love;
On barren sweets they feed, and feeding starve;
I wail their lot, but will not other prove.
My sheephook is wanhope, which all upholds;
My weeds, desire, cut out in endless folds.
What wool my sheep shall bear, while thus they live,
In you it is, you must the judgement give.

(Pub. 1590)

149

O sweet woods, the delight of solitariness,
O how much I do like your solitariness!
Where man's mind hath a freed consideration
Of goodness to receive lovely direction;
Where senses do behold th'order of heavenly host,
And wise thoughts do behold what the creator is.
Contemplation here holdeth his only seat,
Bounded with no limits, borne with a wing of hope,
Climbs even unto the stars; Nature is under it.
Nought disturbs thy quiet; all to thy service yield; 10

148
5 wanhope] despair

Each sight draws on a thought, thought mother of
  science;
Sweet birds kindly do grant harmony unto thee;
Fair trees' shade is enough fortification,
Nor danger to thyself if be not in thyself.

O sweet woods, the delight of solitariness,
O how much I do like your solitariness!
Here no treason is hid, veiled in innocence,
Nor envy's snaky eye finds any harbour here,
Nor flatterers' venomous insinuations,
Nor cunning humorists' puddled opinions, 20
Nor courteous ruin of proffered usury,
Nor time prattled away, cradle of ignorance,
Nor causeless duty, nor cumber of arrogance;
Nor trifling title of vanity dazzleth us,
Nor golden manacles stand for a paradise.
Here wrong's name is unheard; slander a monster is.
Keep thy sprite from abuse, here no abuse doth haunt.
What man grafts in a tree dissimulation?

O sweet woods, the delight of solitariness,
O how well I do like your solitariness! 30
Yet dear soil, if a soul closed in a mansion
As sweet as violets, fair as a lily is,
Straight as a cedar, a voice stains the canary birds,
Whose shade safety doth hold, danger avoideth her;
Such wisdom, that in her lives speculation;
Such goodness, that in her simplicity triumphs;
Where envy's snaky eye winketh or else dieth;
Slander wants a pretext, flattery gone beyond;
O, if such a one have bent to a lonely life
Her steps, glad we receive, glad we receive her eyes, 40
And think not she doth hurt our solitariness:
For such company decks such solitariness.

(1593)

150

My true love hath my heart, and I have his,
By just exchange one for the other given.
I hold his dear, and mine he cannot miss:
There never was a better bargain driven.

20 cunning humorists' puddled opinions] clever crackpots' muzzy notions
31 mansion] dwelling-place
33 a voice stains] with a voice that eclipses
37 winketh] is closed

His heart in me keeps me and him in one;
My heart in him his thoughts and senses guides;
He loves my heart, for once it was his own;
I cherish his, because in me it bides.
His heart his wound received from my sight;
My heart was wounded with his wounded heart; 10
For as from me on him his hurt did light,
So still, methought, in me his hurt did smart;
Both equal hurt, in this change sought our bliss:
My true love hath my heart, and I have his.

(1593)

151

WHY dost thou haste away,
O Titan fair, the giver of the day?
Is it to carry news
To Western wights, what stars in East appear?
Or dost thou think that here
Is left a sun, whose beams thy place may use?
Yet stay, and well peruse
What be her gifts, that make her equal thee;
Bend all thy light to see
In earthly clothes enclosed a heavenly spark. 10
Thy running course cannot such beauties mark;
No, no; thy motions be
Hastened from us with bar of shadow dark,
Because that thou, the author of our sight,
Disdain'st we see thee stained with other's light.

(1593)

152

Strephon. Ye goat-herd gods, that love the grassy mountains;
Ye nymphs, which haunt the springs in pleasant valleys;
Ye satyrs, joyed with free and quiet forests;
Vouchsafe your silent ears to plaining music,
Which to my woes gives still an early morning,
And draws the dolour on till weary evening.

Klaius. O Mercury, foregoer to the evening;
O heavenly huntress of the savage mountains;

151
4 wights] people
15 stained] eclipsed, made pale

152
1 goat-herd gods] the fauns
8 heavenly huntress] Diana

O lovely star, entitled of the morning;
While that my voice doth fill these woeful valleys, 10
Vouchsafe your silent ears to plaining music,
Which oft hath echo tired in secret forests.

Strephon. I, that was once free-burgess of the forests,
Where shade from sun and sport I sought in evening;
I, that was once esteemed for pleasant music,
Am banished now among the monstrous mountains
Of huge despair, and foul affliction's valleys;
Am grown a scrich-owl to myself each morning.

Klaius. I, that was once delighted every morning,
Hunting the wild inhabiters of forests; 20
I, that was once the music of these valleys.
So darkened am, that all my day is evening;
Heart-broken so, that molehills seem high mountains,
And fill the vales with cries instead of music.

Strephon. Long since, alas, my deadly swannish music
Hath made itself a crier of the morning,
And hath with wailing strength climbed highest mountains;
Long since my thoughts more desert be than forests;
Long since I see my joys come to their evening,
And state thrown down to over-trodden valleys. 30

Klaius. Long since the happy dwellers of these valleys
Have prayed me leave my strange exclaiming music,
Which troubles their day's work, and joys of evening;
Long since I hate the night, more hate the morning;
Long since my thoughts chase me like beasts in forests,
And make me wish myself laid under mountains.

Strephon. Me seems I see the high and stately mountains
Transform themselves to low dejected valleys;
Me seems I hear, in these ill-changed forests,
The nightingales do learn of owls their music; 40
Me seems I feel the comfort of the morning
Turned to the mortal serene of an evening.

9 lovely star] Venus
25 swannish] i.e. like a swan song (before death)
42 mortal serene] poisonous nocturnal dew

Klaius. Me seems I see a filthy cloudy evening
As soon as sun begins to climb the mountains;
Me seems I feel a noisome scent, the morning,
When I do smell the flowers of these valleys;
Me seems I hear, when I do hear sweet music,
The dreadful cries of murdered men in forests.

Strephon. I wish to fire the trees of all these forests;
I give the sun a last farewell each evening; 50
I curse the fiddling finders-out of music;
With envy I do hate the lofty mountains,
And with despite despise the humble valleys;
I do detest night, evening, day and morning.

Klaius. Curse to myself my prayer is, the morning;
My fire is more than can be made with forests;
My state more base than are the basest valleys;
I wish no evening more to see, each evening;
Shamed I hate myself in sight of mountains,
And stop mine ears, lest I grow mad with music. 60

Strephon. For she, whose parts maintained a perfect music,
Whose beauties shined more than the blushing morning;
Who much did pass in state the stately mountains,
In straightness passed the cedars of the forests,
Hath cast me, wretch, into eternal evening,
By taking her two suns from these dark valleys.

Klaius. For she, compared with whom the Alps are valleys;
She, whose least word brings from the spheres their music;
At whose approach the sun rose in the evening;
Who, where she went, bare in her forehead morning, 70
Is gone, is gone, from these our spoiled forests,
Turning to deserts our best pastured mountains.

Strephon. These mountains witness shall, so shall these valleys;

Klaius. These forests eke, made wretched by our music.
Our morning hymn is this, and song at evening.

(1590)

from *Certain Sonnets*

153 *Ring out your bells*

RING out your bells, let mourning shows be spread,
For love is dead:
All love is dead, infected
With plague of deep disdain,
Worth, as nought worth, rejected,
And faith fair scorn doth gain.
From so ungrateful fancy,
From such a female franzy,
From them that use men thus:
Good lord, deliver us. 10

Weep, neighbours, weep: do you not hear it said
That love is dead?
His death-bed peacock's folly,
His winding-sheet is shame,
His will false-seeming holy,
His sole executor blame.
From so ungrateful fancy,
From such a female franzy,
From them that use men thus:
Good lord, deliver us. 20

Let dirge be sung, and trentals rightly read,
For love is dead.
Sir wrong his tomb ordaineth,
My mistress' marble heart,
Which epitaph containeth:
'Her eyes were once his dart.'
From so ungrateful fancy,
From such a female franzy,
From them that use men thus:
Good lord, deliver us. 30

Alas, I lie: rage hath this error bred;
Love is not dead.
Love is not dead, but sleepeth
In her unmatched mind,
Where she his counsel keepeth
Till due desert she find.

21 trentals] masses for the dead

Therefore from so vile fancy,
To call such wit a franzy
Who love can temper thus:
Good lord, deliver us. 40

(Wr. 1591 or earlier; pub. 1598)

from *Astrophil and Stella*

154 LOVING in truth, and fain in verse my love to show,
That she (dear she) might take some pleasure of my pain;
Pleasure might cause her read, reading might make her know,
Knowledge might pity win, and pity grace obtain;
I sought fit words to paint the blackest face of woe,
Studying inventions fine, her wits to entertain;
Oft turning others' leaves, to see if thence would flow
Some fresh and fruitful showers upon my sunburnt brain.
But words came halting forth, wanting invention's stay;
Invention, Nature's child, fled step-dame study's blows; 10
And others' feet still seemed but strangers in my way.
Thus great with child to speak, and helpless in my throes,
Biting my truant pen, beating myself for spite,
'Fool,' said my muse to me, 'look in thy heart and write.'

155 LET dainty wits cry on the sisters nine,
That bravely masked, their fancies may be told;
Or Pindar's apes, flaunt they in phrases fine,
Enam'lling with pied flowers their thoughts of gold;
Or else let them in statelier glory shine,
Ennobling new-found tropes with problems old;
Or with strange similes enrich each line,
Of herbs or beasts, which Ind or Afric hold.
For me, in sooth, no muse but one I know;
Phrases and problems from my reach do grow, 10
And strange things cost too dear for my poor sprites.
How then? Even thus: in Stella's face I read
What love and beauty be; then all my deed
But copying is, what in her Nature writes.

156 IT is most true, that eyes are formed to serve
The inward light; and that the heavenly part
Ought to be king, from whose rules who do swerve,
Rebels to Nature, strive for their own smart.
It is most true what we call Cupid's dart
An image is which for ourselves we carve;
And, fools, adore in temple of our heart,
Till that good god make church and churchmen starve.
True, that true beauty virtue is indeed,
Whereof this beauty can be but a shade, 10
Which elements with mortal mixture breed;
True, that on earth we are but pilgrims made,
And should in soul up to our country move;
True; and yet true that I must Stella love.

157 SOME lovers speak, when they their muses entertain,
Of hopes begot by fear, of wot not what desires,
Of force of heavenly beams, infusing hellish pain,
Of living deaths, dear wounds, fair storms and freezing fires.
Some one his song in Jove, and Jove's strange tales, attires,
Broidered with bulls and swans, powdered with golden rain.
Another, humbler, wit to shepherd's pipe retires,
Yet hiding royal blood full oft in rural vein.
To some a sweetest plaint a sweetest style affords,
While tears pour out his ink, and sighs breathe out his words, 10
His paper, pale despair, and pain his pen doth move.
I can speak what I feel, and feel as much as they,
But think that all the map of my state I display,
When trembling voice brings forth, that I do Stella love.

158 ALAS, have I not pain enough, my friend,
Upon whose breast a fiercer gripe doth tire
Than did on him who first stale down the fire,
While Love on me doth all his quiver spend,
But with your rhubarb words you must contend

157
2 wot not what] I know not what
6 Broidered] embroidered
powdered] spangled

158
2 gripe] vulture
tire] prey
5 rhubarb] purgative

To grieve me worse, in saying that desire
Doth plunge my well-formed soul even in the mire
Of sinful thoughts, which do in ruin end?
If that be sin, which doth the manners frame,
Well stayed with truth in word, and faith of deed, 10
Ready of wit, and fearing nought but shame:
If that be sin, which in fixed hearts doth breed
A loathing of all loose unchastity:
Then love is sin, and let me sinful be.

159

YOU that do search for every purling spring
Which from the ribs of old Parnassus flows;
And every flower, not sweet perhaps, which grows
Near thereabouts, into your poesy wring;
You that do dictionary's method bring
Into your rhymes, running in rattling rows;
You that poor Petrarch's long-deceased woes
With new-born sighs and denizened wit do sing:
You take wrong ways, those far-fet helps be such
As do bewray a want of inward touch: 10
And sure at length stol'n goods do come to light.
But if, both for your love and skill, your name
You seek to nurse at fullest breasts of fame,
Stella behold, and then begin to endite.

160

WITH what sharp checks I in myself am shent
When into reason's audit I do go,
And by just counts myself a bankrupt know
Of all those goods, which heaven to me hath lent,
Unable quite to pay even nature's rent,
Which unto it by birthright I do owe:
And which is worse, no good excuse can show,
But that my wealth I have most idly spent.
My youth doth waste, my knowledge brings forth toys,
My wit doth strive those passions to defend 10
Which for reward spoil it with vain annoys.
I see my course to lose myself doth bend:
I see, and yet no greater sorrow take
Than that I lose no more for Stella's sake.

160 shent] shamed
1 checks] rebukes

161

On Cupid's bow how are my heart-strings bent,
That see my wrack, and yet embrace the same!
When most I glory, then I feel most shame:
I willing run, yet while I run, repent.
My best wits still their own disgrace invent;
My very ink turns straight to Stella's name;
And yet my words, as them my pen doth frame,
Avise themselves that they are vainly spent.
For though she pass all things, yet what is all
That unto me, who fare like him that both 10
Looks to the skies, and in a ditch doth fall?
O let me prop my mind, yet in his growth,
And not in nature for best fruits unfit.
'Scholar,' saith Love, 'bend hitherward your wit.'

162

Fly, fly, my friends, I have my death wound, fly;
See there that boy, that murd'ring boy I say,
Who like a thief hid in dark bush doth lie,
Till bloody bullet get him wrongful prey.
So tyrant he no fitter place could spy,
Nor so fair level in so secret stay
As that sweet black which veils the heavenly eye;
There himself with his shot he close doth lay.
Poor passenger, pass now thereby I did;
And stayed, pleased with the prospect of the place, 10
While that black hue from me the bad guest hid:
But straight I saw motions of lightning grace,
And then descried the glist'ring of his dart:
But ere I could fly thence, it pierced my heart.

163

Your words, my friend, right healthful caustics, blame
My young mind marred, whom love doth windlass so
That mine own writings like bad servants show,
My wits, quick in vain thoughts, in virtue lame;
That Plato I read for nought, but if he tame
Such coltish gyres; that to my birth I owe
Nobler desires, lest else that friendly foe,
Great expectation, wear a train of shame.

2 wrack] ruin
8 Avise] inform
9 pass] surpass

162
6 level] aim
7 black] pupil of the eye
9 passenger] passer-by

163
2 windlass] ensnare
6 coltish gyres] youthful gyrations

For since mad March great promise made of me,
If now the May of my years much decline, 10
What can be hoped my harvest time will be?
Sure you say well; your wisdom's golden mine
Dig deep with learning's spade; now tell me this,
Hath this world aught so fair as Stella is?

164

THE curious wits, seeing dull pensiveness
Bewray itself in my long settled eyes,
Whence these same fumes of melancholy rise
With idle pains, and missing aim, do guess.
Some, that know how my spring I did address,
Deem that my muse some fruit of knowledge plies;
Others, because the Prince my service tries,
Think that I think state errors to redress.
But harder judges judge ambition's rage,
Scourge of itself, still climbing slippery place, 10
Holds my young brain captived in golden cage.
O fools, or over-wise: alas, the race
Of all my thoughts hath neither stop nor start
But only Stella's eyes and Stella's heart.

165

BECAUSE I oft, in dark abstracted guise,
Seem most alone in greatest company,
With dearth of words, or answers quite awry,
To them that would make speech of speech arise,
They deem, and of that doom the rumour flies,
That poison foul of bubbling pride doth lie
So in my dwelling breast, that only I
Fawn on myself, and others do despise.
Yet pride, I think, doth not my soul possess,
Which looks too oft in his unflatt'ring glass; 10
But one worse fault, ambition, I confess,
That makes me oft my best friends overpass,
Unseen, unheard, while thought to highest place
Bends all his powers, even unto Stella's grace.

166

YOU that with allegory's curious frame
Of others' children changelings use to make,
With me those pains, for God's sake, do not take;
I list not dig so deep for brazen fame.

When I say 'Stella', I do mean the same
  Princess of beauty, for whose only sake
  The reins of love I love, though never slake,
And joy therein, though nations count it shame.
  I beg no subject to use eloquence,
Nor in hid ways to guide philosophy. 10
Look at my hands for no such quintessence,
But know that I, in pure simplicity,
  Breathe out the flames which burn within my heart,
  Love only reading unto me this art.

167 WHETHER the Turkish new moon minded be
  To fill his horns this year on Christian coast;
  How Pole's right king means, without leave of host,
To warm with ill-made fire cold Muscovy;
If French can yet three parts in one agree;
  What now the Dutch in their full diets boast;
  How Holland hearts, now so good towns be lost,
Trust in the pleasing shade of Orange tree;
  How Ulster likes of that same golden bit
Wherewith my father once made it half tame; 10
If in the Scottish court be welt'ring yet;
These questions busy wits to me do frame.
  I, cumbered with good manners, answer do,
  But know not how, for still I think of you.

168 WITH how sad steps, O moon, thou climb'st the skies;
  How silently, and with how wan a face.
  What, may it be that even in heav'nly place
That busy archer his sharp arrows tries?
Sure, if that long with love acquainted eyes
  Can judge of love, thou feel'st a lover's case;
  I read it in thy looks; thy languished grace
To me, that feel the like, thy state descries.
  Then even of fellowship, O moon, tell me,
Is constant love deemed there but want of wit? 10
Are beauties there as proud as here they be?
Do they above love to be loved, and yet
  Those lovers scorn whom that love doth possess?
  Do they call virtue there ungratefulness?

7 slake] slacken
167
3 Pole's] Poland's
4 Muscovy] Russia
6 the Dutch] the Germans
diets] councils
11 welt'ring] (political) turbulence

169 COME sleep, O sleep, the certain knot of peace,
The baiting place of wit, the balm of woe,
The poor man's wealth, the prisoner's release,
The indifferent judge between the high and low;
With shield of proof shield me from out the prease
Of those fierce darts despair at me doth throw:
O make in me those civil wars to cease;
I will good tribute pay, if thou do so.
Take thou of me sweet pillows, sweetest bed,
A chamber deaf to noise, and blind to light; 10
A rosy garland, and a weary head;
And if these things, as being thine by right,
Move not thy heavy grace, thou shalt in me,
Livelier than elsewhere, Stella's image see.

170 AS good to write, as for to lie and groan.
O Stella dear, how much thy power hath wrought,
That hast my mind, none of the basest, brought
My still kept course, while others sleep, to moan.
Alas, if from the height of virtue's throne
Thou canst vouchsafe the influence of a thought
Upon a wretch, that long thy grace hath sought;
Weigh then how I by thee am overthrown:
And then, think thus: although thy beauty be
Made manifest by such a victory, 10
Yet noblest conquerors do wrecks avoid.
Since then thou hast so far subdued me,
That in my heart I offer still to thee,
O, do not let thy temple be destroyed.

171 STELLA oft sees the very face of woe
Painted in my beclouded stormy face;
But cannot skill to pity my disgrace,
Not though thereof the cause herself she know;
Yet hearing late a fable, which did show
Of lovers never known a grievous case,
Pity thereof gat in her breast such place
That, from that sea derived, tears' springs did flow.

2 baiting place] resting place
4 indifferent] impartial
5 of proof] impregnable
prease] press, onslaught

171
3 cannot skill] is unable

Alas, if fancy drawn by imaged things,
Though false, yet with free scope more grace doth breed 10
Than servant's wrack, where new doubts honour brings;
Then think, my dear, that you in me do read
Of lover's ruin some sad tragedy:
I am not I, pity the tale of me.

172 In martial sports I had my cunning tried,
And yet to break more staves did me address,
While with the people's shouts, I must confess,
Youth, luck and praise even filled my veins with pride;
When Cupid, having me, his slave, descried
In Mars's livery, prancing in the press:
'What now, sir fool,' said he; 'I would no less,
Look here, I say.' I looked, and Stella spied,
Who hard by made a window send forth light.
My heart then quaked, then dazzled were mine eyes, 10
One hand forgot to rule, th'other to fight;
Nor trumpet's sound I heard, nor friendly cries;
My foe came on, and beat the air for me,
Till that her blush taught me my shame to see.

173 Because I breathe not love to every one,
Nor do not use set colours for to wear,
Nor nourish special locks of vowed hair,
Nor give each speech a full point of a groan,
The courtly nymphs, acquainted with the moan
Of them, who in their lips love's standard bear:
'What, he?' say they of me, 'now I dare swear,
He cannot love; no, no, let him alone.'
And think so still, so Stella know my mind.
Profess indeed I do not Cupid's art; 10
But you fair maids, at length this true shall find,
That his right badge is but worn in the heart;
Dumb swans, not chattering pies, do lovers prove;
They love indeed, who quake to say they love.

174 Who will in fairest book of nature know
How virtue may best lodged in beauty be,
Let him but learn of love to read in thee,
Stella, those fair lines which true goodness show.

172
2 staves] staffs used in the tilt or tournament
12 trumpet's sound] i.e. signal to start
173
13 pies] magpies

There shall he find all vices' overthrow,
  Not by rude force, but sweetest sovereignty
  Of reason, from whose light those night-birds fly,
That inward sun in thine eyes shineth so.
  And not content to be perfection's heir
Thy self, dost strive all minds that way to move, 10
Who mark in thee what is in thee most fair;
So while thy beauty draws the heart to love,
  As fast thy virtue bends that love to good.
  But ah, desire still cries: 'Give me some food.'

175

HAVE I caught my heavenly jewel
Teaching sleep most fair to be?
Now will I teach her that she,
When she wakes, is too too cruel.

Since sweet sleep her eyes hath charmed,
The two only darts of Love,
Now will I with that boy prove
Some play, while he is disarmed.

Her tongue waking still refuseth,
Giving frankly niggard 'no'; 10
Now will I attempt to know
What 'no' her tongue sleeping useth.

See, the hand which, waking, guardeth,
Sleeping, grants a free resort;
Now will I invade the fort;
Cowards Love with loss rewardeth.

But, O fool, think of the danger
Of her just and high disdain;
Now will I, alas, refrain;
Love fears nothing else but anger. 20

Yet those lips so sweetly swelling
Do invite a stealing kiss:
Now will I but venture this;
Who will read, must first learn spelling.

O sweet kiss—but ah, she is waking,
Louring beauty chastens me;
Now will I away hence flee;
Fool, more fool, for no more taking.

176

I NEVER drank of Aganippe well,
Nor ever did in shade of Tempe sit;
And muses scorn with vulgar brains to dwell;
Poor layman I, for sacred rites unfit.
  Some do I hear of poet's fury tell,
But (God wot) wot not what they mean by it;
And this I swear, by blackest brook of hell,
I am no pick-purse of another's wit.
  How falls it then, that with so smooth an ease
My thoughts I speak, and what I speak doth flow 10
In verse, and that my verse best wits doth please?
Guess we the cause: 'What, is it thus?' Fie, no;
  'Or so?' Much less. 'How then?' Sure, thus it is:
  My lips are sweet, inspired with Stella's kiss.

177

OF all the kings that ever here did reign,
Edward, named fourth, as first in praise I name;
Not for his fair outside, nor well lined brain,
Although less gifts imp feathers oft on fame;
  Nor that he could, young-wise, wise-valiant, frame
His sire's revenge, joined with a kingdom's gain;
And gained by Mars, could yet mad Mars so tame,
That balance weighed what sword did late obtain;
  Nor that he made the flower-de-luce so 'fraid,
Though strongly hedged, of bloody lion's paws, 10
That witty Lewis to him a tribute paid;
Nor this, nor that, nor any such small cause;
  But only for this worthy knight durst prove
  To lose his crown rather than fail his love.

178

ONLY joy, now here you are,
Fit to hear and ease my care;
Let my whispering voice obtain
Sweet reward for sharpest pain:
Take me to thee and thee to me.
'No, no, no, no, my dear, let be.'

Night hath closed all in her cloak,
Twinkling stars love-thoughts provoke;
Danger hence good care doth keep;
Jealousy itself doth sleep: 10

177
4 imp] engraft
8 balance] justice
9 flower-de-luce] France
10 bloody lion] Scotland
11 witty Lewis] cunning Louis XI

Take me to thee and thee to me.
'No, no, no, no, my dear, let be.'

Better place can no man find
Cupid's yoke to loose or bind;
These sweet flowers on fine bed too
Us in their best language woo:
Take me to thee and thee to me.
'No, no, no, no, my dear, let be.'

This small light the moon bestows
Serves thy beams but to disclose, 20
So to raise my hap more high;
Fear not else, none can us spy:
Take me to thee and thee to me.
'No, no, no, no, my dear, let be.'

That you heard was but a mouse;
Dumb sleep holdeth all the house;
Yet asleep, methinks, they say,
Young folks, take time while you may:
Take me to thee and thee to me.
'No, no, no, no, my dear, let be.' 30

Niggard time threats, if we miss
This large offer of our bliss
Long stay ere he grant the same;
Sweet then, while each thing doth frame:
Take me to thee and thee to me.
'No, no, no, no, my dear, let be.'

Your fair mother is abed,
Candles out, and curtains spread;
She thinks you do letters write;
Write, but first let me endite: 40
Take me to thee and thee to me.
'No, no, no, no, my dear, let be.'

Sweet, alas, why strive you thus?
Concord better fitteth us.
Leave to Mars the force of hands,
Your power in your beauty stands:
Take me to thee and thee to me.
'No, no, no, no, my dear, let be.'

Woe to me, and do you swear
Me to hate, but I forbear? 50
Cursed be my destinies all,
That brought me so high, to fall;
Soon with my death I will please thee.
'No, no, no, no, my dear, let be.'

179

In a grove most rich of shade,
Where birds wanton music made,
May then young his pied weeds showing,
New perfumed with flowers fresh growing,

Astrophil with Stella sweet
Did for mutual comfort meet;
Both within themselves oppressed,
But each in the other blessed.

Him great harms had taught much care:
Her fair neck a foul yoke bare; 10
But her sight his cares did banish,
In his sight her yoke did vanish.

Wept they had, alas the while;
But now tears themselves did smile,
While their eyes, by love directed,
Interchangeably reflected.

Sigh they did; but now betwixt
Sighs of woes were glad sighs mixed,
With arms crossed, yet testifying
Restless rest, and living dying. 20

Their ears hungry of each word,
Which the dear tongue would afford,
But their tongues restrained from walking,
Till their hearts had ended talking.

But when their tongues could not speak
Love itself did silence break;
Love did set his lips asunder,
Thus to speak in love and wonder:

19 arms crossed] a sign of sorrow

'Stella, sovereign of my joy,
Fair triumpher of annoy, 30
Stella, star of heavenly fire,
Stella, lodestar of desire;

Stella, in whose shining eyes
Are the lights of Cupid's skies;
Whose beams, where they once are darted,
Love therewith is straight imparted;

Stella, whose voice when it speaks
Senses all asunder breaks;
Stella, whose voice when it singeth
Angels to acquaintance bringeth; 40

Stella, in whose body is
Writ each character of bliss;
Whose face all, all beauty passeth,
Save thy mind, which yet surpasseth:

Grant, O grant—but speech, alas,
Fails me, fearing on to pass;
Grant—O me, what am I saying?
But no fault there is in praying:

Grant, O dear, on knees I pray'—
(Knees on ground he then did stay) 50
'That not I, but since I love you,
Time and place for me may move you.

Never season was more fit,
Never room more apt for it;
Smiling air allows my reason;
These birds sing, "Now use the season";

This small wind, which so sweet is,
See how it the leaves doth kiss,
Each tree in his best attiring
Sense of love to love inspiring. 60

Love makes earth the water drink,
Love to earth makes water sink;
And if dumb things be so witty,
Shall a heavenly grace want pity?'

54 room] place

There his hands in their speech fain
Would have made tongue's language plain:
But her hands, his hands repelling,
Gave repulse, all grace excelling.

Then she spake; her speech was such
As not ears, but heart did touch; 70
While such wise she love denied,
As yet love she signified.

'Astrophil,' said she, 'my love
Cease in these effects to prove:
Now be still; yet still believe me,
Thy grief more than death would grieve me.

If that any thought in me
Can taste comfort but of thee,
Let me, fed with hopeless anguish,
Joyless, hopeless, endless languish. 80

If those eyes you praised be
Half so dear as you to me,
Let me home return, stark blinded
Of those eyes, and blinder minded.

If to secret of my heart
I do any wish impart
Where thou art not foremost placed,
Be both wish and I defaced.

If more may be said, I say:
All my bliss in thee I lay; 90
If thou love, my love content thee,
For all love, all faith is meant thee.

Trust me, while I thee deny,
In my self the smart I try;
Tyrant honour thus doth use thee;
Stella's self might not refuse thee.

Therefore, dear, this no more move,
Lest, though I leave not thy love,
Which too deep in me is framed,
I should blush when thou art named.' 100

71 such wise] in such a way 74 prove] test 94 try] experience

Therewithal away she went,
Leaving him so passion-rent
With what she had done and spoken,
That therewith my song is broken.

180

Go, my flock, go get you hence,
Seek a better place of feeding,
Where you may have some defence
From the storms in my breast breeding,
And showers from my eyes proceeding.

Leave a wretch, in whom all woe
Can abide to keep no measure;
Merry flock, such one forego,
Unto whom mirth is displeasure,
Only rich in mischief's treasure. 10

Yet, alas, before you go,
Hear your woeful master's story,
Which to stones I else would show:
Sorrow only then hath glory,
When 'tis excellently sorry.

Stella, fiercest shepherdess,
Fiercest, but yet fairest ever;
Stella, whom, O heavens, still bless,
Though against me she persever,
Though I bliss inherit never; 20

Stella hath refused me,
Stella, who more love hath proved
In this caitiff heart to be
Than can in good ewes be moved
Toward lambkins best beloved.

Stella hath refused me;
Astrophil, that so well served,
In this pleasant spring must see,
While in pride flowers be preserved,
Himself only winter-starved. 30

Why, alas, doth she then swear
That she loveth me so dearly,
Seeing me so long to bear
Coals of love, that burn so clearly,
And yet leave me helpless merely?

Is that love? Forsooth, I trow,
If I saw my good dog grieved,
And a help for him did know,
My love should not be believed
But he were by me relieved. 40

No, she hates me, wellaway,
Feigning love somewhat, to please me;
For she knows, if she display
All her hate, death soon would seize me,
And of hideous torments ease me.

Then adieu, dear flock, adieu:
But alas, if in your straying
Heavenly Stella meet with you,
Tell her, in your piteous blaying,
Her poor slave's unjust decaying. 50

181

STELLA, think not that I by verse seek fame;
Who seek, who hope, who love, who live, but thee:
Thine eyes my pride, thy lips my history;
If thou praise not, all other praise is shame.
Nor so ambitious am I, as to frame
A nest for my young praise in laurel tree;
In truth I swear, I wish not there should be
Graved in mine epitaph a poet's name:
Ne if I would, could I just title make,
That any laud to me thereof should grow, 10
Without my plumes from others' wings I take.
For nothing from my wit or will doth flow,
Since all my words thy beauty doth endite,
And love doth hold my hand, and makes me write.

182

BE your words made, good sir, of Indian ware,
That you allow me them by so small rate?
Or do you cutted Spartans imitate?
Or do you mean my tender ears to spare,
That to my questions you so total are?
When I demand of Phoenix Stella's state,
You say, forsooth, you left her well of late.

49 blaying] bleating

182
3 cutted] laconic, clipped of speech
5 total] brief

O God, think you that satisfies my care?
  I would know whether she did sit or walk,
How clothed, how waited on? Sighed she or smiled? 10
Whereof, with whom, how often did she talk?
With what pastime, time's journey she beguiled?
  If her lips deigned to sweeten my poor name?
  Say all, and all well said, still say the same.

183

WHEN far-spent night persuades each mortal eye,
  To whom nor art nor nature granteth light,
  To lay his then mark-wanting shafts of sight,
Closed with their quivers, in sleep's armoury;
With windows ope then most my mind doth lie,
  Viewing the shape of darkness and delight,
  Takes in that sad hue, which with the inward night
Of his mazed powers keeps perfect harmony.
  But when birds charm, and that sweet air, which is
Morn's messenger, with rose-enamelled skies, 10
Calls each wight to salute the flower of bliss:
In tomb of lids then buried are mine eyes,
  Forced by their lord, who is ashamed to find
  Such light in sense, with such a darkened mind.

184

'WHO is it that this dark night
Underneath my window plaineth?'
It is one that from thy sight
Being, ah, exiled, disdaineth
Every other vulgar light.

'Why, alas, and are you he?
Be not yet those fancies changed?'
Dear, when you find change in me,
Though from me you be estranged,
Let my change to ruin be. 10

'Well, in absence this will die;
Leave to see, and leave to wonder.'
Absence sure will help, if I
Can learn, how myself to sunder
From what in my heart doth lie.

183
8 mazed] confused 9 charm] sing together 14 light in sense] outward light

'But time will these thoughts remove;
Time doth work what no man knoweth.'
Time doth as the subject prove;
With time still the affection groweth
In the faithful turtle dove. 20

'What if you new beauties see,
Will they not stir new affection?'
I will think they pictures be,
Image-like of saint's perfection,
Poorly counterfeiting thee.

'But your reason's purest light
Bids you leave such minds to nourish.'
Dear, do reason no such spite;
Never doth thy beauty flourish
More than in my reason's sight. 30

'But the wrongs love bears, will make
Love at length leave undertaking.'
No, the more men do it shake
In a ground of so firm making
Deeper still they drive the stake.

'Peace, I think that some give ear;
Come no more, lest I get anger.'
Bliss, I will my bliss forbear,
Fearing, sweet, you to endanger,
But my soul shall harbour there. 40

'Well, be gone, be gone, I say,
Lest that Argus' eyes perceive you.'
O, unjust is fortune's sway,
Which can make me thus to leave you,
And from louts to run away.

(Wr. 1582; pub. 1591)

from *The Psalms of David Translated into English Verse*

185

Psalm 13

1. HOW long, O lord, shall I forgotten be?
What, ever?
How long wilt thou thy hidden face from me
Dissever?

2. How long shall I consult with careful sprite
 In anguish?
 How long shall I with foes' triumphant might
 Thus languish?

3. Behold me, lord; let to thy hearing creep
 My crying; 10
 Nay, give me eyes, and light, lest that I sleep
 In dying.

4. Lest my foe brag, that in my ruin he
 Prevailed,
 And at my fall they joy that, troublous, me
 Assailed.

5. No, no; I trust on thee, and joy in thy
 Great pity:
 Still therefore of thy graces shall be my
 Song's ditty. 20

(Pub. 1823)

SIR EDWARD DYER
d. 1607

186 *Prometheus, when first from heaven high*

PROMETHEUS, when first from heaven high
 He brought down fire, ere then on earth not seen,
Fond of delight, a satyr standing by
 Gave it a kiss, as it like sweet had been.
Feeling forthwith the outward burning power,
 Wood with the smart, with shouts and shrieking still,
He sought his ease in river, field and bower,
 But for the time his grief went with him still.
So silly I, with that unwonted sight,
 In human shape an angel from above, 10
Feeding mine eyes, th'impression there did light,
 That since I run and rest as pleaseth love.
  The difference is, the satyr's lips, my heart;
  He for a while, I evermore have smart.

(Pub. 1598)

186
6 Wood] mad

187 *In praise of a contented mind*

My mind to me a kingdom is.
  Such perfect joy therein I find
That it excels all other bliss
  That world affords or grows by kind.
    Though much I want which most men have,
    Yet still my mind forbids to crave.

No princely pomp, no wealthy store,
  No force to win the victory,
No wily wit to salve a sore,
  No shape to feed each gazing eye; 10
    To none of these I yield as thrall,
    For why my mind doth serve for all.

I see how plenty suffers oft,
  And hasty climbers soon do fall;
I see that those that are aloft
  Mishap doth threaten most of all;
    They get with toil, they keep with fear:
    Such cares my mind could never bear.

Content I live, this is my stay,
  I seek no more than may suffice; 20
I press to bear no haughty sway;
  Look what I lack my mind supplies.
    Lo thus I triumph like a king,
    Content with that my mind doth bring.

Some have too much, yet still do crave;
  I little have, and seek no more.
They are but poor, though much they have,
  And I am rich with little store.
    They poor, I rich; they beg, I give;
    They lack, I leave; they pine, I live. 30

I laugh not at another's loss;
  I grudge not at another's gain;
No worldly waves my mind can toss;
  My state at one doth still remain.
    I fear no foe nor fawning friend;
    I loathe not life, nor dread my end.

Some weigh their pleasure by their lust,
  Their wisdom by their rage of will;
Their treasure is their only trust,
  And cloaked craft their store of skill: 40
    But all the pleasure that I find
    Is to maintain a quiet mind.

My wealth is health and perfect ease,
  My conscience clear my chief defence;
I neither seek by bribes to please,
  Nor by desert to breed offence.
    Thus do I live; thus will I die.
    Would all did so as well as I.

(Pub. 1588)

ANONYMOUS

188

The lowest trees have tops

THE lowest trees have tops, the ant her gall,
  The fly her spleen, the little spark his heat;
Hairs cast their shadows, though they be but small,
  And bees have stings, although they be not great.
Seas have their source, and so have shallow springs,
And love is love, in beggars and in kings.

The ermine hath the fairest skin on earth,
  Yet doth she choose the weasel for her peer;
The panther hath a sweet perfumed breath,
  Yet doth she suffer apes to draw her near. 10
No flower more fresh than is the damask rose,
Yet next her side the nettle often grows.

Where waters smoothest run, deep'st are the fords,
  The dial stirs, though none perceive it move;
The fairest faith is in the sweetest words,
  The turtles sing not love, and yet they love.
True hearts have eyes and ears, no tongues to speak,
They hear and see, and sigh, and then they break.

(Pub. 1602)

188
2 spleen] courage, passion

HUMPHREY GIFFORD

*fl. c.*1580

189 *For Soldiers*

YE buds of Brutus' land,
Courageous youth, now play your parts:
Unto your tackle stand,
Abide the brunt with valiant hearts.
For news is carried to and fro
That we must forth to warfare go:
Men muster now in every place,
And soldiers are prest forth apace.
Faint not, spend blood,
To do your Queen and country good! 10
Fair words, good pay,
Will make men cast all care away.

The time of war is come:
Prepare your corslet, spear and shield.
Methinks I hear the drum
Strike doleful marches to the field.
Tantaria, tantara! Ye trumpets sound,
Which makes our hearts with joy abound!
The roaring guns are heard afar,
And every thing denounceth war. 20
Serve God, stand stout:
Bold courage brings this gear about.
Fear not; forth run:
Faint heart fair lady never won.

Ye curious carpet knights,
That spend the time in sport and play,
Abroad and see new sights!
Your country's cause calls you away.
Do not to make your ladies game
Bring blemish to your worthy name. 30
Away to field to win renown,
With courage beat your enemies down!

1 Brutus' land] Britain
3 tackle] arms, weapons
8 prest] enlisted, levied
14 corslet] body-armour
20 denounceth] proclaims, announces
22 gear] business, matter
25 curious] cautious, fastidious
carpet knights] stay-at-home soldiers

Stout hearts gain praise,
When dastards sail in slander's seas:
Hap what hap shall,
We sure shall die but once for all.

'Alarm!' methinks they cry,
Be packing, mates; begone with speed!
Our foes are very nigh:
Shame have that man that shrinks at need. 40
Unto it boldly let us stand,
God will give right the upper hand.
Our cause is good, we need not doubt:
In sign of courage give a shout!
March forth, be strong:
Good hap will come ere it be long.
Shrink not, fight well,
For lusty lads must bear the bell.

All you that will shun evil
Must dwell in warfare every day. 50
The world, the flesh and devil
Always do seek our soul's decay.
Strive with these foes with all your might,
So shall you fight a worthy fight.
That conquest doth deserve most praise
Where vice do yield to virtue's ways.
Beat down foul sin,
A worthy crown then shall ye win.
If ye live well,
In heaven with Christ our souls shall dwell. 60

(1580)

190 *In the praise of music*

THE books of Ovid's changed shapes
A story strange do tell,
How Orpheus to fetch his wife
Made voyage unto hell.
Who having passed old Charon's boat
Unto a palace came,
Where dwelt the prince of damned sprites,
Which Pluto had to name.

48 bear the bell] take the first place, be the best
52 decay] destruction

190
1 The book] i.e. Ovid's *Metamorphoses*
8 to name] for name

When Orpheus was once arrived
  Before the regal throne, 10
He played on harp, and sang so sweet,
  As moved them all to moan.
At sound of his melodious tunes
  The very souls did mourn;
Ixion with his whirling wheel
  Stood still, and would not turn;
And Tantalus would not assay
  The fleeting floods to taste;
The sisters with their hollow sieves
  For water made no haste; 20
The greedy vultures that are feigned
  On Titius' heart to gnaw
Left off to feed, and stood amazed
  When Orpheus they saw;
And Sisyphus which rolls the stone
  Against a mighty hill,
Whiles that his music did endure
  Gave ear, and sat him still.
The Furies eke which at no time
  Were seen to weep before, 30
Were moved to moan his heavy hap,
  And shed of tears great store.
If music with her notes divine
  So great remorse can move,
I deem that man bereft of wits
  Which music will not love.
She with her silver sounding tunes
  Revives man's dulled sprites;
She feeds the ear, she fills the heart,
  With choice of rare delights. 40
Her sugared descant doth withdraw
  Thy mind from earthly toys,
And makes thee feel within thy breast
  A taste of heavenly joys.
The planets and celestial parts
  Sweet harmony contain,
Of which if creatures were deprived
  This world could not remain.
It is no doubt the very deed
  Of golden melody 50
That neighbours do together live
  In love and unity.

34 remorse] pity, tender feeling

Where man and wife agrees in one,
  Sweet music doth abound;
But when such strings begin to jar,
  Unpleasant is the sound.
Amongst all sorts of harmony
  None doth so well accord
As when we live in perfect fear
  And favour of the Lord. 60
Who grant unto us sinful wights
  Sufficient power and might,
According to his mercy great
  To tune this string aright.

(1580)

RICHARD STANYHURST
1547–1618

from *The First Four Books of Virgil his Æneis*

191 [*Polyphemus*]

HEERE we doe not lynger; thee vowd sollemnitye finnisht,
Vp we gad, owt spredding oure sayls and make to the seaward:
Al creeks mistrustful with Greekish countrye refusing.
Hercules his dwelling (yf bruite bee truelye reported)
Wee se, Tarent named, to which heunlye Lacinia fronteth,
And Caulons castels we doe spy, with Scylla the wreckmake.
Then far of vplandish we doe view thee fird Sicil Ætna.
And a seabelch grounting on rough rocks rapfulye frapping
Was hard; with ramping bounce clapping neer to the seacoast
Fierce the waters ruffle, thee sands with wroght flud ar hoysed.
  Quod father Anchises, heere loa that scuruye Charybdis.
Theese stoans king Helenus, theese ragd rocks rustye fore vttred.
Hence hye, my deere feloes, duck the oars, and stick to the tacklings.
  Thus sayd he, then swiftly this his heast thee coompanye practise.
First thee pilot Palinure thee steerd ship wrigs to the lifthand;
Right so to thee same boord thee maysters al wrye the vessels.
Vp we fle too skyward with wild fluds hautye, then vnder
Wee duck too bottom with waues contrarye repressed.

61 wights] creatures
191
1 thee] the
3 refusing] avoiding
9 hard] heard
10 hoysed] raised, stirred up
13 hye] hasten
15 lifthand] lefthand

Thus thrise in oure diuing thee rocks moste horribly roared:
And thrise in oure mounting to the stars thee surges vs heaued. 20
Thee winds and soonbeams vs, poore souls weerye, refused,
And to soyl of Cyclops with wandring iournye we roamed.
A large roade fenced from rough ventositye blustring.
But neere ioynctlye brayeth with rufflerye rumboled Ætna.
Soomtyme owt yt balcketh from bulck clowds grimlye bedymmed.
Lyke fyerd pitche skorching, or flash flame sulphurus heating:
Flownce to the stars towring thee fire, lyke a pellet, is hurled,
Ragd rocks vp raking: and guts of mounten yrented
From roote vp hee iogleth: stoans hudge slag molten he rowseth:
With route snort grumbling, in bottom flash furye kendling. 30
Men say that Enceladus with bolt haulf blasted here harbrouth,
Dingd with this squising and massiue burthen of Ætna,
Which pres on hym nayled from broached chymnye stil heateth.
As oft as the giant his broyld syds croompeled altreth,
So oft Sicil al shiuereth, there with flaks smoakye be sparckled.
That night in forrest to vs pouke bugs gastlye be tendred.
Thee cause wee find not, for noise phantastical offred.
Thee stars imparted no light, thee welken is heauye:
And the moon enshryned with closet clowdye remayned.
Thee morning brightnesse dooth luster in east seat Eöus, 40
And night shades moysturs glittring Aurora repealeth.
When that on a suddeyn we behold a windbeaten hard shrimp,
With lanck wan visadge, with rags iags patcherye clowted,
His fists too the skyward rearing: heere wee stood amazed.
A meigre leane rake with a long berd goatlyke; aparrayld
In shrub weeds thorny: by his byrth a Grecian holden,
One that too Troy broyls whillon from his countrye repayred.
When the skrag had marcked far a loof thee Troian atyring,
And Troian weapons, in steps he stutted, apaled,
And fixt his footing; at leingth with desperat offer 50
Too the shore hee neered, theese speeches merciful vttring.
By stars I craue you, by the ayre, by the celical houshold,
Hoyse me hence (O Troians), too sum oother countrye me whirrye.
Playnelye to speake algats, for a Greeke my self I do knowledge,

23 roade] haven, roadstead
24 rufflerye] uproar
25 balcketh] blackens
bulck] massy
30 route] bellow
36 pouke bugs] bugbears, malignant spectres
42 shrimp] little man
43 clowted] clothed
44 too] to
47 whillon] once, formerly
48 skrag] skinny creature
a loof] at a distance
atyring] dress
49 stutted] stumbled
51 merciful] i.e. begging mercy
52 celical] heavenly
53 whirrye] carry swiftly
54 algats] whatever may happen

And that I too Troytowne with purposed enmitye sayled.
If this my trespasse now claymeth duelye reuengment
Plunge me deepe in the waters, and lodge me in Neptun his harboure.
If mens hands slea mee, such mannish slaughter I wish for.
  Thus sayd he, downe kneeling, and oure feete mournefuly clasping.
Then we hym desyred first too discoouer his ofspring, 60
After too manifest this his hard and destenye bitter.
My father Anchises gaue his hand to the wretch on a suddeyn,
And with al a pardon, with saulfe protection, offred.
Thee captiue, shaking of feare, too parlye thus entred.
  Borne I was in the Itacan countrey, mate of haples Vlisses,
Named Achæmenides, my syre also cald Adamastus,
A good honest poore man (would we in that penurye lasted)
Sent me toe your Troywars; at last my coompanye skared
From this countrye cruel, dyd posting leaue me behynde theym,
In Cyclops kennel, thee laystow dirtye, the foule den. 70
In this grislye palaice, in forme and quantitye mightye,
Palpable and groaping darcknesse with murther aboundeth.
Hee doth in al mischiefe surpasse, hee mounts to the sky top.
(Al the heunly feloship from the earth such a monster abandon)
Hard he is too be viewed, too se hym no person abydeth.
Thee blud with the entrayls of men, by hym slaughtred, he gnaweth.
And of my feloes I saw that a couple he grapled
On ground sow groueling, and theym with villenye crusshed,
At flint hard dasshing, thee goare blood spowteth of eeche syde,
And swyms in the thrashold, I saw flesh bluddye toe slauer, 80
When the cob had maunged the gobets foule garbaged haulfe quick.
Yeet got he not shotfree, this butcherye quighted Vlisses:
In which doughtye peril the Ithacan moste wiselye bethoght hym.
For the vnsauerye rakhel with collops bludred yfrancked,
With chuffe chaffe wynesops lyke a gourd bourrachoe replennisht,
His nodil in crossewise wresting downe droups to the growndward,
In belche galp vometing with dead sleape snortye the collops,
Raw with wyne soused, we doe pray toe supernal assemblye,
Round with al embaying thee muffe maffe loller; eke hastlye
With toole sharp poincted wee boarde and perced his oane light, 90

64 shaking of] shaking off
68 skared] in terrified flight
69 posting] in haste
70 laystow] place where refuse and dung are thrown
81 cob] huge man, monster
84 collops] lumps of flesh
yfrancked] crammed
85 chuffe chaffe wynesops] thick lumps of bread soaked in wine
gourd bourrachoe] leather wine bag or bottle
86 nodil] head
87 galp] gaping
88 Raw] suffering indigestion
soused] pickled
89 embaying] surrounding
thee muffe maffe loller] the lolling (i.e. unconscious) brute
90 his oane light] his one eye

That stood in his lowring front gloommish malleted onlye,
Lyke Greekish tergat glistring, or Phœbus his hornebeams.
Thus the death of feloes on a lout wee gladlye reuenged.
But se ye flee caytiefs, hy ye hence, cut swiftlye the cables.
Pack fro the shoare.
For such as in prison thee great Polyphemus is holden,
His sheepflocks foddring, from dugs mylck thriftelye squising,
Thee lyke heere in mountayns doo randge in number an hundred,
That bee cursd Cyclopes in naming vsual highted.
Thee moone three seasons her passadge orbical eended 100
Sence I heere in forrest and cabbans gastlye dyd harboure,
With bestes fel saluadge: and in caues stoanye Cyclopes
Dayly I se, theire trampling and yelling hellish abhorring.
My self I dieted with sloas, and thinlye with hawthorns,
With mast, and with roots of eeche herb I swadgde my great hunger.
I pryed al quarters, and first this nauye to shoare ward
Swift, I scryed, sayling too which my self I remitted,
Of what condicion, what country so eauer yt had beene.
Now tis sufficient that I skape fro this horribil Island.
Mee rather extinguish with soom blud murther or oother. 110
Scant had he thus spoaken: when that from mountenus hil toppe
All wee see the giaunt, with his hole flock lowbylyke hagling,
Namde the shepeherd Polyphem, to the wel knowne sea syd aproching.
A fowle fog monster, great swad, depriued of eyesight.
His fists and stalcking are propt with trunck of a pynetreė.
His flock hym doe folow, this charge hym chieflye reioyceth,
In grief al his coomfort; on neck his whistle is hanged.
When that too the seasyde thee swayne Longolius hobbled,
Hee rinst in the water thee drosse from his late bored eyelyd.
His tusk grimlye gnashing, in seas far waltred, he groyleth: 120
Scantly doo the water surmounting reache toe the shoulders.
But we being feared, from that coast hastlye remooued,
And with vs embarcked thee Greekish suitur, as amplye
Hid due request merited; wee chopt of softlye the cables.
Swift wee sweepe the seafroth with nimble lustilad oare striefe.
Thee noise he perceaued, then he turning warelye listeth,
But when he considerd, that wee preuented his handling,
And that from foloing oure ships thee fluds hye reuockt hym,

91 gloommish] somewhat gloomy
malleted] fixed as if by hammering
92 hornebeams] beams of light issuing from the head like horns
99 highted] called
101 cabbans] i.e. cabins, huts, lairs
105 swadgde] assuaged
112 lowbylyke] (i.e. looby-like) loutishly, clumsily
hagling] advancing with difficulty
114 fog] gross, bloated
swad] lout, clodhopper
120 waltred] tossed, rolled about
groyleth] moves, makes his way
124 chopt of] chopped off
127 preuented] escaped
handling] laying hands (on us)
128 reuockt] restrained

Loud the lowbye brayed with belling monsterus eccho:
Thee water hee shaketh, with his owt cryes Italye trembleth, 130
And with a thick thundring thee fyerde fordge Ætna rebounded.
Then runs from mountayns and woods thee rownseual helswarme
Of Cyclopan lurdens to the shoars in coompanye clustring.
Far we se theym distaunt: vs grimly and vaynely beholding.
Vp to the sky reatching, thee breetherne swish swash of Ætna.
A folck moaste fulsoom, for sight moste fitlye resembling
Trees of loftye cipers, with thickned multitud oakroas,
Or Ioues great forrest, or woods of mightye Diana.
Feare thear vs enforced with posting speedines headlong
Too swap of oure cables, and fal to the seas at auenture. 140
But yeet king Helenus iumptwixt Scylla and the Charybdis
For to sayl vs monished, with no great dangerus hazard.
Yeet we wer ons mynded, backward thee nauye to mayster.
Heere loa behold Boreas from bouch of north blo Pelorus
Oure ships ful chargeth, thee quick rocks stoanye we passed:
And great Pantagia, and Megarus with Tapsus his Island.
Theese soyls fore wandred to oure men were truelye related
By poore Achæmenides, mate too thee luckles Vlisses.

(1582)

THOMAS WATSON

c.1557–1592

192 *My love is past*

YE captive souls of blindfold Cyprian's boat,
Mark with advice in what estate ye stand:
Your boatman never whistles merry note,
And Folly keeping stern, still puts from land,
And makes a sport to toss you to and fro
Twixt sighing winds and surging waves of woe.

129 lowbye] (i.e. looby) lout, clown
belling] bellowing
132 rownseual] gigantic, huge
133 lurdens] dull-witted boors
135 swish swash] braggarts, swaggerers
136 fulsoom] disgusting
137 cipers] cypresses
139 posting] hurrying
140 swap of] cut off
141 iumptwixt] exactly between
144 bouch] mouth
145 quick] living

192
1 Cyprian's] belonging to Cupid
4 keeping stern] steering
still] continually, ever

On Beauty's rock she runs you at her will,
And holds you in suspense twixt hope and fear,
Where dying oft, yet are you living still,
But such a life as death much better were. 10
Be therefore circumspect, and follow me,
When chance or change of manners sets you free.

Beware how you return to seas again.
Hang up your votive tables in the choir
Of Cupid's church, in witness of the pain
You suffer now by forced fond desire.
Then hang your throughwet garments on the wall,
And sing with me that love is mixed with gall.

(1582)

ANONYMOUS

193 *Verses made by a Catholic in praise of Campion that was executed at Tyburn for treason, as is made known by the Proclamation*

WHY do I use my paper, ink, and pen,
And call my wits to counsel what to say?
Such memories were made for mortal men.
I speak of saints whose names shall not decay.
And angels' trump were fitter for to sound
Their glorious death if such on earth were found.

Pardon my want, I offer nought but will;
Their register remaineth safe above.
Campion exceeds the compass of my skill.
Yet let me use the measure of my love, 10
And give me leave in base and lowly verse
His high attempts in England to rehearse.

14 votive tables] inscribed panels anciently hung in a temple in fulfilment of a vow, e.g. after deliverance from shipwreck (cf. Horace, *Odes* I.5)
17 throughwet] drenched

193
9 Campion] Edmund Campion, along with Ralph Sherwin and Alexander Bryan was executed on 1 December 1581

He came by vow, the cause to conquer sin;
His armour, prayer; the Word, his targe and shield;
His comfort, heaven; his spoil, our souls to win;
The devil, his foe; the wicked world, his field;
His triumph, joy; his wage, eternal bliss;
His captain, Christ, which ever blessed is.

From ease to pain, from honour to disgrace,
From love to hate, to danger being well, 20
From safe abode to fear in every place,
Contemning death, to save our souls from hell,
Our new apostle coming to restore
The faith which Austin planted here before.

His natures's flowers were mixed with herbs of grace,
His mild behaviour tempered well with skill;
A lowly mind possessed a learned place,
A sugared speech a rare and virtuous will;
A saintlike man was set on earth below
The seed of truth in erring hearts to sow. 30

With tongue and pen the truth he taught and wrote,
By force whereof they came to Christ apace;
But when it pleased God it was his lot
He should be thralled, He lent him so much grace
His patience then did work as much, or more,
Than had his heavenly speeches done before.

His fare was hard, yet mild and sweet his cheer,
His prison close, yet free and loose his mind,
His torture great, yet small or none his fear,
His offers large, yet nothing could him blind. 40
O constant man, O mind, O virtue strange,
Whom want nor woe, nor fear nor hope could change!

From rack in Tower they brought him to dispute,
Bookless, alone, to answer all that came.
But Christ gave grace; he did them all confute
So sweetly there in glory of His name
That even the adverse part were forced to say
That Campion's cause did bear the bell away.

24 Austin] St Augustine of Canterbury
34 thralled] imprisoned
48 bear the bell away] win the victory

This foil enraged the minds of some so far
They thought it best to take his life away, 50
Because they saw he would their matter mar,
And leave them shortly nought at all to say.
  Traitor he was with many a silly sleight,
  Yet packed a jury that cried guilty straight.

Religion there was treason to the Queen,
Preaching of penance, war against the land;
Priests were such dangerous men as had not been;
Prayers and beads were fight and force of hand;
  Cases of conscience, bane unto the state:
  So blind is error, so false witness, hate. 60

And yet behold, these lambs are drawn to die;
Treason proclaimed, the Queen is put in fear.
Out upon Satan! Fie, malice, fie!
Speak'st thou to those that did the guiltless hear?
  Can humble souls departing now to Christ
  Protest untrue? Avaunt, foul fiend, thou li'st!

My sovereign liege, behold your subjects' end:
Your secret foes do misinform your grace;
Who for your cause their holy lives would spend,
As traitors die—a rare and monstrous case. 70
  The bloody wolf condemns the harmless sheep
  Before the dog, the while the shepherds sleep.

England look up: thy soil is stained with blood.
Thou hast made martyrs many of thine own.
If thou have grace, their death will do thee good;
The seed will take that in such blood is sown,
  And Campion's learning, fertile so before,
  Thus watered to must needs of force be more.

Repent thee, Eliot, of thy Judas kiss:
I wish thy penance, not thy desperate end. 80
Let Norton think, which now in prison is,
To whom he said he was not Caesar's friend,
  And let the judge consider well in fear
  That Pilate washed his hands and was not clear.

49 foil] defeat
79 Eliot] George Eliot (a witness against Campion)
81 Norton] Thomas Norton (official censor of Catholics in England)

The witness false, Sled, Munday, and the rest,
That had your slanders noted in your book,
Confess your fault beforehand it were best,
Lest God do find it written when he look
  In dreadful doom upon the souls of men:
  It will be late alas to mend it then. 90

You bloody jury, Lee and all th'eleven,
Take heed your verdict which was given in haste
Do not exclude you from the joys of heaven
And cause you rue it when the time is past,
  And every one whose malice caused him say
  'Crucifige!' dread the terror of that day.

Fond Elderton, call in thy foolish rhymes,
Thy scurril ballads are too bad to sell;
Let good men rest, and mend thyself betimes,
Confess in prose thou hast not metred well. 100
  Or if thy folly cannot choose but feign,
  Write alehouse toys, blaspheme not in thy vein.

Remember ye that would oppress the cause,
The Church is Christ's, His honour cannot die,
Though hell herself revest her grisly jaws
And join in league with schism and heresy;
  Though craft devise, and cruel rage oppress,
  Yet still will write and martyrdom confess.

Ye thought, perhaps, when learned Campion dies,
His pen must cease, his sugared tongue be still. 110
But you forget how loud his death it cries,
How far beyond the sound of tongue or quill.
  You did not know how rare and great a good
  It was to write those precious gifts in blood.

He living spake to them that present were;
His writings took their censure of the view.
Now fame reports his learning far and near,
And now his death confirms their doctrine true.
  His virtues now are written in the skies
  And often read with holy inward eyes. 120

85 Sled, Munday] Charles Sled, Anthony Munday (witnesses against Campion)
89 doom] judgement
91 Lee] William Lee (foreman of the jury)
96 'Crucifige'] crucify
97 Elderton] William Elderton (well-known writer of ballads)
102 toys] trifles
105 revest] (?) array, make ready
110 sugared] eloquent
116 censure] judgement

All Europe wonders at so rare a man.
England was filled with rumour of his end,
And London most, for it was present then
When constantly three saints their lives did spend.
  The streets, the steps, the stones you hauled them by
  Proclaims the cause wherefore these martyrs die.

The Tower doth tell the truth he did defend;
The bar bears witness of his guiltless mind;
Tyburn did try he made a patient end;
On every gate his martyrdom we find. 130
  In vain ye wrought that would obscure his name,
  For heaven and earth will still record the same.

Your sentence wrong pronounced of him here
Exempts him from the Judgment now to come.
O happy he that is not judged there!
God grant me too to have an earthly doom!
  Your witness false and lewdly taken in
  Doth cause he is not now accused of sin.

His prison now the city of the King,
His rack and torture, joys and heavenly bliss; 140
For men's reproach, with angels he doth sing
A sacred song that everlasting is.
  For shame but short and loss of small renown
  He purchased hath an ever during crown.

His quartered limbs shall join with joy again,
And rise a body brighter than the sun.
Your blinded malice tortured him in vain;
For every wrench some glory hath been won.
  And every drop of blood that he did spend
  Hath reaped a joy that never shall have end. 150

Can dreary death then daunt our deeds or pain?
Is't ling'ring life we fear to lose or ease?
No, no: such death procureth life again:
'Tis only God we tremble to displease,
  Who kills but once and ever still we die,
  Whose hot revenge torments eternally.

122 rumour] talk, report
129 try] prove
131 that would] i.e. what would
136 doom] judgement
137 lewdly] ignorantly
  taken in] believed, accepted
151 dreary] bloody, cruel

We cannot fear a mortal torment, we;
This martyr's blood hath moistened all our hearts;
Whose parted quarters when we chance to see
We learn to play the constant Christian's parts. 160
His head doth speak, and heavenly precepts give
How that we look, should frame ourselves to live.

His youth instructs us how to spend our days;
His flying bids us how to banish sin;
His strait profession shows the narrow ways
Which they must walk that look to enter in.
His home return by danger and distress
Emboldens us our conscience to profess.

His hurdle draws us with him to the cross;
His speeches there provoketh us to die; 170
His death doth say his life is but our loss;
His martyred blood from heaven to us doth cry.
His first and last and all agree in this,
To show the way that leadeth unto bliss.

Blessed be God who lent him so much grace!
Thanked be Christ that blest his martyr so!
Happy is he that sees his master's face!
Cursed are they that thought to work his woe!
Bounden we be to give eternal praise
To Jesus' name who such a saint did raise! 180

(Pub. 1583)

194 [*Hymn to the Virgin*]

FLOWER of roses, angels' joy,
Tower of David, Ark of Noy,
First of saints whose true protecting
Of the young and weak in sprite
Makes my soul these lines endite
To thy throne her plaint directing;

Orphan child alone I lie,
Childlike to thee I cry,

165 strait] strict, rigorous
profession] monastic or priestly order; declaration, promise

194
2 Noy] Noah
4 sprite] spirit
5 endite] write

Queen of Heaven, used to cherish;
Eyes of grace, behold I fall; 10
Ears of pity, hear my call
Lest in swadling clouts I perish.

Hide the greatness of each fault;
My desert, if there be aught,
By thy merits be enlarged
That the debts wherein I fall,
Paying nought but owing all,
By thy prayer be discharged.

Pray to Him whose shape I bear,
By thy love, thy care, thy fear, 20
By thy glorious birth and breeding,
That though our sins touch the sky
Yet his mercies mount more high,
All his other works exceeding.

Tell Him that in strength'ning me
With His grace he graceth thee,
Every little one defending;
Tell Him that I cloy thine ears
With the cry of childish tears
From his footstool still ascending. 30

Hear my cries and grant me aid,
Perfect mother, perfect maid;
Hear my cries to thee addressed;
From my plaints turn not thy face,
Humble and yet full of grace,
Pure, untouched, for ever blessed.

(Pub. 1960)

THOMAS GILBART

*fl. c.*1583

195 *A declaration of the death of John Lewes, a most detestable and obstinate heretic, burned at Norwich, the xviii day of September, 1583. About three of the clock in the afternoon.*

SHALL silence shroud such sin
As Satan seems to show
Even in his imps, in these our days
That all men might it know?

No, no, it cannot be;
But such as love the Lord,
With heart and voice, will him confess
And to his word accord.

And do not as this devil did,
Though shape of man he bare: 10
Denying Christ, did silence keep
At death, devoid of care.

Yet did this wretch, most wickedly
(John Lewes, who to name),
Full boldly speak, and brutishly
God's glory to defame,

In presence of those persons which
Were learned, wise, and grave,
That wished in heart, with weeping tears,
Repentance he would crave. 20

But he, despising reverence
To prince or any state,
Not them regards, but used terms
As each had been his mate.

For he did thou each wight the which
With him had any talk.
Thus did his tongue most devilishly
With defamy still walk.

3 imps] children
24 mate] fellow-worker
25 thou] speak familiarly or contemptuously to (by using 'thou' instead of 'you') wight] person
28 defamy] defamation

But when that no persuasion might
  Procure him to relent, 30
Then Judgment did, by Justice right,
  Unto his death consent.

That he should burned be to death,
  This Justice did award.
Now mark what after did ensue,
  And thereto have regard.

The time then of his death being come—
  Which was the eighteen day
Of September, in eighty-three—
  This wretch wrought his decay. 40

For when he to the place was brought
  Where he his life should end,
He forced was a time to stay,
  A sermon to perpend.

The which was preached by the Dean
  Of Norwich, in such wise,
Which well might move each sinful soul
  From seat of sin to rise.

He, like a tender father, did
  Give documents most pure 50
Unto this wretch as to his child,
  From ill him to procure.

But all in vain, this varlet vild
  His doctrine did detest;
For when he spake of Christ, God's Son,
  He made thereat a jest.

And smilingly his face would turn
  From preachers present there,
Which argued that he never stood
  Of God or man in fear. 60

40 wrought] brought about, contrived
decay] destruction
44 perpend] consider, ponder
53 vild] vile
54 detest] denounce, execrate; hate, abhor

When that the sermon drew to end,
Then did the Dean desire
Him that he would fall on his knees
And God's mercy require.

But still he stood as any stone,
Not lifting hand or eye
Unto the heavens, which showed his heart
To God was nothing nigh.

The shrieve, then, strikes him on the breast,
Wishing him to return; 70
Yea, gentlewomen two or three,
Before he went to burn,

Would seem to pull him on his knees
His sins for to confess;
But he full stoutly stood therein,
Not meaning nothing less.

From preaching place unto the stake
They straight did him convey,
Where preachers two or three him willed
Unto the Lord to pray, 80

And Christ our Saviour to confess
Both God and man to be;
That soul and body, by true faith
In him, might be set free

From Satan, who had him in hold.
But he not this regard,
As countenance his did show full plain,
For why no word was hard

That he did speak; but like a dog
Did end his days with shame, 90
Not bending knee, hand, heart, or tongue,
To glorify God's name.

64 require] beg
69 shrieve] sheriff
86 regard] regarded, took notice of
88 For why] because
hard] heard

For though that divers preachers than,
Both godly, grave, and wise,
Did hope (in heart) to win this man,
Yet all would not suffice.

For not one word that they could get,
What so they did or said,
Till one that was right earnest set
By these words him assayed: 100

'If that thou dost not Jesus Christ
God's only Son confess,
Both God and Man, and hope in him
For thy salvation doubtless,

'As sure as now thou shalt be burnt
Before us here at stake,
So sure in hell thou shalt be burnt
In that infernal lake.'

Quoth he, 'Thou liest', and no more words
At all this caitiff said; 110
Nor no repentant sign would show,
Which made us all dismayed.

And when the fire did compass him
About on every side,
The people looked he then would speak,
And therefore loud they cried:

'Now call on Christ to save thy soul;
Now trust in Christ his death.'
But all in vain; no words he spake,
But thus yields up his breath. 120

Oh woeful state, oh danger deep,
That he was drowned in!
Oh grant us, God, for Christ his sake,
We fall not in such sin.

And we that think we stand in faith
So firm, Lord let it be
To Thee, Thy Son, and Holy Ghost—
One God in Persons three.

(1583)

93 than] then
98 What so] whatever
110 caitiff] wretch, vile man
115 looked] expected

ANONYMOUS

196 *A new courtly sonnet of the Lady Greensleeves*

GREENSLEEVES *was all my joy,*
Greensleeves was my delight;
Greensleeves was my heart of gold,
And who but Lady Greensleeves.

Alas, my love, ye do me wrong
To cast me off discourteously;
And I have loved you so long,
Delighting in your company.
Greensleeves was all my joy, &c.

I have been ready at your hand, 10
To grant whatever you would crave;
I have both waged life and land,
Your love and goodwill for to have.
Greensleeves was all my joy, &c.

I bought thee kerchers to thy head,
That were wrought fine and gallantly;
I kept thee both at board and bed,
Which cost my purse well favouredly.
Greensleeves was all my joy, &c.

I bought thee petticoats of the best, 20
The cloth so fine as fine might be;
I gave thee jewels for thy chest,
And all this cost I spent on thee.
Greensleeves was all my joy, &c.

Thy smock of silk, both fair and white,
With gold embroidered gorgeously;
Thy petticoat of sendal right;
And thus I bought thee gladly.
Greensleeves was all my joy, &c.

Thy girdle of gold so red, 30
With pearls bedecked sumptuously;
The like no other lasses had,
And yet thou wouldst not love me.
Greensleeves was all my joy, &c.

12 waged] risked 27 sendal] fine silk

Thy purse and eke thy gay gilt knives,
Thy pincase gallant to the eye;
No better wore the burgess wives,
And yet thou wouldst not love me.
Greensleeves was all my joy, &c.

Thy crimson stockings all of silk, 40
With gold all wrought above the knee;
Thy pumps as white as was the milk,
And yet thou wouldst not love me.
Greensleeves was all my joy, &c.

Thy gown was of the grassy green,
Thy sleeves of satin hanging by,
Which made thee be our harvest queen,
And yet thou wouldst not love me.
Greensleeves was all my joy, &c.

Thy garters fringed with the gold, 50
And silver aglets hanging by,
Which made thee blithe for to behold,
And yet thou wouldst not love me.
Greensleeves was all my joy, &c.

My gayest gelding I thee gave,
To ride wherever liked thee;
No lady ever was so brave,
And yet thou wouldst not love me.
Greensleeves was all my joy, &c.

My men were clothed all in green, 60
And they did ever wait on thee;
All this was gallant to be seen,
And yet thou wouldst not love me.
Greensleeves was all my joy, &c.

They set thee up, they took thee down,
They served thee with humility;
Thy foot might not once touch the ground,
And yet thou wouldst not love me.
Greensleeves was all my joy, &c.

57 brave] finely dressed

For every morning when thou rose, 70
I sent thee dainties orderly,
To cheer thy stomach from all woes,
And yet thou wouldst not love me.
Greensleeves was all my joy, &c.

Thou couldst desire no earthly thing
But still thou hadst it readily;
Thy music still to play and sing,
And yet thou wouldst not love me.
Greensleeves was all my joy, &c.

And who did pay for all this gear 80
That thou didst spend when pleased thee?
Even I that am rejected here,
And thou disdain'st to love me.
Greensleeves was all my joy, &c.

Well, I will pray to God on high,
That thou my constancy mayst see,
And that yet once before I die,
Thou wilt vouchsafe to love me.
Greensleeves was all my joy, &c.

Greensleeves, now farewell! adieu! 90
God I pray to prosper thee;
For I am still thy lover true.
Come once again and love me.
Greensleeves was all my joy,
Greensleeves was my delight;
Greensleeves was my heart of gold,
And who but Lady Greensleeves.

(1584)

197 *A Nosegay*

A NOSEGAY lacking flowers fresh
to you now I do send,
Desiring you to look thereon
when that you may intend:
For flowers fresh begin to fade,
and Boreas in the field
Even with his hard congealed frost
now better flowers doth yield.

6 Boreas] north wind

But if that winter could have sprung
  a sweeter flower than this, 10
I would have sent it presently
  to you withouten miss.
Accept this then as time doth serve;
  be thankful for the same.
Despise it not, but keep it well,
  and mark each flower his name.

Lavender is for lovers true,
  which evermore be fain,
Desiring always for to have
  some pleasure for their pain. 20
And when that they obtained have
  the love that they require,
then have they all their perfect joy,
  and quenched is the fire.

Rosemary is for remembrance
  between us day and night,
Wishing that I might always have
  you present in my sight.
And when I cannot have
  as I have said before, 30
Then Cupid with his deadly dart
  doth wound my heart full sore.

Sage is for sustenance
  that should man's life sustain,
For I do still lie languishing
  continually in pain,
And shall do still until I die,
  except thou favour show;
My pain and all my grievous smart
  full well you do it know. 40

Fennel is for flatterers;
  an ill thing it is sure.
But I have always meant truly,
  with constant heart most pure;
And will continue in the same
  as long as life doth last,
Still hoping for a joyful day
  when all our pains be past.

11 presently] at once   18 fain] glad   22 require] beg

Violet is for faithfulness,
which in me shall abide; 50
Hoping likewise that from your heart
you will not let it slide,
And will continue in the same
as you have now begun,
And then for ever to abide:
then you my heart have won.

Thyme is to try me,
as each be tried must;
Letting you know while life doth last
I will not be unjust; 60
And if I should I would to God
to hell my soul should bear,
And eke also that Beelzebub
with teeth he should me tear.

Roses is to rule me
with reason as you will,
For to be still obedient
your mind for to fulfil;
And thereto will not disagree
in nothing that you say, 70
But will content your mind truly
in all things that I may.

Gillyflowers is for gentleness
which in me shall remain,
Hoping that no sedition shall
depart our hearts in twain.
As soon the sun shall lose his course,
the moon against her kind
Shall have no light, if that I do
once put you from my mind. 80

Carnations is for graciousness;
mark that now by the way;
Have no regard to flatterers
nor pass not what they say.
For they will come with lying tales
your ears for to fulfil.
In any case do you consent
nothing unto their will.

76 depart] separate, divide   84 pass] care   86 fulfil] fill
78 kind] nature

Marigolds is for marriage
  that would our minds suffice, 90
Lest that suspicion of us twain
  by any means should rise.
As for my part, I do not care;
  myself I will still use
That all the women in the world
  for you I will refuse.

Pennyroyal is to print your love
  so deep within my heart,
That when you look this nosegay on
  my pain you may impart; 100
And when that you have read the same
  consider well my woe:
Think ye then how to recompense
  even him that loves you so.

Cowslips is for counsel
  for secrets us between,
That none but you and I alone
  should know the thing we mean.
And if you will thus wisely do
  as I think to be best, 110
Then have you surely won the field
  and set my heart at rest.

I pray you keep this nosegay well,
  and set by it some store;
And thus farewell, the gods thee guide
  both now and evermore:
Not as the common sort do use,
  to set it in your breast,
That when the smell is gone away
  on ground he takes his rest. 120

(1584)

97 Pennyroyal] species of mint
100 impart] share, partake
105 counsel] private or confidential matter
111 field] i.e. battle, contest
114 set . . . store] value it greatly

JOHN LYLY
*c.*1554–1606

from *Campaspe*

198

Granichus. O, FOR a bowl of fat Canary,
Rich Palermo, sparkling Sherry,
Some nectar else, from Juno's dairy;
O, these draughts would make us merry.

Psyllus. O, for a wench (I deal in faces,
And in other daintier things);
Tickled am I with her embraces,
Fine dancing in such fairy rings.

Manes. O, for a plump fat leg of mutton,
Veal, lamb, capon, pig, and coney: 10
None is happy but a glutton,
None an ass but who wants money.

Chorus. Wines (indeed) and girls are good,
But brave victuals feast the blood;
For wenches, wine, and lusty cheer,
Jove would leap down to surfeit here.

199

Apelles. CUPID and my Campaspe played
At cards for kisses, Cupid paid;
He stakes his quiver, bow, and arrows,
His mother's doves, and team of sparrows;
Loses them too; then, down he throws
The coral of his lip, the rose
Growing on 's cheek (but none knows how);
With these, the crystal of his brow,
And then the dimple of his chin:
All these did my Campaspe win. 10
At last, he set her both his eyes;
She won, and Cupid blind did rise.
O Love, has she done this to thee?
What shall (alas) become of me?

200
Trico.

WHAT bird so sings, yet so does wail?
O, tis the ravished nightingale.
Jug, Jug, Jug, Jug, Tereu, she cries,
And still her woes at midnight rise.
Brave prick song! Who is 't now we hear?
None but the lark so shrill and clear;
How at heaven's gates she claps her wings,
The morn not waking till she sings.
Hark, hark, with what a pretty throat
Poor Robin Redbreast tunes his note;
Hark how the jolly cuckoos sing
Cuckoo, to welcome in the spring,
Cuckoo, to welcome in the spring.

(Wr. *c.*1580; pub. 1632)

from *Sapho and Phao*

201
Sapho.

O cruel Love, on thee I lay
My curse, which shall strike blind the day:
Never may sleep with velvet hand
Charm thine eyes with sacred wand;
Thy jailers shall be hopes and fears;
Thy prison-mates, groans, sighs, and tears;
Thy play to wear out weary times,
Fantastic passions, vows, and rhymes;
Thy bread be frowns, thy drink be gall,
Such as when I on Phao call; 10
The bed thou liest on be despair;
Thy sleep, fond dreams; thy dreams, long care;
Hope (like thy fool) at thy bed's head
Mock thee, till madness strike thee dead;
As Phao, thou dost me, with thy proud eyes.
In thee poor Sapho lives, for thee she dies.

The Song in making of the Arrows

202
Vulcan.

MY shag-hair Cyclops, come, let's ply
Our Lemnian hammers lustily.
By my wife's sparrows,
I swear these arrows
Shall singing fly

Through many a wanton's eye.
These headed are with golden blisses,
These silver ones feathered with kisses,
But this of lead
Strikes a clown dead, 10
When in a dance
He falls in a trance
To see his black-brow lass not buss him,
And then whines out for death t'untruss him.
So, so, our work being done let's play,
Holiday (boys), cry holiday!

(Wr. *c.*1582; pub. 1632)

from *Endimion*

203

Watch. STAND! Who goes there?
We charge you, appear
Fore our constable here
(In the name of the Man in the Moon).
To us billmen relate
Why you stagger so late,
And how you come drunk so soon.
Pages. What are ye (scabs)?
Watch. The Watch.
This the constable.
Pages. A patch.
Constable. Knock 'em down unless they all stand. 10
If any run away,
'Tis the old watchman's play,
To reach him a bill of his hand.
Pages. O, gentlemen, hold,
Your gowns freeze with cold,
And your rotten teeth dance in your head.
Epiton. Wine, nothing shall cost ye.
Samias. Nor huge fires to roast ye.
Dares. Then soberly let us be led.
Constable. Come, my brown bills we'll roar, 20
Bounce loud at tavern door,
Omnes. And i' th' morning steal all to bed.

13 buss] kiss

203
5 billmen] watchmen armed with a bill (broadsword, halberd)
9 patch] fool
20 brown bills] watchmen
21 Bounce] knock

204

Omnes. PINCH him, pinch him black and blue;
Saucy mortals must not view
What the queen of stars is doing,
Nor pry into our fairy wooing.

1 *Fairy.* Pinch him blue.

2 *Fairy.* And pinch him black.

3 *Fairy.* Let him not lack
Sharp nails to pinch him blue and red,
Till sleep has rocked his addle head.

4 *Fairy.* For the trespass he hath done, 10
Spots o'er all his flesh shall run.
Kiss Endimion, kiss his eyes,
Then to our midnight hay-de-guys.

(Wr. *c.*1585; pub. 1632)

from *Midas*

205

Apollo. MY Daphne's hair is twisted gold,
Bright stars a-piece her eyes do hold,
My Daphne's brow enthrones the Graces,
My Daphne's beauty stains all faces,
On Daphne's cheek grow rose and cherry,
On Daphne's lip a sweeter berry;
Daphne's snowy hand but touched does melt,
And then no heavenlier warmth is felt;
My Daphne's voice tunes all the spheres,
My Daphne's music charms all ears. 10
Fond am I thus to sing her praise;
These glories now are turned to bays.

206

Pan. PAN'S Syrinx was a girl indeed,
Though now she's turned into a reed.
From that dear reed Pan's pipe does come,
A pipe that strikes Apollo dumb;
Nor flute, nor lute, nor gittern can
So chant it as the pipe of Pan;

13 hay-de-guys] dances

205
4 stains] eclipses

Cross-gartered swains, and dairy girls,
With faces smug, and round as pearls,
When Pan's shrill pipe begins to play,
With dancing wear out night and day; 10
The bagpipe's drone his hum lays by,
When Pan sounds up his minstrelsy;
His minstrelsy! O base! This quill,
Which at my mouth with wind I fill,
Puts me in mind, though her I miss,
That still my Syrinx' lips I kiss.

207

Pipenetta.

'LAS, how long shall I
And my maidenhead lie
In a cold bed all the night long?
I cannot abide it,
Yet away cannot chide it,
Though I find it does me some wrong.

Can any one tell
Where this fine thing does dwell,
That carries nor form nor fashion?
It both heats and cools, 10
'Tis a bauble for fools,
Yet catched at in every nation.

Say a maid were so crossed
As to see this toy lost,
Cannot hue and cry fetch it again?
'Las, no, for 'tis driven
Nor to hell nor to heaven;
When 'tis found, 'tis lost even then.

208

All.

SING to Apollo, God of Day,
Whose golden beams with morning play,
And make her eyes so brightly shine,
Aurora's face is called divine.
Sing to Phoebus, and that throne
Of diamonds which he sits upon;
Iô, paeans let us sing
To physic's and to poesy's king.

8 smug] smooth, sleek **207** 14 toy] trifle 8 physic] medicine

Crown all his altars with bright fire,
Laurels bind about his lyre, 10
A Daphnean coronet for his head,
The Muses dance about his bed,
When on his ravishing lute he plays;
Strew his temple round with bays.
Iô, paeans let us sing
To the glittering Delian king.

(Wr. 1589; pub. 1632)

FULKE GREVILLE, LORD BROOKE
1554–1628

from *Caelica*

209 THE world, that all contains, is ever moving;
The stars within their spheres for ever turned;
Nature, the queen of change, to change is loving,
And form to matter new is still adjourned.
Fortune, our fancy-god, to vary liketh;
Place is not bound to things within it placed;
The present time upon time passed striketh;
With Phoebus' wandering course the earth is graced.
The air still moves, and by its moving cleareth;
The fire up ascends, and planets feedeth; 10
The water passeth on, and all lets weareth;
The earth stands still, yet change of changes breedeth.
Her plants, which summer ripes, in winter fade;
Each creature in unconstant mother lieth;
Man made of earth, and for whom earth is made,
Still dying lives, and living ever dieth.
Only, like fate, sweet Myra never varies,
Yet in her eyes the doom of all change carries.

(Pub. 1633)

209
1 world] universe
4 adjourned] moved, transferred
11 lets] obstacles, bounds
14 mother] i.e. the earth.

210

I WITH whose colours Myra dressed her head,
I, that ware posies of her own hand making,
I, that mine own name in the chimneys read
By Myra finely wrought ere I was waking:
Must I look on, in hope time coming may
With change bring back my turn again to play?

I, that on Sunday at the church-stile found
A garland sweet, with true-love knots in flowers,
Which I to wear about mine arm was bound
That each of us might know that all was ours: 10
Must I now lead an idle life in wishes?
And follow Cupid for his loaves and fishes?

I, that did wear the ring her mother left,
I, for whose love she gloried to be blamed,
I, with whose eyes her eyes committed theft,
I, who did make her blush when I was named;
Must I lose ring, flowers, blush, theft and go naked,
Watching with sighs, till dead love be awaked?

I, that when drowsy Argus fell asleep,
Like Jealousy o'erwatched with desire, 20
Was even warned modesty to keep,
While her breath, speaking, kindled nature's fire:
Must I look on a-cold, while others warm them?
Do Vulcan's brothers in such fine nets arm them?

Was it for this that I might Myra see
Washing the water with her beauties, white?
Yet would she never write her love to me;
Thinks wit of change while thoughts are in delight?
Mad girls must safely love, as they may leave,
No man can print a kiss, lines may deceive. 30

(Pub. 1633)

211

ALL my senses, like beacon's flame,
Gave alarum to desire
To take arms in Cynthia's name,
And set all my thoughts on fire:

Fury's wit persuaded me,
Happy love was hazard's heir,
Cupid did best shoot and see
In the night where smooth is fair;
Up I start believing well
To see if Cynthia were awake; 10
Wonders I saw, who can tell?
And thus unto myself I spake:
'Sweet God Cupid where am I,
That by pale Diana's light
Such rich beauties do espy,
As harm our senses with delight?
Am I borne up to the skies?
See where Jove and Venus shine,
Showing in her heavenly eyes
That desire is divine: 20
Look where lies the milken way,
Way unto that dainty throne,
Where while all the gods would play,
Vulcan thinks to dwell alone.'
I gave reins to this conceit,
Hope went on the wheels of lust:
Fancy's scales are false of weight,
Thoughts take thought that go of trust.
I stepped forth to touch the sky,
I a god by Cupid dreams, 30
Cynthia who did naked lie
Runs away like silver streams,
Leaving hollow banks behind,
Who can neither forward move,
Nor, if rivers be unkind,
Turn away or leave to love.
There stand I, like Arctic pole,
Where Sol passeth o'er the line,
Mourning my benighted soul,
Which so loseth light divine. 40
There stand I like men that preach
From the execution place,
At their death content to teach
All the world with their disgrace:
He that lets his Cynthia lie,
Naked on a bed of play,
To say prayers ere she die,
Teacheth time to run away.

38 the line] the equator

Let no love-desiring heart,
In the stars go seek his fate, 50
Love is only Nature's art,
Wonder hinders love and hate.
None can well behold with eyes,
But what underneath him lies.

(Pub. 1633)

212

WHEN all this All doth pass from age to age,
And revolution in a circle turn,
Then heavenly justice doth appear like rage,
The caves do roar, the very seas do burn,
Glory grows dark, the sun becomes a night,
And makes this great world feel a greater might.

When love doth change his seat from heart to heart,
And worth about the wheel of fortune goes,
Grace is diseased, desert seems overthwart,
Vows are forlorn, and truth doth credit lose, 10
Chance then gives law, desire must be wise,
And look more ways than one, or lose her eyes.

My age of joy is past, of woe begun,
Absence my presence is, strangeness my grace,
With them that walk against me, is my sun:
The wheel is turned, I hold the lowest place.
What can be good to me since my love is,
To do me harm, content to do amiss?

(Pub. 1633)

213

LOVE is the peace, whereto all thoughts do strive,
Done and begun with all our powers in one:
The first and last in us that is alive,
End of the good, and therewith pleased alone.
Perfection's spirit, goddess of the mind,
Passed through hope, desire, grief and fear,
A simple goodness in the flesh refined,
Which of the joys to come doth witness bear.

212
1 All] universe
2 revolution] time taken for the universe to complete a full cycle
6 world] universe
9 overthwart] obstructed
10 forlorn] forsaken

213
8 to come] after death

Constant, because it sees no cause to vary,
A quintessence of passions overthrown, 10
Raised above all that change of objects carry,
A nature by no other nature known:
For glory's of eternity a frame,
That by all bodies else obscures her name.

(Pub. 1633)

214

THE earth with thunder torn, with fire blasted,
With waters drowned, with windy palsy shaken
Cannot for this with heaven be distasted;
Since thunder, rain and winds from earth are taken:
Man torn with love, with inward furies blasted,
Drowned with despair, with fleshly lustings shaken,
Cannot for this with heaven be distasted;
Love, fury, lustings out of man are taken.

Then man, endure thyself, those clouds will vanish;
Life is a top which whipping Sorrow driveth; 10
Wisdom must bear what our flesh cannot banish,
The humble lead, the stubborn bootless striveth:
Or man, forsake thyself, to heaven turn thee,
Her flames enlighten nature, never burn thee.

(Pub. 1633)

215

WHEN as man's life, the light of human lust,
In socket of his earthly lantern burns,
That all this glory unto ashes must,
And generation to corruption turns;
Then fond desires that only fear their end,
Do vainly wish for life, but to amend.

But when this life is from the body fled,
To see itself in that eternal glass,
Where time doth end, and thoughts accuse the dead,
Where all to come, is one with all that was; 10
Then living men ask how he left his breath,
That while he lived never thought of death.

(Pub. 1633)

13 frame] form, emanation

214
3 distasted] offended

216 MAN, dream no more of curious mysteries,
As what was here before the world was made,
The first man's life, the state of Paradise,
Where Heaven is, or Hell's eternal shade,
For God's works are like him, all infinite;
And curious search, but crafty sin's delight.

The flood that did, and dreadful fire that shall,
Drown, and burn up the malice of the earth,
The divers tongues, and Babylon's downfall,
Are nothing to the man's renewed birth; 10
First, let the Law plough up thy wicked heart,
That Christ may come, and all these types depart.

When thou hast swept the house that all is clear,
When thou the dust hast shaken from thy feet,
When God's all-might doth in thy flesh appear,
Then seas with streams above thy sky do meet;
For goodness only doth God comprehend,
Knows what was first, and what shall be the end.

(Pub. 1633)

217 ETERNAL Truth, almighty, infinite,
Only exiled from man's fleshly heart,
Where ignorance and disobedience fight,
In hell and sin, which shall have greatest part:
When thy sweet mercy opens forth the light,
Of grace which giveth eyes unto the blind,
And with the Law even ploughest up our sprite
To faith, wherein flesh may salvation find;
Thou bidd'st us pray, and we do pray to thee,
But as to power and God without us placed, 10
Thinking a wish may wear out vanity,
Or habits be by miracles defaced.
One thought to God we give, the rest to sin,
Quickly unbent is all desire of good,
True words pass out, but have no being within,
We pray to Christ, yet help to shed his blood;

216
12 types] events of Old Testament history, prefiguring some thing revealed in the new dispensation

For while we say *Believe*, and feel it not,
Promise amends, and yet despair in it,
Hear Sodom judged, and go not out with Lot,
Make Law and Gospel riddles of the wit: 20
We with the Jews even Christ still crucify,
As not yet come to our impiety.

(Pub. 1633)

218

WRAPT up, O Lord, in Man's degeneration,
The glories of thy truth, thy joys eternal,
Reflect upon my soul dark desolation,
And ugly prospects o'er the sprites infernal.
Lord, I have sinned, and mine iniquity
Deserves this hell; yet Lord deliver me.

Thy power and mercy never comprehended,
Rest lively imaged in my conscience wounded;
Mercy to grace, and power to fear extended,
Both infinite, and I in both confounded; 10
Lord, I have sinned, and mine iniquity,
Deserves this hell, yet Lord deliver me.

If from this depth of sin, this hellish grave,
And fatal absence from my Saviour's glory,
I could implore his mercy, who can save,
And for my sins, not pains of sin, be sorry;
Lord, from this horror of iniquity
And hellish grave, thou wouldst deliver me.

(Pub. 1633)

219

DOWN in the depth of mine iniquity,
That ugly centre of infernal spirits
Where each sin feels her own deformity
In these peculiar torments she inherits,
Deprived of human graces and divine,
Even there appears this saving God of mine.

And in this fatal mirror of transgression
Shows man, as fruit of his degeneration,
The error's ugly infinite impression,
Which bears the faithless down to desperation; 10
Deprived of human graces and divine,
Even there appears this saving God of mine;

22 As] who has

In power and truth, almighty and eternal,
Which on the sin reflects strange desolation,
With glory scourging all the sprites infernal,
And uncreated hell with unprivation;
  Deprived of human graces, not divine,
  Even there appears this saving God of mine.

For on this spiritual cross condemned lying,
To pains infernal by eternal doom, 20
I see my Saviour for the same sins dying,
And from that hell I feared, to free me, come;
  Deprived of human graces, not divine,
  Thus hath his death raised up this soul of mine.

(Pub. 1633)

220 THREE things there be in man's opinion dear,
Fame, many friends, and Fortune's dignities:
False visions all, which in our sense appear,
To sanctify desire's idolatries.

For what is Fortune, but a wat'ry glass?
Whose crystal forehead wants a steely back,
Where rain and storms bear all away that was,
Whose shape alike both depths and shallows wrack.

Fame again, which from blinding power takes light,
Both Caesar's shadow is, and Cato's friend, 10
The child of humour, not allied to right,
Living by oft exchange of winged end.

And many friends, false strength of feeble mind,
Betraying equals, as true slaves to might;
Like echoes still send voices down the wind,
But never in adversity find right.

Then man, though virtue of extremities
The middle be, and so hath two to one,
By place and nature constant enemies,
And against both these no strength but her own, 20
  Yet quit thou for her, friends, fame, Fortune's throne;
  Devils, there many be, and Gods but one.

(Pub. 1633)

16 unprivation] absolute power and goodness

220
21 quit thou for her] i.e. for virtue

221

SION lies waste, and thy Jerusalem,
O Lord, is fall'n to utter desolation,
Against thy prophets and thy holy men,
The sin hath wrought a fatal combination,
Prophaned thy name, thy worship overthrown,
And made thee, living Lord, a God unknown.

Thy powerful laws, thy wonders of creation,
Thy word incarnate, glorious heaven, dark hell,
Lie shadowed under man's degeneration,
Thy Christ still crucified for doing well. 10
Impiety, O Lord, sits on thy throne,
Which makes thee, living light, a God unknown.

Man's superstition hath thy truths entombed,
His atheism again her pomps defaceth,
That sensual unsatiable vast womb
Of thy seen church, thy unseen church disgraceth;
There lives no truth with them that seem thine own,
Which makes thee, living Lord, a God unknown.

Yet unto thee, Lord (mirror of transgression)
We who for earthly idols have forsaken 20
Thy heavenly image (sinless pure impression)
And so in nets of vanity lie taken,
All desolate implore that to thine own,
Lord, thou no longer live a God unknown.

Yet, Lord, let Israel's plagues not be eternal,
Nor sin for ever cloud thy sacred mountains,
Nor with false flames spiritual but infernal,
Dry up thy mercy's ever springing fountains.
Rather, sweet Jesus, fill up time and come,
To yield the sin her everlasting doom. 30

(Pub. 1633)

from *Mustapha*

222

[*Chorus of Priests*]

O WEARISOME condition of humanity!
Born under one law, to another bound;
Vainly begot, and yet forbidden vanity;
Created sick, commanded to be sound.
What meaneth Nature by these diverse laws?
Passion and Reason self-division cause.

Is it the mark or majesty of power
  To make offences that it may forgive?
Nature herself doth her own self deflower
  To hate those errors she herself doth give. 10
    For how should man think that he may not do
    If Nature did not fail, and punish too?

Tyrant to others, to her self unjust,
  Only commands things difficult and hard;
Forbids us all things which it knows is lust,
  Makes easy pains unpossible reward.
    If Nature did not take delight in blood,
    She would have made more easy ways to good.

We that are bound by vows and by promotion,
  With pomp of holy sacrifice and rites, 20
To teach belief in good and still devotion,
  To preach of Heaven's wonders and delights:
    Yet when each of us in his own heart looks,
    He finds the God there far unlike his books.

(Pub. 1609)

SIR WALTER RALEGH
*c.*1552–1618

223 *Praised be Diana's fair and harmless light*

PRAISED be Diana's fair and harmless light,
  Praised be the dews, wherewith she moists the ground,
Praised be her beams, the glory of the night,
  Praised be her power, by which all powers abound.

Praised be her nymphs, with whom she decks the woods,
  Praised be her knights, in whom true honour lives,
Praised be that force, by which she moves the floods;
  Let that Diana shine, which all these gives.

In heaven Queen she is among the spheres,
  In ay she mistress-like makes all things pure, 10
Eternity in her oft change she bears,
  She beauty is, by her the fair endure.

223
10 In ay] for ever

Time wears her not, she doth his chariot guide,
Mortality below her orb is placed,
By her the virtue of the stars down slide,
In her is virtue's perfect image cast.

A knowledge pure it is her worth to know,
With Circes let them dwell that think not so.

(Pub. 1593)

224 *Like truthless dreams*

LIKE truthless dreams, so are my joys expired,
And past return are all my dandled days;
My love misled, and fancy quite retired,
Of all which past, the sorrow only stays.

My lost delights, now clean from sight of land,
Have left me all alone in unknown ways;
My mind to woe, my life in fortune's hand,
Of all which past, the sorrow only stays.

As in a country strange without companion,
I only wail the wrong of death's delays, 10
Whose sweet spring spent, whose summer well nigh done,
Of all which past, the sorrow only stays;

Whom care forewarns, ere age and winter cold,
To haste me hence, to find my fortune's fold.

(Pub. 1593)

225 *Like to a hermit poor*

LIKE to a hermit poor in place obscure,
I mean to spend my days of endless doubt,
To wail such woes as time cannot recure,
Where none but Love shall ever find me out.

My food shall be of care and sorrow made,
My drink nought else but tears fall'n from mine eyes,
And for my light in such obscured shade
The flames shall serve which from my heart arise.

A gown of gray my body shall attire,
My staff of broken hope whereon I'll stay; 10
Of late repentance linked with long desire
The couch is framed whereon my limbs I'll lay.

And at my gate despair shall linger still,
To let in death when Love and Fortune will.

(Pub. 1593)

226 *Conceit begotten by the Eyes*

CONCEIT begotten by the eyes
Is quickly born, and quickly dies,
For while it seeks our hearts to have,
Meanwhile there reason makes his grave;
For many things the eyes approve,
Which yet the heart doth seldom love.

For as the seeds in springtime sown
Die in the ground ere they be grown,
Such is conceit, whose rooting fails,
As child that in the cradle quails, 10
Or else within the mother's womb
Hath his beginning, and his tomb.

Affection follows Fortune's wheels,
And soon is shaken from her heels;
For following beauty or estate,
Her liking still is turned to hate;
For all affections have their change,
And fancy only loves to range.

Desire himself runs out of breath,
And getting, doth but gain his death; 20
Desire, nor reason hath, nor rest,
And blind doth seldom choose the best;
Desire attained is not desire,
But as the cinders of the fire.

As ships in ports desired are drowned,
As fruit, once ripe, then falls to ground,
As flies that seek for flames are brought
To cinders by the flames they sought;
So fond desire when it attains,
The life expires, the woe remains. 30

226
1 conceit] judgement, understanding
13 Affection] passion, lust

And yet some poets fain would prove
Affection to be perfect love,
And that desire is of that kind,
No less a passion of the mind.
As if wild beasts and men did seek
To like, to love, to choose alike.

(Pub. 1602)

227 *Sir Walter Ralegh to the Queen*

OUR passions are most like to floods and streams;
The shallow murmur, but the deep are dumb.
So when affections yield discourse, it seems
The bottom is but shallow whence they come.
They that are rich in words must needs discover
That they are poor in that which makes a lover.

Wrong not, dear empress of my heart,
The merit of true passion,
With thinking that he feels no smart
That sues for no compassion; 10
Since, if my plaints serve not to prove
The conquest of your beauty,
It comes not from defect of love,
But from excess of duty.

For knowing that I sue to serve
A saint of such perfection,
As all desire, but none deserve,
A place in her affection,
I rather choose to want relief
Than venture the revealing; 20
When glory recommends the grief,
Despair distrusts the healing.

Thus those desires that aim too high
For any mortal lover,
When reason cannot make them die,
Discretion will them cover.
Yet when discretion doth bereave
The plaints that they should utter,
Then your discretion may perceive
That silence is a suitor. 30

227
21 glory] vainglory

Silence in love bewrays more woe
  Than words, though ne'er so witty;
A beggar that is dumb, ye know,
  Deserveth double pity.
Then misconceive not (dearest heart)
  My true, though secret passion;
He smarteth most that hides his smart,
  And sues for no compassion.

(Wr. early 1590s (?); pub. 1655)

228 *As you came from the holy land*

As you came from the holy land
  Of Walsinghame,
Met you not with my true love
  By the way as you came?

How shall I know your true love,
  That have met many one
As I went to the holy land,
  That have come, that have gone?

She is neither white nor brown,
  But as the heavens fair, 10
There is none hath a form so divine
  In the earth or the air.

Such an one did I meet, good sir,
  Such an angel-like face,
Who like a queen, like a nymph, did appear
  By her gait, by her grace.

She hath left me here all alone,
  All alone as unknown,
Who sometimes did me lead with herself,
  And me loved as her own. 20

What's the cause that she leaves you alone
  And a new way doth take,
Who loved you once as her own
  And her joy did you make?

31 bewrays] reveals, makes known

228
19 sometimes] sometime, once
lead with herself] accompany

I have loved her all my youth,
  But now old as you see;
Love likes not the falling fruit
  From the withered tree.

Know that Love is a careless child,
  And forgets promise past; 30
He is blind, he is deaf when he list
  And in faith never fast.

His desire is a dureless content
  And a trustless joy;
He is won with a world of despair
  And is lost with a toy.

Of womenkind such indeed is the love
  Or the word love abused,
Under which many childish desires
  And conceits are excused. 40

But true love is a durable fire
  In the mind ever burning;
Never sick, never old, never dead,
  From itself never turning.

(Pub. 1631(?), 1678)

229 *If all the world and love were young*

If all the world and love were young,
And truth in every shepherd's tongue,
These pretty pleasures might me move
To live with thee and be thy love.

Time drives the flocks from field to fold,
When rivers rage and rocks grow cold,
And Philomel becometh dumb;
The rest complain of cares to come.

32 fast] sure, firm
33 dureless] not lasting, transient
36 toy] trifle, idle fancy
44 turning] changing

229
1 If all the world] (see Marlowe's poem 'Come live with me and be my love', to which this is a reply)
7 Philomel] the nightingale

The flowers do fade, and wanton fields
To wayward winter reckoning yields; 10
A honey tongue, a heart of gall,
Is fancy's spring, but sorrow's fall.

Thy gowns, thy shoes, thy beds of roses,
Thy cap, thy kirtle, and thy posies
Soon break, soon wither, soon forgotten,
In folly ripe, in reason rotten.

Thy belt of straw and ivy buds,
Thy coral clasps and amber studs,
All these in me no means can move
To come to thee and be thy love. 20

But could youth last and love still breed,
Had joys no date nor age no need,
Then these delights my mind might move
To live with thee and be thy love.

(Pub. 1600)

230 *Sir Walter Ralegh to his son*

THREE things there be that prosper up apace,
And flourish, whilst they grow asunder far,
But on a day, they meet all in one place,
And when they meet they one another mar.
And they be these: the wood, the weed, the wag.
The wood is that, which makes the gallow tree;
The weed is that, which strings the hangman's bag;
The wag, my pretty knave, betokeneth thee.
Mark well, dear boy, whilst these assemble not,
Green springs the tree, hemp grows, the wag is wild; 10
But when they meet, it makes the timber rot,
It frets the halter, and it chokes the child.
Then bless thee, and beware, and let us pray,
We part not with thee at this meeting day.

(Pub. 1870)

14 kirtle] skirt, outer petticoat

22 date] end

231 *Farewell false love*

FAREWELL false love, the oracle of lies,
  A mortal foe and enemy to rest,
An envious boy, from whom all cares arise,
  A bastard vile, a beast with rage possessed,
    A way of error, a temple full of treason,
    In all effects contrary unto reason.

A poisoned serpent covered all with flowers,
  Mother of sighs, and murderer of repose,
A sea of sorrows from whence are drawn such showers
  As moisture lend to every grief that grows, 10
    A school of guile, a net of deep deceit,
    A gilded hook, that holds a poisoned bait.

A fortress foiled, which reason did defend,
  A Syren song, a fever of the mind,
A maze wherein affection finds no end,
  A ranging cloud that runs before the wind,
    A substance like the shadow of the sun,
    A goal of grief for which the wisest run.

A quenchless fire, a nurse of trembling fear,
  A path that leads to peril and mishap, 20
A true retreat of sorrow and despair,
  An idle boy that sleeps in pleasure's lap,
    A deep mistrust of that which certain seems,
    A hope of that which reason doubtful deems.

Sith then thy trains my younger years betrayed
  And for my faith ingratitude I find,
And sith repentance hath my wrongs bewrayed
  Whose course was ever contrary to kind,
    False love, desire, and beauty frail adieu!
    Dead is the root whence all these fancies grew. 30

(1588)

3 envious] malicious
15 affection] passion, feeling
25 sith] since
27 bewrayed] revealed
28 kind] nature

232

A Vision upon this Conceit of 'The Faerie Queene'

METHOUGHT I saw the grave, where Laura lay,
  Within that temple, where the vestal flame
Was wont to burn, and passing by that way,
  To see that buried dust of living fame,
Whose tomb fair Love and fairer Virtue kept,
  All suddenly I saw the Faerie Queene;
At whose approach the soul of Petrarch wept,
  And from thenceforth those graces were not seen,
For they this Queen attended, in whose stead
  Oblivion laid him down on Laura's hearse. 10
Hereat the hardest stones were seen to bleed,
  And groans of buried ghosts the heavens did pierce;
    Where Homer's spright did tremble all for grief,
    And cursed th'access of that celestial thief.

(1590)

233

The Lie

GO, soul, the body's guest,
  Upon a thankless arrant;
Fear not to touch the best;
  The truth shall be thy warrant.
    Go, since I needs must die,
    And give the world the lie.

Say to the court, it glows
  And shines like rotten wood;
Say to the church, it shows
  What's good, and doth no good: 10
    If church and court reply,
    Then give them both the lie.

Tell potentates, they live
  Acting by others' action,
Not loved unless they give,
  Not strong but by affection.
    If potentates reply,
    Give potentates the lie.

233
2 arrant] errand   16 affection] passion, lust

Tell men of high condition
  That manage the estate, 20
Their purpose is ambition,
  Their practice only hate:
    And if they once reply,
    Then give them all the lie.

Tell them that brave it most,
  They beg for more by spending,
Who, in their greatest cost,
  Seek nothing but commending:
    And if they make reply,
    Then give them all the lie. 30

Tell zeal it wants devotion;
  Tell love it is but lust;
Tell time it metes but motion;
  Tell flesh it is but dust:
    And wish them not reply,
    For thou must give the lie.

Tell age it daily wasteth;
  Tell honour how it alters;
Tell beauty how she blasteth;
  Tell favour how it falters: 40
    And as they shall reply,
    Give every one the lie.

Tell wit how much it wrangles
  In tickle points of niceness;
Tell wisdom she entangles
  Herself in over-wiseness:
    And when they do reply,
    Straight give them both the lie.

Tell physic of her boldness;
  Tell skill it is prevention; 50
Tell charity of coldness;
  Tell law it is contention:
    And as they do reply,
    So give them still the lie.

25 brave it] dress extravagantly   33 metes] measures   49 physic] medicine

Tell fortune of her blindness;
  Tell nature of decay;
Tell friendship of unkindness;
  Tell justice of delay:
    And if they will reply,
    Then give them all the lie. 60

Tell arts they have no soundness,
  But vary by esteeming;
Tell schools they want profoundness,
  And stand too much on seeming:
    If arts and schools reply,
    Give arts and schools the lie.

Tell faith it's fled the city;
  Tell how the country erreth;
Tell, manhood shakes off pity;
  Tell, virtue least preferreth: 70
    And if they do reply,
    Spare not to give the lie.

So when thou hast, as I
  Commanded thee, done blabbing,
Although to give the lie
  Deserves no less than stabbing,
    Stab at thee he that will,
    No stab thy soul can kill.

(Pub. 1608)

234 *Fortune hath taken thee away, my love*

FORTUNE hath taken thee away, my love,
My life's soul, and my soul's heaven above;
Fortune hath taken thee away, my princess,
My only light, and my true fancy's mistress.

Fortune hath taken all away from me,
Fortune hath taken all by taking thee;
Dead to all joy, I only live to woe,
So fortune now becomes my mortal foe.

In vain, mine eyes, in vain you waste your tears,
In vain my sighs do smoke forth my despairs, 10
In vain you search the earth and heaven above,
In vain you search, for fortune rules in love.

Thus now I leave my love in fortune's hands,
Thus now I leave my love in fortune's bands,
And only love the sorrows due to me;
Sorrow henceforth it shall my princess be.

I joy in this, that fortune conquers kings;
Fortune, that rules on earth and earthly things,
Hath taken my love in spite of Cupid's might;
So blind a dame did never Cupid right. 20

With wisdom's eyes had but blind fortune seen,
Then had my love my love for ever been;
But love farewell, though fortune conquer thee:
No fortune base shall ever alter me.

(Pub. 1960)

235 *The Ocean to Cynthia*

SUFFICETH it to you, my joys interred,
In simple words that I my woes complain,
You that then died when first my fancy erred,
Joys under dust that never live again.

If to the living were my Muse addressed,
Or did my mind her own spirit still inhold,
Were not my living passion so repressed
As to the dead the dead did these unfold,

Some sweeter words, some more becoming verse,
Should witness my mishap in higher kind; 10
But my love's wounds, my fancy in the hearse,
The idea but resting of a wasted mind,

The blossoms fallen, the sap gone from the tree,
The broken monuments of my great desires;
From these so lost what may th'affections be?
What heat in cinders of extinguished fires?

Lost in the mud of those high-flowing streams,
Which through more fairer fields their courses bend,
Slain with self-thoughts, amazed in fearful dreams,
Woes without date, discomforts without end, 20

12 resting] remaining 20 date] termination

From fruitful trees I gather withered leaves,
And glean the broken ears with miser's hand;
Who sometime did enjoy the weighty sheaves,
I seek fair flowers amid the brinish sand.

All in the shade, even in the fair sun days,
Under those healthless trees I sit alone,
Where joyful birds sing neither lovely lays,
Nor Philomen recounts her direful moan.

No feeding flocks, no shepherds' company,
That might renew my dolorous conceit, 30
While happy then, while love and fantasy
Confined my thoughts on that fair flock to wait;

No pleasing streams fast to the ocean wending,
The messengers sometimes of my great woe;
But all on earth, as from the cold storms bending,
Shrink from my thoughts in high heavens and below.

O hopeful love, my object, and invention,
O true desire, the spur of my conceit,
O worthiest spirit, my mind's impulsion,
O eyes transpersant, my affection's bait; 40

O princely form, my fancy's adamant,
Divine conceit, my pain's acceptance,
O all in one oh, heaven on earth transparent,
The seat of joys and love's abundance!

Out of that mass of miracles my Muse
Gathered those flowers, to her pure senses pleasing;
Out of her eyes (the store of joys) did choose
Equal delights, my sorrows counterpeising.

Her regal looks my rigorous sighs suppressed;
Small drops of joys sweetened great worlds of woes; 50
One gladsome day a thousand cares redressed.
Whom Love defends, what fortune overthrows?

When she did well, what did there else amiss?
When she did ill, what empires could have pleased?
No other power effecting woe or bliss,
She gave, she took, she wounded, she appeased.

41 adamant] magnet

The honour of her love, love still devising,
  Wounding my mind with contrary conceit,
Transferred itself sometime to her aspiring,
  Sometime the trumpet of her thought's retreat. 60

To seek new worlds for gold, for praise, for glory,
  To try desire, to try love severed far,
When I was gone, she sent her memory,
  More strong than were ten thousand ships of war,

To call me back, to leave great honour's thought,
  To leave my friends, my fortune, my attempt,
To leave the purpose I so long had sought,
  And hold both cares and comforts in contempt.

Such heat in ice, such fire in frost remained,
  Such trust in doubt, such comfort in despair; 70
Much like the gentle lamb, though lately weaned,
  Plays with the dug, though finds no comfort there.

But as a body, violently slain,
  Retaineth warmth although the spirit be gone,
And by a power in nature moves again,
  Till it be laid below the fatal stone;

Or as the earth, even in cold winter days,
  Left for a time by her life-giving sun,
Doth by the power remaining of his rays
  Produce some green, though not as it hath done; 80

Or as a wheel, forced by the falling stream,
  Although the course be turned some other way,
Doth for a time go round upon the beam,
  Till, wanting strength to move, it stands at stay;

So my forsaken heart, my withered mind,
  Widow of all the joys it once possessed,
My hopes clean out of sight with forced wind,
  To kingdoms strange, to lands far-off, addressed,

Alone, forsaken, friendless, on the shore,
  With many wounds, with death's cold pangs embraced, 90
Writes in the dust, as one that could no more,
  Whom love, and time, and fortune, had defaced,

Of things so great, so long, so manifold,
  With means so weak, the soul even then departing,
The weal, the woe, the passages of old,
  And worlds of thoughts described by one last sighing;

As if, when after Phoebus is descended,
  And leaves a light much like the past day's dawning,
And, every toil and labour wholly ended,
  Each living creature draweth to his resting, 100

We should begin by such a parting light
  To write the story of all ages past,
And end the same before th'approaching night.

Such is again the labour of my mind,
  Whose shroud, by sorrow woven now to end,
Hath seen that ever shining sun declined,
  So many years that so could not descend,

But that the eyes of my mind held her beams
  In every part transferred by love's swift thought;
Far off or near, in waking or in dreams, 110
  Imagination strong their lustre brought.

Such force her angel-like appearance had
  To master distance, time, or cruelty;
Such art to grieve, and after to make glad;
  Such fear in love, such love in majesty.

My weary limbs her memory embalmed;
  My darkest ways her eyes make clear as day.
What storms so great but Cynthia's beams appeased?
  What rage so fierce, that love could not allay?

Twelve years entire I wasted in this war, 120
  Twelve years of my most happy younger days;
But I in them, and they now wasted are,
  'Of all which past the sorrow only stays'.

So wrote I once, and my mishap foretold,
  My mind still feeling sorrowful success,
Even as before a storm the marble cold
  Doth by moist tears tempestuous times express.

116 embalmed] anointed and soothed

125 sorrowful success] the sadness to come

So felt my heavy mind my harms at hand,
  Which my vain thought in vain sought to recure;
At middle day my sun seemed under land, 130
  When any little cloud did it obscure.

And as the icicles in a winter's day,
  Whenas the sun shines with unwonted warm,

So did my joys melt into secret tears,
  So did my heart dissolve in wasting drops;
And as the season of the year outwears,
  And heaps of snow from off the mountain tops

With sudden streams the valleys overflow,
  So did the time draw on my more despair;
Then floods of sorrow and whole seas of woe 140
  The banks of all my hope did overbear,

And drowned my mind in depths of misery.
  Sometime I died, sometime I was distract,
My soul the stage of fancy's tragedy;
  Then furious madness, where true reason lacked,

Wrote what it would, and scourged mine own conceit.
  O heavy heart, who can thee witness bear?
What tongue, what pen, could thy tormenting treat,
  But thine own mourning thoughts which present were?

What stranger mind believe the meanest part? 150
  What altered sense conceive the weakest woe,
That tare, that rent, that pierced thy sad heart?
  And as a man distract, with treble might,

Bound in strong chains doth strive and rage in vain,
  Till, tired and breathless, he is forced to rest,
Finds by contention but increase of pain,
  And fiery heat inflamed in swollen breast;

So did my mind in change of passion
  From woe to wrath, from wrath return to woe,
Struggling in vain from love's subjection. 160

Therefore, all lifeless and all helpless bound,
  My fainting spirits sunk, and heart appaled,
My joys and hopes lay bleeding on the ground,
  That not long since the highest heaven scaled.

I hated life and cursed destiny;
  The thoughts of passed times, like flames of hell,
Kindled afresh within my memory
  The many dear achievements that befell

In those prime years and infancy of love,
  Which to describe were but to die in writing; 170
Ah, those I sought, but vainly, to remove,
  And vainly shall, by which I perish living.

And though strong reason hold before mine eyes
  The images and forms of worlds past,
Teaching the cause why all those flames that rise
  From forms external can no longer last,

Than that those seeming beauties hold in prime
  Love's ground, his essence, and his empery,
All slaves to age, and vassals unto time,
  Of which repentance writes the tragedy. 180

But this my heart's desire could not conceive,
  Whose love outflew the fastest flying time,
A beauty that can easily deceive
  Th'arrest of years, and creeping age outclimb,

A spring of beauties which time ripeth not,
  Time that but works on frail mortality,
A sweetness which woe's wrongs outwipeth not,
  Whom love hath chose for his divinity,

A vestal fire that burns but never wasteth,
  That loseth nought by giving light to all, 190
That endless shines each where, and endless lasteth,
  Blossoms of pride that can nor vade nor fall.

These were those marvellous perfections,
  The parents of my sorrow and my envy,
Most deathful and most violent infections;
  These be the tyrants that in fetters tie

162 appaled] grew pale   171 remove] move again   192 vade] fade

Their wounded vassals, yet nor kill nor cure,
  But glory in their lasting misery,
That, as her beauties would, our woes should dure;
  These be the effects of powerful empery. 200

Yet have these wonders want, which want compassion;
  Yet hath her mind some marks of human race;
Yet will she be a woman for a fashion,
  So doth she please her virtues to deface.

And like as that immortal power doth seat
  An element of waters, to allay
The fiery sunbeams that on earth do beat,
  And temper by cold night the heat of day,

So hath perfection, which begat her mind,
  Added thereto a change of fantasy, 210
And left her the affections of her kind,
  Yet free from every evil but cruelty.

But leave her praise; speak thou of nought but woe;
  Write on the tale that Sorrow bids thee tell;
Strive to forget, and care no more to know
  Thy cares are known, by knowing those too well.

Describe her now as she appears to thee,
  Not as she did appear in days fordone;
In love, those things that were no more may be,
  For fancy seldom ends where it begun. 220

And as a stream by strong hand bounded in
  From nature's course where it did sometime run,
By some small rent or loose part doth begin
  To find escape, till it a way hath won;

Doth then all unawares in sunder tear
  The forced bounds, and, raging, run at large
In th'ancient channels as they wonted were;
  Such is of women's love the careful charge,

Held and maintained with multitude of woes;
  Of long erections such the sudden fall. 230
One hour diverts, one instant overthrows,
  For which our life's, for which our fortune's, thrall.

228 careful] causing sorrow

So many years those joys have dearly bought,
  Of which when our fond hopes do most assure,
All is dissolved; our labours come to nought,
  Nor any mark thereof there doth endure;

No more than, when small drops of rain do fall
  Upon the parched ground by heat updried,
No cooling moisture is perceived at all,
  Nor any show or sign of wet doth bide. 240

But as the fields, clothed with leaves and flowers,
  The banks of roses smelling precious sweet,
Have but their beauty's date and timely hours,
  And then defaced by winter's cold and sleet,

So far as neither fruit nor form of flower
  Stays for a witness what such branches bare,
But as time gave, time did again devour,
  And change our rising joy to falling care;

So of affection which our youth presented.
  When she that from the sun reaves power and light, 250
Did but decline her beams as discontented,
  Converting sweetest days to saddest night,

All droops, all dies, all trodden under dust,
  The person, place, and passages forgotten,
The hardest steel eaten with softest rust,
  The firm and solid tree both rent and rotten.

Those thoughts, so full of pleasure and content,
  That in our absence were affection's food,
Are razed out and from the fancy rent,
  In highest grace and heart's dear care that stood, 260

Are cast for prey to hatred and to scorn;
  Our dearest treasures and our heart's true joys,
The tokens hung on breast and kindly worn,
  Are now elsewhere disposed or held for toys,

And those which then our jealousy removed,
  And others for our sakes then valued dear,
The one forgot, the rest are dear beloved,
  When all of ours doth strange or vild appear.

254 passages] confidences, amorous relations 265 removed] moved

Those streams seem standing puddles, which before
We saw our beauties in, so were they clear; 270
Belphebe's course is now observed no more;

That fair resemblance weareth out of date;
Our ocean seas are but tempestuous waves,
And all things base, that blessed were of late.

And as a field, wherein the stubble stands
Of harvest past the ploughman's eye offends,
He tills again, or tears them up with hands,
And throws to fire as foiled and fruitless ends,

And takes delight another seed to sow;
So doth the mind root up all wonted thought, 280
And scorns the care of our remaining woes;
The sorrows, which themselves for us have wrought,

Are burnt to cinders by new kindled fires;
The ashes are dispersed into the air;
The sighs, the groans of all our past desires
Are clean outworn, as things that never were.

With youth is dead the hope of love's return,
Who looks not back to hear our after cries;
Where he is not, he laughs at those that mourn;
Whence he is gone, he scorns the mind that dies; 290

When he is absent, he believes no words;
When reason speaks, he careless stops his ears;
Whom he hath left, he never grace affords,
But bathes his wings in our lamenting tears.

Unlasting passion, soon outworn conceit,
Whereon I built, and on so dureless trust!
My mind had wounds, I dare not say deceit,
Where I resolved her promise was not just.

Sorrow was my revenge and woe my hate;
I powerless was to alter my desire; 300
My love is not of time or bound to date;
My heart's internal heat and living fire

275 stands] stalks

Would not, or could, be quenched with sudden showers;
  My bound respect was not confined to days,
My vowed faith not set to ended hours;
  I love the bearing and not bearing sprays

Which now to others do their sweetness send,
  Th'incarnate, snow-driven white, and purest azure,
Who from high heaven doth on their fields descend,
  Filling their barns with grain, and towers with treasure. 310

Erring or never erring, such is Love
  As, while it lasteth, scorns th'accompt of those
Seeking but self contentment to improve,
  And hides, if any be, his inward woes,

And will not know, while he knows his own passion,
  The often and unjust perseverance
In deeds of love and state, and every action
  From that first day and year of their joy's entrance.

But I, unblessed and ill born creature,
  That did embrace the dust her body bearing, 320
That loved her both by fancy and by nature,
  That drew, even with the milk in my first sucking,

Affection from the parent's breast that bare me,
  Have found her as a stranger so severe,
Improving my mishap in each degree.
  But love was gone; so would I my life were!

A queen she was to me, no more Belphebe,
  A lion then, no more a milk-white dove;
A prisoner in her breast I could not be;
  She did untie the gentle chains of love. 330

Love was no more the love of hiding . . .

All trespass and mischance for her own glory.
  It had been such; it was still for the elect;
But I must be th'example in love's story;
  This was of all forepast the sad effect.

308 incarnate] pink   309 Who] which

But thou, my weary soul and heavy thought,
  Made by her love a burden to my being,
Dost know my error never was forethought,
  Or ever could proceed from sense of loving.

Of other cause if then it had proceeding, 340
  I leave th'excuse, sith judgement hath been given;
The limbs divided, sundered, and a-bleeding,
  Cannot complain the sentence was uneven.

This did that nature's wonder, virtue's choice,
  The only paragon of time's begetting,
Divine in words, angelical in voice,
  That spring of joys, that flower of love's own setting,

The Idea remaining of those golden ages,
  That beauty, braving heaven's and earth embalming,
Which after worthless worlds but play on stages; 350
  Such didst thou her long since describe, yet sighing

That thy unable spirit could not find aught
  In heaven's beauties or in earth's delight,
For likeness fit to satisfy thy thought.
  But what hath it availed thee so to write?

She cares not for thy praise, who knows not theirs;
  It's now an idle labour, and a tale
Told out of time, that dulls the hearer's ears,
  A merchandise whereof there is no sale.

Leave them, or lay them up with thy despairs. 360
  She hath resolved, and judged thee long ago.
Thy lines are now a murmuring to her ears,
  Like to a falling stream, which, passing slow,

Is wont to nourish sleep and quietness.
  So shall thy painful labours be perused,
And draw on rest, which sometime had regard;
  But those her cares thy errors have excused;

Thy days fordone have had their day's reward.
  So her hard heart, so her estranged mind,
In which above the heavens I once reposed; 370
  So to thy error have her ears inclined,

349 braving] challenging

350 after worthless worlds] i.e. worthless after-worlds

And have forgotten all thy past deserving,
  Holding in mind but only thine offence;
And only now affecteth thy depraving,
  And thinks all vain that pleadeth thy defence.

Yet greater fancy beauty never bred;
  A more desire the heart-blood never nourished;
Her sweetness an affection never fed,
  Which more in any age hath ever flourished.

The mind and virtue never have begotten 380
  A firmer love, since love on earth had power,
A love obscured, but cannot be forgotten,
  Too great and strong for time's jaws to devour,

Containing such a faith as ages wound not,
  Care, wakeful ever of her good estate,
Fear, dreading loss, which sighs and joys not,
  A memory of the joys her grace begat,

A lasting gratefulness for those comforts past,
  Of which the cordial sweetness cannot die.
These thoughts, knit up by faith, shall ever last, 390
  These time assays, but never can untie,

Whose life once lived in her pearl-like breast,
  Whose joys were drawn but from her happiness,
Whose heart's high pleasure, and whose mind's true rest,
  Proceeded from her fortune's blessedness;

Who was intentive, wakeful, and dismayed
  In fears, in dreams, in feverous jealousy;
Who long in silence served, and obeyed
  With secret heart and hidden loyalty,

Which never change to sad adversity, 400
  Which never age, or nature's overthrow,
Which never sickness or deformity,
  Which never wasting care or wearing woe,

If subject unto these she could have been . . .

Which never words or wits malicious,
  Which never honour's bait, or world's fame,
Achieved by attempts adventurous,
  Or aught beneath the sun or heaven's frame,

396 intentive] attentive, heedful

Can so dissolve, dissever, or destroy,
  The essential love of no frail parts compounded, 410
Though of the same now buried be the joy,
  The hope, the comfort, and the sweetness ended,

But that the thoughts and memories of these
  Work a relapse of passion, and remain
Of my sad heart the sorrow-sucking bees;
  The wrongs received, the scorns, persuade in vain.

And though these medicines work desire to end,
  And are in others the true cure of liking,
The salves that heal love's wounds, and do amend
  Consuming woe, and slake our hearty sighing, 420

They work not so in thy mind's long disease;
  External fancy time alone recureth,
All whose effects do wear away with ease.
  Love of delight, while such delight endureth;

Stays by the pleasure, but no longer stays . . .

But in my mind so is her love inclosed,
  And is thereof not only the best part
But into it the essence is disposed.
  O love (the more my woe), to it thou art

Even as the moisture in each plant that grows; 430
  Even as the sun unto the frozen ground;
Even as the sweetness to th'incarnate rose;
  Even as the centre in each perfect round;

As water to the fish, to men as air,
  As heat to fire, as light unto the sun;
O love, it is but vain to say thou were;
  Ages and times cannot thy power outrun.

Thou art the soul of that unhappy mind
  Which, being by nature made an idle thought,
Begun even then to take immortal kind, 440
  When first her virtues in thy spirits wrought.

420 hearty sighing] sighing from the heart

432 incarnate] pink, crimson

From thee therefore that mover cannot move,
  Because it is become thy cause of being;
Whatever error may obscure that love,
  Whatever frail effect of mortal living,

Whatever passion from distempered heart,
  What absence, time, or injuries effect,
What faithless friends or deep dissembled art
  Present to feed her most unkind suspect.

Yet as the air in deep caves underground 450
  Is strongly drawn when violent heat hath rent
Great clefts therein, till moisture do abound,
  And then the same, imprisoned and up-pent,

Breaks out in earthquakes tearing all asunder;
  So, in the centre of my cloven heart,
My heart, to whom her beauties were such wonder,
  Lies the sharp poisoned head of that love's dart,

Which, till all break and all dissolve to dust,
  Thence drawn it cannot be, or therein known.
There, mixed with my heart-blood, the fretting rust 460
  The better part hath eaten and outgrown.

But what of those or these, or what of aught
  Of that which was, or that which is, to treat?
What I possess is but the same I sought;
  My love was false, my labours were deceit.

Nor less than such they are esteemed to be;
  A fraud bought at the price of many woes;
A guile, whereof the profits unto me—
  Could it be thought premeditate for those?

Witness those withered leaves left on the tree, 470
  The sorrow-worn face, the pensive mind.
The external shows what may th'internal be;
  Cold care hath bitten both the root and vind.

But stay, my thoughts, make end, give fortune way.
  Harsh is the voice of woe and sorrow's sound;
Complaints cure not, and tears do but allay
  Griefs for a time, which after more abound.

472 external] exterior   internal] interior   473 vind] vine

To seek for moisture in th'Arabian sands
  Is but a loss of labour and of rest.
The links which time did break of hearty bands 480

Words cannot knit, or wailings make anew.
  Seek not the sun in clouds when it is set.
On highest mountains, where those cedars grew,
  Against whose banks the troubled ocean beat,

And were the marks to find thy hoped port,
  Into a soil far off themselves remove.
On Sestus' shore, Leander's late resort,
  Hero hath left no lamp to guide her love.

Thou lookest for light in vain, and storms arise;
  She sleeps thy death, that erst thy danger sighed, 490
Strive then no more; bow down thy weary eyes,
  Eyes which to all these woes thy heart have guided.

She is gone, she is lost, she is found, she is ever fair.
  Sorrow draws weakly, where love draws not too;
Woe's cries sound nothing, but only in love's ear;
  Do then by dying what life cannot do.

Unfold thy flocks and leave them to the fields,
  To feed on hills, or dales, where likes them best,
Of what the summer or the spring time yields,
  For love and time hath given thee leave to rest. 500

Thy heart which was their fold, now in decay,
  By often storms and winter's many blasts,
All torn and rent becomes misfortune's prey;
  False hope, my shepherd's staff, now age hath brast.

My pipe, which love's own hand gave my desire
  To sing her praises and my woe upon,
Despair hath often threatened to the fire,
  As vain to keep now all the rest are gone.

Thus home I draw, as death's long night draws on;
  Yet, every foot, old thoughts turn back mine eyes. 510
Constraint me guides, as old age draws a stone
  Against the hill, which over-weighty lies

480 hearty] located in the heart

504 brast] broken

For feeble arms or wasted strength to move;
  My steps are backward, gazing on my loss,
My mind's affection and my soul's sole love,
  Not mixed with fancy's chaff or fortune's dross.

To God I leave it, who first gave it me,
  And I her gave, and she returned again,
As it was hers; so let his mercies be
  Of my last comforts the essential mean. 520

But be it so or not, th'effects are past;
Her love hath end; my woe must ever last.

(Pub. 1870)

Translations from *The History of the World*

236

from *Catullus*

THE sun may set and rise
But we contrariwise
Sleep after our short light
One everlasting night.

237

from *Euripides*

HEAVEN and Earth one form did bear;
But when disjoined once they were
  From mutual embraces,
All things to light appeared then,
Of trees, birds, beasts, fishes, and men
  The still-remaining races.

238

from *Ausonius*

I AM that Dido which thou here dost see,
Cunningly framed in beauteous imagery.
Like this I was, but had not such a soul
As Maro feigned, incestuous and foul.
Æneas never with his Trojan host
Beheld my face, or landed on this coast;
But flying proud Iarbas' villainy,
Not moved by furious love or jealousy,

I did with weapons chaste, to save my fame,
Make way for death untimely, ere it came. 10
This was my end; but first I built a town,
Revenged my husband's death, lived with renown.
Why didst thou stir up Virgil, envious Muse,
Falsely my name and honour to abuse?
Readers, believe historians, not those
Which to the world Jove's thefts and vice expose.
Poets are liars, and for verses' sake
Will make the Gods of human crimes partake.

(Pub. 1614)

239 *What is our life?*

WHAT is our life? A play of passion,
Our mirth the music of division.
Our mother's wombs the tiring-houses be,
Where we are dressed for this short comedy.
Heaven the judicious sharp spectator is,
That sits and marks still who doth act amiss.
Our graves that hide us from the searching sun
Are like drawn curtains when the play is done.
Thus march we, playing, to our latest rest,
Only we die in earnest, that's no jest. 10

(Pub. 1612)

240 *Verses made the night before he died*

EVEN such is Time, which takes in trust
Our youth, our joys, and all we have,
And pays us but with age and dust;
Who in the dark and silent grave,
When we have wandered all our ways,
Shuts up the story of our days:
And from which earth, and grave, and dust,
The Lord shall raise me up, I trust.

(Pub. 1618)

13 envious] malicious

239
2 division] a florid melodic passage
3 tiring-houses] dressing-rooms

241 *On the snuff of a candle, the night before he died*

COWARDS fear to die, but courage stout,
Rather than live in snuff, will be put out.

(Pub. 1651)

ATTRIBUTED TO SIR WALTER RALEGH

242 *To his love when he had obtained her*

NOW, Serena, be not coy;
Since we freely may enjoy
Sweet embraces, such delights
As will shorten tedious nights,
Think that beauty will not stay
With you always, but away;
And that tyrannising face
That now holds such perfect grace,
Will both changed and ruined be;
So frail is all things as we see, 10
So subject unto conquering Time.
Then gather flowers in their prime,
Let them not fall and perish so.
Nature her bounties did bestow
On us that we might use them; And
'Tis coldness not to understand
What she and youth and form persuade
With opportunity, that's made
As we could wish it. Let's then meet
Often with amorous lips, and greet 20
Each other till our wanton kisses
In number pass the days Ulysses
Consumed in travel, and the stars
That look upon our peaceful wars
With envious lustre. If this store
Will not suffice, we'll number o'er
The same again, until we find
No number left to call to mind
 And show our plenty. They are poor
 That can count all they have and more. 30

(Pub. 1931)

2 snuff] candle-end

242
17 form] proper behaviour, decorum

SIR ARTHUR GORGES

1557–1625

243 [*Dialogue from* Desportes]

Mistress. TELL me, my heart, how wilt thou do
Now I must part from thy sweet sight.

Servant. Even as a body without soul,
Or as the eyes that want their light.

Mistress. But shall this parting of us two
Our minds likewise asunder set?

Servant. No, no; that heart loves not in deed
That can in absence so forget.

Mistress. Of all the griefs that absence brings
What one dost thou most grievous find? 10

Servant. Lest you with change of time and place
Should likewise change your friendly mind.

Mistress. In using of such doubtful speech
My constancy you do disprove.

Servant. Not so; but rather I do show
With how great care and fear I love.

Mistress. But yet methinks my tried faith
Should put thee out of all such fear.

Servant. Alas, who would not live in dread
To hazard that they hold most dear? 20

Mistress. And would you much aggrieved be
If other liking me possessed?

Servant. Yea, all as much as I would joy
To know that still you loved me best.

Mistress. Tell me, my dear, at this farewell
What kind of passion dost thou try?

Servant. As one that had his sentence given
And knew that he the death must die.

Mistress. Beware, for if thou die with grief
Then must your love end with your breath. 30

Servant. Not so; my love rests with my soul
And is not subject unto death.

(Wr. probably before 1590; pub. 1953)

CHIDIOCK TICHBORNE
d. 1586

244 *Tichborne's Elegy*

Written with his own hand in the Tower before his execution

MY prime of youth is but a frost of cares,
My feast of joy is but a dish of pain,
My crop of corn is but a field of tares,
And all my good is but vain hope of gain.
The day is past, and yet I saw no sun,
And now I live, and now my life is done.

My tale was heard and yet it was not told,
My fruit is fallen and yet my leaves are green;
My youth is spent and yet I am not old,
I saw the world and yet I was not seen. 10
My thread is cut and yet it is not spun,
And now I live, and now my life is done.

I sought my death and found it in my womb,
I looked for life and saw it was a shade;
I trod the earth and knew it was my tomb,
And now I die, and now I was but made.
My glass is full, and now my glass is run,
And now I live, and now my life is done.

(1586)

ROBERT SOUTHWELL SJ

1561–1595

245 *The Burning Babe*

As I in hoary winter's night
  Stood shivering in the snow,
Surprised I was with sudden heat,
  Which made my heart to glow.

And lifting up a fearful eye
  To view what fire was near,
A pretty Babe all burning bright
  Did in the air appear;

Who, scorched with excessive heat,
  Such floods of tears did shed, 10
As though his floods should quench his flames
  Which with his tears were fed.

'Alas', quoth he, 'but newly born
  In fiery heats I fry,
Yet none approach to warm their hearts,
  Or feel my fire, but I.

'My faultless breast the furnace is,
  The fuel wounding thorns;
Love is the fire, and sighs the smoke,
  The ashes, shame and scorns; 20

'The fuel Justice layeth on,
  And Mercy blows the coals;
The metal in this furnace wrought
  Are men's defiled souls:

'For which, as now on fire I am
  To work them to their good,
So will I melt into a bath
  To wash them in my blood.'

With this he vanished out of sight
  And swiftly shrunk away, 30
And straight I called unto mind
  That it was Christmas day.

(Wr. 1586–92; pub. 1602)

246 *New Prince, New Pomp*

BEHOLD, a silly tender Babe
  In freezing winter night
In homely manger trembling lies,
  Alas, a piteous sight!

The inns are full; no man will yield
  This little pilgrim bed,
But forced he is with silly beasts
  In crib to shroud his head.

Despise him not for lying there.
  First, what he is inquire; 10
An orient pearl is often found
  In depth of dirty mire.

Weigh not his crib, his wooden dish,
  Nor beasts that by him feed;
Weigh not his Mother's poor attire,
  Nor Joseph's simple weed.

This stable is a Prince's court,
  This crib his chair of state;
The beasts are parcel of his pomp,
  The wooden dish his plate. 20

The persons in that poor attire
  His royal liveries wear;
The Prince himself is come from heaven;
  This pomp is prized there.

With joy approach, O Christian wight,
  Do homage to thy King;
And highly praise his humble pomp,
  Which he from heaven doth bring.

(Wr. 1586–92; pub. 1602)

1 silly] helpless 19 parcel] part

247 *A Vale of Tears*

A VALE there is enwrapped with dreadful shades
Which thick of mourning pines shrouds from the sun,
Where hanging clifts yield short and dumpish glades,
And snowy flood with broken streams doth run;

Where eye-room is from rocks to cloudy sky,
From thence to dales with stony ruins strowed,
Then to the crushed waters' frothy fry,
Which tumbleth from the tops where snow is thowed;

Where ears of other sound can have no choice
But various blustring of the stubborn wind 10
In trees, in caves, in straits with divers noise,
Which now doth hiss, now howl, now roar by kind;

Where waters wrestle with encount'ring stones
That break their streams and turn them into foam;
The hollow clouds full fraught with thund'ring groans
With hideous thumps discharge their pregnant womb.

And in the horror of this fearful quire
Consists the music of this doleful place:
All pleasant birds their tunes from thence retire,
Where none but heavy notes have any grace. 20

Resort there is of none but pilgrim wights,
That pass with trembling foot and panting heart;
With terror cast in cold and shivering frights,
They judge the place to terror framed by art.

Yet Nature's work it is, of art untouched,
So strait indeed, so vast unto the eye,
With such disordered order strangely couched,
And so with pleasing horror low and high,

That who it views must needs remain aghast,
Much at the work, more at the Maker's might, 30
And muse how Nature such a plot could cast,
Where nothing seemed wrong, yet nothing right:

3 clifts] cliffs
dumpish] melancholy
5 eye-room] scope afforded to the eye, prospect
7 fry] turbulence
21 wights] creatures, persons
24 framed] contrived
26 strait] difficult to traverse

A place for mated minds, an only bower,
Where every thing doth soothe a dumpish mood.
Earth lies forlorn; the cloudy sky doth lower;
The wind here weeps, here sighs, here cries aloud.

The struggling flood between the marble groans,
Then roaring beats upon the craggy sides;
A little off, amidst the pebble stones,
With bubbling streams and purling noise it glides. 40

The pines thick-set, high-grown, and ever green,
Still clothe the place with sad and mourning veil.
Here gaping cliff, there mossy plain is seen;
Here hope doth spring, and there again doth quail.

Huge massy stones that hang by tickle stay
Still threaten fall, and seem to hang in fear;
Some withered trees ashamed of their decay,
Beset with green, are forced gray coats to wear.

Here crystal springs crept out of secret vein
Straight find some envious hole that hides their grace. 50
Here seared tufts lament the want of rain;
There thunder-rack gives terror to the place.

All pangs and heavy passions here may find
A thousand motives suitly to their griefs,
To feed the sorrows of their troubled mind,
And chase away Dame Pleasure's vain reliefs.

To plaining thoughts this vale a rest may be,
To which from worldly joys they may retire;
Where sorrow springs from water, stone and tree,
Where every thing with mourners doth conspire. 60

Set here, my soul, main streams of tears afloat;
Here all thy sinful foils alone recount;
Of solemn tunes make thou the doleful'st note
That to thy ditties dolour may amount.

33 mated] confounded, amazed
only] lonely, solitary
34 soothe] show, declare
45 tickle] precarious
48 Beset with] surrounded with
50 envious] malicious, spiteful
52 thunder-rack] storm clouds (or perhaps 'thunder-claps'?)
54 motives] things that rouse the feelings, incitements
suitly] in accordance with
61 set . . . afloat] overflow with
62 foils] (?) lapses, weaknesses
64 That] so that
amount] mount up, increase

When Echo doth repeat thy plainful cries,
Think that the very stones thy sins bewray,
And now accuse thee with their sad replies,
As heaven and earth shall in the latter day.

Let former faults be fuel of the fire
For grief in limbeck of thy heart to still 70
Thy pensive thoughts and dumps of thy desire
And vapour tears up to thy eyes at will.

Let tears to tunes and pains to plaints be prest,
And let this be the bourdon of thy song:
Come, deep remorse, possess my sinful breast;
Delights adieu, I harboured you too long.

(Wr. 1586–92; pub. 1595)

248 *Decease release*

Dum morior orior

THE pounded spice both taste and scent doth please;
In fading smoke the force doth incense show;
The perished kernel springeth with increase;
The lopped tree doth best and soonest grow.

God's spice I was and pounding was my due;
In fading breath my incense savoured best;
Death was the mean my kernel to renew;
By lopping shot I up to heavenly rest.

Some things more perfect are in their decay,
Like spark that going out gives clearest light. 10
Such was my hap, whose doleful dying day
Began my joy and termed fortune's spite.

Alive a Queen, now dead I am a Saint;
Once Mary called, my name now Martyr is.
From earthly reign debarred by restraint,
In lieu whereof I reign in heavenly bliss.

66 bewray] reveal, divulge
68 latter day] Day of Judgement
70 limbeck] alembic (the vessel used in the process of distillation)
still] distil
71 dumps] low spirits, dejection
72 vapour] cause to rise up in the form of vapour
73 prest] ready (to be changed)
74 bourdon] low accompaniment to a melody

248
Dum morior orior] dying, I live (rise)
12 termed] ended
14 Mary] Mary Queen of Scots

My life my grief, my death hath wrought my joy;
  My friends my foil, my foes my weal procured.
My speedy death hath shortened long annoy,
  And loss of life an endless life assured. 20

My scaffold was the bed where ease I found,
  The block a pillow of eternal rest.
My headman cast me in a blissful swound;
  His axe cut off my cares from cumbered breast.

Rue not my death; rejoice at my repose.
  It was no death to me but to my woe.
The bud was opened to let out the rose,
  The chains unloosed to let the captive go.

A prince by birth, a prisoner by mishap,
  From crown to cross, from throne to thrall I fell; 30
My right my ruth, my titles wrought my trap,
  My weal my woe, my worldly heaven my hell.

By death from prisoner to a prince enhanced,
  From cross to crown, from thrall to throne again,
My ruth my right, my trap my style advanced,
  From woe to weal, from hell to heavenly reign.

(Wr. 1586–92; pub. 1817)

249 *Man's civil war*

My hovering thoughts would fly to heaven
  And quiet nestle in the sky;
Fain would my ship in virtue's shore
  Without remove at anchor lie.

But mounting thoughts are hailed down
  With heavy poise of mortal load,
And blustering storms deny my ship
  In virtue's haven secure abode.

30 thrall] captivity, misery
31 ruth] matter or occasion of sorrow
trap] snare, pitfall
32 weal] happiness, well-being
33 enhanced] raised

249
6 poise] weight

When inward eye to heavenly sights
  Doth draw my longing heart's desire, 10
The world with jesses of delights
  Would to her perch my thoughts retire.

Fond fancy's trains to pleasures lure,
  Though reason stiffly do repine.
Though wisdom woo me to the saint,
  Yet sense would win me to the shrine.

Where reason loathes, there fancy loves,
  And overrules the captive will;
Foes senses are to virtue's lore:
  They draw the wit their wish to fill. 20

Need craves consent of soul to sense,
  Yet divers bents breed civil fray.
Hard hap where halves must disagree,
  Or truce of halves the whole betray!

O cruel fight where fighting friend
  With love doth kill a favouring foe,
Where peace with sense is war with God,
  And self-delight the seed of woe.

Dame Pleasure's drugs are steeped in sin:
  Their sugared taste doth breed annoy. 30
O fickle sense, beware her gin,
  Sell not thy soul for brittle joy!

(Wr. 1586–92; pub. 1595)

250 *Look home*

RETIRED thoughts enjoy their own delights,
  As beauty doth in self-beholding eye;
Man's mind a mirror is of heavenly sights,
  A brief wherein all marvels summed lie;
    Of fairest forms and sweetest shapes the store,
    Most graceful all, yet thought may grace them more.

11 jesses] the harness on the legs of a hawk
22 divers bents] opposed aims

250
4 brief] epitome, summary

The mind a creature is, yet can create,
To nature's patterns adding higher skill.
Of finest works wit better could the state,
If force of wit had equal power of will. 10
Device of man in working hath no end:
What thought can think another thought can mend.

Man's soul of endless beauties image is,
Drawn by the work of endless skill and might.
This skilful might gave many sparks of bliss,
And to discern this bliss a native light.
To frame God's image as his worths required,
His might, his skill, his word and will conspired.

All that he had his image should present;
All that it should present he could afford; 20
To that he could afford his will was bent;
His will was followed with performing word.
Let this suffice, by this conceive the rest:
He should, he could, he would, he did the best.

(Wr. 1586–92; pub. 1595)

251 *Times Go by Turns*

THE lopped tree in time may grow again,
Most naked plants renew both fruit and flower;
The sorriest wight may find release of pain,
The driest soil suck in some moistening shower.
Times go by turns, and chances change by course,
From foul to fair, from better hap to worse.

The sea of Fortune doth not ever flow,
She draws her favours to the lowest ebb.
Her tides hath equal times to come and go,
Her loom doth weave the fine and coarsest web. 10
No joy so great but runneth to an end,
No hap so hard but may in fine amend.

9 better could] could improve, make better

251
3 wight] person
12 in fine] at last, in the end

Not always fall of leaf, nor ever spring,
No endless night, yet not eternal day;
The saddest birds a season find to sing,
The roughest storm a calm may soon allay.
Thus, with succeeding turns, God tempereth all,
That man may hope to rise, yet fear to fall.

A chance may win that by mischance was lost;
The net, that holds no great, takes little fish; 20
In some things all, in all things none are crossed;
Few all they need, but none have all they wish.
Unmeddled joys here to no man befall;
Who least, hath some; who most, hath never all.

(Wr. 1586–92; pub. 1595)

252 *Loss in Delays*

SHUN delays, they breed remorse;
Take thy time while time doth serve thee;
Creeping snails have weakest force,
Fly their fault lest thou repent thee.
Good is best when soonest wrought,
Lingered labours come to nought.

Hoist up sail while gale doth last,
Tide and wind stay no man's pleasure;
Seek not time when time is past,
Sober speed is wisdom's leisure. 10
After-wits are dearly bought,
Let thy fore-wit guide thy thought.

Time wears all his locks before,
Take thy hold upon his forehead;
When he flies he turns no more,
And behind his scalp is naked.
Works adjourned have many stays,
Long demurs breed new delays.

Seek thy salve while sore is green,
Festered wounds ask deeper lancing; 20
After-cures are seldom seen,
Often sought, scarce ever chancing.

19 that by mischance] i.e. what by mischance
23 Unmeddled] unmixed

252
12 fore-wit] foresight, prudence
19 salve] healing ointment
green] fresh

Time and place give best advice,
Out of season, out of price.

Crush the serpent in the head,
  Break ill eggs ere they be hatched;
Kill bad chickens in the tread,
  Fligg they hardly can be catched.
In the rising stifle ill,
Lest it grow against thy will. 30

Drops do pierce the stubborn flint,
  Not by force but often falling;
Custom kills with feeble dint,
  More by use than strength prevailing.
Single sands have little weight,
Many make a drowning freight.

Tender twigs are bent with ease,
  Aged trees do break with bending;
Young desires make little prease,
  Growth doth make them past amending. 40
Happy man, that soon doth knock
Babble babes against the rock!

(Wr. 1586–92; pub. 1595)

253 *Content and Rich*

I DWELL in Grace's court,
  Enriched with Virtue's rights;
Faith guides my wit, Love leads my will,
  Hope all my mind delights.

In lowly vales I mount
  To pleasure's highest pitch;
My silly shroud true honour brings;
  My poor estate is rich.

My conscience is my crown,
  Contented thoughts my rest; 10
My heart is happy in itself;
  My bliss is in my breast.

27 tread] action of the male bird in coition, i.e. moment of conception
28 Fligg] fledged, ready to fly
39 prease] trouble, difficulty
42 Babble] Babylonian (cf. Ps. 137: 8, 9)
253
7 silly shroud] simple clothing

Enough, I reckon wealth;
  A mean the surest lot,
That lies too high for base contempt,
  Too low for envy's shot.

My wishes are but few,
  All easy to fulfil;
I make the limits of my power
  The bonds unto my will. 20

I have no hopes but one,
  Which is of heavenly reign;
Effects attained, or not desired,
  All lower hopes refrain.

I feel no care of coin;
  Well-doing is my wealth;
My mind to me an empire is,
  While grace affordeth health.

I clip high-climbing thoughts,
  The wings of swelling pride; 30
Their fall is worst, that from the height
  Of greatest honour slide.

Sith sails of largest size
  The storm doth soonest tear,
I bear so low and small a sail
  As freeth me from fear.

I wrestle not with rage,
  While fury's flame doth burn;
It is in vain to stop the stream
  Until the tide doth turn. 40

But when the flame is out,
  And ebbing wrath doth end,
I turn a late enraged foe
  Into a quiet friend.

And taught with often proof,
  A tempered calm I find
To be most solace to itself,
  Best cure for angry mind.

29 clip] cut (wings) short

Spare diet is my fare,
  My clothes more fit than fine; 50
I know I feed and clothe a foe
  That pampered would repine.

I envy not their hap,
  Whom favour doth advance;
I take no pleasure in their pain,
  That have less happy chance.

To rise by other's fall
  I deem a losing gain;
All states with others' ruins built
  To ruin run amain. 60

No change of Fortune's calms
  Can cast my comforts down;
When Fortune smiles, I smile to think
  How quickly she will frown.

And when in froward mood
  She proves an angry foe,
Small gain I found to let her come,
  Less loss to let her go.

(Wr. 1586–92; pub. 1595)

ANONYMOUS

254 *Upon the Image of Death*

BEFORE my face the picture hangs,
  That daily should put me in mind
Of those cold qualms and bitter pangs,
  That shortly I am like to find:
    But yet, alas, full little I
    Do think hereon that I must die.

I often look upon a face
  Most ugly, grisly, bare, and thin;
I often view the hollow place,
  Where eyes and nose had sometimes been; 10
    I see the bones across that lie,
    Yet little think that I must die.

254
65 froward] perverse, refractory

I read the label underneath,
  That telleth me whereto I must;
I see the sentence eke that saith
  'Remember, man, that thou art dust!'
    But yet, alas, but seldom I
    Do think indeed that I must die.

Continually at my bed's head
  A hearse doth hang, which doth me tell, 20
That I ere morning may be dead,
  Though now I feel myself full well:
    But yet, alas, for all this, I
    Have little mind that I must die.

The gown which I do use to wear,
  The knife wherewith I cut my meat,
And eke that old and ancient chair
  Which is my only usual seat:
    All these do tell me I must die,
    And yet my life amend not I. 30

My ancestors are turned to clay,
  And many of my mates are gone;
My youngers daily drop away,
  And can I think to scape alone?
    No, no, I know that I must die,
    And yet my life amend not I.

Not Solomon, for all his wit,
  Nor Samson, though he were so strong,
No king nor person ever yet
  Could scape, but death laid him along: 40
    Wherefore I know that I must die,
    And yet my life amend not I.

Though all the East did quake to hear
  Of Alexander's dreadful name,
And all the West did likewise fear
  To hear of Julius Caesar's fame,
    Yet both by death in dust now lie.
    Who then can scape, but he must die?

If none can scape death's dreadful dart,
  If rich and poor his beck obey, 50
If strong, if wise, if all do smart,
  Then I to scape shall have no way.
    Oh grant me grace, O God, that I
    My life may mend, sith I must die.

(Pub. 1595)

[*Songs set by William Byrd*]

255

I JOY not in no earthly bliss;
  I force not Croesus' wealth a straw;
For care I know not what it is;
  I fear not Fortune's fatal law.
My mind is such as may not move
For beauty bright, nor force of love.

I wish but what I have at will;
  I wander not to seek for more;
I like the plain, I climb no hill;
  In greatest storms I sit on shore, 10
And laugh at them that toil in vain
To get what must be lost again.

I kiss not where I wish to kill;
  I feign not love where most I hate;
I break no sleep to win my will;
  I wait not at the mighty's gate.
I scorn no poor, nor fear no rich,
I feel no want, nor have too much.

The court and cart I like nor loathe;
  Extremes are counted worst of all; 20
The golden mean between them both
  Doth surest sit and fear no fall.
This is my choice; forwhy I find
No wealth is like the quiet mind.

(1588)

256

WHAT pleasure have great princes
  More dainty to their choice,
Than herdmen wild, who careless
  In quiet life rejoice,
And Fortune's fate not fearing
Sing sweet in summer morning.

255
2 force not] care not for
23 forwhy] because

Their dealings plain and rightful
Are void of all deceit;
They never know how spiteful
It is to kneel and wait 10
On favourite presumptuous
Whose pride is vain and sumptuous.

All day their flocks each tendeth,
At night they take their rest,
More quiet than who sendeth
His ship into the East,
Where gold and pearl are plenty,
But getting very dainty.

For lawyers and their pleading
They 'steem it not a straw; 20
They think that honest meaning
Is of itself a law;
Where conscience judgeth plainly
They spend no money vainly.

O happy who thus liveth,
Not caring much for gold,
With clothing which sufficeth
To keep him from the cold.
Though poor and plain his diet,
Yet merry it is and quiet. 30

(1588)

257 CONSTANT Penelope sends to thee, careless Ulysses.
Write not again, but come, sweet mate, thyself to revive me.
Troy we do much envy, we desolate lost ladies of Greece;
Not Priamus, nor yet all Troy can us recompense make.
Oh, that he had, when he first took shipping to Lacedaemon,
That adulter I mean, had been o'erwhelmed with waters.
Then had I not lain now all alone, thus quivering for cold,
Nor used this complaint, nor have thought the day to be so long.

(1588)

9 spiteful] shameful, vexing
18 dainty] rare

257
1 careless] unconcerned, free of care
3 envy] (stress on second syllable) feel hostile towards
6 adulter] adulterer

from *Six Idillia . . . chosen out of . . . Theocritus*

258 *Cyclops*

O NICIAS, there is no other remedy for love,
With ointing, or with sprinkling on, that ever I could prove,
Beside the Muses nine. This pleasant medicine of the mind
Grows among men; and seems but light, yet very hard to find:
As well I wote you know; who are in physic such a leech,
And of the Muses so beloved. The cause of this my speech
A Cyclops is, who lived here with us right wealthily;
That ancient Polyphem, when first he loved Galatee
When, with a bristled beard, his chin and cheeks first clothed were.
He loved her not with roses, apples, or with curled hair; 10
But with the Furies' rage. All other things he little plied.
Full often to their fold, from pastures green, without a guide,
His sheep returned home: when all the while he singing lay
In honour of his love, and on the shore consumed away
From morning until night; sick of the wound, fast by the heart,
Which mighty Venus gave, and in his liver stuck the dart.
For which, this remedy he found, that sitting oftentimes
Upon a rock and looking on the sea, he sang these rhymes:

'O Galatea fair, why dost thou shun thy lover true?
More tender than a lamb, more white than cheese when it is new, 20
More wanton than a calf, more sharp than grapes unripe, I find.
You use to come when pleasant sleep, my senses all do bind:
But you are gone again when pleasant sleep doth leave mine eye;
And as a sheep you run, that on the plain a wolf doth spy.

'I then began to love thee, Galatee, when first of all
You, with my mother, came to gather leaves of crowtoe small
Upon our hill; when I, as usher, squired you all the way.
Nor when I saw thee first, nor afterwards, nor at this day,
Since then could I refrain: but you, by Jove, nought set thereby.

'But well I know, fair Nymph, the very cause why you thus fly. 30
Because upon my front, one only brow, with bristles strong
From one ear to the other ear is stretched all along:
'Neath which, one eye; and on my lips, a hugy nose, there stands.
Yet I, this such a one, a thousand sheep feed on these lands;
And pleasant milk I drink, which from the strouting bags is pressed.
Nor want I cheese in summer, nor in autumn of the best,

5 physic] medicine
leech] physician, doctor
26 crowtoe] hyacinth
35 strouting] swelling, bulging

Nor yet in winter time. My cheese racks ever laden are;
And better can I pipe than any Cyclops may compare.
O apple sweet of thee, and of myself I use to sing,
And that at midnight oft. For thee, eleven fawns up I bring, 40
All great with young: and four bears' whelps I nourish up for thee.
But come thou hither first, and thou shalt have them all of me.
And let the bluish-coloured sea beat on the shore so nigh,
The night with me in cave, thou shalt consume more pleasantly.
There are the shady bays, and there tall cypress trees do sprout:
And there is ivy black, and fertile vines are all about.
Cool water there I have, distilled of the whitest snow,
A drink divine, which out of woody Etna mount doth flow.
In these respects, who in the sea and waves would rather be?

'But if I seem as yet too rough and savage unto thee, 50
Great store of oaken wood I have, and never-quenched fire;
And I can well endure my soul to burn with thy desire,
With this my only eye, than which I nothing think more trim:
Now woe is me, my mother bore me not with fins to swim,
That I might dive to thee; that I thy dainty hand might kiss,
If lips thou wouldst not let. Then would I lilies bring iwis,
And tender poppy-toe that bears a top like rattles red,
And these in summer time: but other are in winter bred,
So that I cannot bring them all at once. Now certainly
I'll learn to swim of some or other stranger passing by, 60
That I may know what pleasure 'tis in waters deep to dwell.

'Come forth, fair Galatee, and once got out, forget thee well
(As I do, sitting on this rock) home to return again.
But feed my sheep with me, and for to milk them take the pain.
And cheese to press, and in the milk the rennet sharp to strain.
My mother only wrongeth me; and her I blame, for she
Spake never yet to thee one good, or lovely, word of me:
And that, although she daily sees how I away do pine.
But I will say, "My head and feet do ache," that she may whine,
And sorrow at the heart: because my heart with grief is swoll'n. 70

'O Cyclops, Cyclops, whither is thy wit and reason flown?
If thou would'st baskets make; and cut down browsing from the tree,
And bring it to thy lambs, a great deal wiser thou should'st be.
Go, coy some present nymph. Why dost thou follow flying wind?
Perhaps another Galatee, and fairer, thou shalt find.

56 iwis] truly
58 other] others
65 rennet] a kind of apple
72 browsing] young shoots and twigs
74 coy] blandish, court

For many maidens in the evening tide with me will play,
And all do sweetly laugh, when I stand heark'ning what they say:
And I somebody seem, and in the earth do bear a sway.'

Thus Polyphemus singing, fed his raging love of old;
Wherein he sweeter did, than had he sent her sums of gold. 80

(1588)

259 *Neatherd*

EUNICA scorned me, when her I would have sweetly kissed
And railing at me said, 'Go with a mischief, where thou list!
Think'st thou, a wretched neatherd, me to kiss? I have no will
After the country guise to smouch! Of city lips I skill!
My lovely mouth, so much as in thy dream thou shalt not touch.
How dost thou look! How dost thou talk! How playest thou the slouch!
How daintily thou speak'st! What courting words thou bringest out!
How soft a beard thou hast! How fair thy locks hang round about!
Thy lips are like a sick man's lips! thy hands, so black they be!
And rankly thou dost smell! Away, lest thou defilest me!' 10
Having thus said, she spattered on her bosom twice or thrice;
And, still beholding me from top to toe in scornful wise,
She muttered with her lips; and with her eyes she looked aside,
And of her beauty wondrous coy she was; her mouth she wryed,
And proudly mocked me to my face. My blood boiled in each vein,
And red I wox for grief as doth the rose with dewy rain.
Thus leaving me, away she flung. Since when, it vexeth me
That I should be so scorned of such a filthy drab as she.
'Ye shepherds, tell me true, am not I as fair as any swan?
Hath of a sudden any god made me another man? 20
For well I wot, before a comely grace in me did shine,
Like ivy round about a tree, and decked this beard of mine.
My crisped locks, like parsley, on my temples wont to spread;
And on my eyebrows black a milk-white forehead glistered:
More seemly were mine eyes than are Minerva's eyes, I know.
My mouth for sweetness passed cheese; and from my mouth did flow
A voice more sweet than honeycombs. Sweet is my roundelay
When on the whistle, flute, or pipe, or cornet I do play.
And all the women on our hills do say that I am fair,
And all do love me well: but these that breathe the city air 30
Did never love me yet. And why? The cause is this I know.
That I a neatherd am. They hear not how in vales below,

3 neatherd] cowherd
4 smouch] kiss
skill] have knowledge
6 slouch] lout, clown
16 wox] grew
23 crisped] curled
28 cornet] horn

Fair Bacchus kept a herd of beasts. Nor can these nice ones tell
How Venus, raving for a neatherd's love, with him did dwell
Upon the hills of Phrygia; and how she loved again
Adonis in the woods, and mourned in woods when he was slain.
Who was Endymion? Was he not a neatherd? Yet the Moon
Did love this neatherd so, that from the heavens descending soon,
She came to Latmos grove where with the dainty lad she lay.
And Rhea, thou a neatherd dost bewail! and thou, all day, 40
O mighty Jupiter, but for a shepherd's boy didst stray!
Eunica only, deigned not a neatherd for to love:
Better forsooth than Cybel, Venus, or the Moon above!
And Venus thou hereafter must not love thy fair Adone
In city, nor on hill, but all the night must sleep alone!'

(1588)

260 *Adonis*

WHEN Venus first did see
Adonis dead to be;
With woeful tattered hair
And cheeks so wan and sear,
The winged Loves she bade,
The boar should straight be had.
Forthwith like birds they fly,
And through the wood they hie;
The woeful beast they find,
And him with cords they bind. 10
One with a rope before
Doth lead the captive boar:
Another on his back
Doth make his bow to crack.
The beast went wretchedly,
For Venus horribly
He feared; who thus him curst:
'Of all the beasts the worst,
Didst thou this thigh so wound?
Didst thou my love confound?' 20
The beast thus spake in fear
'Venus, to thee I swear!
By thee, and husband thine,
And by these bands of mine,
And by these hunters all,
Thy husband fair and tall
I minded not to kill.
But, as an image still,

I him beheld for love:
Which made me forward shove 30
His thigh, that naked was;
Thinking to kiss, alas,
And that hath hurt me thus.
'Wherefore these teeth, Venus,
Or punish, or cut out:
Why bear I in my snout
These needless teeth about!
If these may not suffice,
Cut off my chaps likewise!'
To ruth he Venus moves, 40
And she commands the Loves
His bands for to untie.
After he came not nigh
The wood; but at her will
He followed Venus still.
And coming to the fire,
He burnt up his desire.

(1588)

LODOWICK BRYSKETT (LODOVICO BRUSCHETTO)

1546–1612

261 *A Pastoral Eclogue upon the death of Sir Philip Sidney Knight*

Lycon. Colin.

Lycon. Colin, well fits thy sad cheer this sad stound,
This woeful stound, wherein all things complain
This great mishap, this grievous loss of ours.
Hear'st thou the Orown?—how with hollow sound
He slides away, and murmuring doth plain,
And seems to say unto the fading flowers,
Along his banks, unto the bared trees,
'Philisides is dead.' Up, jolly swain,

261
Lycon] (Bryskett)
Colin] (Edmund Spenser)
1 stound] time; time of pain
4 Orown] Irish river
5 plain] complain, lament
8 jolly] gallant, brave

Thou that with skill canst tune a doleful lay,
Help him to mourn. My heart with grief doth freeze, 10
Hoarse is my voice with crying, else a part
Sure would I bear, though rude. But as I may,
With sobs and sighs I second will thy song,
And so express the sorrows of my heart.

Colin. Ah Lycon, Lycon, what needs skill to teach
A grieved mind pour forth his plaints? How long
Hath the poor turtle gone to school (ween'st thou?)
To learn to mourn her lost make? No, no, each
Creature by nature can tell how to wail.
Seest not these flocks, how sad they wander now? 20
Seemeth their leaders bell their bleating tunes
In doleful sound. Like him, not one doth fail
With hanging head to show a heavy cheer.
What bird (I pray thee) hast thou seen, that prunes
Himself of late? Did any cheerful note
Come to thine ears, or gladsome sight appear
Unto thine eyes, since that same fatal hour?
Hath not the air put on his mourning coat,
And testified his grief with flowing tears?
Sith then, it seemeth each thing to his power 30
Doth us invite to make a sad consort.
Come, let us join our mournful song with theirs.
Grief will indite, and sorrow will enforce
Thy voice, and Echo will our words report.

Lycon. Though my rude rhymes ill with thy verses frame,
That others far excel, yet will I force
Myself to answer thee the best I can,
And honour my base words with his high name.
But if my plaints annoy thee where thou sit
In secret shade or cave, vouchsafe (O Pan) 40
To pardon me, and hear this hard constraint
With patience while I sing, and pity it.
And eke, ye rural Muses, that do dwell
In these wild woods, if ever piteous plaint
We did indite, or taught a woeful mind
With words of pure affect his grief to tell,
Instruct me now. Now, Colin, then go on,
And I will follow thee, though far behind.

17 turtle] turtle-dove
ween'st] know'st
18 make] mate
21 bell] loudly utter
30 Sith] since
33 indite] dictate, put into words
34 report] repeat, send back
35 frame] suit, fit
41 constraint] affliction, distress
46 affect] feeling

Colin. Philisides is dead. O harmful death,
O deadly harm! Unhappy Albion, 50
When shalt thou see among thy shepheards all
Any so sage, so perfect? Whom unneath
Envy could touch for virtuous life and skill;
Courteous, valiant, and liberal.
Behold the sacred Pales, where with hair
Untrussed she sits, in shade of yonder hill.
And her fair face bent sadly down doth send
A flood of tears to bathe the earth; and there
Doth call the heavens despiteful, envious;
Cruel his fate, that made so short an end 60
Of that same life, well worthy to have been
Prolonged with many years, happy and famous.
The Nymphs and Oreads her round about
Do sit lamenting on the grassy green,
And with shrill cries, beating their whitest breasts,
Accuse the direful dart that death sent out
To give the fatal stroke. The stars they blame,
That deaf or careless seem at their request.
The pleasant shade of stately groves they shun;
They leave their crystal springs, where they wont frame 70
Sweet bowers of myrtle twigs and laurel fair,
To sport themselves free from the scorching sun.
And now the hollow caves where horror dark
Doth dwell, whence banished is the gladsome air,
They seek; and there in mourning spend their time
With wailful tunes, while wolves do howl and bark,
And seem to bear a bourdon to their plaint.

Lycon. Philisides is dead. O doleful rhyme,
Why should my tongue express thee? Who is left
Now to uphold thy hopes when they do faint, 80
Lycon unfortunate? What spiteful fate,
What luckless destiny hath thee bereft
Of thy chief comfort, of thy only stay?
Where is become thy wonted happy state,
(Alas) wherein through many a hill and dale,
Through pleasant woods and many an unknown way
Thou with him yodest; and with him didst scale
The craggy rocks of th'Alps and Appenine,
Still with the Muses sporting, while those beams
Of virtue kindled in his noble breast 90

52 unneath] scarcely
55 Pales] (goddess of flocks and herds)
63 Oreads] mountain nymphs
77 bourdon] low undersong
87 yodest] went
90 kindled] become inflamed, glow

Which after did so gloriously forth shine?
But (woe is me) they now yquenched are
All suddenly, and death hath them oppressed.
Lo, father Neptune with sad countenance,
How he sits mourning on the strond now bare,
Yonder, where th'ocean with his rolling waves
The white feet washeth (wailing this mischance)
Of Dover cliffs. His sacred skirt about
The sea-gods all are set; from their moist caves
All for his comfort gathered there they be. 100
The Thamis rich, the Humber rough and stout,
The fruitful Severn, with the rest are come
To help their lord to mourn, and eke to see
The doleful sight and sad pomp funeral
Of the dead corpse passing through his kingdom.
And all their heads with cypress garlands crowned,
With woeful shrieks salute him, great and small.
Eke wailful Echo, forgetting her dear
Narcissus, their last accents doth resound.

Colin. Philisides is dead. O luckless age, 110
O widow world! O brooks and fountains clear!
O hills, O dales, O woods that oft have rung
With his sweet carolling, which could assuage
The fiercest wrath of tiger or of bear.
Ye Sylvans, Fauns and Satyrs, that among
These thickets oft have danced after his pipe,
Ye Nymphs and Naiades with golden hair
That oft have left your purest crystal springs
To hearken to his lays, that coulden wipe
Away all grief and sorrow from your hearts, 120
Alas who now is left that like him sings?
When shall you hear again like harmony?
So sweet a sound, who to you now imparts?
Lo, where engraved by his hand yet lives
The name of Stella, in yonder bay tree!
Happy name, happy tree; fair may you grow,
And spread your sacred branch, which honour gives
To famous emperors, and poets crown.
Unhappy flock that wander scattered now,
What marvel if through grief ye woxen lean, 130
Forsake your food, and hang your heads adown?
For such a shepherd never shall you guide,
Whose parting hath of weal bereft you clean.

104 pomp] procession 130 woxen] grow 133 weal] well-being

Lycon. Philisides is dead. O happy sprite,
That now in heaven with blessed souls dost bide!
Look down a while from where thou sitst above,
And see how busy shepherds be to indite
Sad songs of grief, their sorrows to declare,
And grateful memory of their kind love.
Behold myself with Colin, gentle swain 140
(Whose learned Muse thou cherished most whylere),
Where we thy name recording, seek to ease
The inward torment and tormenting pain
That thy departure to us both hath bred;
Ne can each other's sorrow yet appease.
Behold the fountains now left desolate,
And withered grass with cypress boughs bespread;
Behold these flowers which on thy grave we strew,
Which, faded, show the givers' faded state
(Though eke they show their fervent zeal and pure), 150
Whose only comfort on thy welfare grew.
Whose prayers importune shall the heavens for aye,
That to thy ashes rest they may assure:
That learned shepherds honour may thy name
With yearly praises, and the Nymphs alway
Thy tomb may deck with fresh and sweetest flowers,
And that for ever may endure thy fame.

Colin. The sun (lo) hastened hath his face to steep
In western waves, and th'air with stormy showers
Warns us to drive homeward our silly sheep. 160
Lycon, let's rise, and take of them good keep.

(Pub. 1595)

ANONYMOUS

262 *Like to a ring without a finger*

LIKE to a ring without a finger,
Or a bell without a ringer,
Like a horse was never ridden,
Or a feast and no guest bidden,

137 indite] compose
141 whylere] formerly
142 recording] repeating; calling to mind
158 steep] bathe
159 air] wind
160 silly] simple, helpless
161 keep] care

Like a well without a bucket,
Or a rose if no man pluck it,
Just such as these may she be said
That lives, not loves, but dies a maid.

The ring, if worn, the finger decks;
The bell pulled by the ringer speaks; 10
The horse does ease if he be ridden;
The feast doth please if guest be bidden;
The bucket draws the water forth;
The rose, when plucked, is still most worth:
Such is the virgin in my eyes
That lives, loves, marries ere she dies.

Like a stock not grafted on,
Or like a lute not played upon,
Like a jack without a weight,
Or a bark without a freight, 20
Like a lock without a key,
Or like a candle in the day,
Just such as these may she be said
That lives, not loves, but dies a maid.

The graffed stock doth bear best fruit;
There's music in the fingered lute;
The weight doth make the clock go ready;
The freight doth make the bark go steady;
The key the lock doth open right;
A candle's useful in the night: 30
Such is the virgin in mine eyes
That lives, loves, marries ere she dies.

Like a call without a Non-sir,
Or a question and no answer,
Like a ship was never rigged,
Or a mine was never digged,
Like a wound without a tent,
Or a box without a scent,
Just such as these may she be said
That lives, not loves, but dies a maid. 40

The Non-sir doth obey the call;
The question answered pleaseth all;

25 graffed] grafted
33 Non-sir] i.e. 'Anon, sir!', the tapster's cry of 'Coming!'
37 tent] a probe or dressing for a wound

Who rigs a ship sails with the wind;
Who digs a mine doth treasure find;
The wound by wholesome tent hath ease;
The box perfumed the senses please:
  Such is the virgin in mine eyes
  That lives, loves, marries ere she dies.

Like marrow-bone was never broken,
Or commendation and no token, 50
Like a fort and none to win it,
Or like the moon and no man in it,
Like a school without a teacher,
Or like a pulpit and no preacher,
  Just such as these may she be said
  That lives, not loves, but dies a maid.

The broken marrow-bone is sweet;
The token doth adorn the greet;
There's triumph in the fort being won;
The man rides glorious in the moon; 60
The school is by the teacher skilled;
The pulpit by the preacher filled:
  Such is the virgin in mine eyes
  That lives, loves, marries ere she dies.

Like a cage without a bird,
Or a thing too long deferred,
Like the gold was never tried,
Or the ground unoccupied,
Like a house that's not possessed,
Or the book was never pressed, 70
  Just such as these may she be said
  That lives, ne'er loves, but dies a maid.

The bird in cage doth sweetly sing;
Due season proffers everything;
The gold that's tried from dross is pured;
There's profit in the ground manured;
The house is by possession graced;
The book, when pressed, is then embraced:
  Such is the virgin in mine eyes
  That lives, loves, marries ere she dies. 80

(Pub. 1951)

58 greet] great 67 tried] purified 70 pressed] printed

ROBERT GREENE
1558–1592

263 [*Phillis and Coridon*]

PHILLIS kept sheep along the western plains,
  And Coridon did feed his flocks hard by:
This shepherd was the flower of all the swains,
  That traced the downs of fruitful Thessaly,
    And Phillis, that did far her flocks surpass
    In silver hue, was thought a bonny lass.

A bonny lass, quaint in her country 'tire,
  Was lovely Phillis, Coridon swore so;
Her locks, her looks, did set the swain on fire.
  He left his lambs, and he began to woo, 10
    He looked, he sithed, he courted with a kiss:
    No better could the silly swad than this.

He little knew to paint a tale of love;
  Shepherds can fancy, but they cannot say:
Phillis gan smile, and wily thought to prove,
  What uncouth grief poor Coridon did pay;
    She asked him how his flocks or he did fare,
    Yet pensive thus his sighs did tell his care.

The shepherd blushed when Phillis questioned so,
  And swore by Pan it was not for his flocks: 20
' 'Tis love, fair Phillis, breedeth all this woe:
  My thoughts are trapped within thy lovely locks,
    Thine eye hath pierced, thy face hath set on fire.
    Fair Phillis kindleth Coridon's desire'

'Can shepherds love?', said Phillis to the swain.
  'Such saints as Phillis,' Coridon replied.
'Men, when they lust, can many fancies feign,'
  Said Phillis. This not Coridon denied,
    That lust had lies. 'But love,' quoth he, 'says truth.
    Thy shepherd loves; then, Phillis, what ensueth?' 30

11 sithed] sighed 12 swad] clown

Phillis was won, she blushed and hung the head;
  The swain stepped to, and cheered her with a kiss:
With faith, with troth, they struck the matter dead;
  So used they when men thought not amiss:
    This love begun and ended both in one;
    Phillis was loved, and she liked Coridon.

(1588)

264 *Weep not, my wanton*

Weep not, my wanton, smile upon my knee;
When thou art old there's grief enough for thee.
    Mother's wag, pretty boy,
    Father's sorrow, father's joy.
    When thy father first did see
    Such a boy by him and me,
    He was glad, I was woe:
    Fortune changed made him so,
    When he left his pretty boy,
    Last his sorrow, first his joy. 10

Weep not, my wanton, smile upon my knee;
When thou art old there's grief enough for thee.
    Streaming tears that never stint,
    Like pearl drops from a flint,
    Fell by course from his eyes,
    That one another's place supplies:
    Thus he grieved in every part,
    Tears of blood fell from his heart,
    When he left his pretty boy,
    Father's sorrow, father's joy. 20

Weep not, my wanton, smile upon my knee;
When thou art old there's grief enough for thee.
    The wanton smiled, father wept;
    Mother cried, baby leapt;
    More he crowed, more we cried;
    Nature could not sorrow hide.
    He must go, he must kiss
    Child and mother, baby bliss;
    For he left his pretty boy,
    Father's sorrow, father's joy. 30

Weep not, my wanton, smile upon my knee;
When thou art old there's grief enough for thee.

(1589)

265 *The Description of the Shepherd and His Wife*

It was near a thicky shade
That broad leaves of beech had made,
Joining all their tops so nigh
That scarce Phoebus in could pry
To see if lovers in the thick
Could dally with a wanton trick,
Where sat the swain and his wife,
Sporting in that pleasing life
That Corydon commendeth so
All other lives to overgo. 10
He and she did sit and keep
Flocks of kids and folds of sheep;
He upon his pipe did play,
She tuned voice unto his lay.
And for you might her huswife know,
Voice did sing and fingers sew.
He was young, his coat was green,
With welts of white, seamed between,
Turned over with a flap,
That breast and bosom in did wrap; 20
Skirts side and plighted free
Seemly hanging to his knee.
A whittle with a silver chape;
Cloak was russet, and the cape
Served for a bonnet oft,
To shroud him from the wet aloft.
A leather scrip of colour red,
With a button on the head;
A bottle full of country whig
By the shepherd's side did ligge; 30
And in a little bush hard by
There the shepherd's dog did lie,
Who while his master gan to sleep
Well could watch both kids and sheep.
The shepherd was a frolic swain,
For though his 'parel was but plain,
Yet doon the authors soothly say
His colour was both fresh and gay;

18 welts] strips
23 whittle] knife
chape] metal plate of scabbard or sheath
27 scrip] bag
29 whig] whey or buttermilk
30 ligge] lie

And in their writs plain discuss
Fairer was not Tityrus, 40
Nor Menalcas whom they call
The alderliefest swain of all.
Seeming to him was his wife,
Both in line and in life:
Fair she was as fair might be,
Like the roses on the tree;
Buxom, blithe, and young, I ween,
Beauteous like a summer's queen,
For her cheeks were ruddy hued,
As if lilies were imbrued 50
With drops of blood to make the white
Please the eye with more delight.
Love did lie within her eyes
In ambush for some wanton prize.
A liefer lass than this had been
Corydon had never seen.
Nor was Phillis, that fair may,
Half so gaudy or so gay:
She wore a chaplet on her head;
Her cassock was of scarlet red; 60
Long and large as straight as bent;
Her middle was both small and gent;
A neck as white as whales bone,
Compassed with a lace of stone.
Fine she was and fair she was,
Brighter than the brightest glass.
Such a shepherd's wife as she
Was not more in Thessaly.

(1590)

266 [*The Shepherd's Wife's Song*]

Ah, what is love? It is a pretty thing,
As sweet unto a shepherd as a king,
And sweeter too;
For kings have cares that wait upon a crown,
And cares can make the sweetest love to frown.
Ah then, ah then,
If country loves such sweet desires gain,
What lady would not love a shepherd swain?

39 discuss] declare
42 alderliefest] dearest
43 Seeming] suitable, beseeming
57 may] maid
61 bent] reed, stiff grass
62 small] slender

His flocks are folded, he comes home at night,
As merry as a king in his delight, 10
    And merrier too;
For kings bethink them what the state require,
Where shepherds careless carol by the fire.
    Ah then, ah then,
If country loves such sweet desires gain,
What lady would not love a shepherd swain?

He kisseth first, then sits as blithe to eat
His cream and curds, as doth the king his meat,
    And blither too;
For kings have often fears when they do sup, 20
Where shepherds dread no poison in their cup.
    Ah then, ah then,
If country loves such sweet desires gain,
What lady would not love a shepherd swain?

To bed he goes, as wanton then I ween,
As is a king in dalliance with a queen,
    More wanton too;
For kings have many griefs affects to move,
Where shepherds have no greater grief than love.
    Ah then, ah then, 30
If country loves such sweet desires gain,
What lady would not love a shepherd swain?

Upon his couch of straw he sleeps as sound,
As doth the king upon his bed of down,
    More sounder too;
For cares cause kings full oft their sleep to spill,
Where weary shepherds lie and snort their fill.
    Ah then, ah then,
If country loves such sweet desires gain,
What lady would not love a shepherd swain? 40

Thus with his wife he spends the year as blithe,
As doth the king at every tide or sithe,
    And blither too;
For kings have war and broils to take in hand,
Where shepherds laugh, and love upon the land.
    Ah then, ah then,
If country loves such sweet desires gain,
What lady would not love a shepherd swain?

(1590)

42 sithe] time

267 [*A Night Visitor*]

CUPID abroad was lated in the night;
His wings were wet with ranging in the rain.
Harbour he sought; to me he took his flight:
To dry his plumes I heard the boy complain,
I oped the door and granted his desire;
I rose myself, and made the wag a fire.

Looking more narrow by the fire's flame,
I spied his quiver hanging by his back.
Doubting the boy might my misfortune frame,
I would have gone for fear of further wrack. 10
But what I drad, did me poor wretch betide:
For forth he drew an arrow from his side.

He pierced the quick, and I began to start,
A pleasing wound but that it was too high;
His shaft procured a sharp yet sugared smart.
Away he flew, for why his wings were dry,
But left the arrow sticking in my breast,
That sore I grieved I welcomed such a guest.

(Prob. wr. 1590; pub. 1599)

268 *The Palmer*

DOWN the valley gan he track,
Bag and bottle at his back,
In a surcoat all of gray,
Such wear palmers on the way,
When with scrip and staff they see
Jesus' grave on Calvary;
A hat of straw like a swain,
Shelter for the sun and rain,
With a scallop shell before;
Sandals on his feet he wore; 10
Legs were bare, arms unclad:
Such attire this palmer had.
His face fair like Titan's shine;
Gray and buxom were his eyne,

267
9 Doubting] fearing
16 for why] because

268
Palmer] pilgrim
5 scrip] wallet, small bag
14 buxom] bright, lively

Whereout dropped pearls of sorrow:
Such sweet tears Love doth borrow,
When in outward dews he plains
Heart's distress that lovers pains.
Ruby lips, cherry cheeks;
Such rare mixture Venus seeks, 20
When to keep her damsels quiet
Beauty sets them down their diet.
Adon was not thought more fair:
Curled locks of amber hair,
Locks where Love did sit and twine
Nets to snare the gazer's eyne:
Such a palmer ne'er was seen,
'Less Love himself had palmer been.
Yet for all he was so quaint,
Sorrow did his visage taint. 30
Midst the riches of his face
Grief deciphered high disgrace.
Every step strained a tear;
Sudden sighs shewed his fear;
And yet his fear by his sight
Ended in a strange delight;
That his passions did approve
Weeds and sorrow were for love.

(1590)

269 *Old Menalcas on a day*

OLD Menalcas on a day,
As in field this shepherd lay,
Tuning of his oaten pipe,
Which he hit with many a stripe,
Said to Coridon that he
Once was young and full of glee:
'Blithe and wanton was I then;
Such desires follow men.
As I lay and kept my sheep,
Came the God that hateth sleep, 10
Clad in armour all of fire,
Hand in hand with Queen Desire;
And with a dart that wounded nigh,
Pierced my heart as I did lie;

29 quaint] handsome, elegant

32 deciphered] revealed

269
4 stripe] blow

That when I woke I gan swear,
Phillis' beauty palm did bear.
Up I start, forth went I,
With her face to feed mine eye:
There I saw Desire sit,
That my heart with love had hit, 20
Laying forth bright beauty's hooks
To entrap my gazing looks.
Love I did and gan to woo,
Pray and sigh; all would not do:
Women, when they take the toy,
Covet to be counted coy.
Coy she was, and I gan court,
She thought love was but a sport.
Profound hell was in my thought,
Such a pain Desire had wrought, 30
That I sued with sighs and tears.
Still ingrate she stopped her ears,
Till my youth I had spent.
Last a passion of Repent
Told me flat that Desire
Was a brand of loves fire,
Which consumeth men in thrall,
Virtue, youth, wit, and all.
At this saw back I start,
Beat Desire from my heart, 40
Shook off love and made an oath,
To be enemy to both.
Old I was when thus I fled
Such fond toys as cloyed my head.
But this I learned at Virtue's gate,
The way to good is never late.'

(1590)

270 *Deceiving World*

DECEIVING world, that with alluring toys
Hast made my life the subject of thy scorn,
And scornest now to lend thy fading joys,
To length my life, whom friends have left forlorn,
How well are they that die ere they be born;
And never see thy sleights, which few men shun
Till unawares they helpless are undone.

39 saw] saying, maxim

Oft have I sung of Love and of his fire,
  But now I find that poet was advised
Which made full feasts increasers of desire, 10
  And proves weak love was with the poor despised;
  For when the life with food is not sufficed,
    What thought of love, what motion of delight,
    What pleasance can proceed from such a wight?

Witness my want, the murderer of my wit.
  My ravished sense, of wonted fury reft,
Wants such conceit, as should in poems fit
  Set down the sorrow wherein I am left.
  But therefore have high heavens their gifts bereft,
    Because so long they lent them me to use, 20
    And I so long their bounty did abuse.

O, that a year were granted me to live,
  And for that year my former wits restored,
What rules of life, what counsel would I give!
  How should my sin with sorrow be deplored!
  But I must die of every man abhorred.
    Time loosely spent will not again be won;
    My time is loosely spent, and I undone.

(1592)

271 *The Description of Sir Geoffrey Chaucer*

His stature was not very tall,
Lean he was, his legs were small,
Hosed within a stock of red,
A buttoned bonnet on his head,
From under which did hang, I ween,
Silver hairs both bright and sheen.
His beard was white, trimmed round,
His countenance blithe and merry found.
A sleeveless jacket large and wide,
With many plights and skirts side, 10
Of water camlet did he wear;
A whittle by his belt he bare,

271
2 small] slender
3 stock] stocking
10 side] long
11 water camlet] material made of goat-hair with a wavy surface
12 whittle] knife, dagger

His shoes were corned, broad before,
His inkhorn at his side he wore,
And in his hand he bore a book.
Thus did this ancient poet look.

(1592)

WILLIAM WARNER

c. 1558–1609

from *Albion's England*

272 *A Tale of the beginning of Friars and Cloisterers*

A MERRY mate amongst the rest, of cloisterers thus told.
'This cloist'ring and fat feeding of religion is not old',
Quoth he. 'Not long since was a man that did his devoir give
To kill the passions of his flesh, and did in penance live,
And, though beloved of the King, he lived by his sweat,
Affirming men that would not work unworthy for to eat.
He told the erring their amiss, and taught them to amend.
He counselled the comfortless, and all his days did spend
In prayer and in poverty. Amongst his doings well
High ways he mended, doing which this accident befell. 10
'A dozen thieves to have been hanged were led this hermit by,
To whom he went, exhorting them as Christian men to die.
So penitent they were, and he so pitiful (poor man)
As to the King for pardon of the prisoners he ran:
Which got, he gave it them. But this proviso did he add,
That they should ever work as he. They grant, poor souls, and
glad.
He got them gowns of country grey and hoods for rain and cold,
And hempen girdles which, besides themselves, might burthens
hold,
Pickaxe and spade; and hard to work the covent fell together.
With robes and ropes and every tool for every work and weather 20
So did they toil as thereabout no causey was unwrought.
Wherefore new labours for his men the holy hermit sought;

13 corned] pointed

272

2 cloist'ring] living in a convent or monastery

3 devoir] duty, best endeavour

19 covent] convent, company of 'religious' persons

21 causey] causeway

But at departure prayed them to fast, to watch, and pray,
And live remote from worldly men; and goeth so his way.
'The holy thieves (for now in them had custom wrought content)
Could much of Scripture and, indeed, did heartily repent.
But when the country folk did hear of these same men devout,
Religiously they haunt their cells, and lastly brought about
That from the woods to buildings brave they wound the hermit's crew,
Who was from found-out work returned, and their aposta knew. 30
He going to the stately place did find in every dish
Fat beef and brewis, and great store of dainty fowl and fish.
Who seeing their saturity, and practising to win
His pupils thence, "Excess", he said, "doth work access to sin.
Who fareth finest doth but feed, and over-feedeth oft.
Who sleepeth softest doth but sleep, and sometimes over-soft.
Who clads him trimmest is but clad. The fairest is but fair;
And all but live—yea, if so long, yet not with lesser care
Than forms, backs, bones and bellies that more homely nourished are.
Learn freedom and felicity. Hawks flying where they list 40
Be kindlier and more sound than hawks best tended on the fist."
Thus preached he promised abstinence, and bids them come away.
No haft but good: well were they, and so well as they would stay.
The godly hermit, when all means in vain he did perceive,
Departing said, "I found you knaves, and knaves I do you leave."
'Hence', said this merry fellow (if the merriment be true),
'That cloist'ring, friars' clothing, and a covent's number grew.'
This heard a simple northern-man, no friend to monk, or friar,
Or preaching limmer, for his speech disclosed thus his ire:
'A foul ill on their weasands, for the carls gar sic a din 50
That more we member of their japes than mend us of our sin.
At Yule we wonten gambol, dance, to carol and to sing,
To have good spiced sew, and roast, and plum-pie for a king.
At Fast's-eve pan-puffs; Gang-tide gaits did aly masses bring.

26 Could much] knew much
30 aposta] apostasy, betrayal of their principles
32 brewis] thick broth, with meat and bread
33 saturity] fulness, repletion
practising] intending, planning
43 haft] (?) permanent abode
48 simple] poor, humble, of low rank
49 limmer] rogue, scoundrel
50 weasands] throats
carls] villains, base fellows
gar sic a din] make such an uproar
51 member] remember
japes] tricks, frauds, deceptions
52 Yule] Christmas
wonten] were accustomed to
53 sew] pottage, broth
plum-pie] pie with raisins and currants; mince-pie
54 Fast's-eve] evening or day before the fast; Shrove-Tuesday
pan-puffs] pancakes
Gang-tide] Rogation time; the three days preceding Ascension Day or Holy Thursday
gaits] walks, 'rounds'
aly masses] feasts with ale-drinking

At Paske begun our morris, and ere Pentecost our May:
Tho Robin Hood, lile John, Friar Tuck and Marian deftly play,
And lord and lady gang to kirk with lads and lasses gay.
Fra mass and e'ensong sa gud cheer and glee on ery green
As, save our wakes twixt emes and sibs, like gam was never seen.
At Baptis-day with ale and cakes bout bonfires neighbours stood; 60
At Martilmas wa turned a crab, thilk told of Robin Hood,
Till after long time mirk, when blessed were windows, dares, and lights,
And pails were filled, and hathes were swept, gainst fairies, elves and sprites.
Rock, and Plough Mondays games sal gang, with saint-feasts and kirk sights.
I's tell ye, clerks erst racked not of purpoe ne of pall:
Ilk yeoman fed moe poor toom wambes than gentiles now in hall.
Yea, ledge they ne'er sa Haly Writ, thilk tide was greater wrang
Than heretoforne: tho words had sooth, na writing now so strang.
I is na wizard, yet I drad it will be warse ere lang.

55 Paske] Easter
morris] morris-dancing
Pentecost] Whitsuntide
May] festivities associated with May-day
56 Tho] then
lile] little
deftly] skilfully, handsomely
play] perform, take their parts in the play
57 gang] go
kirk] church
58 glee] sport, play
ery] every
59 wakes] vigils or eves of a festival; festivals with dancing
emes] uncles; friends, gossips
sibs] kinsfolk, relatives
gam] fun
60 Baptis-day] St John Baptist's Day (24 June)
61 Martilmas] Martinmas, the feast of St Martin (11 November)
wa] one
turned a crab] roasted crab-apples by rotating them over a fire
thilk] this one
62 mirk] darkness
blessed] i.e. committed to God's blessing
dares] doors
lights] openings in walls to admit light
63 hathes] hearths
64 Rock] i.e. the day after Twelfth Day, when women resumed their spinning which had been interrupted by Christmas; rock=distaff
Plough Monday] the first Monday after Epiphany, when the beginning of the ploughing season was celebrated with a procession of disguised ploughmen and boys
sal gang] shall proceed, take their course
65 clerks] members of the clergy; scholars, those able to read and write
erst] formerly
racked] recked, i.e. cared, heeded
purpoe] purple (colour of Kings)
pall] rich vestment (e.g. woollen garment worn by the Pope)
66 Ilk] the same
moe] more
toom wambes] empty bellies
gentiles] gentles, gentry
67 ledge] allege, i.e. cite, quote
thilk tide] this time
68 heretoforne] before this time, before now
tho] then
sooth] truth
strang] strange, unfamiliar, foreign
69 wizard] wise man, prophet

Belive doon lither kirkmen reave the crop, and we the tithe, 70
And mickle bookish ben they gif they tache our lakins blithe.
Some egg us sla the Prince and show a bullock fra the Pape,
Whilk, gif it guds the sawle, I's sure the crag gangs till the rape.
Sic votion guiles the people, sa but sild gud Princes scape.
Sa teend our King his life, and sung is *Requiem* for the monk:
Gud King, God rest thy sawle, but fiends reave him bath sawle and
trunk.'
Such talk was long on foot, and still was quittance tale for tale.

(1589)

SIR HENRY LEE
1530–1610

273 [*Farewell to the Court*]

HIS golden locks time hath to silver turned
(O time too swift, O swiftness never ceasing!);
His youth gainst time and age hath ever spurned,
But spurned in vain: youth waneth by increasing.
Beauty, strength, youth are flowers but fading seen;
Duty, faith, love are roots, and ever green.

His helmet now shall make a hive for bees,
And lover's sonnets turn to holy psalms;
A man-at-arms must now serve on his knees,
And feed on prayers, which are age's alms. 10
But though from court to cottage he depart,
His saint is sure of his unspotted heart.

70 Belive] quickly, soon
lither] bad, wicked
reave] steal, plunder
71 mickle] greatly
gif] if
tache] teach
lakins] (?) children, babies
blithe] cheerfully, kindly
72 egg us] incited us to
sla] slay
bullock] papal bull (exculpating the murderer)
Pape] Pope
73 Whilk] which
guds] benefits
crag] neck
gangs till the rape] goes to the rope (ends on the gallows)
74 votion] devotion, religious display
sild] seldom
75 teend] tend, i.e. look after, guard

And when he saddest sits in homely cell,
He'll teach his swains this carol for a song:
'Blest be the hearts that wish my sovereign well,
Curst be the souls that think her any wrong.'
Goddess, allow this aged man his right,
To be your beadsman now that was your knight.

(Pub. 1590)

THOMAS LODGE
1558–1625

274

Love in my bosom like a bee

LOVE in my bosom like a bee
Doth suck his sweet;
Now with his wings he plays with me,
Now with his feet.
Within mine eyes he makes his nest,
His bed amidst my tender breast;
My kisses are his daily feast,
And yet he robs me of my rest.
Ah, wanton, will ye?

And if I sleep, then percheth he 10
With pretty flight,
And makes his pillow of my knee
The livelong night.
Strike I my lute, he tunes the string;
He music plays if so I sing;
He lends me every lovely thing;
Yet cruel he my heart doth sting.
Whist, wanton, still ye!

Else I with roses every day
Will whip you hence, 20
And bind you, when you long to play,
For your offence.
I'll shut mine eyes to keep you in,
I'll make you fast it for your sin,
I'll count your power not worth a pin.
Alas! what hereby shall I win
If he gainsay me?

What if I beat the wanton boy
With many a rod?
He will repay me with annoy, 30
Because a god.
Then sit thou safely on my knee,
And let thy bower my bosom be;
Lurk in mine eyes, I like of thee.
O Cupid, so thou pity me,
Spare not, but play thee!

(1590)

275 *Love guards the roses of thy lips*

LOVE guards the roses of thy lips
And flies about them like a bee;
If I approach he forward skips,
And if I kiss he stingeth me.

Love in thine eyes doth build his bower,
And sleeps within their pretty shine;
And if I look the boy will lour,
And from their orbs shoots shafts divine.

Love works thy heart within his fire,
And in my tears doth firm the same; 10
And if I tempt it will retire,
And of my plaints doth make a game.

Love, let me cull her choicest flowers;
And pity me, and calm her eye;
Make soft her heart, dissolve her lours;
Then will I praise thy deity.

But if thou do not, Love, I'll truly serve her,
In spite of thee, and by firm faith deserve her.

(1593)

276 *My Phillis hath the morning sun*

MY Phillis hath the morning sun
At first to look upon her;
And Phillis hath morn-waking birds
Her risings for to honour.

My Phillis hath prime-feathered flowers
  That smile when she treads on them;
And Phillis hath a gallant flock
  That leaps since she doth own them.
But Phillis hath so hard a heart—
  Alas that she should have it!— 10
As yields no mercy to desert,
  Nor grace to those that crave it.
  Sweet sun, when thou lookest on,
  Pray her regard my moan;
  Sweet birds, when you sing to her,
  To yield some pity, woo her;
  Sweet flowers, wheneas she treads on,
  Tell her, her beauty deads one:
And if in life her love she nill agree me,
Pray her, before I die she will come see me. 20

(1593)

277 *The Shepherd's Sorrow, being disdained in love*

MUSES, help me; sorrow swarmeth,
  Eyes are fraught with seas of languish:
Hapless hope my solace harmeth,
  Mind's repast is bitter anguish.

Eye of day regarded never,
  Certain trust in world untrusty:
Flattering hope beguileth ever,
  Weary old, and wanton lusty.

Dawn of day beholds enthroned
  Fortune's darling proud and dreadless: 10
Darksome night doth hear him moaned,
  Who before was rich and needless.

Rob the sphere of lines united,
  Make a sudden void in nature:
Force the day to be benighted,
  Reave the cause of time and creature,

Ere the world will cease to vary;
  This I weep for, this I sorrow:
Muses, if you please to tarry,
  Further help I mean to borrow. 20

277
12 needless] not in want

Courted once by Fortune's favour,
  Compassed now with envy's curses:
All my thoughts of sorrows savour,
  Hopes run fleeting like the sources.

Aye me, wanton scorn hath maimed
  All the joys my heart enjoyed:
Thoughts their thinking have disclaimed,
  Hate my hopes hath quite annoyed.

Scant regard my weal hath scanted,
  Looking coy hath forced my louring: 30
Nothing liked, where nothing wanted,
  Weds mine eyes to ceaseless show'ring.

Former love was once admired,
  Present favour is estranged:
Loathed the pleasure long desired,
  Thus both men and thoughts are changed.

Lovely swain with lucky speeding,
  Once, but now no more so friended:
Thou my flocks hast had in feeding
  From the morn till day was ended. 40

Drink and fodder, food and folding,
  Had my lambs and ewes together:
I with them was still beholding,
  Both in warmth and winter weather.

Now they languish, since refused,
  Ewes and lambs are pained with pining:
I with ewes and lambs confused,
  All unto our deaths declining.

Silence, leave thy cave obscured,
  Deign a doleful swain to tender: 50
Though disdains I have endured,
  Yet I am no deep offender.

Philip's son can with his finger
  Hide his scar, it is so little:
Little sin a day to linger,
  Wise men wander in a tittle.

24 sources] springs, fountains
37 speeding] fortune
53 Philip's son] Alexander the Great

Trifles yet my swain have turned,
  Though my son he never showeth:
Though I weep, I am not mourned,
  Though I want, no pity groweth. 60

Yet for pity, love my Muses,
  Gentle silence be their cover:
They must leave their wonted uses,
  Since I leave to be a lover.

They shall live with thee enclosed,
  I will loathe my pen and paper:
Art shall never be supposed
  Sloth shall quench the watching taper.

Kiss them, silence, kiss them kindly,
  Though I leave them, yet I love them: 70
Though my wit have led them blindly,
  Yet my swain did once approve them.

I will travel soils removed,
  Night and morning never merry:
Thou shalt harbour that I loved,
  I will love that makes me weary.

If perchance the shepherd strayeth,
  In thy walks and shades unhaunted:
Tell the teen my heart betrayeth,
  How neglect my joys have daunted. 80

(Pub. 1593)

278 [*Animal Weather-forecasting*]

SIR, laugh no more at Pliny and the rest,
Who in their public writings do protest
That birds and beasts (by natural respects
And motions) judge of subsequent effects:
For I will prove that creatures, being dumb,
Have some foreknowledge of events to come.

73 soils removed] far-off countries 79 teen] sorrow

'How prove you that?' I hear some Momus cry.
Thus (gentle sir) by good Philosophy.
First, brutish beasts, who are possessed of nought
But fantasy, to ordinate their thought, 10
And wanting reason's light (which men alone
Partake to help imagination),
It followeth that their fantasies do move
And imitate impressions from above;
And therefore often by the motion
Of birds and beasts some certain things are known.
Hereon the Stagirite (with judgment deep)
Discourseth in his book of watch and sleep:
That some imprudent are most provident—
He meaneth beasts, in reason indigent; 20
Where natheless their intellective parts
(Nothing affected with care-killing hearts,
But desert, as it were, and void of all)
Seem with their manners half-connatural.
For proof, the bitter stings of fleas and flies,
The slime-bred frogs, their harsh reports and cries,
Foresignify and prove a following rain.
'How prove you that?' cries Momus once again.
Why this, dull dunce. The moist and stormy time
Fitting the frogs that dwell in wet and slime, 30
Makes them by natural instinct to croak,
Because ensuing rains the spleen provoke.
And too, the fleas and flies in their degree,
By their attracted moist humidity,
Drawn from a certain virtue elative
Whence rain his generation doth derive,
Seek more than their accustomed nutriment.
So cocks in season inconvenient
That often crow, and asses that do rub
And chafe their hanging ears against a shrub, 40
A following rain do truly prophesy;
And this the reason in philosophy:
The cock, whose dryness by the heat was fed,
By moisture seeks the same extinguished;
The ass with vapours caused by the rain,
The humours then abounding in his brain,

7 Momus] fool
8 Philosophy] natural philosophy (study of nature)
10 ordinate] order, regulate
17 the Stagirite] Aristotle
24 connatural] innate
32 spleen] impulse
33 too] also
35 elative] that raises, elevates
38 season] time

Engendereth an itching in his head.
What need I more? He that hath Virgil read
(Were he as Cato crooked and precise),
Would grant that birds and beasts were weather-wise. 50
But if some misbelieving lad there be
That scorns herein to judge, and join with me,
This pain I do enjoin him for his sins:
When porpoise beat the sea with eager fins,
And beasts more greedily do chaw their cud,
And cormorants seek shore and fly the flood,
And birds do booze them in the pleasant springs,
And crows do ceaseless cry and beat their wings,
That cloakless in a champian he were set
Till to the skin he thoroughly be wet. 60

(1595)

ANONYMOUS

279 [*Hopeless desire soon withers and dies*]

THOUGH naked trees seem dead to sight
When winter winds doth keenly blow,
Yet if the root maintain her right
The spring their hidden life will show.
But if the root be dead and dry,
No marvel though the branches die.

While hope did live within my breast
No winter storm could kill desire,
But now disdain hath hope oppressed
Dead is the root, dead is the spire. 10
Hope was the root, the spire was love,
No sap beneath, no life above.

And as we see the rootless stock
Retain some sap, and spring a while,
Yet quickly prove a lifeless block
Because the root doth life beguile,
So lives desire which hope hath left,
As twilight shines when sun is rest.

(Pub. 1602)

49 crooked] perverse, cantankerous
precise] rigorous, punctilious
53 pain] punishment, penalty
enjoin] impose on
57 booze them] drink copiously
59 champian] expanse of level open country

279
10 spire] tree-top

280 *Were I as base*

WERE I as base as is the lowly plain,
And you (my love) as high as heaven above,
Yet should the thoughts of me your humble swain
Ascend to heaven, in honour of my love.
Were I as high as heaven above the plain,
And you (my love) as humble and as low
As are the deepest bottoms of the main,
Whereso'er you were with you my love should go.
Were you the earth (dear love) and I the skies,
My love should shine on you like to the sun, 10
And look upon you with ten thousand eyes,
Till heaven waxed blind, and till the world were dun.
Whereso'er I am, below or else above you,
Whereso'er you are, my heart shall truly love you.

(Pub. 1602)

281 [*Anacreon, Ode 3*]

OF late, what time the Bear turned round
At midnight in her wonted way,
And men of all sorts slept full sound,
O'ercome with labour of the day,

The God of Love came to my door,
And took the ring and knocked it hard.
'Who's there' quoth I, 'that knocks so sore?
You break my sleep! My dreams are marred!'

'A little boy forsooth', quoth he,
'Dung-wet with rain this moonless night'. 10
With that methought it pitied me;
I ope the door, and candle light.

And straight a little boy I spied,
A winged boy with shafts and bow.
I took him to the fire side,
And set him down to warm him so.

12 dun] dark

281
10 Dung-wet] wet through
11 it pitied me] I was moved to pity

His little hands in mine I strain,
To rub and warm them therewithal.
Out of his locks I crush the rain,
From which the drops apace down fall. 20

At last, when he was waxen warm,
'Now let me try my bow', quoth he;
'I fear my string hath caught some harm,
And wet, will prove too slack for me.'

He said, and bent his bow, and shot,
And wightly hit me in the heart.
The wound was sore and raging hot,
The heat like fury wreaks my smart.

'Mine host', quoth he, 'my string is well',
And laughed, so that he leapt again. 30
'Look to your wound for fear it swell,
Your heart may hap to feel the pain.'

(Pub. 1602)

FRANCIS TREGIAN
1548–1608

282 [*An imprisoned recusant writes to his wife*]

My wont is not to write in verse,
You know, good wife, I wisse,
Wherefore you may well bear with me,
Though now I write amiss.
For lack of ink the candle coal,
For pen a pin I use,
The which also I may allege
In part of my excuse:
For said it is of many men
And such as are no fools, 10
A workman is but little worth
If he do want his tools.

26 wightly] at once

282
2 I wisse] indeed, truly
7 allege] plead

And what although my vein in verse
  Be not as Maro's was,
Yet may such lines as Francis frames
  To his one Mary pass.

What I should send I know not well,
  But sure I am of this,
The doleful mind restored to mirth
  By perfect prayer is. 20
Let prayer be your practice, wife,
  Let prayer be your play,
Let prayer be your staple of trust,
  Let prayer be your stay.
Let prayer be your castle strong,
  Let prayer be your fort,
Let prayer be your place of rest,
  Let prayer be your port.

I know not what to send you, wife,
  I know not what to say; 30
I know not in this world a mean
  Whereby so well you may
Appease your grief, procure relief,
  And eke all ill resist,
As prayer and to meditate
  Upon the life of Christ.
My keeper knocks at door who comes
  To see his hawks in mew,
Wherefore, good wife, I must make short,
  Farewell, sweet spouse, adieu. 40
Farewell the anchor of my hope,
  Farewell my stay of life,
Farewell my poor Penelope,
  Farewell my faithful wife.
Bless in my name my little babes,
  God send them all good hap,
And bless withal that little babe
  That lieth in your lap.

Farewell again thou lamp of light,
  Vicegerent of my heart. 50
He that takes leave so oft, I think,
  He likes not to depart.

(Pub. 1938)

14 Maro] Virgil   38 mew] cage

SIR JOHN HARINGTON
1560–1612

from *'Orlando Furioso' in English Heroical Verse*

283 *[The beginning of Orlando's madness]*

THUS much he prayed, and thence away he went
To seek out Mandricard but found him not,
And (for the day now more than half was spent,
The sun and season waxing somewhat hot)
A shady grove he found, and there he meant
To take some ease but found small ease God wot:
He thinks his thirst and heat a while to swage
But found that set him in worse heat and rage,

For looking all about the grove, behold,
In sundry places fair engrav'n he sees 10
Her name whose love he more esteems than gold
By her own hand in barks of divers trees.
This was the place wherein before I told
Medoro used to pay his surgeon's fees,
Where she to boast of that that was her shame
Used oft to write hers and Medoro's name,

And then with true-love knots and pretty posies
(To show how she to him by love was knit)
Her inward thoughts by outward words discloses,
In her much love to show her little wit. 20
Orlando knew the hand and yet supposes
It was not she that had such posies writ,
And to beguile him self, 'Tush, tush', quoth he,
'There may be more Angelicas than she.

'Yea, but I know too well that pretty hand;
Oft hath she sent me letters of her writing'.
Then he bethinks how she might understand
His name and love by that same new inditing,

1 he prayed] i.e. Orlando
6 wot] knows
7 swage] assuage
8 that] what
11 Her name] i.e. Angelica's (whom Orlando loves)

And how it might be done long time he scanned,
With this fond thought so fondly him delighting. 30
Thus with small hope, much fear, all malcontent,
In these and such conceits the time he spent,

And ay the more he seeks out of his thought
To drive this fancy still it doth increase,
Even as a bird that is with birdlime caught
Doth beat her wings and strives and doth not cease
Until she hath herself all overwrought
And quite entangled in the slimy grease.
Thus on went he till him the way did bring
Unto a shady cave and pleasant spring. 40

This was a place wherein above the rest
This loving pair, leaving their homely host,
Spent time in sports that may not be expressed;
Here in the parching heat they tarried most,
And here Medore (that thought himself most blest)
Wrote certain verses as in way of boast
Which in his language doubtless sounded pretty,
And thus I turn them to an English ditty:

'Ye pleasant plants, green herbs, and waters fair,
And cave with smell and grateful shadow mixed, 50
Where sweet Angelica, daughter and heir
Of Galafronne, on whom in vain were fixed
Full many hearts, with me did oft repair
Alone and naked lay mine arms betwixt,
I, poor Medore, can yield but praise and thanks
For these great pleasures found amid your banks;

'And pray each lord whom Cupid holds in prey,
Each knight, each dame, and every one beside,
Of gentle or mean sort that pass this way,
As fancy or his fortune shall him guide, 60
That to the plants, herbs, spring, and cave he say
"Long may the sun and moon maintain your pride
And the fair crew of nymphs make such purveyance
As hither come no herds to your annoyance".'

29 scanned] examined minutely
30 fond] (1) loving; (2) foolish
32 conceits] notions
33 ay] ever, still
59 sort] rank

It written was there in th'Arabian tongue,
Which tongue Orlando perfect understood
As having learnt it when he was but young
And oft the skill thereof had done him good,
But at this time it him so deeply stung,
It had been well that he it never could, 70
And yet we see to know men still are glad,
And yet we see much knowledge makes men mad.

Twice, thrice, yea five times he doth read the rime,
And though he saw and knew the meaning plain,
Yet that his love was guilty of such crime
He will not let it sink into his brain.
Oft he perused it, and every time
It doth increase his sharp tormenting pain,
And ay the more he on the matter mused
The more his wits and senses were confused. 80

Even then was he of wit wellnigh bestraught,
So quite he was given over unto grief,
(And sure if we believe as proof hath taught
This torture is of all the rest the chief)
His sprite was dead, his courage quailed with thought;
He doth despair and look for no relief,
And sorrow did his senses so surprise
That words his tongue and tears forsook his eyes.

The raging pang remained still within
That would have burst out all at once too fast: 90
Even so we see the water tarry in
A bottle little mouthed and big in waist
That though you topsy-turvy turn the brim
The liquor bides behind with too much haste
And with the striving oft is in such taking
As scant a man can get it out with shaking.

At last he comes unto himself anew
And in his mind another way doth frame
That that which there was written was not true
But writ of spite his lady to defame, 100

70 could] knew
81 bestraught] bereft
83 proof] experience
85 sprite] spirit
thought] depression, dejection

Or to that end that he the same might view
And so his heart with jealousy enflame.
'Well, be't who list', quoth he, 'I see this clearly,
He hath her hand resembled passing nearly.'

With this small hope, with this poor little spark
He doth somedeal revive his troubled sprite,
And, for it was now late and waxed dark,
He seeks some place where he may lie that night.
At last he hears a noise of dogs that bark;
He smells some smoke and sees some candle-light; 110
He takes his inn with will to sleep not eat
As filled with grief and with none other meat;

But lo, his hap was at that house to host
Where fair Angelica had lain before
And where her name on every door and post
With true-love knots was joined to Medore.
That knot, his name, whom he detested most,
Was in his eye and thought still evermore.
He dares not ask nor once the matter touch
For knowing more of that he knows too much, 120

But vain it was himself so to beguile,
For why his host unasked by and by
That saw his guest sit there so sad the while
And thinks to put him from his dumps thereby
Beginneth plain without all fraud or guile,
Without concealing truth or adding lie,
To tell that tale to him without regard
Which divers had before with pleasure heard:

As thus, how at Angelica's request
He holp unto his house to bring Medore, 130
Who then was sorely wounded in his breast,
And she with surgery did heal his sore;
But, while with her own hands the wound she dressed,
Blind Cupid wounded her as much or more,
That when her skill and herbs had cured her patient
Her cureless wound in love made her unpatient.

106 somedeal] somewhat 122 For why] because 124 dumps] low spirits

So that, admit she were the greatest queen
  Of fame and living in those easter parts,
Yet so with fancy she was overseen
  To marry with a page of mean desarts. 140
'Thus Love', quoth he, 'will have his godhead seen
  In famous queens' and highest princes' hearts'.
    This said, to end the tale, he showed the jewel
    That she had given him which Orlando knew well.

This tale, and chiefly this same last conclusion,
  Was even a hatchet to cut off all hope
When love had after many a vain collusion
  Now for his farewell lent him such a rope
To hang himself and drown him in confusion,
  Yet fain he would deny his sorrow scope; 150
    And though a while to show it he forbears,
    It breaketh out at last in sighs and tears,

And as it were enforced he gives the rein
  To raging grief upon his bed alone.
His eyes do shed a very shower of rain
  With many a scalding sigh and bitter groan.
He slept as much as if he had then lain
  Upon a bed of thorns and stuffed with stone,
    And as he lay thereon and could not rest him,
    The bed it self gave matter to molest him. 160

'Ah wretch I am', thus to himself he said,
  'Shall I once hope to take repose and rest me
In that same house, yea even in that same bed
  Where my ungrateful love so lewdly dressed me?
Nay, let me first an hundred times be dead,
  First wolves devour and vultures shall digest me'.
    Straight up he starts, and on he puts his clothes,
    And leaves the house, so much the bed he loathes.

He leaves his host nor once doth take his leave;
  He fared so ill he bids not them farewell. 170
He leaves the town, his servants he doth leave,
  He rides, but where he rides he cannot tell;

138 easter] eastern
139 overseen] made rash
147 collusion] trick
164 dressed] treated

And when alone himself he doth perceive,
  To weep and wail, nay even to howl and yell,
    He doth not cease, to give his grief a vent,
    That inwardly so sore did him torment.

The day, the night, to him were both aleek;
  Abroad upon the cold bare earth he lies;
No sleep, no food he takes nor none would seek;
  All sustenance he to himself denies. 180
Thus he began and ended half the week,
  And he himself doth marvel whence his eyes
    Are fed so long with such a spring of water,
    And to himself thus reasons on the matter:

'No, no, these be no tears that now I shed,
  These be no tears, nor can tears run so rife,
But fire of frenzy drawth up to my head
  My vital humour that should keep my life.
This stream will never cease till I be dead.
  Then welcome death and end my fatal strife. 190
    No comfort in this life my woe can minish
    But thou who canst both life and sorrow finish.

These are not sighs, for sighs some respite have;
  My gripes, my pangs no respite do permit.
The blindfold boy made me a seeing slave
  When from her eyes my heart he first did hit.
Now all enflamed I burn, I rage and rave,
  And in the midst of flame consume no whit.
    Love sitting in my heart, a master cruel,
    Blows with his wings, feeds with his will the fuel. 200

I am not I the man that erst I was:
  Orlando; he is buried and dead.
His most ungrateful love—ah foolish lass!—
  Hath killed Orlando and cut off his head.
I am his ghost that up and down must pass,
  In this tormenting hell for ever led,
    To be a fearful sample and a just
    To all such fools as put in love their trust'.

177 aleek] alike 191 minish] diminish

Thus wand'ring still in ways that have no way
He happed again to light upon the cave 210
Where (in remembrance of their pleasant play)
Medoro did that epigram engrave.
To see the stones again his woes display
And her ill name and his ill hap deprave
Did on the sudden all his sense enrage
With hate, with fury, with revenge and rage.

Straightways he draweth forth his fatal blade
And hews the stones: to heaven the shivers flee.
Accursed was that fountain, cave, and shade,
The arbour and the flowers and every tree. 220
Orlando of all places havoc made
Where he those names together joined may see;
Yea, to the spring he did perpetual hurt
By filling it with leaves, boughs, stones, and dirt.

And having done this foolish frantic feat,
He lays him downe all weary on the ground,
Distempered in his body with much heat,
In mind with pains that no tongue can expound.
Three days he doth not sleep nor drink nor eat,
But lay with open eyes as in a sound. 230
The fourth with rage and not with reason waked,
He rents his clothes and runs about stark naked.

His helmet here he flings, his pouldrons there,
He casts away his curats and his shield.
His sword he throws away he cares not where;
He scatters all his armour in the field.
No rag about his body he doth bear
As might from cold or might from shame him shield,
And save he left behind this fatal blade,
No doubt he had therewith great havoc made. 240

But his surpassing force did so exceed
All common men that neither sword nor bill
Nor any other weapon he did need:
Mere strength sufficed him to do what he will.

212 epigram] inscription
230 sound] swoon
233 pouldrons] shoulder-plates
234 curats] cuirass, i.e. body-armour

He roots up trees as one would root a weed,
  And even as birders laying nets with skill
    Pare slender thorns away with easy strokes,
    So he did play with ashes, elms and oaks.

The herdman and the shepherds that did hear
  The hideous noise and unacquainted sound 250
With fear and wonder great approached near
  To see and know what was hereof the ground;
But now I must cut off this treatise here
  Lest this my book do grow beyond his bound;
    And if you take some pleasure in this text,
    I will go forward with it in my next.

(1591)

284 [*Astolfo recovers Orlando's wits*]

SOON after, he a crystal stream espying,
  From foot to head he washed himself therein.
Then up he gets him on his courser flying
  And of the air he more and more doth win,
Affecting heaven, all earthly thoughts defying.
  As fishes cut the liquid stream with fin,
    So cutteth he the air, and doth not stop
    Till he was come unto that mountain's top.

This hill nigh touched the circle of the moon.
  The top was all a fruitful pleasant field 10
And light at night as ours is here at noon;
  The sweetest place that ever man beheld
(There would I dwell if God gave me my boon);
  The soil there of most fragrant flowers did yield,
    Like rubies, gold, pearls, sapphires, topaz stones,
    Chrysolites, diamonds, jacinths for the nonce.

The trees that there did grow were ever green;
  The fruits that thereon grew were never fading;
The sundry coloured birds did sit between
  And sing most sweet, the fruitful boughs them shading; 20

284
1 He] (i.e. Duke Astolfo)
3 courser] winged horse
5 Affecting] aspiring to
9 circle] sphere
13 boon] request, wish
16 for the nonce] on purpose (used as a metrical tag)

The rivers clear as crystal to be seen;
  The fragrant smell the sense and soul invading
    With air so temperate and so delightsome
    As all the place beside was clear and lightsome.

Amid the plain a palace passing fair
  There stood above conceit of mortal men,
Built of great height into the clearest air,
  And was in circuit twenty mile and ten.
To this fair palace the Duke did straight repair,
  And viewing all that goodly country then 30
    He thought this world, compared with that palace,
    A dunghill vile or prison void of solace.

But when as nearer to the place he came,
  He was amazed at the wondrous sight.
The wall was all one precious stone, the same,
  And than the carbuncle more sanguine bright.
O workman rare, O most stupendious frame,
  What Daedalus of this had oversight!
    Peace, ye that wont to praise the wonders seven:
    Those earthly kings made, this the King of Heaven. 40

Now while the Duke his eyes with wonder fed,
  Behold a fair old man in th'entry stood,
Whose gown was white but yet his jacket red,
  The tone was snow, the tother looked as blood.
His beard was long and white, so was his head;
  His count'nance was so grave, his grace so good,
    A man thereby might at first sight suspect
    He was a saint and one of God's elect.

He coming to the Duke with cheerful face,
  Who now alighted was for rev'rence sake: 50
'Bold Baron', said the saint, 'by special grace
  That suffered wast this voyage strange to make
And to arrive at this most blessed place
  Not knowing why thou didst this journey take,
    Yet know that not without the will celestial
    Thou comest here to paradise terrestrial.

24 lightsome] luminous
26 conceit] conception
36 carbuncle] red-coloured precious stone
sanguine] blood-red
44 tone . . . tother] the one . . . the other

'The cause you come a journey of such length
  Is here of me to learn what must be done
That Charles and Holy Church may now at length
  Be freed that erst were well nigh overrun. 60
Wherefore impute it not to thine own strength
  Nor to thy courage, nor thy wit, my son,
    For neither could thy horn nor winged steed
    Without God's help stand thee in any stead.

But at more leisure hereof we will reason
  And more at large I mind with you to speak.
Now with some meat refresh you, as is reason,
  Lest fasting long may make your stomach weak.
Our fruits', said he, 'be never out of season.'
  The Duke rejoiced much and marvelled eke, 70
    Then chief when by his speeches and his coat
    He knew 'twas he that the fourth Gospel wrote,

That holy John whom Christ did hold so dear
  That others thought he death should never see,
Though in the Gospel it appears not clear,
  But thus he said: 'What if it pleased me,
O Peter, that thy fellow tarry here
  Until my coming, what is that to thee?'
    So though our Saviour not directly spake it,
    Yet sure it was so every one did take it. 80

He here assumed was in happy hour
  Whereas before Enoch the Patriarch was
And where the Prophet bides of mighty power
  That in the fiery coach did thither pass.
These three in that so happy sacred bower
  In high felicity their days did pass
    Where in such sort to stand they are allowed
    Till Christ return upon the burning cloud.

These saints him welcome to that sacred seat
  And to a stately lodging him they brought, 90
And for his horse likewise ordained meat;
  And then the Duke himself by them was taught

59 Charles] i.e. Charlemagne, the commander of the Christian army
60 erst] formerly
63 horn] (Astolfo's horn, which has the power to terrify his enemies)
64 stand . . . stead] be of use to you
81 assumed] received up into heaven
82 Whereas] where
83 Prophet] i.e. Ezekiel

The dainty fruits of Paradise to eat,
  So delicate in taste as sure he thought
    Our first two parents were to be excused
    That for such fruit obedience they refused.

Now when the Duke had nature satisfied
  With meat and drink and with his due repose
(For there were lodgings fair and all beside
  That needful for man's use man can suppose), 100
He gets up early in the morning tide
  What time with us alow the sun arose,
    But ere that he from out his lodging moved
    Came that disciple whom our Saviour loved,

And by the hand the Duke abroad he led
  And said some things to him I may not name;
But in the end, I think, 'My son', he said,
  'Although that you from France so lately came,
You little know how those in France have sped.
  There your Orlando quite is out of frame, 110
    For God his sin most sharply now rewardeth,
    Who most doth punish whom he most regardeth.

'Know that the champion, your Orlando, whom
  God so great strength and so great courage gave
And so rare grace that from his mother's womb
  By force of steel his skin no hurt might have
To th'end that he might fight for his own home
  And those that hold the Christian faith to save,
    As Samson erst enabled was to stand
    Against Philistines for the Hebrew land; 120

'This your Orlando hath been so ungrate,
  For so great grace received, unto his maker,
That when his country was in weakest state
  And needed succour most, he did forsake her
For love (O woeful love that breeds God's hate!),
  To woo a pagan wench with mind to take her,
    And to such sin this love did him entice
    He would have killed his kinsman once or twice.

101 tide] time   102 alow] below   110 frame] order

'For this same cause doth mighty God permit
  Him mad to run with belly bare and breast, 130
And so to daze his reason and his wit
  He knows not others and himself knows least.
So in times past Our Lord did deem it fit
  To turn the King of Babel to a beast,
    In which estate he seven whole years did pass
    And like an ox did feed on hay and grass.

'But, for the paladin's offence is not
  So great as was the King of Babel's crime,
The mighty Lord of mercy doth allot
  Unto his punishment a shorter time. 140
Twelve weeks in all he must remain a sot;
  And for this cause you suffered were to climb
    To this high place that here you may be taught
    How to his wits Orlando may be brought.

'Here you shall learn to work the feat, I warrant;
  But yet, before you can be fully sped
Of this your great but not forethought-on arrant,
  You must with me a more strange way be led
Up to the planet that of all stars errant
  Is nearest us; when she comes overhead, 150
    Then will I bring you where the medicine lies
    That you must have to make Orlando wise.'

Thus all that day they spent in divers talk,
  With solace great as never wanteth there.
But when the sun began this earth to baulk
  And pass into the tother hemisphere,
Then they prepared to fetch a further walk,
  And straight the fiery charret that did bear
    Elias when he up to heaven was carried
    Was ready in a trice and for them tarried. 160

Four horses fierce, as red as flaming fire,
  Th'Apostle doth into the charret set,
Which when he framed had to his desire,
  Astolfo in the car by him he set.

138 Babel] Babylon, i.e. Nebuchadnezzar
146 sped] discharged
147 arrant] errand, mission
149 errant] wandering
152 wise] sane
155 baulk] pass by
158 charret] chariot
159 Elias] i.e. Elijah

Then up they went, and still ascending higher
Above the fiery region they did get
Whose nature so th'Apostle then did turn
That though they went through fire they did not burn.

I say although the fire were wondrous hot,
Yet in their passage they no heat did feel, 170
So that it burned them nor offends them not.
Thence to the moon he guides the running wheel.
The moon was like a glass all void of spot,
Or like a piece of purely burnished steel,
And looked, although to us it seemed so small,
Well nigh as big as earth and sea and all.

Here had Astolfo cause of double wonder:
One, that that region seemeth there so wide
That unto us that are so far asunder
Seems but a little circle, and beside, 180
That to behold the ground that lay him under
A man had need to have been sharply eyed
And bend his brows and mark ev'n all they might,
It seemed so small, now chiefly wanting light.

'Twere infinite to tell what wondrous things
He saw that passed ours not few degrees.
What towns, what hills, what rivers, and what springs,
What dales, what palaces, what goodly trees—
But to be short, at last his guide him brings
Unto a goodly valley where he sees 190
A mighty mass of things strangely confused:
Things that on earth were lost or were abused.

A storehouse strange, that what on earth is lost
By fault, by time, by fortune, there is found,
And like a merchandise is there engrossed
In stranger sort than I can well expound.
Nor speak I sole of wealth or things of cost
In which blind fortune's power doth most abound,
But ev'n of things quite out of fortune's power,
Which wilfully we waste each day and hour. 200

167 turn] change 186 passed] surpassed 195 engrossed] collected

The precious time that fools misspend in play,
  The vain attempts that never take effect,
The vows that sinners make and never pay,
  The counsels wise that careless men neglect,
The fond desires that lead us oft astray,
  The praises that with pride the heart infect,
    And all we lose with folly and misspending
    May there be found unto this place ascending.

Now as Astolfo by those regions passed
  He asked many questions of his guide; 210
And as he on tone side his eye did cast,
  A wondrous hill of bladders he espied,
And he was told they had been in time past
  The pompous crowns and sceptres full of pride
    Of monarchs of Assyria and of Greece
    Of which now scantly there is left a piece.

He saw great store of baited hooks with gold,
  And those were gifts that foolish men prepared
To give to princes covetous and old
  With fondest hope of future vain reward. 220
Then were there ropes all in sweet garlands rolled,
  And those were all false flatteries he heard.
    Then heard he crickets' songs like to the verses
    The servant in his master's praise rehearses.

There did he see fond loves that men pursue
  To look like golden gyves with stones all set;
Then things like eagles' talents he did view,
  Those offices that favourites do get;
Then saw he bellows large that much wind blew,
  Large promises that lords make and forget 230
    Unto their Ganymedes in flower of youth—
    But after, nought but beggary ensu'th.

He saw great cities seated in fair places
  That overthrown quite topsy-turvy stood:
He asked, and learned the cause of their defaces
  Was treason that doth never turn to good.

226 gyves] fetters
227 talents] talons
231 Ganymedes] boys, catamites
235 defaces] defacement, destruction

He saw foul serpents with fair women's faces,
Of coiners and of thieves the cursed brood.
He saw fine glasses all in pieces broken,
Of service lost in court a woeful token. 240

Of mingled broth he saw a mighty mass
That to no use all spilt on ground did lie.
He asked his teacher, and he heard it was
The fruitless alms that men give when they die.
Then by a fair green mountain he did pass
That once smelt sweet but now it stinks perdie:
This was that gift (be't said without offence)
That Constantine gave Sylvester long since.

Of bird-limed rods he saw no little store—
And these, O ladies fair, your beauties be. 250
I do omit ten thousand things and more
Like unto these that there the Duke did see,
For all that here is lost, there evermore
Is kept and thither in a trice doth flee.
Howbeit more or less there was no folly,
For still that here with us remaineth wholly.

He saw some of his own lost time and deeds,
But yet he knew them not to be his own,
They seemed to him disguised in so strange weeds
Till his instructor made them better known. 260
But last the thing which no man thinks he needs,
Yet each man needeth most, to him was shown:
By name, man's wit, which here we leese so fast
As that one substance all the other passed.

It seemed to be a body moist and soft
And apt to mount by every exhalation,
And when it hither mounted was aloft
It there was kept in pots of such a fashion
As we call jars, where oil is kept in oft.
The Duke beheld with no small admiration 270
The jars of wit, amongst which one had writ
Upon the side thereof, 'Orlando's wit'.

246 perdie] by God (polite oath)
249 store] abundance
263 wit] understanding, sanity
leese] lose

This vessel bigger was than all the rest;
  And every vessel had engraven with art
His name that erst the wit therein possessed.
  There of his own the Duke did find a part,
And much he mused and much himself he blessed
  To see some names of men of great desart
    That think they have great store of wit and boast it,
    And here it plain appeared they quite had lost it. 280

Some lose their wit with love, some with ambition,
  Some running to the sea great wealth to get,
Some following lords and men of high condition,
  And some in fair jewels rich and costly set;
One hath desire to prove a rare magician,
  And some with poetry their wit forget;
    Another thinks to be an alchemist
    Till all be spent and he his number missed.

Astolfo takes his own before he goes,
  For so th'Evangelist did him permit. 290
He set the vessel's mouth but to his nose,
  And to his place he snuffed up all his wit.
Long after, wise he lived as Turpin shows
  Until one fault he after did commit,
    By name, the love of one fair northern lass,
    Sent up his wit unto the place it was.

The vessel where Orlando's wit was closed
  Astolfo took, and thence with him did bear.
It was far heavier than he had supposed,
  So great a quantity of wit was there; 300
But yet ere back their journey they disposed
  The holy Prophet brought Astolfo where
    A palace (seldom seen by mortal man)
    Was placed, by which a thick dark river ran.

(1591)

285 *Of Treason*

TREASON doth never prosper, what's the reason?
For if it prosper, none dare call it treason.

(Pub. 1615)

278 desart] deserving   301 disposed] directed

286 *Of the wars in Ireland*

I PRAISE the speech, but cannot now abide it,
That war is sweet to those that have not tried it:
For I have proved it now, and plainly see 't,
It is so sweet, it maketh all things sweet.
At home Canary wines and Greek grow loathsome;
Here milk is nectar, water tasteth toothsome.
There without baked, roast, boiled, it is no cheer;
Biscuit we like, and bonny clabo here.
There we complain of one rare-roasted chick;
Here viler meat, worse cooked, ne'er makes me sick. 10
At home in silken sparvers, beds of down,
We scant can rest, but still toss up and down;
Here I can sleep, a saddle to my pillow,
A hedge the curtain, canopy a willow.
There if a child but cry, Oh what a spite!
Here we can brook three larums in one night.
There homely rooms must be perfumed with roses;
Here match and powder ne'er offends our noses.
There from a storm of rain we run like pullets;
Here we stand fast against a shower of bullets. 20
Lo then how greatly their opinions err
That think there is no great delight in war.
But yet for this, sweet war, I'll be thy debtor:
I shall for ever love my home the better.

(Wr. 1599; pub. 1615)

287 *A Groom of the Chamber's Religion in King Henry the eighth's time*

ONE of King Henry's favourites began
To move the King one day to take a man
Whom of his chamber he might make a groom.
'Soft', said the King, 'before I grant that room,

1 the speech] i.e. Erasmus's adage 'Dulce bellum inexpertis'
8 bonny clabo] sour buttermilk
9 rare-roasted] underdone
11 sparvers] canopies at the top of the bed
12 scant] scarce
16 larums] calls to arms; sudden attacks
19 pullets] chickens

287
A Groom of the Chamber] i.e. an officer of the Royal Household
4 Soft] wait; not so fast
room] office, post

It is a question not to be neglected
How he in his religion stands affected.'
'For his religion', answered then the minion,
'I do not certain know what's his opinion.
But sure he may, talking with men of learning,
Conform himself in less than ten days' warning.' 10

(Pub. 1618)

288 *Sir John Raynsford's Confession*

RAYNSFORD, a knight, fit to have served King Arthur,
And in Queen Mary's days a demi-martyr:
For though both then, before, and since he turned,
Yet sure, *per ignem hanc*, he might be burned—
This knight agreed with those of that profession,
And went, as others did, to make confession.
Among some peccadillos he confessed
That same sweet sin that some but deem a jest,
And told how, by good help of bawds and varlets,
Within twelve months he had six times twelve harlots. 10
The priest, that at the tale was half astonished,
With grave and ghostly counsel him admonished
To fast and pray, to drive away that devil
That was to him causer of so great evil,
That the lewd spirit of Lechery, no question,
Stirred up his lust, with many a lewd suggestion:
'A filthy fiend', said he, 'most foul and odious,
Named, as appears, in holy writs, Asmodeus.'
Thus, with some penance that was ne'er performed,
Away went that same knight, smally reformed. 20
Soon after this, ensued religious change,
That in the Church bred alteration strange,
And Raynsford, with the rest, followed the stream.
The priest went roving round about the realm.
This priest in clothes disguised himself did hide,
Yet Raynsford three years after him had spied,
And laid unto his charge, and sorely pressed him
To tell if 'twere not he that had confessed him.
The priest, though this knight's words did sore him daunt,
Yet what he could not well deny, did grant, 30

288
3 turned] adopted a different form of religion; was converted
4 *per ignem hanc*] through this fire
7 peccadillos] venial sins or faults

And prayed him not to punish, or control,
That he had done for safety of his soul.
'No, knave', quoth he, 'I will not harm procure thee,
Upon my worship here I do assure thee.
I only needs must laugh at thy great folly
That wouldst persuade with me to be so holy;
To chastise mine own flesh, to fast and pray,
To drive the spirit of Lechery away.
Sounds, foolish knave, I fasted not, nor prayed,
Yet is that spirit quite gone from me', he said. 40
'If thou couldst help me to that spirit again,
Thou shouldst a hundred pound have for thy pain.
  That lusty Lord of Lechery, Asmodeus,
  That thou call'st odious, I do count commodious.'

(Pub. 1618)

ANONYMOUS

289 *The Passionate Man's Pilgrimage*

GIVE me my scallop-shell of quiet,
My staff of faith to walk upon,
My scrip of joy, immortal diet,
My bottle of salvation,
My gown of glory, hope's true gage,
And thus I'll take my pilgrimage.

Blood must be my body's balmer,
No other balm will there be given,
Whilst my soul like a white palmer
Travels to the land of heaven, 10
Over the silver mountains,
Where spring the nectar fountains;
And there I'll kiss
The bowl of bliss,
And drink my eternal fill
On every milken hill.
My soul will be a-dry before,
But after it, will ne'er thirst more.

31 control] rebuke
32 That] what
39 Sounds] Zounds (i.e. by God's wounds)
44 commodious] beneficial; serviceable

And by the happy blissful way
More peaceful pilgrims I shall see, 20
That have shook off their gowns of clay
And go apparelled fresh like me.
I'll bring them first
To slake their thirst,
And then to taste those nectar suckets,
At the clear wells
Where sweetness dwells,
Drawn up by saints in crystal buckets.

And when our bottles and all we
Are filled with immortality, 30
Then the holy paths we'll travel,
Strewed with rubies thick as gravel,
Ceilings of diamonds, sapphire floors,
High walls of coral and pearl bowers.

From thence to heaven's bribeless hall
Where no corrupted voices brawl,
No conscience molten into gold,
Nor forged accusers bought and sold,
No cause deferred, nor vain-spent journey,
For there Christ is the King's Attorney, 40
Who pleads for all without degrees,
And he hath angels, but no fees.

When the grand twelve million jury
Of our sins and sinful fury
'Gainst our souls black verdicts give,
Christ pleads his death, and then we live.
Be thou my speaker, taintless pleader,
Unblotted lawyer, true proceeder;
Thou movest salvation even for alms,
Not with a bribed lawyer's palms. 50

And this is my eternal plea
To him that made heaven, earth, and sea:
Seeing my flesh must die so soon,
And want a head to dine next noon,
Just at the stroke when my veins start and spread,
Set on my soul an everlasting head.
Then am I ready, like a palmer fit,
To tread those blest paths which before I writ.

(Pub. 1604)

25 suckets] sweetmeats

42 angels] word-play on 'angel' meaning 'coin'

HENRY CONSTABLE
1562–1613

290 *To the Marquess of Piscat's Soul, endued in her life time with infinite perfections, as her divine poems do testify*

SWEET soul, which now with heavenly songs dost tell
  Thy dear Redeemer's glory and his praise,
  No marvel though thy skilful Muse assays
The songs of other souls there to excel:
For thou didst learn to sing divinely well
  Long time before thy fair and glittering rays
  Increased the light of heaven, for even thy lays
Most heavenly were when thou on earth didst dwell.
When thou didst on the earth sing poet-wise,
  Angels in heaven prayed for thy company; 10
And now thou sing'st with angels in the skies
  Shall not all poets praise thy memory?
    And to thy name shall not their works give fame
    Whenas their works be sweetened by thy name?

(Pub. 1595)

291 *To Sir Philip Sidney's Soul*

GIVE pardon, blessed soul, to my bold cries,
  If they, importune, interrupt thy song,
  Which now with joyful notes thou sing'st among
The angel-choristers of th'heavenly skies.
Give pardon eke dear soul, to my slow eyes,
  That since I saw thee it is now so long,
  And yet the tears which unto thee belong
To thee as yet they did not sacrifice.
I did not know that thou wert dead before;
  I did not feel the grief I did sustain; 10
The greater stroke astonisheth the more;
  Astonishment takes from us sense of pain.
    I stood amazed when others' tears begun,
    And now begin to weep when they have done.

(Pub. 1595)

Marquess of Piscat] Vittoria Colonna, marchioness of Pescara (1490–1547)

291
2 importune] importunate

292 *To God the Holy Ghost*

ETERNAL sprite, which art in heaven the love
With which God and His Son each other kiss,
And who, to show who God's beloved is,
The shape, and wings, took'st of a loving dove:
When Christ ascending sent thee from above
In fiery tongues, thou cam'st down unto his
That skill in uttering heavenly mysteries
By heat of zeal both faith and love might move.
True God of Love, from whom all true love springs,
Bestow upon my love thy wings and fire; 10
My soul a spirit is, and with thy wings
May like an angel fly from earth's desire,
And with thy fire a heart enflamed may bear,
And in thy sight a seraphin appear.

(Pub. 1815)

293 *To the Blessed Sacrament*

WHEN thee (O holy sacrificed Lamb)
In severed signs I white and liquid see:
As on thy body slain I think on thee,
Which pale by shedding of thy blood became;
And when again I do behold the same
Veiled in white to be received of me,
Thou seemest in thy sindon wrapped to be
Like to a corse, whose monument I am.
Buried in me, unto my soul appear
Prisoned in earth, and banished from thy sight, 10
Like our forefathers, who in limbo were.
Clear thou my thoughts, as thou didst give the light;
And as thou others freed from purging fire,
Quench in my heart the flames of bad desire.

(Pub. 1815)

292
1 sprite] spirit
14 seraphin] seraph

293
2 severed signs] the bread and wine of the eucharist
7 sindon] shroud
8 corse] corpse

294 *To Our Blessed Lady*

In that, O Queen of queens, thy birth was free
From guilt, which others doth of grace bereave,
When in their mother's womb they life receive,
God as his sole-born daughter loved thee.
To match thee like thy birth's nobility,
He thee his Spirit for thy spouse did leave,
Of whom thou didst his only Son conceive,
And so wast linked to all the Trinity.
Cease then, O queens, who earthly crowns do wear,
To glory in the pomp of worldly things. 10
If men such high respect unto you bear,
Which daughters, wives and mothers are of kings,
What honour should unto that Queen be done,
Who had your God for father, spouse and son?

(Pub. 1815)

295 *To St John Baptist*

As Anne, long barren, mother did become
Of him who last was judge in Israel,
Thou last of prophets born, like Samuel
Didst from a womb past hope of issue come.
His mother silent spake; thy father dumb,
Recovering speech, God's wonder did foretell.
He after death a prophet was in hell,
And thou unborn within thy mother's womb.
He did anoint the king, whom God did take
From charge of sheep to rule His chosen land; 10
But that high King who heaven and earth did make
Received a holier liquor from thy hand,
When God His flock in human shape did feed,
As Israel's king kept his in shepherd's weed.

(Pub. 1815)

295
5 silent spake] cf. I Samuel 1: 12–13
thy father dumb] cf. Luke 1: 18–22, 64
7 prophet was in hell] cf. I Samuel 28: 11–19
8 unborn] cf. Luke 1: 41–4
9 anoint the king] I Samuel 16: 11–13
12 holier liquor] cf. Matthew 3

296 *To St Peter and St Paul*

HE that for fear his master did deny
And at a maiden's voice amazed stood,
The mightiest monarch of the earth withstood,
And on his master's Cross rejoiced to die.
He whose blind zeal did rage with cruelty,
And helped to shed the first of martyrs' blood,
By light from heaven his blindness understood,
And with the chief apostle slain doth lie.
O three times happy two! O golden pair,
Who with your blood did lay the Church's ground 10
Within the fatal town which twins did found,
And settled there the Hebrew fisher's chair,
Where first the Latin shepherd raised his throne,
And since the world and Church were ruled by one.

(Pub. 1815)

297 *To St Mary Magdalen*

FOR few nights' solace in delicious bed
Where heat of lust did kindle flames of hell,
Thou nak'd on naked rock in desert cell
Lay thirty years, and tears of grief did shed.
But for that time thy heart there sorrowed
Thou now in heaven eternally dost dwell,
And for each tear which from thine eyes then fell,
A sea of pleasure now is rendered.
If short delights entice my heart to stray,
Let me by thy long penance learn to know 10
How dear I should for trifling pleasures pay;
And if I virtue's rough beginning shun,
Let thy eternal joys unto me show
What high rewards by little pain is won.

(Pub. 1815)

11 fatal town] Rome, founded by Romulus and Remus

292
4 thirty years] i.e. her (legendary) penance at Marseilles

298 *To St Mary Magdalen*

BLESSED offender, who thyself hast tried
How far a sinner differs from a saint,
Join thy wet eyes with tears of my complaint,
While I sigh for that grave for which thou cried.
No longer let my sinful soul abide
In fever of thy first desires faint,
But let that love, which last thy heart did taint
With pangs of thy repentance, pierce my side.
So shall my soul no foolish virgin be
With empty lamp, but like a Magdalen bear 10
For ointment box a breast with oil of grace;
And so the zeal which then shall burn in me
May make my heart like to a lamp appear
And in my spouse's palace give me place.

(Pub. 1815)

299 *To St Margaret*

FAIR amazon of heaven, who took'st in hand
St Michael and St George to imitate,
And for a tyrant's love transformed to hate
Wast for thy lily faith retained in band:
Alone on foot, and with thy naked hand,
Thou did'st like Michael and his host, and that
For which on horse armed George we celebrate;
Whilst thou, like them, a dragon did'st withstand.
Behold my soul shut in my body's jail,
The which the drake of hell gapes to devour. 10
Teach me (O Virgin) how thou did'st prevail.
Virginity, thou say'st, was all thy aid.
Give me then purity instead of power,
And let my soul, maid chaste, pass for a maid.

(Pub. 1815)

1 tried] proved by experience

299
4 retained] detained band] imprisonment
10 drake] dragon

MARY HERBERT, COUNTESS OF PEMBROKE

1561–1621

300 *Psalm 52*

TYRANT, why swell'st thou thus,
  Of mischief vaunting?
Since help from God to us
  Is never wanting.

Lewd lies thy tongue contrives,
  Loud lies it soundeth;
Sharper than sharpest knives
  With lies it woundeth.

Falsehood thy wit approves,
  All truth rejected: 10
Thy will all vices loves,
  Virtue neglected.

Not words from cursed thee,
  But gulfs are poured;
Gulfs wherein daily be
  Good men devoured.

Think'st thou to bear it so?
  God shall displace thee;
God shall thee overthrow,
  Crush thee, deface thee. 20

The just shall fearing see
  These fearful chances,
And laughing shoot at thee
  With scornful glances.

Lo, lo, the wretched wight,
  Who God disdaining,
His mischief made his might,
  His guard his gaining.

I as an olive tree
  Still green shall flourish: 30
God's house the soil shall be
  My roots to nourish.

My trust on his true love
  Truly attending,
Shall never thence remove,
  Never see ending.

Thee will I honour still,
  Lord, for this justice;
There fix my hopes I will
  Where thy saints' trust is. 40

Thy saints trust in thy name,
  Therein they joy them:
Protected by the same,
  Naught can annoy them.

(Wr. prob. before 1599;
pub. 1823)

301 *Psalm 58*

AND call ye this to utter what is just,
  You that of justice hold the sovereign throne?
And call ye this to yield, O sons of dust,
  To wronged brethren every man his own?
O no: it is your long malicious will
  Now to the world to make by practice known,
With whose oppression you the balance fill,
  Just to yourselves, indifferent else to none.

But what could they, who even in birth declined,
  From truth and right to lies and injuries? 10
To show the venom of their cancered mind
  The adder's image scarcely can suffice;
Nay scarce the aspic may with them contend,
  On whom the charmer all in vain applies
His skilful'st spells: ay missing of his end,
  While she self-deaf and unaffected lies.

Lord, crack their teeth; Lord, crush these lions' jaws,
  So let them sink as water in the sand.
When deadly bow their aiming fury draws,
  Shiver the shaft ere past the shooter's hand. 20
So make them melt as the dis-housed snail
  Or as the embryo, whose vital band
Breaks ere it holds, and formless eyes do fail
  To see the sun, though brought to lightful land.

301
13 aspic] asp, snake

O let their brood, a brood of springing thorns,
  Be by untimely rooting overthrown,
Ere bushes waxed they push with pricking horns,
  As fruits yet green are oft by tempest blown.
The good with gladness this revenge shall see,
  And bathe his feet in blood of wicked one; 30
While all shall say: the just rewarded be;
  There is a God that carves to each his own.

(Wr. probably before 1599; pub. 1823)

302 *Psalm 59*

SAVE me from such as me assail:
    Let not my foes,
O God, against my life prevail;
    Save me from those
Who make a trade of cursed wrong
And, bred in blood, for blood do long.

Of these one sort do seek by sleight
    My overthrow;
The stronger part with open might
    Against me go, 10
And yet thou, God, my witness be
From all offence my soul is free.

But what if I from fault am free?
    Yet they are bent
To band and stand against poor me,
    Poor innocent.
Rise, God, and see how these things go,
And rescue me from instant woe.

Rise, God of armies, mighty God
    Of Israel,
Look on them all who spread abroad 20
    On earth do dwell,
And let thy hand no longer spare
Such as of malice wicked are.

27 waxed] grown

When golden sun in west doth set,
    Returned again,
As hounds that howl their food to get,
    They run amain
The city through from street to street,
With hungry maw some prey to meet. 30

Night elder grown, their fittest day,
    They babbling prate,
How my lost life extinguish may
    Their deadly hate.
They prate and babble void of fear,
For, tush, say they, who now can hear?

Even thou canst hear, and hearing scorn,
    All that they say;
For them (if not by thee upborne)
    What props do stay? 40
Then will I, as they wait for me,
On God my fortress, wait on thee.

Thou ever me with thy free grace
    Prevented hast;
With thee my prayer shall take place
    Ere from me passed;
And I shall see who me doth hate
Beyond my wish in woeful state.

For fear my people it forget
    Slay not outright 50
But scatter them and so them set
    In open sight,
That by thy might they may be known,
Disgraced, debased, and overthrown.

No witness of their wickedness
    I need produce
But their own lips, fit to express
    Each vile abuse:
In cursing proud, proud when they lie,
O let them dear such pride a-buy. 60

28 amain] violently, in large numbers 44 Prevented] anticipated

At length in rage consume them so
    That nought remain;
Let them all being quite forego,
    And make it plain
That God, who Jacob's rule upholds,
Rules all all-bearing earth enfolds.

Now thus they fare: when sun doth set,
    Returned again,
As hounds that howl their food to get,
    They run amain   70
The city through from street to street
With hungry maw some prey to meet.

Abroad they range, and hunt apace
    Now that, now this,
As famine trails a hungry trace;
    And though they miss,
Yet will they not to kennel hie,
But all the night at bay do lie.

But I will of thy goodness sing
    And of thy might,   80
When early sun again doth bring
    His cheerful light;
For thou, my refuge and my fort,
In all distress dost me support.

My strength doth of thy strength depend.
    To thee I sing:
Thou art my fort, me to defend.
    My God, my king,
To thee I owe, and thy free grace,
That free I rest in fearless place.   90

(Wr. probably before 1599; pub. 1823)

303 *Psalm 73*

It is most true that God to Israel,
  I mean to men of undefiled hearts,
  Is only good, and nought but good imparts.
Most true, I see, albe almost I fell
  From right conceit into a crooked mind;
  And from this truth with straying steps declined.

For lo, my boiling breast did chafe and swell
  When first I saw the wicked proudly stand,
  Prevailing still in all they took in hand.
And sure no sickness dwelleth where they dwell: 10
  Nay, so they guarded are with health and might,
  It seems of them death dares not claim his right.

They seem as privileged from others' pain:
  The scourging plagues, which on their neighbours fall,
  Torment not them, nay touch them not at all.
Therefore with pride, as with a gorgeous chain,
  Their swelling necks encompassed they bear,
  All clothed in wrong, as if a robe it were:
So fat become, that fatness doth constrain
  Their eyes to swell; and if they think on aught, 20
  Their thought they have, yea have beyond their thought.
They wanton grow, and in malicious vein
  Talking of wrong, pronounce as from the skies.
  So high a pitch their proud presumption flies!

Nay heaven itself, high heaven escapes not free
  From their base mouths; and in their common talk
  Their tongues no less than all the earth do walk.
Wherefore even godly men, when so they see
  Their horn of plenty freshly flowing still,
  Leaning to them, bend from their better will; 30
And thus, they reasons frame: how can it be
  That God doth understand? that he doth know,
  Who sits in heaven, how earthly matters go?
See here the godless crew, while godly we
  Unhappy pine, all happiness possess:
  Their riches more, our wealth still growing less.

Nay even within my self, my self did say:
  In vain my heart I purge, my hands in vain
  In cleanness washed I keep from filthy stain,
Since thus afflictions scourge me every day; 40
  Since never a day from early east is sent
  But brings my pain, my check, my chastisement.
And shall I then these thoughts in words bewray?
  O let me, Lord, give never such offence
  To children thine that rest in thy defence.
So then I turned my thoughts another way,
  Sounding, if I this secret's depth might find;
  But cumbrous clouds my inward sight did blind.

Until at length nigh weary of the chase,
  Unto thy house I did my steps direct: 50
  There, lo, I learned what end did these expect,
And what? but that in high but slippery place
  Thou didst them set: whence when they least of all
  To fall did fear, they fell with headlong fall.
For how are they in less than moment's space
  With ruin overthrown? with frightful fear
  Consumed so clean, as if they never were?
Right as a dream, which waking doth deface.
  So, Lord, most vain thou dost their fancies make,
  When thou dost them from careless sleep awake. 60

Then for what purpose was it? to what end
  For me to fume with malcontented heart,
  Tormenting so in me each inward part?
I was a fool (I can it not defend),
  So quite deprived of understanding might
  That as a beast I bare me in thy sight.
But as I was, yet did I still attend,
  Still follow thee, by whose upholding hand,
  When most I slide, yet still upright I stand.
Then guide me still, then still upon me spend 70
  The treasures of thy sure advice, until
  Thou take me hence into thy glory's hill.

O what is he will teach me climb the skies?
  With thee, thee good, thee goodness to remain?
  No good on earth doth my desires detain.
Often my mind and oft my body tries
  Their weak defects: but thou, my God, thou art
  My endless lot and fortress of my heart.
The faithless fugitives who thee despise
  Shall perish all, they all shall be undone, 80
  Who leaving thee to whorish idols run.
But as for me, nought better in my eyes
  Than cleave to God, my hopes in him to place,
  To sing his works while breath shall give me space.

(Wr. prob. before 1599; pub. 1823)

304

Psalm 134

YOU that Jehovah's servants are,
Whose careful watch, whose watchful care,
Within his house are spent;
Say thus with one assent:
Jehovah's name be praised.
Then let your hands be raised
To holiest place,
Where holiest grace
Doth ay
Remain: 10
And say
Again,
Jehovah's name be praised.
Say last unto the company
Who tarrying make
Their leave to take:
All blessings you accompany,
For him in plenty showered
Whom Sion holds embowered,
Who heaven and earth of nought hath raised. 20

(Wr. prob. before 1599; pub. 1823)

305

Psalm 139

O LORD, in me there lieth nought
But to thy search revealed lies,
For when I sit
Thou markest it;
No less thou notest when I rise;
Yea, closest closet of my thought
Hath open windows to thine eyes.

Thou walkest with me when I walk;
When to my bed for rest I go,
I find thee there, 10
And everywhere:
Not youngest thought in me doth grow,
No, not one word I cast to talk
But yet unuttered thou dost know.

If forth I march, thou goest before,
  If back I turn, thou com'st behind:
    So forth nor back
    Thy guard I lack,
  Nay on me too, thy hand I find.
Well I thy wisdom may adore, 20
  But never reach with earthy mind.

To shun thy notice, leave thine eye,
  O whither might I take my way?
    To starry sphere?
    Thy throne is there.
  To dead men's undelightsome stay?
There is thy walk, and there to lie
  Unknown, in vain I should assay.

O sun, whom light nor flight can match,
  Suppose thy lightful flightful wings 30
    Thou lend to me,
    And I could flee
  As far as thee the evening brings:
Even led to west he would me catch,
  Nor should I lurk with western things.

Do thou thy best, O secret night,
  In sable veil to cover me:
    Thy sable veil
    Shall vainly fail;
  With day unmasked my night shall be, 40
For night is day, and darkness light,
  O father of all lights, to thee.

Each inmost piece in me is thine:
  While yet I in my mother dwelt,
    All that me clad
    From thee I had.
  Thou in my frame hast strangely dealt:
Needs in my praise thy works must shine
  So inly them my thoughts have felt.

Thou, how my back was beam-wise laid, 50
  And raft'ring of my ribs, dost know;
    Know'st every point
    Of bone and joint,
  How to this whole these parts did grow,
In brave embroid'ry fair arrayed,
  Though wrought in shop both dark and low.

Nay fashionless, ere form I took,
  Thy all and more beholding eye
    My shapeless shape
    Could not escape: 60
  All these time framed successively
Ere one had being, in the book
  Of thy foresight enrolled did lie.

My God, how I these studies prize,
  That do thy hidden workings show!
    Whose sum is such
    No sum so much,
  Nay, summed as sand they sumless grow.
I lie to sleep, from sleep I rise,
  Yet still in thought with thee I go. 70

My God, if thou but one wouldst kill,
  Then straight would leave my further chase
    This cursed brood
    Inured to blood,
  Whose graceless taunts at thy disgrace
Have aimed oft; and hating still
  Would with proud lies thy truth outface.

Hate not I them, who thee do hate?
  Thine, Lord, I will the censure be.
    Detest I not 80
    The cankered knot
  Whom I against thee banded see?
O Lord, thou know'st in highest rate
  I hate them all as foes to me.

Search me, my God, and prove my heart,
  Examine me, and try my thought;
    And mark in me
    If ought there be
  That hath with cause their anger wrought.
If not (as not) my life's each part, 90
  Lord, safely guide from danger brought.

(Wr. prob. before 1599; pub. 1823)

CHRISTOPHER MARLOWE
1564–1593

from *All Ovid's Elegies*

306

Elegia IV

Amicam, qua arte, quibusve nutibus in cæna, presente viro, uti debeat, admonet

THY husband to a banquet goes with me,
Pray God it may his latest supper be.
Shall I sit gazing as a bashful guest,
While others touch the damsel I love best?
Wilt lying under him, his bosom clip?
About thy neck shall he at pleasure skip?
Marvel not, though the fair bride did incite
The drunken Centaurs to a sudden fight.
I am no half horse, nor in woods I dwell,
Yet scarce my hands from thee contain I well. 10
But how thou should'st behave thyself now know:
Nor let the winds away my warnings blow.
Before thy husband come, though I not see
What may be done, yet there before him be.
Lie with him gently, when his limbs he spread
Upon the bed, but on my foot first tread.
View me, my becks, and speaking countenance;
Take, and receive each secret amorous glance.
Words without voice shall on my eyebrows sit,
Lines thou shalt read in wine by my hand writ. 20
When our lascivious toys come in thy mind,
Thy rosy cheeks be to thy thumb inclin'd.
If aught of me thou speak'st in inward thought,
Let thy soft finger to thy ear be brought.
When I (my light) do or say aught that please thee,
Turn round thy gold ring, as it were to ease thee.
Strike on the board like them that pray for evil,
When thou dost wish thy husband at the devil.
What wine he fills thee, wisely will him drink,
Ask thou the boy, what thou enough dost think. 30

Amicam . . . admonet] He tells his mistress how to behave when they, and her husband, are at a dinner-party
5 clip] embrace
15 Lie with him] i.e. on the couch on which guests at Roman dinner-parties reclined
21 toys] amorous play

When thou hast tasted, I will take the cup,
And where thou drink'st, on that part I will sup.
If he gives thee what first himself did taste,
Even in his face his offered gobbets cast.
Let not thy neck by his vile arms be pressed,
Nor lean thy soft head on his boistrous breast.
Thy bosom's roseate buds let him not finger;
Chiefly on thy lips let not his lips linger.
If thou giv'st kisses, I shall all disclose,
Say they are mine, and hands on thee impose. 40
Yet this I'll see, but if thy gown aught cover,
Suspicious fear in all my veins will hover.
Mingle not thighs, nor to his leg join thine,
Nor thy soft foot with his hard foot combine.
I have been wanton, therefore am perplexed,
And with mistrust of the like measure vexed.
I and my wench oft under clothes did lurk,
When pleasure moved us to our sweetest work.
Do not thou so; but throw thy mantle hence
Lest I should think thee guilty of offence. 50
Entreat thy husband drink, but do not kiss,
And while he drinks, to add more do not miss;
If he lies down with wine and sleep oppressed,
The thing and place shall counsel us the rest.
When to go homewards we rise all along,
Have care to walk in middle of the throng.
There will I find thee or be found by thee,
There touch whatever thou canst touch of me.
Aye me, I warn what profits some few hours,
But we must part, when heaven with black night lours. 60
At night thy husband clips thee: I will weep
And to the doors sight of thyself will keep:
Then will he kiss thee, and not only kiss,
But force thee give him my stol'n honey bliss.
Constrained against thy will give it the peasant,
Forbear sweet words, and be your sport unpleasant.
To him I pray it no delight may bring,
Or if it do, to thee no joy thence spring.
But, though this night thy fortune be to try it,
To me to-morrow constantly deny it. 70

(Pub. before 1599)

36 boistrous] massive, rough
45 perplexed] involved in doubt, suspicious
61 clips] embraces
69 try] experience

307

Elegia V

Corinnæ concubitus

IN summer's heat, and mid-time of the day,
To rest my limbs upon a bed I lay;
One window shut, the other open stood,
Which gave such light as twinkles in a wood,
Like twilight glimpse at setting of the sun,
Or night being past, and yet not day begun.
Such light to shamefast maidens must be shown
Where they may sport, and seem to be unknown.
Then came Corinna in a long loose gown,
Her white neck hid with tresses hanging down, 10
Resembling fair Semiramis going to bed,
Or Lais of a thousand wooers sped.
I snatched her gown; being thin, the harm was small;
Yet strived she to be covered therewithal;
And striving thus as one that would be cast,
Betrayed herself, and yielded at the last.
Stark naked as she stood before mine eye,
Not one wen in her body could I spy.
What arms and shoulders did I touch and see,
How apt her breasts were to be pressed by me! 20
How smooth a belly under her waist saw I,
How large a leg, and what a lusty thigh!
To leave the rest, all liked me passing well;
I clinged her naked body, down she fell;
Judge you the rest: being tired she bade me kiss.
Jove send me more such afternoons as this.

(Pub. before 1599)

308

Elegia XIII

Ad Auroram, ne properet

NOW o'er the sea from her old love comes she
That draws the day from heaven's cold axle-tree.
Aurora, whither slid'st thou? down again,
And birds for Memnon yearly shall be slain.

Corinnæ concubitus] going to bed with Corinna
12 sped] enjoyed
15 cast] overcome
18 wen] blemish

308
Ad . . . properet] To the Dawn, that she should not be in a hurry
4 Memnon] son of Aurora (the 'dawn') and Tithonus

Now in her tender arms I sweetly bide,
If ever, now well lies she by my side.
The air is cold, and sleep is sweetest now,
And birds send forth shrill notes from every bough:
Whither runn'st thou, that men and women love not?
Hold in thy rosy horses that they move not. 10
Ere thou rise, stars teach seamen where to sail,
But when thou com'st, they of their courses fail.
Poor travellers though tired, rise at thy sight,
And soldiers make them ready to the fight.
The painful hind by thee to field is sent;
Slow oxen early in the yoke are pent.
Thou cozen'st boys of sleep, and dost betray them
To pedants that with cruel lashes pay them.
Thou mak'st the surety to the lawyer run,
That with one word hath nigh himself undone. 20
The lawyer and the client hate thy view,
Both whom thou raisest up to toil anew.
By thy means women of their rest are barred,
Thou set'st their labouring hands to spin and card.
All could I bear; but that the wench should rise
Who can endure, save him with whom none lies?
How oft wished I night would not give thee place,
Nor morning stars shun thy uprising face.
How oft that either wind would break thy coach,
Or steeds might fall, forced with thick clouds' approach. 30
Whither goest thou, hateful nymph? Memnon the elf
Received his coal-black colour from thyself.
Say that thy love with Cephalus were not known,
Then thinkest thou thy loose life is not shown?
Would Tithon might but talk of thee awhile,
Not one in heaven should be more base and vile.
Thou leav'st his bed, because he's faint through age,
And early mount'st thy hateful carriage;
But held'st thou in thy arms some Cephalus,
Then would'st thou cry, 'Stay night, and run not thus.' 40
Dost punish me, because years make him wane?
I did not bid thee wed an aged swain.
The moon sleeps with Endymion every day,
Thou art as fair as she, then kiss and play.
Jove, that thou should'st not haste but wait his leisure,
Made two nights one to finish up his pleasure.

15 painful hind] long-suffering labourer
18 pedants] schoolmasters
31 elf] mischievous child

I chid no more; she blushed, and therefore heard me,
Yet lingered not the day, but morning scared me.

(Pub. before 1599)

309 Elegia XV

Ad invidos, quod fama poetarum sit perennis

ENVY, why carp'st thou my time is spent so ill?
And term'st my works fruits of an idle quill?
Or that unlike the line from whence I sprung,
War's dusty honours are refus'd, being young?
Nor that I study not the brawling laws,
Nor set my voice to sale in every cause?
Thy scope is mortal; mine, eternal fame,
That all the world may ever chant my name.
Homer shall live while Tenedos stands and Ide,
Or into sea swift Simois doth slide. 10
Ascræus lives while grapes with new wine swell,
Or men with crooked sickles corn down fell.
The world shall of Callimachus ever speak,
His art excelled, although his wit was weak.
For ever lasts high Sophocles' proud vein;
With sun and moon Aratus shall remain.
While bondmen cheat, fathers be hard, bawds whorish,
And strumpets flatter, shall Menander flourish.
Rude Ennius, and Plautus full of wit,
Are both in Fame's eternal legend writ. 20
What age of Varro's name shall not be told,
And Jason's Argos, and the fleece of gold?
Lofty Lucretius shall live that hour
That nature shall dissolve this earthly bower.
Æneas' war and Tityrus shall be read,
While Rome of all the conquered world is head.
Till Cupid's bow and fiery shafts be broken,
Thy verses, sweet Tibullus, shall be spoken:
And Gallus shall be known from east to west,
So shall Lycoris whom he loved best. 30
Therefore when flint and iron wear away,
Verse is immortal and shall ne'er decay.
To verse let kings give place, and kingly shows,
And banks o'er which gold-bearing Tagus flows.

309
Ad invidos . . . perennis] To those who maliciously envy the immortality of poets
7 scope] aim, purpose
19 wit] genius

Let base-conceited wits admire vile things,
Fair Phœbus lead me to the Muses' springs.
About my head be quivering myrtle wound,
And in sad lovers' heads let me be found.
The living, not the dead, can envy bite,
For after death all men receive their right. 40
Then though death rakes my bones in funeral fire,
I'll live, and as he pulls me down mount higher.

(Pub. before 1599)

310 *The Passionate Shepherd to his Love*

COME live with me, and be my love,
And we will all the pleasures prove,
That valleys, groves, hills and fields,
Woods, or steepy mountain yields.

And we will sit upon the rocks,
Seeing the shepherds feed their flocks
By shallow rivers, to whose falls
Melodious birds sing madrigals.

And I will make thee beds of roses,
And a thousand fragrant posies, 10
A cap of flowers and a kirtle
Embroidered all with leaves of myrtle.

A gown made of the finest wool
Which from our pretty lambs we pull,
Fair lined slippers for the cold,
With buckles of the purest gold;

A belt of straw and ivy-buds,
With coral clasps and amber studs,
And if these pleasures may thee move,
Come live with me, and be my love. 20

The shepherd swains shall dance and sing
For thy delight each May-morning,
If these delights thy mind may move;
Then live with me, and be my love.

(Pub. 1599)

310
2 prove] experience

from *Lucan's First Book, translated line for line*

311 [*The causes of the civil war*]

ALL great things crush themselves; such end the gods
Allot the height of honour; men so strong
By land and sea no foreign force could ruin.
O Rome, thyself art cause of all these evils,
Thyself thus shivered out to three men's shares:
Dire league of partners in a kingdom last not.
O faintly-join'd friends, with ambition blind,
Why join you force to share the world betwixt you?
While th'earth the sea, and air the earth sustains,
While Titan strives against the world's swift course, 10
Or Cynthia, night's queen, waits upon the day,
Shall never faith be found in fellow kings.
Dominion cannot suffer partnership.
This need no foreign proof nor far-fet story:
Rome's infant walls were steeped in brother's blood;
Nor then was land or sea, to breed such hate;
A town with one poor church set them at odds.
Caesar's and Pompey's jarring love soon ended,
'Twas peace against their wills; betwixt them both
Stepped Crassus in; even as the slender Isthmus, 20
Betwixt the Aegæan and the Ionian sea,
Keeps each from other, but being worn away,
They both burst out, and each encounter other:
So whenas Crassus' wretched death, who stayed them,
Had filled Assyrian Carra's walls with blood,
His loss made way for Roman outrages.
Parthians, y'afflict us more than ye suppose;
Being conquered, we are plagued with civil war.
Swords share our empire: Fortune, that made Rome
Govern the earth, the sea, the world itself, 30
Would not admit two lords; for Julia,
Snatched hence by cruel Fates, with ominous howls,
Bare down to hell her son, the pledge of peace,
And all bands of that death-presaging alliance.
Julia, had heaven given thee longer life,
Thou hadst restrained thy headstrong husband's rage,
Yea, and thy father too, and, swords thrown down,
Made all shake hands, as once the Sabines did;

8 share] share out, divide

31 Julia] Caesar's daughter and Pompey's wife

Thy death broke amity, and trained to war
These captains emulous of each other's glory. 40
Thou feared'st, great Pompey, that late deeds would dim
Old triumphs, and that Caesar's conquering France
Would dash the wreath thou wear'st for pirates' wrack.
Thee war's use stirred, and thoughts that always scorned
A second place; Pompey could bide no equal,
Nor Caesar no superior: which of both
Had justest cause, unlawful 'tis to judge.
Each side had great partakers; Caesar's cause
The gods abetted, Cato liked the other.
Both differed much: Pompey was struck in years, 50
And by long rest forgot to manage arms,
And being popular sought by liberal gifts
To gain the light unstable commons' love,
And joyed to hear his theatre's applause:
He lived secure, boasting his former deeds,
And thought his name sufficient to uphold him;
Like to a tall oak in a fruitful field,
Bearing old spoils and conquerors' monuments,
Who though his root be weak, and his own weight
Keep him within the ground, his arms all bare, 60
His body, not his boughs, send forth a shade;
Though every blast it nod, and seem to fall,
When all the woods about stand bolt upright,
Yet he alone is held in reverence.
Caesar's renown for war was less, he restless,
Shaming to strive but where he did subdue;
When ire or hope provoked, heady, and bold;
At all times charging home, and making havoc;
Urging his fortune, trusting in the gods,
Destroying what withstood his proud desires, 70
And glad when blood and ruin made him way:
So thunder, which the wind tears from the clouds,
With crack of riven air and hideous sound
Filling the world, leaps out and throws forth fire,
Affrights poor fearful men, and blasts their eyes
With overthwarting flames, and raging shoots
Alongst the air, and, nought resisting it,
Falls, and returns, and shivers where it lights.
Such humours stirred them up: but this war's seed
Was even the same that wrecks all great dominions. 80
When Fortune made us lords of all, wealth flowed,

39 trained] drew, induced
44 Thee] i.e. Caesar
55 secure] over-confident
58 monuments] memorials

And then we grew licentious and rude;
The soldiers' prey and rapine brought in riot;
Men took delight in jewels, houses, plate,
And scorned old sparing diet, and ware robes
Too light for women; Poverty (who hatched
Rome's greatest wits) was loathed, and all the world
Ransacked for gold, which breeds the world decay;
And then large limits had their butting lands;
The ground which Curius and Camillus tilled 90
Was stretched unto the fields of hinds unknown.
Again, this people could not brook calm peace;
Them freedom without war might not suffice;
Quarrels were rife, greedy desire, still poor,
Did vile deeds; then 'twas worth the price of blood,
And deemed renown, to spoil their native town;
Force mastered right, the strongest governed all.
Hence came it that th'edicts were over-ruled,
That laws were broke, tribunes with consuls strove,
Sale made of offices, and people's voices 100
Bought by themselves and sold, and every year
Frauds and corruption in the Field of Mars;
Hence interest and devouring usury sprang,
Faith's breach, and hence came war, to most men welcome.

(Pub. 1600)

312 [*Caesar summons his forces from Gaul*]

WHEN Caesar saw his army prone to war,
And Fates so bent, lest sloth and long delay
Might cross him, he withdrew his troops from France,
And in all quarters musters men for Rome.
They by Lemannus' nook forsook their tents;
They whom the Lingones foiled with painted spears,
Under the rocks by crooked Vogesus;
And many came from shallow Isara,
Who, running long, falls in a greater flood,
And, ere he sees the sea, loseth his name; 10
The yellow Ruthens left their garrisons;
Mild Atax glad it bears not Roman boats,
And frontier Varus that the camp is far,
Sent aid; so did Alcides' port, whose seas
Eat hollow rocks, and where the north-west wind
Nor Zephyr rules not, but the north alone

89 butting] boundary

Turmoils the coast, and enterance forbids;
And others came from that uncertain shore
Which is nor sea, nor land, but ofttimes both,
And changeth as the ocean ebbs and flows; 20
Whether the sea rolled always from that point
Whence the wind blows, still forced to and fro;
Or that the wandering main follow the moon;
Or flaming Titan (feeding on the deep)
Pulls them aloft, and makes the surge kiss heaven,
Philosophers, look you; for unto me,
Thou cause, whate'er thou be whom God assigns
This great effect, art hid. They came that dwell
By Nemes' fields, and banks of Satirus,
Where Tarbel's winding shores embrace the sea; 30
The Santons that rejoice in Caesar's love;
Those of Bituriges, and light Axon pikes;
And they of Rhene and Leuca, cunning darters,
And Sequana that well could manage steeds;
The Belgians apt to govern British cars;
Th'Averni too, which boldly feign themselves
The Romans' brethren, sprung of Ilian race;
The stubborn Nervians stained with Cotta's blood;
And Vangions who, like those of Sarmata,
Wear open slops; and fierce Batavians, 40
Whom trumpets' clang incites; and those that dwell
By Cinga's stream, and where swift Rhodanus
Drives Araris to sea; they near the hills,
Under whose hoary rocks Gebenna hangs;
And, Trevier, thou being glad that wars are past thee;
And you, late-shorn Ligurians, who were wont
In large-spread hair to exceed the rest of France;
And where to Hesus and fell Mercury
They offer human flesh, and where Jove seems
Bloody like Dian, whom the Scythians serve. 50
And you, French Bardi, whose immortal pens
Renown the valiant souls slain in your wars,
Sit safe at home and chant sweet poesy.
And, Druides, you now in peace renew
Your barbarous customs and sinister rites;
In unfelled woods and sacred groves you dwell,
And only gods and heavenly powers you know,
Or only know you nothing. For you hold
That souls pass not to silent Erebus
Or Pluto's bloodless kingdom, but elsewhere 60

40 open slops] loose breeches

Resume a body; so (if truth you sing)
Death brings long life. Doubtless these northern men,
Whom death, the greatest of all fears, affright not,
Are blest by such sweet error; this makes them
Run on the sword's point and desire to die,
And shame to spare life which being lost is won.
You likewise, that repuls'd the Cayc foe,
March towards Rome; and you, fierce men of Rhene,
Leaving your country open to the spoil.

(Pub. 1600)

313 *Hero and Leander*

On Hellespont, guilty of true love's blood,
In view and opposite two cities stood,
Seaborderers, disjoined by Neptune's might;
The one Abydos, the other Sestos hight.
At Sestos, Hero dwelt; Hero the fair,
Whom young Apollo courted for her hair,
And offered as a dower his burning throne,
Where she should sit for men to gaze upon.
The outside of her garments were of lawn,
The lining purple silk, with gilt stars drawn; 10
Her wide sleeves green, and bordered with a grove,
Where Venus in her naked glory strove
To please the careless and disdainful eyes
Of proud Adonis, that before her lies;
Her kirtle blue, whereon was many a stain,
Made with the blood of wretched lovers slain.
Upon her head she ware a myrtle wreath,
From whence her veil reached to the ground beneath.
Her veil was artificial flowers and leaves,
Whose workmanship both man and beast deceives. 20
Many would praise the sweet smell as she passed,
When 'twas the odour which her breath forth cast;
And there for honey bees have sought in vain,
And beat from thence, have lighted there again.
About her neck hung chains of pebble-stone,
Which, lightened by her neck, like diamonds shone.
She ware no gloves, for neither sun nor wind
Would burn or parch her hands, but to her mind
Or warm or cool them, for they took delight
To play upon those hands, they were so white. 30

313
4 hight] called
26 lightened] brightened

Buskins of shells all silvered used she,
And branched with blushing coral to the knee,
Where sparrows perched, of hollow pearl and gold,
Such as the world would wonder to behold:
Those with sweet water oft her handmaid fills,
Which, as she went, would chirrup through the bills.
Some say, for her the fairest Cupid pined,
And, looking in her face, was strooken blind.
But this is true, so like was one the other,
As he imagined Hero was his mother; 40
And oftentimes into her bosom flew,
About her naked neck his bare arms threw,
And laid his childish head upon her breast,
And with still panting rocked, there took his rest.
So lovely fair was Hero, Venus' nun,
As Nature wept, thinking she was undone,
Because she took more from her than she left,
And of such wondrous beauty her bereft;
Therefore, in sign her treasure suffered wrack,
Since Hero's time hath half the world been black. 50
Amorous Leander, beautiful and young,
(Whose tragedy divine Musaeus sung)
Dwelt at Abydos; since him, dwelt there none
For whom succeeding times make greater moan.
His dangling tresses that were never shorn,
Had they been cut and unto Colchos borne,
Would have allured the vent'rous youth of Greece
To hazard more than for the Golden Fleece.
Fair Cynthia wished his arms might be her sphere;
Grief makes her pale, because she moves not there. 60
His body was as straight as Circe's wand;
Jove might have sipped out nectar from his hand.
Even as delicious meat is to the taste,
So was his neck in touching, and surpassed
The white of Pelops' shoulder. I could tell ye
How smooth his breast was, and how white his belly,
And whose immortal fingers did imprint
That heavenly path, with many a curious dint,
That runs along his back; but my rude pen
Can hardly blazon forth the loves of men, 70
Much less of powerful gods; let it suffice
That my slack muse sings of Leander's eyes,
Those orient cheeks and lips, exceeding his
That leapt into the water for a kiss

36 went] walked 44 still] continual 72 slack] feeble

Of his own shadow, and despising many,
Died ere he could enjoy the love of any.
Had wild Hippolytus Leander seen,
Enamoured of his beauty had he been;
His presence made the rudest peasant melt,
That in the vast uplandish country dwelt; 80
The barbarous Thracian soldier, moved with nought,
Was moved with him, and for his favour sought.
Some swore he was a maid in man's attire,
For in his looks were all that men desire,
A pleasant smiling cheek, a speaking eye,
A brow for love to banquet royally;
And such as knew he was a man, would say,
'Leander, thou art made for amorous play;
Why art thou not in love, and loved of all?
Though thou be fair, yet be not thine own thrall.' 90
The men of wealthy Sestos, every year
(For his sake whom their goddess held so dear,
Rose-cheeked Adonis), kept a solemn feast.
Thither resorted many a wandering guest
To meet their loves; such as had none at all,
Came lovers home from this great festival.
For every street, like to a firmament,
Glistered with breathing stars, who, where they went,
Frighted the melancholy earth, which deemed
Eternal heaven to burn, for so it seemed 100
As if another Phaeton had got
The guidance of the sun's rich chariot.
But, far above the loveliest, Hero shined,
And stole away th'enchanted gazer's mind;
For like sea-nymphs' inveigling harmony,
So was her beauty to the standers by.
Nor that night-wandering pale and watery star
(When yawning dragons draw her thirling car
From Latmos' mount up to the gloomy sky,
Where, crowned with blazing light and majesty, 110
She proudly sits) more over-rules the flood,
Than she the hearts of those that near her stood.
Even as, when gaudy nymphs pursue the chase,
Wretched Ixion's shaggy-footed race,
Incensed with savage heat, gallop amain
From steep pine-bearing mountains to the plain;
So ran the people forth to gaze upon her,
And all that viewed her were enamoured on her.

80 uplandish] coarsely rustic
108 thirling] shooting through the air
114 shaggy-footed race] i.e. the centaurs

And as in fury of a dreadful fight,
Their fellows being slain or put to flight, 120
Poor soldiers stand with fear of death dead-strooken,
So at her presence all, surprised and tooken,
Await the sentence of her scornful eyes;
He whom she favours lives, the other dies.
There might you see one sigh, another rage,
And some, their violent passions to assuage,
Compile sharp satires; but alas, too late,
For faithful love will never turn to hate.
And many, seeing great princes were denied,
Pined as they went, and thinking on her, died. 130
On this feast day, oh, cursed day and hour!
Went Hero thorough Sestos, from her tower
To Venus' temple, where unhappily,
As after chanced, they did each other spy.
So fair a church as this had Venus none;
The walls were of discoloured jasper stone,
Wherein was Proteus carved, and o'erhead
A lively vine of green sea-agate spread,
Where by one hand light-headed Bacchus hung,
And with the other wine from grapes out-wrung. 140
Of crystal shining fair the pavement was;
The town of Sestos called it Venus' glass.
There might you see the gods in sundry shapes,
Committing heady riots, incest, rapes:
For know that underneath this radiant floor
Was Danae's statue in a brazen tower;
Jove slyly stealing from his sister's bed
To dally with Idalian Ganymede,
And for his love Europa bellowing loud,
And tumbling with the rainbow in a cloud; 150
Blood-quaffing Mars heaving the iron net
Which limping Vulcan and his Cyclops set;
Love kindling fire to burn such towns as Troy;
Sylvanus weeping for the lovely boy
That now is turned into a cypress tree,
Under whose shade the wood-gods love to be.
And in the midst a silver altar stood;
There Hero sacrificing turtles' blood,
Vailed to the ground, veiling her eyelids close,
And modestly they opened as she rose: 160
Thence flew love's arrow with the golden head,
And thus Leander was enamoured.

136 discoloured] variously coloured   159 Vailed] sank

Stone-still he stood, and evermore he gazed,
Till with the fire that from his countenance blazed
Relenting Hero's gentle heart was strook;
Such force and virtue hath an amorous look.
It lies not in our power to love or hate,
For will in us is over-ruled by fate.
When two are stripped, long ere the course begin,
We wish that one should lose, the other win; 170
And one especially do we affect
Of two gold ingots, like in each respect.
The reason no man knows; let it suffice,
What we behold is censured by our eyes.
Where both deliberate, the love is slight;
Who ever loved, that loved not at first sight?
He kneeled, but unto her devoutly prayed;
Chaste Hero to herself thus softly said:
'Were I the saint he worships, I would hear him';
And as she spake these words, came somewhat near him. 180
He started up; she blushed as one ashamed;
Wherewith Leander much more was inflamed.
He touched her hand; in touching it she trembled;
Love deeply grounded hardly is dissembled.
These lovers parlied by the touch of hands;
True love is mute, and oft amazed stands.
Thus while dumb signs their yielding hearts entangled,
The air with sparks of living fire was spangled,
And night, deep drenched in misty Acheron,
Heaved up her head, and half the world upon 190
Breathed darkness forth (dark night is Cupid's day).
And now begins Leander to display
Love's holy fire with words, with sighs and tears,
Which like sweet music entered Hero's ears;
And yet at every word she turned aside,
And always cut him off as he replied.
At last, like to a bold sharp sophister,
With cheerful hope thus he accosted her:
'Fair creature, let me speak without offence;
I would my rude words had the influence 200
To lead thy thoughts as thy fair looks do mine;
Then shouldst thou be his prisoner who is thine.
Be not unkind and fair; misshapen stuff
Are of behaviour boisterous and rough.

174 censured] judged
184 hardly] with difficulty
197 sophister] manipulator of arguments
203 stuff] i.e. persons (with reference to their physical appearances)

O shun me not, but hear me ere you go;
God knows I cannot force love, as you do.
My words shall be as spotless as my youth,
Full of simplicity and naked truth.
This sacrifice, whose sweet perfume descending
From Venus' altar to your footsteps bending 210
Doth testify that you exceed her far,
To whom you offer, and whose nun you are.
Why should you worship her? her you surpass
As much as sparkling diamonds flaring glass.
A diamond set in lead his worth retains;
A heavenly nymph, beloved of human swains,
Receives no blemish, but ofttimes more grace;
Which makes me hope, although I am but base,
Base in respect of thee, divine and pure,
Dutiful service may thy love procure; 220
And I in duty will excel all other,
As thou in beauty dost exceed Love's mother.
Nor heaven, nor thou, were made to gaze upon;
As heaven preserves all things, so save thou one.
A stately builded ship, well rigged and tall,
The ocean maketh more majestical:
Why vow'st thou then to live in Sestos here,
Who on Love's seas more glorious wouldst appear?
Like untuned golden strings all women are,
Which, long time lie untouched, will harshly jar. 230
Vessels of brass, oft handled, brightly shine;
What difference betwixt the richest mine
And basest mould, but use? for both, not used,
Are of like worth. Then treasure is abused,
When misers keep it; being put to loan,
In time it will return us two for one.
Rich robes themselves and others do adorn;
Neither themselves nor others, if not worn.
Who builds a palace, and rams up the gate,
Shall see it ruinous and desolate. 240
Ah, simple Hero, learn thyself to cherish;
Lone women, like to empty houses, perish.
Less sins the poor rich man that starves himself
In heaping up a mass of drossy pelf,
Than such as you; his golden earth remains,
Which after his decease some other gains;
But this fair gem, sweet in the loss alone,
When you fleet hence, can be bequeathed to none.

219 in respect of] in comparison with

Or if it could, down from th'enamelled sky
All heaven would come to claim this legacy, 250
And with intestine broils the world destroy,
And quite confound Nature's sweet harmony.
Well therefore by the gods decreed it is
We human creatures should enjoy that bliss.
One is no number; maids are nothing, then,
Without the sweet society of men.
Wilt thou live single still? One shalt thou be
Though never-singling Hymen couple thee.
Wild savages, that drink of running springs,
Think water far excels all earthly things, 260
But they that daily taste neat wine, despise it.
Virginity, albeit some highly prize it,
Compared with marriage, had you tried them both,
Differs as much as wine and water doth.
Base bullion for the stamp's sake we allow;
Even so for men's impression do we you,
By which alone, our reverend fathers say,
Women receive perfection every way.
This idol which you term virginity
Is neither essence subject to the eye, 270
No, nor to any one exterior sense;
Nor hath it any place of residence,
Nor is 't of earth or mould celestial,
Or capable of any form at all.
Of that which hath no being do not boast;
Things that are not at all, are never lost.
Men foolishly do call it virtuous;
What virtue is it, that is born with us?
Much less can honour be ascribed thereto;
Honour is purchased by the deeds we do. 280
Believe me, Hero, honour is not won,
Until some honourable deed be done.
Seek you, for chastity, immortal fame,
And know that some have wronged Diana's name?
Whose name is it, if she be false or not,
So she be fair, but some vile tongues will blot?
But you are fair (ay me) so wondrous fair,
So young, so gentle, and so debonair,
As Greece will think, if thus you live alone,
Some one or other keeps you as his own. 290
Then, Hero, hate me not, nor from me fly,
To follow swiftly blasting infamy.

265 allow] praise, commend 273 mould] form, pattern

Perhaps, thy sacred priesthood makes thee loth;
Tell me, to whom mad'st thou that heedless oath?'
'To Venus', answered she, and as she spake,
Forth from those two tralucent cisterns brake
A stream of liquid pearl, which down her face
Made milk-white paths, whereon the gods might trace
To Jove's high court. He thus replied: 'The rites
In which love's beauteous empress most delights 300
Are banquets, Doric music, midnight revel,
Plays, masks, and all that stern age counteth evil.
Thee as a holy idiot doth she scorn,
For thou, in vowing chastity, hast sworn
To rob her name and honour, and thereby
Commit'st a sin far worse than perjury,
Even sacrilege against her deity,
Through regular and formal purity.
To expiate which sin, kiss and shake hands;
Such sacrifice as this Venus demands.' 310
Thereat she smiled, and did deny him so
As put thereby, yet might he hope for mo.
Which makes him quickly reinforce his speech,
And her in humble manner thus beseech:
'Though neither gods nor men may thee deserve,
Yet for her sake whom you have vowed to serve,
Abandon fruitless cold virginity,
The gentle queen of love's sole enemy.
Then shall you most resemble Venus' nun,
When Venus' sweet rites are performed and done. 320
Flint-breasted Pallas joys in single life,
But Pallas and your mistress are at strife.
Love, Hero, then, and be not tyrannous,
But heal the heart, that thou hast wounded thus;
Nor stain thy youthful years with avarice;
Fair fools delight to be accounted nice.
The richest corn dies, if it be not reaped;
Beauty alone is lost, too warily kept.'
These arguments he used, and many more,
Wherewith she yielded, that was won before. 330
Hero's looks yielded, but her words made war;
Women are won when they begin to jar.
Thus having swallowed Cupid's golden hook,
The more she strived, the deeper was she strook;
Yet, evilly feigning anger, strove she still,
And would be thought to grant against her will.

296 tralucent] translucent
298 trace] go, travel
312 put thereby] dismissed by that
mo] more
326 nice] coy, reserved

So having paused a while, at last she said:
'Who taught thee rhetoric to deceive a maid?
Ay me, such words as these should I abhor,
And yet I like them for the orator.' 340
With that Leander stooped to have embraced her,
But from his spreading arms away she cast her,
And thus bespake him: 'Gentle youth, forbear
To touch the sacred garments which I wear.
'Upon a rock, and underneath a hill,
Far from the town, where all is whist and still,
Save that the sea, playing on yellow sand,
Sends forth a rattling murmur to the land,
Whose sound allures the golden Morpheus
In silence of the night to visit us, 350
My turret stands; and there, God knows, I play
With Venus' swans and sparrows all the day.
A dwarfish beldame bears me company,
That hops about the chamber where I lie,
And spends the night, that might be better spent,
In vain discourse and apish merriment.
Come thither.' As she spake this, her tongue tripped.
For unawares 'Come thither' from her slipped;
And suddenly her former colour changed,
And here and there her eyes through anger ranged. 360
And like a planet moving several ways
At one self instant, she, poor soul, assays,
Loving, not to love at all, and every part
Strove to resist the motions of her heart:
And hands so pure, so innocent, nay such
As might have made heaven stoop to have a touch,
Did she uphold to Venus, and again
Vowed spotless chastity, but all in vain.
Cupid beats down her prayers with his wings;
Her vows above the empty air he flings; 370
All deep enraged, his sinewy bow he bent,
And shot a shaft that burning from him went;
Wherewith she strooken looked so dolefully,
As made Love sigh to see his tyranny.
And as she wept, her tears to pearl he turned,
And wound them on his arm, and for her mourned.
Then towards the palace of the Destinies,
Laden with languishment and grief, he flies,
And to those stern nymphs humbly made request,
Both might enjoy each other, and be blest. 380

349 Morpheus] sleep

But with a ghastly dreadful countenance,
Threatening a thousand deaths at every glance,
They answered Love, nor would vouchsafe so much
As one poor word, their hate to him was such.
Hearken awhile, and I will tell you why:
Heaven's winged herald, Jove-born Mercury,
The self-same day that he asleep had laid
Enchanted Argus, spied a country maid,
Whose careless hair, instead of pearl t'adorn it,
Glistered with dew, as one that seemed to scorn it: 390
Her breath as fragrant as the morning rose,
Her mind pure, and her tongue untaught to gloze;
Yet proud she was, for lofty pride that dwells
In towered courts is oft in shepherds' cells,
And too too well the fair vermilion knew,
And silver tincture of her cheeks, that drew
The love of every swain. On her this god
Enamoured was, and with his snaky rod
Did charm her nimble feet, and made her stay,
The while upon a hillock down he lay, 400
And sweetly on his pipe began to play,
And with smooth speech her fancy to assay;
Till in his twining arms he locked her fast,
And then he wooed with kisses, and at last,
As shepherds do, her on the ground he laid,
And tumbling in the grass, he often strayed
Beyond the bounds of shame, in being bold
To eye those parts which no eye should behold;
And like an insolent commanding lover,
Boasting his parentage, would needs discover 410
The way to new Elysium: but she,
Whose only dower was her chastity,
Having striven in vain, was now about to cry,
And crave the help of shepherds that were nigh.
Herewith he stayed his fury, and began
To give her leave to rise; away she ran;
After went Mercury, who used such cunning,
As she, to hear his tale, left off her running.
Maids are not won by brutish force and might,
But speeches full of pleasure and delight. 420
And knowing Hermes courted her, was glad
That she such loveliness and beauty had
As could provoke his liking, yet was mute,
And neither would deny nor grant his suit.

392 gloze] prevaricate, flatter

Still vowed he love; she wanting no excuse
To feed him with delays, as women use,
Or thirsting after immortality—
All women are ambitious naturally—
Imposed upon her lover such a task
As he ought not perform, nor yet she ask. 430
A draught of flowing nectar she requested,
Wherewith the king of gods and men is feasted.
He, ready to accomplish what she willed,
Stole some from Hebe (Hebe Jove's cup filled)
And gave it to his simple rustic love;
Which being known (as what is hid from Jove?)
He inly stormed, and waxed more furious
Than for the fire filched by Prometheus,
And thrusts him down from heaven: he wandering here,
In mournful terms, with sad and heavy cheer, 440
Complained to Cupid. Cupid, for his sake,
To be revenged on Jove did undertake;
And those on whom heaven, earth, and hell relies,
I mean the adamantine Destinies,
He wounds with love, and forced them equally
To dote upon deceitful Mercury.
They offered him the deadly fatal knife,
That shears the slender threads of human life;
At his fair feathered feet the engines laid
Which th'earth from ugly Chaos' den upweighed: 450
These he regarded not, but did entreat
That Jove, usurper of his father's seat,
Might presently be banished into hell,
And aged Saturn in Olympus dwell.
They granted what he craved, and once again
Saturn and Ops began their golden reign.
Murder, rape, war, lust, and treachery
Were with Jove closed in Stygian empery.
But long this blessed time continued not;
As soon as he his wished purpose got, 460
He, reckless of his promise, did despise
The love of th'everlasting Destinies.
They seeing it, both Love and him abhorred,
And Jupiter unto his place restored.
And but that Learning, in despite of Fate,
Will mount aloft, and enter heaven gate,
And to the seat of Jove itself advance,
Hermes had slept in hell with Ignorance;
Yet as a punishment they added this,
That he and Poverty should always kiss. 470

And to this day is every scholar poor;
Gross gold from them runs headlong to the boor.
Likewise, the angry sisters, thus deluded,
To venge themselves on Hermes, have concluded
That Midas' brood shall sit in Honour's chair,
To which the Muses' sons are only heir;
And fruitful wits that in aspiring are,
Shall, discontent, run into regions far;
And few great lords in virtuous deeds shall joy,
But be surprised with every garish toy; 480
And still enrich the lofty servile clown,
Who with encroaching guile keeps learning down.
Then muse not Cupid's suit no better sped,
Seeing in their loves the Fates were injured.
  By this, sad Hero, with love unacquainted,
Viewing Leander's face, fell down and fainted.
He kissed her and breathed life into her lips,
Wherewith as one displeased, away she trips.
Yet as she went, full often looked behind,
And many poor excuses did she find 490
To linger by the way, and once she stayed
And would have turned again, but was afraid,
In offering parley, to be counted light.
So on she goes, and in her idle flight,
Her painted fan of curled plumes let fall,
Thinking to train Leander therewithal.
He, being a novice, knew not what she meant,
But stayed, and after her a letter sent,
Which joyful Hero answered in such sort,
As he had hope to scale the beauteous fort, 500
Wherein the liberal graces locked their wealth,
And therefore to her tower he got by stealth.
Wide open stood the door, he need not climb;
And she herself before the pointed time
Had spread the board, with roses strewed the room,
And oft looked out, and mused he did not come.
At last he came; O, who can tell the greeting
These greedy lovers had at their first meeting.
He asked, she gave, and nothing was denied;
Both to each other quickly were affied. 510
Look how their hands, so were their hearts united,
And what he did she willingly requited.
Sweet are the kisses, the embracements sweet,
When like desires and affections meet;

481 clown] rustic boor   504 pointed] appointed   510 affied] bound in faith, betrothed

For from the earth to heaven is Cupid raised,
Where fancy is in equal balance paised.
Yet she this rashness suddenly repented,
And turned aside, and to herself lamented,
As if her name and honour had been wronged
By being possessed of him for whom she longed; 520
Ay, and she wished, albeit not from her heart,
That he would leave her turret and depart.
The mirthful god of amorous pleasure smiled
To see how he this captive nymph beguiled;
For hitherto he did but fan the fire,
And kept it down that it might mount the higher.
Now waxed she jealous lest his love abated,
Fearing her own thoughts made her to be hated.
Therefore unto him hastily she goes,
And like light Salmacis, her body throws 530
Upon his bosom, where with yielding eyes
She offers up herself a sacrifice,
To slake his anger if he were displeased.
O, what god would not therewith be appeased?
Like Aesop's cock, this jewel he enjoyed,
And as a brother with his sister toyed,
Supposing nothing else was to be done,
Now he her favour and good will had won.
But know you not that creatures wanting sense
By nature have a mutual appetence, 540
And wanting organs to advance a step,
Moved by love's force, unto each other leap?
Much more in subjects having intellect
Some hidden influence breeds like effect.
Albeit Leander, rude in love and raw,
Long dallying with Hero, nothing saw
That might delight him more, yet he suspected
Some amorous rites or other were neglected.
Therefore unto his body hers he clung;
She, fearing on the rushes to be flung, 550
Strived with redoubled strength; the more she strived,
The more a gentle pleasing heat revived,
Which taught him all that elder lovers know;
And now the same gan so to scorch and glow,
As in plain terms, yet cunningly, he craved it;
Love always makes those eloquent that have it.
She, with a kind of granting, put him by it,
And ever as he thought himself most nigh it,

516 paised] poised

Like to the tree of Tantalus she fled,
And, seeming lavish, saved her maidenhead. 560
Ne'er king more sought to keep his diadem,
Than Hero this inestimable gem.
Above our life we love a steadfast friend,
Yet when a token of great worth we send,
We often kiss it, often look thereon,
And stay the messenger that would be gone;
No marvel then though Hero would not yield
So soon to part from that she dearly held.
Jewels being lost are found again, this never;
'Tis lost but once, and once lost, lost forever. 570
Now had the morn espied her lover's steeds,
Whereat she starts, puts on her purple weeds,
And, red for anger that he stayed so long,
All headlong throws herself the clouds among.
And now Leander, fearing to be missed,
Embraced her suddenly, took leave, and kissed.
Long was he taking leave, and loth to go,
And kissed again, as lovers use to do.
Sad Hero wrung him by the hand and wept,
Saying, 'Let your vows and promises be kept.' 580
Then, standing at the door, she turned about,
As loth to see Leander going out.
And now the sun that through th'horizon peeps,
As pitying these lovers, downward creeps,
So that in silence of the cloudy night,
Though it was morning, did he take his flight.
But what the secret trusty night concealed,
Leander's amorous habit soon revealed;
With Cupid's myrtle was his bonnet crowned,
About his arms the purple riband wound, 590
Wherewith she wreathed her largely spreading hair;
Nor could the youth abstain, but he must wear
The sacred ring wherewith she was endowed,
When first religious chastity she vowed;
Which made his love through Sestos to be known,
And thence unto Abydos sooner blown
Than he could sail, for incorporeal Fame,
Whose weight consists in nothing but her name,
Is swifter than the wind, whose tardy plumes
Are reeking water and dull earthly fumes. 600
Home when he came, he seemed not to be there,
But like exiled air thrust from his sphere,

560 lavish] loose, lascivious   588 habit] clothing, dress

Set in a foreign place; and straight from thence,
Alcides like, by mighty violence
He would have chased away the swelling main,
That him from her unjustly did detain.
Like as the sun in a diameter
Fires and inflames objects removed far,
And heateth kindly, shining laterally,
So beauty sweetly quickens when 'tis nigh, 610
But being separated and removed,
Burns where it cherished, murders where it loved.
Therefore even as an index to a book,
So to his mind was young Leander's look.
O, none but gods have power their love to hide;
Affection by the countenance is descried.
The light of hidden fire itself discovers,
And love that is concealed betrays poor lovers.
His secret flame apparently was seen;
Leander's father knew where he had been, 620
And for the same mildly rebuked his son,
Thinking to quench the sparkles new begun.
But love, resisted once, grows passionate,
And nothing more than counsel lovers hate;
For as a hot proud horse highly disdains
To have his head controlled, but breaks the reins,
Spits forth the ringled bit, and with his hooves
Checks the submissive ground, so he that loves,
The more he is restrained, the worse he fares.
What is it now but mad Leander dares? 630
'O Hero, Hero!' thus he cried full oft,
And then he got him to a rock aloft,
Where having spied her tower, long stared he on 't,
And prayed the narrow toiling Hellespont
To part in twain, that he might come and go;
But still the rising billows answered 'No.'
With that he stripped him to the ivory skin,
And crying, 'Love, I come', leaped lively in.
Whereat the sapphire-visaged god grew proud,
And made his capering Triton sound aloud, 640
Imagining that Ganymede, displeased,
Had left the heavens; therefore on him he seized.
Leander strived; the waves about him wound,
And pulled him to the bottom, where the ground
Was strewed with pearl, and in low coral groves
Sweet singing mermaids sported with their loves

607 in a diameter] i.e. when perpendicular at noon
619 apparently] openly
634 toiling] agitated
639 sapphire-visaged god] Neptune
proud] lustful

On heaps of heavy gold, and took great pleasure
To spurn in careless sort the shipwreck treasure:
For here the stately azure palace stood,
Where kingly Neptune and his train abode. 650
The lusty god embraced him, called him love,
And swore he never should return to Jove.
But when he knew it was not Ganymede,
For under water he was almost dead,
He heaved him up, and looking on his face,
Beat down the bold waves with his triple mace,
Which mounted up, intending to have kissed him,
And fell in drops like tears, because they missed him.
Leander, being up, began to swim,
And looking back, saw Neptune follow him; 660
Whereat aghast, the poor soul gan to cry:
'O let me visit Hero ere I die!'
The god put Helle's bracelet on his arm,
And swore the sea should never do him harm.
He clapped his plump cheeks, and with his tresses played,
And smiling wantonly, his love bewrayed.
He watched his arms, and as they opened wide,
At every stroke betwixt them would he slide,
And steal a kiss, and then run out and dance,
And as he turned, cast many a lustful glance, 670
And throw him gaudy toys to please his eye,
And dive into the water, and there pry
Upon his breast, his thighs, and every limb,
And up again, and close behind him swim,
And talk of love. Leander made reply:
'You are deceived, I am no woman, I.'
Thereat smiled Neptune, and then told a tale,
How that a shepherd, sitting in a vale,
Played with a boy so lovely fair and kind,
As for his love both earth and heaven pined; 680
That of the cooling river durst not drink
Lest water-nymphs should pull him from the brink;
And when he sported in the fragrant lawns,
Goat-footed satyrs and up-staring fauns
Would steal him thence. Ere half this tale was done,
'Ay me!' Leander cried, 'th'enamoured sun,
That now should shine on Thetis' glassy bower,
Descends upon my radiant Hero's tower.
O that these tardy arms of mine were wings!'
And as he spake, upon the waves he springs. 690

666 bewrayed] showed

684 up-staring] shaggy, with hair standing up

Neptune was angry that he gave no ear,
And in his heart revenging malice bare;
He flung at him his mace, but as it went
He called it in, for love made him repent.
The mace returning back, his own hand hit,
As meaning to be venged for darting it.
When this fresh bleeding wound Leander viewed,
His colour went and came, as if he rued
The grief which Neptune felt. In gentle breasts
Relenting thoughts, remorse, and pity rests; 700
And who have hard hearts and obdurate minds
But vicious, hare-brained, and illiterate hinds?
The god, seeing him with pity to be moved,
Thereon concluded that he was beloved.
(Love is too full of faith, too credulous,
With folly and false hope deluding us.)
Wherefore, Leander's fancy to surprise,
To the rich Ocean for gifts he flies.
'Tis wisdom to give much; a gift prevails
When deep persuading oratory fails. 710
By this Leander being near the land
Cast down his weary feet, and felt the sand.
Breathless albeit he were, he rested not
Till to the solitary tower he got,
And knocked and called, at which celestial noise
The longing heart of Hero much more joys
Than nymphs or shepherds when the timbrel rings,
Or crooked dolphin when the sailor sings.
She stayed not for her robes, but straight arose,
And drunk with gladness, to the door she goes; 720
Where seeing a naked man, she screeched for fear,
(Such sights as this to tender maids are rare)
And ran into the dark herself to hide.
Rich jewels in the dark are soonest spied.
Unto her was he led, or rather drawn
By those white limbs which sparkled through the lawn.
The nearer that he came, the more she fled,
And seeking refuge, slipped into her bed.
Whereon Leander sitting, thus began,
Through numbing cold all feeble, faint, and wan: 730
'If not for love, yet, love, for pity sake,
Me in thy bed and maiden bosom take;
At least vouchsafe these arms some little room,
Who, hoping to embrace thee, cheerly swum;

702 hinds] boors, peasants 717 timbrel] tambourine 734 cheerly] gladly

This head was beat with many a churlish billow,
And therefore let it rest upon thy pillow.'
Herewith affrighted Hero shrunk away,
And in her lukewarm place Leander lay,
Whose lively heat, like fire from heaven fet,
Would animate gross clay, and higher set 740
The drooping thoughts of base declining souls,
Than dreary Mars carousing nectar bowls.
His hands he cast upon her like a snare:
She, overcome with shame and sallow fear,
Like chaste Diana, when Actaeon spied her,
Being suddenly betrayed, dived down to hide her;
And as her silver body downward went,
With both her hands she made the bed a tent,
And in her own mind thought herself secure,
O'ercast with dim and darksome coverture. 750
And now she lets him whisper in her ear,
Flatter, entreat, promise, protest, and swear;
Yet ever as he greedily assayed
To touch those dainties, she the harpy played,
And every limb did, as a soldier stout,
Defend the fort and keep the foeman out.
For though the rising ivory mount he scaled,
Which is with azure circling lines empaled,
Much like a globe (a globe may I term this,
By which love sails to regions full of bliss) 760
Yet there with Sisyphus he toiled in vain,
Till gentle parley did the truce obtain.
Wherein Leander on her quivering breast
Breathless spoke something, and sighed out the rest;
Which so prevailed, as he with small ado
Enclosed her in his arms and kissed her too.
And every kiss to her was as a charm,
And to Leander as a fresh alarm,
So that the truce was broke, and she, alas,
(Poor silly maiden) at his mercy was. 770
Love is not full of pity (as men say)
But deaf and cruel where he means to prey.
Even as a bird, which in our hands we wring,
Forth plungeth and oft flutters with her wing,
She trembling strove; this strife of hers (like that
Which made the world) another world begat
Of unknown joy. Treason was in her thought,
And cunningly to yield herself she sought.

739 fet] fetched 742 dreary] drear, bloody 758 empaled] encircled

Seeming not won, yet won she was at length;
In such wars women use but half their strength. 780
Leander now, like Theban Hercules,
Entered the orchard of th'Hesperides,
Whose fruit none rightly can describe but he
That pulls or shakes it from the golden tree.
And now she wished this night were never done,
And sighed to think upon th'approaching sun;
For much it grieved her that the bright daylight
Should know the pleasure of this blessed night,
And them like Mars and Erycine displayed,
Both in each other's arms chained as they laid. 790
Again she knew not how to frame her look,
Or speak to him who in a moment took
That which so long so charily she kept;
And fain by stealth away she would have crept,
And to some corner secretly have gone,
Leaving Leander in the bed alone.
But as her naked feet were whipping out,
He on the sudden clinged her so about,
That mermaid-like unto the floor she slid,
One half appeared, the other half was hid. 800
Thus near the bed she blushing stood upright,
And from her countenance behold ye might
A kind of twilight break, which through the hair,
As from an orient cloud, glims here and there;
And round about the chamber this false morn
Brought forth the day before the day was born.
So Hero's ruddy cheek Hero betrayed,
And her all naked to his sight displayed;
Whence his admiring eyes more pleasure took
Than Dis on heaps of gold fixing his look. 810
By this, Apollo's golden harp began
To sound forth music to the Ocean;
Which watchful Hesperus no sooner heard,
But he the day's bright-bearing car prepared,
And ran before, as harbinger of light,
And with his flaring beams mocked ugly Night
Till she, o'ercome with anguish, shame, and rage,
Danged down to hell her loathsome carriage.

(Pub. 1598)

804 glims] gleams 818 Danged] drove violently

SIR HENRY WOTTON
1568–1639

314 *A Poem written by Sir Henry Wotton in his youth*

O FAITHLESS world, and thy most faithless part,
  A woman's heart!
The true shop of variety, where sits
  Nothing but fits
And fevers of desire, and pangs of love,
  Which toys remove.
Why was she born to please? or I to trust
  Words writ in dust,
Suffering her eyes to govern my despair,
  My pain for air; 10
And fruit of time rewarded with untruth,
  The food of youth?
Untrue she was; yet I believed her eyes,
  Instructed spies,
Till I was taught, that love was but a school
  To breed a fool.
Or sought she more, by triumphs of denial,
  To make a trial
How far her smiles commanded my weakness?
  Yield, and confess! 20
Excuse no more thy folly; but, for cure,
  Blush and endure
As well thy shame as passions that were vain;
  And think, 'tis gain,
To know that love lodged in a woman's breast
  Is but a guest.

(Pub. 1602)

315 [*To John Donne*]

'TIS not a coat of gray or shepherd's life,
  'Tis not in fields or woods remote to live,
That adds or takes from one that peace or strife
  Which to our days such good or ill doth give:
It is the mind that makes the man's estate
For ever happy or unfortunate.

Then first the mind of passions must be free
  Of him that would to happiness aspire,
Whether in princes' palaces to be
  Or whether to his cottage he retire; 10
For our desires that on extremes are bent
Are friends to care and traitors to content.

Nor should we blame our friends though false they be,
  Since there are thousands false for one that's true,
But our own blindness that we cannot see
  To choose the best although they be but few;
For he that every feignèd friend will trust
Proves true to friend but to himself unjust.

The faults we have are they that made our woe;
  Our virtues are the motives of our joy. 20
Then is it vain if we to deserts go
  To seek our bliss or shroud us from annoy.
Our place need not be changed but our will,
For everywhere we may do good or ill.

But this I do not dedicate to thee
  As one that holds himself fit to advise,
Or that my lines to him should precepts be
  That is less ill than I, and much more wise.
But 'tis no harm morality to preach,
For men do often learn when they do teach. 30

(Wr. 1597–8; pub. 1911)

SAMUEL DANIEL

*c.*1563–1619

from *Delia*

316 LOOK, Delia, how we steem the half-blown rose,
  The image of thy blush and summer's honour;
Whilst in her tender green she doth enclose
  That pure sweet beauty Time bestows upon her.

20 motives] inducements; inciting causes
21 deserts] wild, uninhabited regions
25 dedicate] address
28 ill] blameworthy

316

1 steem] estimate, value

No sooner spreads her glory in the air
  But straight her full-blown pride is in declining;
She then is scorned that late adorned the fair:
  So clouds thy beauty after fairest shining.
No April can revive thy withered flowers,
  Whose blooming grace adorns thy glory now; 10
Swift speedy Time, feathered with flying hours,
  Dissolves the beauty of the fairest brow.
    O let not then such riches waste in vain,
    But love whilst that thou mayst be loved again.

317 BUT love whilst that thou mayst be loved again,
  Now whilst thy May hath filled thy lap with flowers,
Now whilst thy beauty bears without a stain;
  Now use thy summer smiles ere winter lours.
And whilst thou spread'st unto the rising sun
  The fairest flower that ever saw the light,
Now joy thy time before thy sweet be done:
  And, Delia, think thy morning must have night,
And that thy brightness sets at length to west,
  When thou wilt close up that which now thou
    showest; 10
And think the same becomes thy fading best,
  Which then shall hide it most and cover lowest.
    Men do not weigh the stalk for that it was,
    When once they find her flower, her glory, pass.

318 WHEN men shall find thy flower, thy glory, pass,
  And thou with careful brow sitting alone
Received hast this message from thy glass,
  That tells thee truth, and says that all is gone,
Fresh shalt thou see in me the wounds thou madest,
  Though spent thy flame, in me the heat remaining;
I that have loved thee thus before thou fadest,
  My faith shall wax, when thou art in thy waning.
The world shall find this miracle in me,
  That fire can burn when all the matter's spent; 10
Then what my faith hath been thyself shalt see,
  And that thou wast unkind thou mayst repent.
    Thou mayst repent, that thou hast scorned my
      tears,
    When winter snows upon thy golden hairs.

14 loved again] loved in return

317
3 bears] endures, lasts

318
2 careful] sorrowful, troubled

319 WHEN winter snows upon thy golden hairs,
And frost of age hath nipped thy flowers near;
When dark shall seem thy day that never clears,
And all lies withered that was held so dear;
Then take this picture which I here present thee,
Limned with a pencil not all unworthy;
Here see the gifts that God and Nature lent thee;
Here read thyself, and what I suffered for thee.
This may remain thy lasting monument,
Which happily posterity may cherish; 10
These colours with thy fading are not spent;
These may remain, when thou and I shall perish.
If they remain, then thou shalt live thereby;
They will remain, and so thou canst not die.

320 THOU canst not die whilst any zeal abound
In feeling hearts that can conceive these lines;
Though thou, a Laura, hast no Petrarch found,
In base attire yet clearly beauty shines.
And I, though born within a colder clime,
Do feel mine inward heat as great, I know it;
He never had more faith, although more rhyme;
I love as well, though he could better show it.
But I may add one feather to thy fame
To help her flight throughout the fairest isle, 10
And if my pen could more enlarge thy name
Then shouldst thou live in an immortal style.
But though that Laura better limned be,
Suffice, thou shalt be loved as well as she.

321 BEAUTY, sweet love, is like the morning dew,
Whose short refresh upon the tender green
Cheers for a time, but till the sun doth show,
And straight 'tis gone as it had never been.
Soon doth it fade that makes the fairest flourish;
Short is the glory of the blushing rose,
The hue which thou so carefully dost nourish,
Yet which at length thou must be forced to lose.

6 pencil] paint-brush

320

2 conceive] understand, comprehend
7 rhyme] poetry, poetical genius

321

2 refresh] refreshment
5 flourish] luxuriance, bloom

When thou, surcharged with burthen of thy years,
Shalt bend thy wrinkles homeward to the earth, 10
When Time hath made a passport for thy fears,
Dated in age, the calends of our death—
But ah, no more! This hath been often told,
And women grieve to think they must be old.

322

CARE-CHARMER sleep, son of the sable night,
Brother to death, in silent darkness born,
Relieve my languish, and restore the light,
With dark forgetting of my cares return.
And let the day be time enough to mourn
The shipwreck of my ill-adventured youth;
Let waking eyes suffice to wail their scorn
Without the torment of the night's untruth.
Cease, dreams, th'imagery of our day-desires,
To model forth the passions of the morrow; 10
Never let rising sun approve you liars,
To add more grief to aggravate my sorrow.
Still let me sleep, embracing clouds in vain,
And never wake to feel the day's disdain.

323

LET others sing of knights and paladins
In aged accents and untimely words;
Paint shadows in imaginary lines,
Which well the reach of their high wits records.
But I must sing of thee, and those fair eyes
Authentic shall my verse in time to come,
When yet th'unborn shall say, 'Lo where she lies,
Whose beauty made him speak that else was dumb.'
These are the arks, the trophies I erect,
That fortify thy name against old age; 10
And these thy sacred virtues must protect
Against the dark, and Time's consuming rage.
Though th'error of my youth they shall discover,
Suffice they show I lived and was thy lover.

(1592)

11 passport] authorization, permission
12 calends] beginning, first taste

322
3 languish] languishing
10 model forth] portray, describe in detail
11 approve] prove

323
1 paladins] champions, chivalric heroes
2 untimely] anachronistic (e.g. Spenser's diction in *The Faerie Queene*)
6 Authentic] authenticate, establish the truthfulness of
9 arks] coffers, caskets
trophies] memorial buildings, monuments

324

Ode

NOW each creature joys the other,
  Passing happy days and hours;
One bird reports unto another
  In the fall of silver showers;
Whilst the earth, our common mother,
  Hath her bosom decked with flowers.

Whilst the greatest torch of heaven
  With bright rays warms Flora's lap,
Making nights and days both even,
  Cheering plants with fresher sap; 10
My field, of flowers quite bereaven,
  Wants refresh of better hap.

Echo, daughter of the Air,
  Babbling guest of rocks and hills,
Knows the name of my fierce fair,
  And sounds the accents of my ills.
Each thing pities my despair,
  Whilst that she her lover kills.

Whilst that she, O cruel maid,
  Doth me and my true love despise, 20
My life's flourish is decayed
  That depended on her eyes:
But her will must be obeyed,
  And well he ends for love who dies.

(1592)

from *The Civil Wars*

325

[*King Richard II is taken into custody*]

A PLACE there is, where proudly raised there stands
  A huge aspiring rock, neighbouring the skies,
Whose surly brow imperiously commands
  The sea his bounds, that at his proud feet lies;
And spurns the waves, that in rebellious bands
  Assault his empire, and against him rise;
    Under whose craggy government there was
    A niggard narrow way for men to pass.

And here, in hidden cliffs, concealed lay
  A troop of armed men, to intercept 10
The unsuspecting King, that had no way
  To free his foot, that into danger stepped.
The dreadful ocean, on the one side, lay:
  The hard-encroaching mountain th'other kept;
    Before him, he beheld his hateful foes:
    Behind him, traitorous enemies enclose.

Environed thus, the earl begins to cheer
  His all-amazed lord, by him betrayed;
Bids him take courage, there's no cause of fear,
  These troops, but there to guard him safe, were laid. 20
To whom the King: 'What needs so many here?
  This is against your oath, my lord', he said.
    But now he sees in what distress he stood:
    To strive, was vain; t'entreat, would do no good.

And therefore on with careful heart he goes;
  Complains (but, to himself), sighs, grieves, and frets;
At Rutland dines, though feeds but on his woes:
  The grief of mind hindered the mind of meats.
For sorrow, shame, and fear, scorn of his foes,
  The thought of what he was, and what now threats, 30
    Then what he should, and now what he hath done,
    Musters confused passions all in one.

To Flint, from thence, unto a restless bed,
  That miserable night he comes conveyed;
Poorly provided, poorly followed,
  Uncourted, unrespected, unobeyed;
Where, if uncertain sleep but hovered
  Over the drooping cares that heavy weighed,
    Millions of figures fantasy presents
    Unto that sorrow wakened grief augments. 40

His new misfortune makes deluding sleep
  Say 'twas not so (false dreams the truth deny).
Wherewith he starts; feels waking cares do creep
  Upon his soul, and gives his dream the lie;
Then sleeps again: and then again, as deep
  Deceits of darkness mock his misery.
    So hard believed was sorrow in her youth
    That he thinks truth was dreams and dreams were truth.

The morning light presents unto his view
(Walking upon a turret of the place) 50
The truth of what he sees is proved too true.
A hundred thousand men, before his face,
Came marching on the shore, which thither drew:
And, more to aggravate his foul disgrace,
Those he had wronged, or done to them despite,
(As if they him upbraid) came first in sight.

There might he see that false forsworn vile crew,
Those shameless agents of unlawful lust,
His pandars, parasites (people untrue
To God and man, unworthy any trust) 60
Pressing unto that fortune that was new,
And with unblushing faces foremost thrust;
As those that still with prosperous fortune sort,
And are as born for court, or made in court.

There he beheld how humbly diligent
New Adulation was to be at hand;
How ready Falsehood stepped; how nimbly went
Base pick-thank Flattery, and prevents Command;
He saw the great obey, the grave consent,
And all with this new-raised aspirer stand; 70
But, which was worst, his own part acted there,
Not by himself; his power, not his, appear.

Which whilst he viewed, the Duke he might perceive
Make towards the castle, to an interview,
Wherefore he did his contemplation leave,
And down into some fitter place withdrew;
Where now he must admit, without his leave,
Him, who before with all submission due
Would have been glad, t'attend, and to prepare
The grace of audience, with respective care. 80

Who now being come in presence of his king
(Whether the sight of majesty did breed
Remorse of what he was encompassing,
Or whether but to formalise his deed)
He kneels him down with some astonishing;
Rose; kneels again (for, craft will still exceed):
Whenas the King approached, put off his hood,
And welcomed him, though wished him little good.

68 prevents] anticipates 80 respective] attentive

To whom the Duke began: 'My Lord, I know
 That both uncalled, and unexpected too, 90
I have presumed in this sort to show
 And seek the right which I am born unto.
Yet pardon I beseech you, and allow
 Of that constraint which drives me thus to do.
  For, since I could not by a fairer course
  Attain mine own, I must use this of force.'

'Well; so it seems, dear cousin', said the King,
 'Though you might have procured it otherwise.
And I am here content, in everything
 To right you, as yourself shall best devise. 100
And God vouchsafe the force that here you bring
 Beget not England greater injuries.'
  And so they part. The Duke made haste from thence.
  It was no place to end this difference.

Straight towards London, in this heat of pride,
 They forward set, as they had fore-decreed;
With whom the captive King constrained must ride,
 Most meanly mounted on a simple steed;
Degraded of all grace and ease beside,
 Thereby neglect of all respect to breed. 110
  For, th'over-spreading pomp of prouder might
  Must darken weakness and debase his right.

Approaching near the City, he was met
 With all the sumptuous shows joy could devise,
Where new-desire to please did not forget
 To pass the usual pomp of former guise.
Striving Applause, as out of prison let,
 Runs on, beyond all bounds, to novelties:
  And voice, and hands, and knees, and all do now
  A strange deformed form of welcome show. 120

And manifold Confusion running greets,
 Shouts, cries, claps hands, thrusts, strives and presses near.
Houses impoverished were, t'enrich the streets,
 And streets left naked, that (unhappy) were
Placed from the sight where Joy with Wonder meets;
 Where all, of all degrees, strive to appear;
  Where divers-speaking Zeal one murmur finds,
  In undistinguished voice to tell their minds.

He that in glory of his former fate
Admiring what he thought could never be, 130
Did feel his blood within salute his state,
And lift up his rejoicing soul, to see
So many hands and hearts congratulate
Th'advancement of his long-desired degree;
When, prodigal of thanks, in passing by,
He resalutes them all, with cheerful eye.

Behind him, all aloof, came pensive on
The unregarded King, that drooping went
Alone, and (but for spite) scarce looked upon.
Judge if he did more envy or lament. 140
See what a wondrous work this day is done,
Which th'image of both fortunes doth present:
In th'one, to show the best of glory's face,
In th'other, worse than worst of all disgrace.

Now Isabel, the young afflicted Queen
(Whose years had never showed her but delights,
Nor lovely eyes before had ever seen
Other than smiling joys and joyful sights;
Born great, matched great, lived great, and ever been
Partaker of the world's best benefits) 150
Had placed herself, hearing her Lord should pass
That way, where she unseen in secret was;

Sick of delay, and longing to behold
Her long-missed love in fearful jeopardies;
To whom, although it had, in sort, been told
Of their proceeding and of his surprise;
Yet thinking they would never be so bold
To lead their Lord in any shameful wise,
But rather would conduct him as their King,
As seeking but the state's re-ordering. 160

And forth she looks, and notes the foremost train,
And grieves to view some there she wished not there;
Seeing the chief not come, stays, looks again;
And yet she sees not him that should appear.
Then back she stands, and then desires as fain
Again to look, to see if he were near.
At length a glittering troop far off she spies,
Perceives the throng, and hears the shouts and cries.

'Lo, yonder now at length he comes', saith she;
'Look, my good women, where he is in sight. 170
Do you not see him? Yonder, that is he,
Mounted on that white courser, all in white,
There where the thronging troops of people be.
I know him by his seat, he sits s'upright.
Lo, now he bows. Dear Lord, with what sweet grace!
How long have I longed to behold that face!

'O what delight my heart takes by mine eye!
I doubt me, when he comes but something near,
I shall set wide the window. What care I
Who doth see me, so him I may see clear?' 180
Thus doth false joy delude her wrongfully
(Sweet lady) in the thing she held so dear.
For, nearer come, she finds she had mistook;
And him she marked was Henry Bullingbrooke.

Then Envy takes the place in her sweet eyes,
Where Sorrow had prepared herself a seat;
And words of wrath, from whence complaints should rise,
Proceed from eager looks and brows that threat.
'Traitor!' saith she, 'Is't thou, that in this wise
To brave thy Lord and King art made so great? 190
And have mine eyes done unto me this wrong,
To look on thee? For this, stayed I so long?

'Ah, have they graced a perjured rebel so?
Well; for their error I will weep them out,
And hate the tongue defiled, that praised my foe,
And loathe the mind that gave me not to doubt.
What! Have I added shame unto my woe?
I'll look no more. Ladies, look you about,
And tell me if my Lord be in this train,
Lest my betraying eyes should err again.' 200

And in this passion turns herself away.
The rest look all, and careful note each wight,
Whilst she, impatient of the least delay,
Demands again: 'And what, not yet in sight?
Where is my Lord? What, gone some other way?
I muse at this. O God, grant all go right!'
Then to the window goes again at last,
And sees the chiefest train of all was past;

178 doubt] fear 185 Envy] hostile feeling, ill-will

And sees not him her soul desired to see.
  And yet hope, spent, makes her not leave to look. 210
At last, her love-quick eyes, which ready be,
  Fastens on one; whom though she never took
Could be her Lord; yet that sad cheer which he
  Then showed, his habit and his woeful look,
    The grace he doth in base attire retain,
    Caused her she could not from his sight refrain.

'What might he be', she said, 'that thus alone
  Rides pensive in this universal joy?
Some I perceive, as well as we, do moan:
  All are not pleased with everything this day. 220
It may be, he laments the wrong is done
  Unto my Lord, and grieves; as well he may.
    Then he is some of ours: and we, of right,
    Must pity him, that pities our sad plight.

'But stay: is't not my Lord himself I see?
  In truth, if 'twere not for his base array,
I verily should think that it were he.
  And yet his baseness doth a grace bewray.
Yet God forbid; let me deceived be,
  And be it not my Lord, although it may. 230
    Let my desire make vows against desire,
    And let my sight approve my sight a liar.

'Let me not see him, but himself: a King.
  For so he left me, so he did remove.
This is not he: this feels some other thing;
  A passion of dislike, or else of love.
O yes; 'tis he: that princely face doth bring
  The evidence of majesty to prove.
    That face I have conferred, which now I see,
    With that within my heart, and they agree.' 240

Thus as she stood assured, and yet in doubt,
  Wishing to see what seen she grieved to see;
Having belief, yet fain would be without;
  Knowing, yet striving not to know 'twas he:
Her heart relenting, yet her heart so stout
  As would not yield to think what was, could be;
    Till, quite condemned by open proof of sight,
    She must confess, or else deny the light.

(1595)

228 bewray] reveal   232 approve] prove   239 conferred] compared

from *Musophilus*

326

Philocosmus

FOND man, Musophilus, that thus dost spend
In an ungainful art thy dearest days,
Tiring thy wits and toiling to no end,
But to attain that idle smoke of praise;
Now when this busy world cannot attend
Th'untimely music of neglected lays.
Other delights than these, other desires,
This wiser profit-seeking age requires.

Musophilus

Friend Philocosmus, I confess indeed
I love this sacred art thou set'st so light, 10
And though it never stand my life in stead,
It is enough it gives myself delight,
The whiles my unafflicted mind doth feed
On no unholy thoughts for benefit.

Be it that my unseasonable song
Come out of time, that fault is in the time,
And I must not do virtue so much wrong
As love her aught the worse for others' crime;
And yet I find some blessed spirits among
That cherish me, and like and grace my rhyme. 20

A gain, that I do more in soul esteem,
Than all the gain of dust the world doth crave;
And if I may attain but to redeem
My name from dissolution and the grave,
I shall have done enough, and better deem
T'have lived to be, than to have died to have.

Short-breathed mortality would yet extend
That span of life so far forth as it may,
And rob her fate, seek to beguile her end
Of some few lingering days of after stay, 30
That all this little all might not descend
Into the dark a universal prey.

Philocosmus] i.e. lover of the world
1 Fond] foolish
Musophilus] i.e. lover of the Muse
11 stand . . . in stead] be of advantage
19 among] at intervals, from time to time

And give our labours yet this poor delight,
  That when our days do end they are not done;
And though we die, we shall not perish quite,
  But live two lives, where others have but one.

Philocosmus

Silly desires of self-abusing man,
  Striving to gain th'inheritance of air,
That, having done the uttermost he can,
  Leaves yet, perhaps, but beggary to his heir. 40
All that great purchase of the breath he wan
  Feeds not his race or makes his house more fair.

And what art thou the better, thus to leave
  A multitude of words to small effect,
Which other times may scorn, and so deceive
  Thy promised name of what thou dost expect?
Besides, some viperous critic may bereave
  Th'opinion of thy worth for some defect,

And get more reputation of his wit,
  By but controlling of some word or sense, 50
Than thou shalt honour for contriving it,
  With all thy travail, care, and diligence;
Being learned now enough to contradict
  And censure others with bold insolence.

Besides, so many so confusedly sing,
  As diverse discords have the music marred,
And in contempt that mystery doth bring,
  That he must sing aloud that will be heard;
And the received opinion of the thing,
  For some unhallowed strings that vildly jarred, 60

Hath so unseasoned now the ears of men,
  That who doth touch the tenour of that vein
Is held but vain, and his unreckoned pen
  The title but of levity doth gain.
A poor, light gain, to recompense their toil,
  That thought to get eternity the while.

50 controlling of] finding fault with
57 mystery] art, skill
60 vildly] vilely

And therefore, leave the left and outworn course
  Of unregarded ways, and labour how
To fit the times with what is most in force;
  Be new with men's affections that are now; 70
Strive not to run an idle counter-course
  Out from the scent of humours men allow.

For not discreetly to compose our parts
  Unto the frame of men, which we must be,
Is to put off ourselves, and make our arts
  Rebels to Nature and society;
Whereby we come to bury our deserts
  In th'obscure grave of singularity.

Musophilus

Do not profane the work of doing well,
  Seduced man, that canst not look so high 80
From out that mist of earth as thou canst tell
  The ways of right, which virtue doth descry,
That overlooks the base, contemptible,
  And low-laid follies of mortality;

Nor mete out truth and right-deserving praise
  By that wrong measure of confusion
The vulgar foot, that never takes his ways
  By reason but by imitation,
Rolling on with the rest, and never weighs
  The course which he should go but what is gone. 90

Well were it with mankind if what the most
  Did like were best, but ignorance will live
By others' square, as by example lost;
  And man to man must th'hand of error give
That none can fall alone at their own cost,
  And all because men judge not, but believe.

For what poor bounds have they whom but th'earth
  bounds!
  What is their end whereto their care attains,
When the thing got relieves not but confounds,
  Having but travail to succeed their pains? 100
What joy hath he of living that propounds
  Affliction but his end and grief his gains?

75 put off] hinder, obstruct   93 square] standard

Gathering encroaching, wresting, joining to,
   Destroying, building, decking, furnishing,
Repairing, altering, and so much ado
   To his soul's toil and body's travailing:
And all this doth he, little knowing who
   Fortune ordains to have th'inheriting.

And his fair house raised high in envy's eye,
   Whose pillars reared perhaps on blood and wrong, 110
The spoils and pillage of iniquity—
   Who can assure it to continue long?
If rage spared not the walls of piety,
   Shall the profanest piles of sin keep strong?

How many proud aspiring palaces
   Have we known made the prey of wrath and pride,
Leveled with the earth, left to forgetfulness,
   Whilst titlers their pretended rights decide,
Or civil tumults, or an orderless
   Order pretending change of some strong side? 120

Then where is that proud title of thy name,
   Written in ice of melting vanity?
Where is thine heir left to possess the same?
   Perhaps not so well as in beggary.
Some thing may rise to be beyond the shame
   Of vile and unregarded poverty.

Which, I confess, although I often strive
   To clothe in the best habit of my skill
In all the fairest colours I can give,
   Yet for all that methinks she looks but ill; 130
I cannot brook that face which dead-alive
   Shows a quick body but a buried will.

Yet oft we see the bars of this restraint
   Holds goodness in, which loose wealth would let fly,
And fruitless riches, barrener than want,
      Brings forth small worth from idle liberty;
   Which when disorders shall again make scant,
      It must refetch her state from poverty.

118 titlers] those who claim a legal title

But yet in all this interchange of all,
  Virtue we see, with her fair grace, stands fast. 140
For what high races hath there come to fall,
  With low disgrace, quite vanished and past,
Since Chaucer lived, who yet lives and yet shall,
  Though (which I grieve to say) but in his last.

Yet what a time hath he wrested from time,
  And won upon the mighty waste of days
Unto th'immortal honour of our clime,
  That by his means came first adorned with bays;
Unto the sacred relics of whose rhyme
  We yet are bound in zeal to offer praise. 150

And could our lines begotten in this age
  Obtain but such a blessed hand of years
And scape the fury of that threatening rage
  Which in confused clouds ghastly appears,
Who would not strain his travails to engage,
  When such true glory should succeed his cares?

But whereas he came planted in the spring,
  And had the sun, before him, of respect,
We set in th'autumn, in the withering,
  And sullen season of a cold defect, 160
Must taste those sour distastes the times do bring,
  Upon the fulness of a cloyed neglect,

Although the stronger constitutions shall
  Wear out th'infection of distempered days,
And come with glory to outlive this fall,
  Recov'ring of another spring of praise,
Cleared from th'oppressing humours wherewithal
  The idle multitude surcharge their lays.

Whenas perhaps the words thou scornest now
  May live, the speaking picture of the mind, 170
The extract of the soul that laboured how
  To leave the image of herself behind,
Wherein posterity that love to know
  The just proportion of our spirits may find.

For these lines are the veins, the arteries,
  And undecaying life-strings of those hearts
That still shall pant, and still shall exercise
  The motion spirit and nature both imparts,
And shall with those alive so sympathise
  As nourished with their powers enjoy their parts. 180

O blessed letters, that combine in one
  All ages past, and make one live with all!
By you we do confer with who are gone,
  And the dead living unto council call;
By you th'unborn shall have communion
  Of what we feel and what doth us befall.

Soul of the world, knowledge, without thee
  What hath the earth that truly glorious is?
Why should our pride make such a stir to be,
  To be forgot? What good is like to this, 190
To do worthy the writing and to write
  Worthy the reading and the world's delight?

And let th'unnatural and wayward race
  Born of one womb with us, but to our shame,
That never read t'observe but to disgrace,
  Raise all the tempest of their power to blame.
That puff of folly never can deface
  The work a happy genius took to frame.

327

Musophilus

SACRED Religion, mother of form and fear,
  How gorgeously sometimes dost thou sit decked!
What pompous vestures do we make thee wear!
  What stately piles we prodigal erect!
How sweet-perfumed thou art, how shining clear!
  How solemnly observed, with what respect!

Another time all plain, and quite threadbare,
  Thou must have all within and nought without;
Sit poorly without light, disrobed, no care
  Of outward grace, to amuse the poor devout; 10
Powerless, unfollowed, scarcely men can spare
  Thee necessary rites to set thee out.

198 happy] felicitous
genius] turn or temper of mind

327
1 form] proper or right behaviour

Either truth, goodness, virtue are not still
  The self same which they are, and always one,
But alter to the project of our will,
  Or we our actions make them wait upon,
Putting them in the livery of our skill,
  And cast them off again when we have done.

You mighty lords, that with respected grace
  Do at the stern of fair example stand, 20
And all the body of this populace
  Guide with the only turning of your hand,
Keep a right course, bear up from all disgrace,
  Observe the point of glory to our land;

Hold up disgraced knowledge from the ground,
  Keep virtue in request, give worth her due,
Let not neglect with barbarous means confound
  So fair a good to bring in night anew.
Be not, O be not accessary found
  Unto her death that must give life to you. 30

Where will you have your virtuous names safe laid,
  In gorgeous tombs, in sacred cells secure?
Do you not see those prostrate heaps betrayed
  Your fathers' bones, and could not keep them sure?
And will you trust deceitful stones fair laid,
  And think they will be to your honour truer?

No, no; unsparing time will proudly send
  A warrant unto wrath that with one frown
Will all these mockeries of vainglory rend,
  And make them as before, ungraced, unknown; 40
Poor idle honours that can ill defend
  Your memories, that cannot keep their own.

And whereto serve that wondrous trophy now
  That on the goodly plain near Wilton stands?
That huge dumb heap, that cannot tell us how,
  Nor what, nor whence it is, nor with whose hands,
Nor for whose glory, it was set to show
  How much our pride mocks that of other lands.

Whereon whenas the gazing passenger
  Hath greedy looked with admiration, 50
And fain would know his birth, and what he were,
  How there erected, and how long agone,
Enquires and asks his fellow traveller
  What he hath heard and his opinion:

And he knows nothing. Then he turns again,
  And looks and sighs, and then admires afresh,
And in himself with sorrow doth complain
  The misery of dark forgetfulness,
Angry with time that nothing should remain
  Our greatest wonders'-wonder to express. 60

328

Philocosmus

BEHOLD how every man, drawn with delight
  Of what he doth, flatters him in his way;
Striving to make his course seem only right
  Doth his own rest and his own thoughts betray;
Imagination bringing bravely dight
  Her pleasing images in best array,

With flattering glasses that must show him fair
  And others foul; his skill and his wit best,
Others seduced, deceived and wrong in their;
  His knowledge right, all ignorant the rest, 10
Not seeing how these minions in the air
  Present a face of things falsely expressed,
And that the glimmering of these errors shown
  Are but a light to let him see his own.

Alas poor Fame, in what a narrow room,
  As an encaged parrot, art thou pent
Here amongst us, when even as good be dumb
  As speak, and to be heard with no attent!
How can you promise of the time to come
  Whenas the present are so negligent? 20

Is this the walk of all your wide renown,
  This little point, this scarce-discerned isle,
Thrust from the world, with whom our speech unknown
  Made never any traffic of our style?
And is this all where all this care is shown,
  T'enchant your fame to last so long a while?

49 passenger] passer-by
55 turns again] turns back

328
2 flatters him] flatters himself
5 bravely dight] richly dressed

And for that happier tongues have won so much,
  Think you to make your barbarous language such?

Poor narrow limits for so mighty pains,
  That cannot promise any foreign vent: 30
And yet if here to all your wondrous veins
  Were generally known, it might content.
But lo, how many reads not, or disdains,
  The labours of the chief and excellent.

How many thousands never heard the name
  Of Sidney or of Spenser, or their books!
And yet brave fellows, and presume of fame,
  And seem to bear down all the world with looks.
What then shall they expect of meaner frame,
  On whose endeavours few or none scarce looks? 40

Do you not see these pamphlets, libels, rhymes?
  These strange confused tumults of the mind
Are grown to be the sickness of these times,
  The great disease inflicted on mankind.
Your virtues, by your follies made your crimes,
  Have issue with your indiscretion joined.

Schools, arts, professions, all in so great store,
  Pass the proportion of the present state,
Where being as great a number as before
  And fewer rooms them to accommodate, 50
It cannot be but they must throng the more,
  And kick, and thrust, and shoulder with debate.

For when the greater wits cannot attain
  Th'expected good, which they account their right,
And yet perceive others to reap that gain
  Of far inferior virtues in their sight,
They present with the sharp of envy's strain
  To wound them with reproaches and despite;
And for these cannot have as well as they,
  They scorn their faith should deign to look that way. 60

Hence discontented sects and schisms arise,
  Hence interwounding controversies spring,
That feed the simple and offend the wise,
  Who know the consequence of cavilling.
Disgrace, that these to others do devise,
  Contempt and scorn on all in th'end doth bring,

50 rooms] jobs, positions   52 debate] strife, conflict

Like scolding wives reckoning each other's fault
  Make standers-by imagine both are naught.

For when to these rare dainties time admits
  All comers, all complexions, all that will, 70
Where none should be let in but choicest wits,
  Whose mild discretion could comport with skill,
For when the place their humour neither fits
  Nor they the place, who can expect but ill?

For being unapt for what they took in hand,
  And for ought else whereto they shall b'addressed,
They even become th'encumbrance of the land
  As out of rank disordering all the rest.
This grace of theirs to seem to understand
  Mars all their grace to do, without their rest. 80

Men find that action is another thing
  Than what they in discoursing papers read.
The world's affairs require in managing
  More arts than those wherein you clerks proceed.
Whilst timorous knowledge stands considering,
  Audacious ignorance hath done the deed.
For who knows most, the more he knows to doubt;
  The least discourse is commonly most stout.

This sweet enchanting knowledge turns you clean
  Out from the fields of natural delight, 90
And makes you hide, unwilling to be seen
  In th'open concourse of a public sight.
This skill wherewith you have so cunning been
  Unsinews all your powers, unmans you quite.

Public society and commerce of men
  Require another grace, another port:
This eloquence, these rhymes, these phrases then,
  Begot in shades, do serve us in no sort.
Th'unmaterial swellings of your pen
  Touch not the spirit that action doth import. 100

A manly style fitted to manly ears
  Best grees with wit, not that which goes so gay,
And commonly the gaudy livery wears
  Of nice corruptions which the times do sway,

102 grees] agrees   104 nice] over-refined, unmanly

And waits on th'humour of his pulse that bears
His passions set to such a pleasing key.
Such dainties serve only for stomachs weak,
For men do foulest when they finest speak.

Yet I do not dislike that in some wise
Be sung the great heroical deserts 110
Of brave renowned spirits, whose exercise
Of worthy deeds may call up others' hearts,
And serve a model for posterities
To fashion them fit for like glorious parts;
But so that all our spirits may tend hereto
To make it not our grace to say, but do.

Musophilus

Much hast thou said, and willingly I hear,
As one that am not so possessed with love
Of what I do, but that I rather bear
An ear to learn than a tongue to disprove. 120
I know men must, as carried in their sphere,
According to their proper motions move.
And that course likes them best which they are on,
Yet truth hath certain bounds, but falsehood none.

I do confess our limits are but small
Compared with all the whole vast earth beside;
All which again, rated to that great All,
Is likewise as a point scarcely descried;
So that in these respects we may this call
A point but of a point where we abide. 130

But if we shall descend from that high stand
Of overlooking contemplation,
And cast our thoughts but to, and not beyond,
This spacious circuit which we tread upon,
We then may estimate our mighty land
A world within a world standing alone;

Where if our fame confined cannot get out,
What, shall we then imagine it is penned
That hath so great a world to walk about,
Whose bounds with her reports have both one end? 140
Why shall we not rather esteem her stout
That farther than her own scorn to extend?

Where being so large a room both to do well
   And eke to hear th'applause of things well done,
That farther if men shall our virtues tell
   We have more mouths but not more merit won.
It doth not greater make that which is laudable;
   The flame is bigger blown, the fire all one.

And for the few that only lend their ear,
   That few is all the world, which with a few   150
Doth ever live, and move, and work and stir.
   This is the heart doth feel; and only know
The rest of all, that only bodies bear,
   Roll up and down, and fill but up the row;

And serve as others' members, not their own,
   The instruments of those that do direct.
Then what disgrace is this not to be known
   To those know not to give themselves respect?
And though they swell with pomp of folly blown,
   They live ungraced, and die but in neglect.   160

And for my part, if only one allow
   The care my labouring spirits take in this,
He is to me a theatre large enow,
   And his applause only sufficient is.
All my respect is bent but to his brow:
   That is my all, and all I am is his.

And if some worthy spirits be pleased too,
   It shall more comfort breed, but not more will.
But what if none? It cannot yet undo
   The love I bear unto this holy skill.   170
This is the thing that I was born to do,
   This is my scene, this part I must fulfil.

329

POWER above powers, O heavenly Eloquence,
   That with the strong rein of commanding words
Dost manage, guide, and master th'eminence
   Of men's affections, more than all their swords,
Shall we not offer to thy excellence
   The richest treasure that our wit affords?

161 allow] approve

Thou that canst do much more with one poor pen,
  Than all the powers of princes can effect;
And draw, divert, dispose and fashion men,
  Better than force or rigour can direct. 10
Should we this ornament of glory then,
  As th'unmaterial fruits of shades, neglect?

Or should we careless come behind the rest
  In power of words, that go before in worth;
Whenas our accents equal to the best
  Is able greater wonders to bring forth?
When all that ever hotter spirits expressed
  Comes bettered by the patience of the north.

And who, in time, knows whither we may vent
  The treasure of our tongue, to what strange shores 20
This gain of our best glory shall be sent,
  T' enrich unknowing nations with our stores?
What worlds in th' yet unformed Occident
  May come refined with th'accents that are ours?

Or who can tell for what great work in hand
  The greatness of our style is now ordained?
What powers it shall bring in, what spirits command?
  What thoughts let out, what humours keep restrained?
What mischief it may powerfully withstand;
  And what fair ends may thereby be attained? 30

And as for Poesy, mother of this force,
  That breeds, brings forth, and nourishes this might,
Teaching it in a loose, yet measured course,
  With comely motions how to go upright;
And fostering it with bountiful discourse,
  Adorns it thus in fashions of delight:

What should I say? since it is well approved
  The speech of Heaven, with whom they have
    commerce,
That only seem out of themselves removed,
  And do with more than human skills converse: 40
Those numbers, wherewith Heaven and Earth are
    moved,
  Show weakness speaks in prose, but power in verse.

(1599)

MICHAEL DRAYTON
1563–1631

from *The Shepherds' Garland*

330

The eighth eclogue

Good Gorbo of the golden world
and Saturn's reign doth tell,
And afterward doth make report,
of bonny Dowsabell.

Motto

SHEPHERD, why creep we in this lowly vein,
As though our Muse no store at all affords,
Whilst others vaunt it with the frolic swain
And strut the stage with reperfumed words?

See how these younkers rave it out in rhyme,
Who make a traffic of their rarest wits, 10
And in Bellona's buskin tread it fine,
Like Bacchus' priests raging in frantic fits.

Those myrtle groves decayed, done grow again,
Their roots refreshed with Helicona's spring,
Whose pleasant shade invites the homely swain
To sit him down and hear the Muses sing.

Then if thy Muse hath spent her wonted zeal,
With ivy twist thy temples shall be crowned,
Or if she dares hoise up top-gallant sail,
Amongst the rest then may she be renowned. 20

Gorbo

My boy, these younkers reachen after fame,
And so done press into the learned troop,
With filed quill to glorify their name,
Which otherwise were penned in shameful coop.

But this high object hath abjected me,
And I must pipe amongst the lowly sort,
Those little herd-grooms who admired to see,
When I by moonshine made the fairies sport.

9 younkers] youngsters

Who dares describe the toils of Hercules
  And puts his hand to fame's eternal pen 30
Must invocate the soul of Hercules,
  Attended with the troops of conquered men.

Who writes of thrice-renowned Theseus,
  A monster-tamer's rare description,
Trophies the jaws of ugly Cerberus,
  And paints out Styx and fiery Acheron.

My Muse may not affect night-charming spells,
  Whose force affects th'Olympic vault to quake,
Nor call those grisly goblins from their cells,
  The ever-damned fry of Limbo lake. 40

And who erects the brave Pyramides
  Of monarchs or renowned warriors,
Need bathe his quill for such attempts as these
  In flowing streams of learned Maro's showers.

For when the great world's conqueror began
  To prove his helmet and his habergeon,
The sweat that from the poets'-god Orpheus ran
  Foretold his prophets had to play upon.

When pens and lances saw the Olympiad prize,
  Those chariot triumphs with the laurel crown, 50
Then gan the worthies' glory first to rise,
  And plumes were vailed to the purple gown.

The gravest censor, sagest senator,
  With wings of justice and religion,
Mounted the top of Nimrod's stately tower,
  Soaring unto that high celestial throne,

Where blessed angels in their heavenly queres,
  Chant anthems with shrill syren harmony,
Tuned to the sound of those air-crowding spheres,
  Which herien their maker's eternity. 60

35 Trophies] makes a trophy of, celebrates
44 Maro's] i.e. Virgil's
45 conqueror] i.e. Alexander the Great
46 habergeon] sleeveless coat of mail
52 vailed] lowered (in deference to)
60 herien] praise, glorify

Those who foretell the times of unborn men
  And future things in foretime augured,
Have slumbered in that spell-god's darkest den,
  Which first inspired his prophesying head.

Sooth-saying Sybils sleepen long agone,
  We have their rede, but few have conned their art;
Welsh-wizard Merlin cleaveth to a stone;
  No oracle more wonders may impart.

The infant age could deftly carol love
  Till greedy thirst of that ambitious honour 70
Drew poet's pen from his sweet lass's glove
  To chant of slaughtering broils and bloody horror.

Then Jove's love-theft was privily descried,
  How he played false play in Amphitrio's bed,
And how Apollo in the mount of Ide
  Gave Oenon physic for her maidenhead.

The tender grass was then the softest bed,
  The pleasant'st shades were deemed the stateliest halls;
No belly-god with Bacchus banqueted,
  Nor painted rags then covered rotten walls. 80

Then simple love with simple virtue weighed,
  Flowers the favours which true faith revealed;
Kindness with kindness was again repaid,
  With sweetest kisses covenants were sealed.

Then beauty's self with herself beautified,
  Scorned painting's parget and the borrowed hair,
Nor monstrous forms deformities did hide,
  Nor foul was varnished with compounded fair.

The purest fleece then covered purest skin,
  For pride as then with Lucifer remained; 90
Deformed fashions now were to begin,
  Nor clothes were yet with poisoned liquor stained.

But when the bowels of the earth were sought,
  And men her golden entrails did espy,
This mischief then into the world was brought,
  This framed the mint which coined our misery.

63 that spell-god's] i.e. Apollo's
66 rede] sayings, lore
conned] studied, learned
86 parget] plaster

Then lofty pines were by ambition hewn,
  And men sea-monsters swam the brackish flood
In wainscot tubs to seek out worlds unknown,
  For certain ill to leave assured good. 100

The startling steed is managed from the field,
  And serves a subject to the rider's laws;
He whom the churlish bit did never wield
  Now feels the curb control his angry jaws.

The hammering Vulcan spent his wasting fire,
  Till he the use of tempering metals found;
His anvil wrought the steeled coat's attire,
  And forged tools to carve the foe-man's wound.

The city-builder then entrenched his towers,
  And walled his wealth within the fenced town, 110
Which afterward in bloody stormy stours
  Kindled that flame which burnt his bulwarks down.

And thus began th'exordium of our woes,
  The fatal dumb-show of our misery;
Here sprang the tree on which our mischief grows,
  The dreary subject of world's tragedy.

Motto

Well, shepherd, well, the golden age is gone,
  Wishes may not revoke that which is past.
It were no wit to make two griefs of one;
  Our proverb saith, nothing can always last. 120

Listen to me, my lovely shepherd's joy,
  And thou shalt hear with mirth and mickle glee
A pretty tale which, when I was a boy,
  My toothless grandame oft hath told to me.

Gorbo

Shepherd, say on, so we may pass the time:
There is no doubt it is some worthy rhyme.

Motto

  Far in the country of Arden,
  There wonned a knight hight Cassamen,
    As bold as Isenbras:
  Fell was he and eager bent, 130
  In battle and in tournament,
    As was the good sir Topas.

111 stours] fights   hight] called   130 Fell] fierce
128 wonned] lived

He had, as antique stories tell,
A daughter cleped Dowsabell,
A maiden fair and free;
And for she was her father's heir,
Full well she was yconned the leir
Of mickle courtesy.

The silk well couth she twist and twine,
And make the fine marchpine, 140
And with the needle work;
And she couth help the priest to say
His matin on a holyday,
And sing a psalm in kirk.

She ware a frock of frolic green,
Might well become a maiden queen,
Which seemly was to see;
A hood to that so neat and fine,
In colour like the columbine,
Ywrought full featously. 150

Her features all as fresh above
As is the grass that grows by Dove,
And lithe as lass of Kent;
Her skin as soft as Lemster wool,
As white as snow on Peakish hull,
Or swan that swims in Trent.

This maiden, in a morn betime,
Went forth when May was in her prime,
To get sweet setywall,
The honey suckle, the harlock, 160
The lily and the lady-smock,
To deck her summer hall.

Thus as she wandered here and there,
Ypicking of the bloomed brere,
She chanced to espy
A shepherd sitting on a bank;
Like Chanticleer he crowed crank,
And piped with merry glee.

134 cleped] called
137 well . . . yconned] knew well
leir] learning, lore
139 couth] could
140 marchpine] marzipan
150 featously] elegantly
155 hull] hill
159 setywall] valerian
160 harlock] unidentified flower
167 crank] cheerfully, lustily

He leered his sheep, as he him list,
When he would whistle in his fist, 170
   To feed about him round;
Whilst he full many a carol sung,
Until the fields and meadows rung
   And that the woods did sound.

In favour this same shepherd swain
Was like the bedlam Tamburlaine,
   Which held proud kings in awe;
But meek he was as lamb mought be,
Ylike that gentle Abel he,
   Whom his lewd brother slaw. 180

This shepherd ware a sheep-gray cloak,
Which was of the finest lock
   That could be cut with shear.
His mittens were of bauzon's skin,
His cockers were of cordiwin,
   His hood of miniver.

His awl and lingel in a thong,
His tar-box on his broad belt hung,
   His breech of Cointree blue;
Full crisp and curled were his locks, 190
His brows as white as Albion rocks,
   So like a lover true.

And piping still he spent the day,
So merry as the popinjay;
   Which liked Dowsabell;
That would she aught, or would she nought,
This lad would never from her thought;
   She in love-longing fell.

At length she tucked up her frock;
White as a lily was her smock; 200
   She drew the shepherd nigh:
But then the shepherd piped a-good,
That all his sheep forsook their food,
   To hear his melody.

169 leered] taught
175 favour] appearance, look
176 bedlam] mad
180 lewd] bad, worthless
184 bauzon's] badger's
185 cockers] boots
   cordiwin] leather
186 miniver] white fur
187 lingel] thread
189 Cointree] Coventry
194 popinjay] parrot

'Thy sheep', quoth she, 'cannot be lean,
That have a jolly shepherd's swain,
The which can pipe so well'.
'Yea, but', saith he, 'their shepherd may,
If piping thus he pine away,
In love of Dowsabell'. 210

'Of love, fond boy, take thou no keep',
Quoth she, 'look well unto thy sheep,
Lest they should hap to stray.'
Quoth he, 'So had I done full well,
Had I not seen fair Dowsabell
Come forth to gather May.'

With that she gan to vail her head,
Her cheeks were like the roses red,
But not a word she said.
With that the shepherd gan to frown, 220
He threw his pretty pipes adown,
And on the ground him laid.

Saith she, 'I may not stay till night,
And leave my summer hall undight,
And all for love of thee.'
'My cote', saith he, 'nor yet my fold,
Shall neither sheep nor shepherd hold,
Except thou favour me.'

Saith she, 'Yet liever I were dead
Than I should lose my maidenhead, 230
And all for love of men.'
Saith he, 'Yet are you too unkind,
If in your heart you cannot find
To love us now and then.

'And I to thee will be as kind
As Colin was to Rosalinde,
Of courtesy the flower.'
'Then will I be as true', quoth she,
'As ever maiden yet might be
Unto her paramour.' 240

With that she bent her snow-white knee,
Down by the shepherd kneeled she,

211 keep] care 217 vail] lower

And him she sweetly kissed.
With that the shepherd whooped for joy:
Quoth he, 'There's never shepherd boy
That ever was so blest.'

Gorbo

Now by my sheep-hook here's a tale alone.
Learn me the same and I will give thee hire.
This were as good as curds for our Joan,
When at a night we sitten by the fire. 250

Motto

Why, gentle hodge, I will not stick for that,
When we two meeten here another day.
But see, whilst we have set us down to chat,
Yon tikes of mine begin to steal away.

And if thou wilt but come unto our green
On Lammas day when as we have our feast,
Thou shalt sit next unto our summer Queen,
And thou shalt be the only welcome guest.

(1593)

from *Idea's Mirror*

331 THE golden sun upon his fiery wheels
The horned Ram doth in his course awake,
And of just length our night and day doth make,
Flinging the Fishes backward with his heels:
Then to the tropic takes his full career,
Trotting his sun-steeds till the palfreys sweat,
Baiting the Lion in his furious heat
Till Virgin's smiles do sound his sweet reteere.
But my fair Planet, who directs me still,
Unkindly such distemperature doth bring, 10
Makes summer winter, autumn in the spring,
Crossing sweet nature by unruly will.
Such is the sun who guides my youthful season,
Whose thwarting course deprives the world of
reason.

(1594)

254 tikes] curs, mongrels

331
5 career] short gallop at full speed
8 sound . . . reteere] sound a retreat (in battle)
9 directs] controls, governs the actions of
10 Unkindly] (1) ungraciously; (2) unnaturally
distemperature] atmospheric disorder or extremity

from *Idea*

332

LOVE, in a humour, played the prodigal,
  And bade my senses to a solemn feast;
Yet more to grace the company withal,
  Invites my heart to be the chiefest guest.
No other drink would serve this glutton's turn
  But precious tears distilling from mine eyne,
Which with my sighs this epicure doth burn,
  Quaffing carouses in this costly wine;
Where, in his cups o'ercome with foul excess,
  Straightways he plays a swagg'ring ruffin's part, 10
And at the banquet, in his drunkenness,
  Slew his dear friend, my kind and truest heart.
    A gentle warning, friends, thus may you see,
    What 'tis to keep a drunkard company.

(1599)

333

AS other men, so I myself do muse
  Why in this sort I wrest invention so,
And why these giddy metaphors I use,
  Leaving the path the greater part do go.
I will resolve you: I am lunatic;
  And ever this in madmen you shall find,
What they last thought of, when the brain grew sick,
  In most distraction they keep that in mind:
Thus talking idly in this bedlam fit,
  Reason and I (you must conceive) are twain. 10
'Tis nine years now since first I lost my wit;
  Bear with me then, though troubled be my brain.
    With diet and correction, men distraught
    (Not too far past) may to their wits be brought.

(1600)

334

AN evil spirit, your beauty, haunts me still,
  Wherewith (alas) I have been long possessed,
Which ceaseth not to tempt me to each ill,
  Nor gives me once but one poor minute's rest.

1 humour] fancy, caprice, whim
6 eyne] eyes
8 carouses] toasts
10 ruffin] ruffian, violent bully

333
10 twain] at variance, estranged

In me it speaks, whether I sleep or wake,
  And when by means to drive it out I try,
With greater torments then it me doth take,
  And tortures me in most extremity.
Before my face it lays down my despairs,
  And hastes me on unto a sudden death, 10
Now tempting me to drown myself in tears,
  And then in sighing to give up my breath.
    Thus am I still provoked to every evil
    By this good wicked spirit, sweet angel devil.

(1599)

335

As Love and I, late harboured in one inn,
  With proverbs thus each other entertain,
'In love there is no lack' thus I begin;
  'Fair words make fools', replieth he again.
'Who spares to speak doth spare to speed', quoth I;
  'As well', say'th he, 'too forward as too slow'.
'Fortune assists the boldest', I reply;
  'A hasty man', quoth he, 'ne'er wanted woe'.
'Labour is light, where love', quoth I, 'doth pay'.
  Say'th he, 'Light burden's heavy, if far borne'. 10
Quoth I, 'The main lost, cast the by away'.
  'You have spun a fair thread', he replies in scorn.
    And having thus awhile each other thwarted,
    Fools as we met, so fools again we parted.

(1600)

336

Truce, gentle Love, a parley now I crave:
  Methinks 'tis long since first these wars begun.
Nor thou, nor I, the better yet can have;
  Bad is the match where neither party won.
I offer free conditions of fair peace:
  My heart for hostage that it shall remain.
Discharge our forces, here let malice cease,
  So for my pledge thou give me pledge again.
Or if no thing but death will serve thy turn,
  Still thirsting for subversion of my state, 10
Do what thou canst, raze, massacre and burn;
  Let the world see the utmost of thy hate.
    I send defiance, since if overthrown,
    Thou vanquishing, the conquest is mine own.

(1599)

335
5 speed] succeed
11 main] in games of hazard, the number called by the 'caster' and then thrown for
by] lesser stake in hazard

337 from *England's Heroical Epistles*

(i)

Queen Katherine to Owen Tudor

JUDGE not a Princess' worth impeached hereby,
That love thus triumphs over majesty;
Nor think less virtue in this royal hand,
That it entreats, and wonted to command:
For in this sort, though humbly now it woo,
The day hath been thou wouldst have kneeled unto.
Nor think that this submission of my state
Proceeds from frailty (rather judge it fate).
Alcides ne'er more fit for war's stern shock
Than when with women spinning at the rock; 10
Never less clouds did Phoebus' glory dim
Than in a clown's shape when he covered him;
Jove's great command was never more obeyed
Than when a satyr's antic parts he played.
He was thy King who sued for love to me,
And she his Queen who sues for love to thee.
When Henry was, my love was only his,
But by his death, it Owen Tudor's is.
My love to Owen, him my Henry giveth;
My love to Henry, in my Owen liveth. 20
Henry wooed me, whilst wars did yet increase;
I woo my Tudor in sweet calms of peace.
To force affection, he did conquest prove;
I come with gentle arguments of love.
Encamped at Melans, in war's hot alarms,
First saw I Henry, clad in princely arms;
At pleasant Windsor first these eyes of mine
My Tudor judged, for wit and shape divine;
Henry abroad, with puissance and with force,
Tudor at home, with courtship and discourse: 30
He then, thou now, I hardly can judge whether
Did like me best, Plantagenet or Tether;
A march, a measure, battle or a dance,
A courtly rapier, or a conquering lance.

Queen Katherine] widow of King Henry V
1 impeached] injured, impaired
4 wonted] was accustomed
10 rock] distaff
14 antic] grotesque, ludicrous
32 Tether] Tudor
33 measure] stately dance

His princely bed hath strengthened my renown,
And on my temples set a double crown;
Which glorious wreath (as Henry's lawful heir)
Henry the Sixth upon his brow doth bear.
At Troy in Champagne he did first enjoy
My bridal rites, to England brought from Troy; 40
In England now that honour thou shalt have,
Which once in Champagne famous Henry gave.
  I seek not wealth, three kingdoms in my power;
If these suffice not, where shall be my dower?
Sad discontent may ever follow her
Which doth base pelf before true love prefer.
If titles still could our affections tie,
What is so great but majesty might buy?
  As I seek thee, so Kings do me desire:
To what they would, thou eas'ly mayst aspire. 50
That sacred fire once warmed my heart before;
The fuel fit, the flame is now the more.
And means to quench it, I in vain do prove:
We may hide treasure, but not hide our love.
And since it is thy fortune (thus) to gain it,
It were too late, nor will I now restrain it.
Nor these great titles vainly will I bring,
Wife, daughter, mother, sister to a King,
Of grandsire, father, husband, son, and brother,
More thou alone to me than all these other. 60
Nor fear, my Tudor, that this love of mine
Should wrong the Gaunt-born, great Lancastrian line,
Or make the English blood, the sun and moon,
Repine at Lorraine, Bourbon, Alençon;
Nor do I think there is such different odds,
They should alone be numbered with the gods:
Of Cadmus' earthly issue reck'ning us,
And they from Jove, Mars, Neptune, Æolus;
Of great Latona's offspring only they,
And we the brats of woeful Niobe. 70
Our famous grandsires (as their own) bestrid
That horse of fame, that god-begotten steed,
Whose bounding hoof ploughed that Boeotian spring,
Where those sweet maids of memory do sing.
I claim not all from Henry, but as well
To be the child of Charles and Isabel:

39 Troy] Troyes
46 pelf] money; 'filthy lucre'
50 aspire] attain, reach
53 prove] experience
72 That horse] Pegasus
73 that Boeotian spring] Hippocrene (sacred to the Muses)

Nor can I think from whence their grief should grow
That by this match they be disparaged so;
When John and Longshanks' issue were affied
And to the Kings of Wales in wedlock tied, 80
Showing the greatness of your blood thereby,
Your race and royal consanguinity:
And Wales as well as haughty England boasts
Of Camelot and all her Pentecosts,
To have precedence in Pendragon's race,
At Arthur's table challenging the place.
  If by the often conquest of your land
They boast the spoils of their victorious hand,
If these our ancient chronicles be true,
They altogether are not free from you. 90
When bloody Rufus sought your utter sack,
Twice ent'ring Wales, yet twice was beaten back;
When famous Cambria washed her in the flood
Made by th'effusion of the English blood,
And oft returned with glorious victory,
From Worcester, Her'ford, Chester, Shrewsbury;
Whose power in every conquest so prevails
As once expulsed the English out of Wales.
  Although my beauty made my country's peace,
And at my bridal former broils did cease, 100
More than his power had not his person been,
I had not come to England as a Queen.
Nor took I Henry to supply my want,
Because in France that time my choice was scant,
When it had robbed all Christendom of men,
And England's flower remained amongst us then:
Gloucester, whose counsels (Nestor-like) assist;
Courageous Bedford, that great martialist;
Clarence, for virtue honoured of his foes;
And York, whose fame yet daily greater grows; 110
Warwick, the pride of Neville's haughty race;
Great Salisbury, so feared in every place;
That valiant Pole, whom no achievement dares;
And Vere, so famous in the Irish wars;
Who, though my self so great a prince were born,
The worst of these, my equal need not scorn.
But Henry's rare perfections, and his parts,
As conqu'ring kingdoms, so he conquered hearts.

78 disparaged] matched unequally in marriage, dishonoured
79 Longshanks] King Edward I
affied] espoused, betrothed
84 Pentecosts] Whitsun festivals
86 challenging] claiming
93 Cambria] Wales, 'Cymru'
113 dares] daunts

As chaste was I to him as Queen might be,
But freed from him, my chaste love vowed to thee. 120
Beauty doth fetch all favour from thy face;
All perfect courtship resteth in thy grace.
If thou discourse, thy lips such accents break
As Love a spirit forth of thee seems to speak.
The British language, which our vowels wants,
And jars so much upon harsh consonants,
Comes with such grace from thy mellifluous tongue
As do the sweet notes of a well-set song,
And runs as smoothly from those lips of thine
As the pure Tuscan from the Florentine; 130
Leaving such seasoned sweetness in the ear
That, the voice past, yet still the sound is there:
In Nisus' tower, as when Apollo lay,
And on his golden viol used to play,
Where senseless stones were with such music drowned
As many years they did retain the sound.
Let not the beams that greatness doth reflect
Amaze thy hopes with timorous respect.
Assure thee, Tudor, majesty can be
As kind in love as can the mean'st degree, 140
And the embraces of a Queen as true
As theirs, which think them much advanced by you;
When in our greatness our affections crave
Those secret joys that other women have.
So I (a Queen) be sovereign in my choice,
Let others fawn upon the public voice;
Or what (by this) can ever hap to thee,
Light in respect, to be beloved of me?
Let peevish worldlings prate of right and wrong;
Leave plaints and pleas to whom they do belong; 150
Let old men speak of chances and events,
And lawyers talk of titles and descents;
Leave fond reports to such as stories tell,
And covenants to those that buy and sell:
Love, my sweet Tudor, that becomes thee best;
And to our good success refer the rest.

125 British] Welsh
137 greatness] exalted rank
138 Amaze] bewilder, stupefy
140 degree] social rank
142 advanced] raised
148 in respect] in comparison
156 success] (1) outcome; (2) (dynastic) succession

(ii)

Owen Tudor to Queen Katherine

When first mine eyes beheld your princely name,
And found from whence this friendly letter came,
As in excess of joy I had forgot
Whether I saw it or I saw it not: 160
My panting heart doth bid mine eyes proceed,
My dazzled eyes invite my tongue to read,
Which wanting their direction dully missed it.
My lips, which should have spoke, were dumb, and kissed it,
And left the paper in my trembling hand,
When all my senses did amazed stand;
Even as a mother coming to her child,
Which from her presence hath been long exiled,
With gentle arms his tender neck doth strain,
Now kissing it, now clipping it again; 170
And yet excessive joy deludes her so,
As still she doubts if this be hers or no.
At length awakened from this pleasing dream,
When passion somewhat left to be extreme,
My longing eyes with their fair object meet,
Where every letter's pleasing, each word sweet.
It was not Henry's conquests, nor his court,
That had the power to win me by report;
Nor was his dreadful terror-striking name
The cause that I from Wales to England came; 180
For Christian Rhodes, and our religion's truth,
To great achievement first had won my youth:
This brave adventure did my valour prove
Before I e'er knew what it was to love.
Nor came I hither by some poor event,
But by th'eternal Destinies' consent;
Whose uncomprised wisdom did foresee
That you in marriage should be linked to me.
By our great Merlin was it not foretold
(Amongst his holy prophecies enrolled) 190
When first he did of Tudor's name divine,
That Kings and Queens should follow in our line?
And that the helm (the Tudors' ancient crest)
Should with the golden fleur-de-luce be dressed;

170 clipping] embracing
185 event] chance occurrence
187 uncomprised] uncomprehended
194 dressed] placed

As that the leek (our country's chief renown)
Should grow with roses in the English crown?
As Charles's daughter, you the lily wear,
As Henry's Queen, the blushing rose you bear.
By France's conquest and by England's oath,
You are the true-made dowager of both; 200
Both in your crown, both in your cheek together,
Join Tether's love to yours, and yours to Tether.
  Then cast no future doubts, nor fear no hate,
When it so long hath been foretold by fate;
And by the all-disposing doom of heaven,
Before our births, we to one bed were given.
No Pallas here, nor Juno is at all,
When I to Venus yield the golden ball;
Nor when the Grecians' wonder I enjoy,
None in revenge to kindle fire in Troy. 210
  And have not strange events divined to us
That in our love we should be prosperous?
When in your presence I was called to dance,
In lofty tricks whilst I myself advance,
And in a turn my footing failed by hap,
Was't not my chance to light into your lap?
Who would not judge it fortune's greatest grace,
Sith he must fall, to fall in such a place?
  His birth from heaven your Tudor not derives,
Nor stands on tiptoe in superlatives, 220
Although the envious English do devise
A thousand jests of our hyperboles;
Nor do I claim that plot by ancient deeds,
Where Phoebus pastures his fire-breathing steeds;
Nor do I boast my god-made grandsire's scars,
Nor giants' trophies in the Titans' wars;
Nor feign my birth (your princely ears to please)
By three nights' getting, as was Hercules;
Nor do I forge my long descent to run
From aged Neptune, or the glorious sun. 230
And yet in Wales, with them that famous be,
Our learned bards do sing my pedigree;
And boast my birth from great Cadwallader,
From old Caer-Septon, in Mount Palador;
And from Eneon's line, the South-Wales King,
By Theodore, the Tudors' name do bring.

203 future doubts] fears for the future
214 advance] raise
218 Sith] since
221 envious] malicious
234 Caer-Septon] 'now called Shaftesbury' (Drayton)

My royal mother's princely stock began
From her great-grandam, fair Gwenellian,
By true descent from Leolin the Great,
As well from North-Wales, as fair Powsland's seat: 240
Though for our princely genealogy
I do not stand to make apology;
Yet who with judgment's true impartial eyes
Shall look from whence our name at first did rise,
Shall find that fortune is to us in debt;
And why not Tudor, as Plantagenet?
Nor that term Croggen, nick-name of disgrace,
Used as a by-word now in every place,
Shall blot our blood, or wrong a Welshman's name,
Which was at first begot with England's shame. 250
Our valiant swords our right did still maintain,
Against that cruel, proud, usurping Dane,
Buckling besides in many dang'rous fights,
With Norways, Swethens, and with Muscovites;
And kept our native language now thus long,
And to this day yet never changed our tongue,
When they which now our nation fain would tame,
Subdued, have lost their country and their name.
Nor ever could the Saxons' swords provoke
Our Briton necks to bear their servile yoke, 260
Where Cambria's pleasant countries bounded be
With swelling Severn and the holy Dee;
And since great Brutus first arrived, have stood,
The only remnant of the Trojan blood.
To every man is not allotted chance,
To boast with Henry to have conquered France;
Yet if my fortunes be thus raised by thee,
This may presage a further good to me,
And our Saint David, in the Britons' right,
May join with George, the sainted English knight; 270
And old Carmarthen, Merlin's famous town,
Not scorned by London, though of such renown.
Ah, would to God, that hour my hopes attend
Were with my wish brought to desired end!
Blame me not, Madam, though I thus desire:
Many there be that after you enquire.
Till now your beauty in night's bosom slept:
What eye durst stir, where awful Henry kept?

239 Leolin] Llewelyn
240 Powsland] Powys (in mid-Wales)
247 Croggen] (see note)
253 Buckling] engaging, grappling
254 Norways . . . Muscovites] Norwegians, Swedes, and Russians
261 countries] regions, lands

Who durst attempt to sail but near the bay
Where that all-conqu'ring great Alcides lay? 280
Your beauty now is set a royal prize,
And Kings repair to cheapen merchandise.
If you but walk to take the breathing air,
Orithia makes me that I Boreas fear;
If to the fire, Jove once in lightning came,
And fair Aegina makes me fear the flame;
If in the sun, then sad suspicion dreams
Phoebus should spread Leucothoe in his beams;
If in a fountain you do cool your blood,
Neptune I fear, which once came in a flood; 290
If with your maids, I dread Apollo's rape,
Who cozened Chion in an old wive's shape;
If you do banquet, Bacchus makes me dread,
Who in a grape Erigone did feed;
And if myself your chamber-door should keep,
Yet fear I Hermes, coming in a sleep.
Pardon (sweet Queen) if I offend in this,
In these delays Love most impatient is;
And youth wants power his hot spleen to suppress,
When hope already banquets in excess. 300
Though Henry's fame in me you shall not find,
Yet that which better shall content your mind.
But only in the title of a King
Was his advantage, in no other thing.
If in his love more pleasure you did take,
Never let Queen trust Briton, for my sake.
Yet judge me not from modesty exempt
That I another Phaeton's charge attempt;
My mind, that thus your favours dare aspire,
Shows that 'tis touched with a celestial fire. 310
If I do fault, the more is beauty's blame,
When she herself is author of the same.
All men to some one quality incline:
Only to love is naturally mine.
Thou art by beauty famous, as by birth,
Ordained by heaven to cheer the drooping earth.
Add faithful love unto your greater state,
And be alike in all things fortunate.
A King might promise more, I not deny,
But yet (by heaven) he loved not more than I. 320

282 cheapen] bid for, offer a price for
306 Briton] Welshman
299 spleen] impetuosity, eagerness

And thus I leave, till time my faith approve.
I cease to write, but never cease to love.

(Pub. 1597; corrected in edns. 1600–37)

ANONYMOUS

338 [*The Ruins of Walsingham*]

In the wracks of Walsingham
  Whom should I choose,
But the Queen of Walsingham
  To be guide to my muse?
Then thou, Prince of Walsingham,
  Grant me to frame
Bitter plaints to rue thy wrong,
  Bitter woe for thy name.
Bitter was it so to see
  The seely sheep 10
Murdered by the ravening wolves
  While the shepherds did sleep.
Bitter was it, O, to view
  The sacred vine
(While gardeners played all close)
  Rooted up by the swine.
Bitter, bitter, O, to behold
  The grass to grow
Where the walls of Walsingham
  So stately did show. 20
Such were the works of Walsingham,
  While she did stand;
Such are the wracks as now do show
  Of that holy land.
Level, level with the ground
  The towers do lie,
Which with their golden glittering tops
  Pierced once to the sky.
Where were gates no gates are now,
  The ways unknown 30
Where the press of peers did pass
  While her fame far was blown.

321 approve] make good, show to be true 10 seely] simple
338
1 wracks] ruins

Owls do shriek where the sweetest hymns
  Lately were sung;
Toads and serpents hold their dens
  Where the palmers did throng.
Weep, weep, O Walsingham,
  Whose days are nights,
Blessings turned to blasphemies,
  Holy deeds to despites. 40
Sin is where Our Lady sat,
  Heaven turned is to hell.
Satan sits where Our Lord did sway:
  Walsingham, O, farewell.

(Pub. 1868)

WILLIAM SHAKESPEARE
1564–1616

from *Venus and Adonis*

339 [*The Death of Adonis*]

Lo, here the gentle lark, weary of rest,
From his moist cabinet mounts up on high,
And wakes the morning, from whose silver breast
The sun ariseth in his majesty;
  Who doth the world so gloriously behold
  That cedar-tops and hills seem burnished gold.

Venus salutes him with this fair good-morrow:
'O thou clear god, and patron of all light,
From whom each lamp and shining star doth borrow
The beauteous influence that makes him bright, 10
  There lives a son that sucked an earthly mother
  May lend thee light, as thou dost lend to other.'

36 palmers] pilgrims
43 sway] rule

339
2 moist cabinet] dewy lodging
8 clear] bright
12 other] others

This said, she hasteth to a myrtle grove,
Musing the morning is so much o'erworn,
And yet she hears no tidings of her love;
She hearkens for his hounds and for his horn.
  Anon she hears them chant it lustily,
  And all in haste she coasteth to the cry.

And as she runs, the bushes in the way
Some catch her by the neck, some kiss her face, 20
Some twine about her thigh to make her stay;
She wildly breaketh from their strict embrace,
  Like a milch doe, whose swelling dugs do ache,
  Hasting to feed her fawn hid in some brake.

By this she hears the hounds are at a bay;
Whereat she starts, like one that spies an adder
Wreathed up in fatal folds just in his way,
The fear whereof doth make him shake and shudder;
  Even so the timorous yelping of the hounds
  Appals her senses and her spirit confounds. 30

For now she knows it is no gentle chase,
But the blunt boar, rough bear, or lion proud,
Because the cry remaineth in one place,
Where fearfully the dogs exclaim aloud.
  Finding their enemy to be so curst,
  They all strain court'sy who shall cope him first.

This dismal cry rings sadly in her ear,
Through which it enters to surprise her heart;
Who, overcome by doubt and bloodless fear,
With cold-pale weakness numbs each feeling part; 40
  Like soldiers, when their captain once doth yield,
  They basely fly and dare not stay the field.

Thus stands she in a trembling ecstasy;
Till, cheering up her senses all dismayed,
She tells them 'tis a causeless fantasy,
And childish error, that they are afraid;
  Bids them leave quaking, bids them fear no more;
  And with that word she spied the hunted boar,

18 coasteth] approaches
22 strict] close, tight
23 milch] giving milk
24 brake] thicket
25 bay] the point when the hunted animal is driven to turn upon its hunters
43 ecstasy] frenzy

Whose frothy mouth, bepainted all with red,
Like milk and blood being mingled both together, 50
A second fear through all her sinews spread,
Which madly hurries her she knows not whither:
  This way she runs, and now she will no further,
  But back retires to rate the boar for murther.

A thousand spleens bear her a thousand ways;
She treads the path that she untreads again;
Her more than haste is mated with delays,
Like the proceedings of a drunken brain,
  Full of respects, yet nought at all respecting,
  In hand with all things, nought at all effecting. 60

Here kennelled in a brake she finds a hound,
And asks the weary caitiff for his master;
And there another licking of his wound,
'Gainst venomed sores the only sovereign plaster;
  And here she meets another sadly scowling,
  To whom she speaks, and he replies with howling.

When he hath ceased his ill-resounding noise,
Another flap-mouthed mourner, black and grim,
Against the welkin volleys out his voice;
Another and another answer him, 70
  Clapping their proud tails to the ground below,
  Shaking their scratched ears, bleeding as they go.

Look how the world's poor people are amazed
At apparitions, signs and prodigies,
Whereon with fearful eyes they long have gazed,
Infusing them with dreadful prophecies;
  So she at these sad signs draws up her breath,
  And, sighing it again, exclaims on Death.

'Hard-favoured tyrant, ugly, meagre, lean,
Hateful divorce of love'—thus chides she Death— 80
'Grim-grinning ghost, earth's worm, what dost thou mean
To stifle beauty and to steal his breath
  Who when he lived, his breath and beauty set
  Gloss on the rose, smell to the violet?

54 rate] scold
55 spleens] sudden impulses
57 mated] checked
59 respects] consideration
62 caitiff] wretch
69 welkin] sky
73 Look how] just as
74 prodigies] portents
79 Hard-favoured] with an ugly face

'If he be dead—O no, it cannot be,
Seeing his beauty, thou shouldst strike at it—
O yes, it may; thou hast no eyes to see,
But hatefully at random dost thou hit.
  Thy mark is feeble age; but thy false dart
  Mistakes that aim, and cleaves an infant's heart. 90

'Hadst thou but bid beware, then he had spoke,
And, hearing him, thy power had lost his power.
The Destinies will curse thee for this stroke;
They bid thee crop a weed, thou pluck'st a flower.
  Love's golden arrow at him should have fled,
  And not Death's ebon dart, to strike him dead.

'Dost thou drink tears, that thou provok'st such weeping?
What may a heavy groan advantage thee?
Why hast thou cast into eternal sleeping
Those eyes that taught all other eyes to see? 100
  Now Nature cares not for thy mortal vigour,
  Since her best work is ruined with thy rigour.'

Here overcome as one full of despair,
She vailed her eyelids, who, like sluices, stopped
The crystal tide that from her two cheeks fair
In the sweet channel of her bosom dropped;
  But through the flood-gates breaks the silver rain,
  And with his strong course opens them again.

O, how her eyes and tears did lend and borrow!
Her eye seen in the tears, tears in her eye; 110
Both crystals, where they viewed each other's sorrow,
Sorrow that friendly sighs sought still to dry;
  But like a stormy day, now wind, now rain,
  Sighs dry her cheeks, tears make them wet again.

Variable passions throng her constant woe,
As striving who should best become her grief;
All entertained, each passion labours so
That every present sorrow seemeth chief,
  But none is best. Then join they all together,
  Like many clouds consulting for foul weather. 120

104 vailed] lowered
sluices] floodgates
111 crystals] magic crystals, in which someone in sympathy with another could see the scene of his distress
120 consulting] plotting, conspiring

By this, far off she hears some huntsman holla;
A nurse's song ne'er pleased her babe so well.
The dire imagination she did follow
This sound of hope doth labour to expel;
  For now reviving joy bids her rejoice,
  And flatters her it is Adonis' voice.

Whereat her tears began to turn their tide,
Being prisoned in her eye like pearls in glass;
Yet sometimes falls an orient drop beside,
Which her cheek melts, as scorning it should pass 130
  To wash the foul face of the sluttish ground,
  Who is but drunken when she seemeth drowned.

O hard-believing love, how strange it seems
Not to believe, and yet too credulous!
Thy weal and woe are both of them extremes;
Despair, and hope, makes thee ridiculous:
  The one doth flatter thee in thoughts unlikely,
  In likely thoughts the other kills thee quickly.

Now she unweaves the web that she hath wrought;
Adonis lives, and Death is not to blame; 140
It was not she that called him all to nought.
Now she adds honours to his hateful name;
  She clepes him king of graves, and grave for kings,
  Imperious supreme of all mortal things.

'No, no,' quoth she, 'sweet Death, I did but jest;
Yet pardon me, I felt a kind of fear
When as I met the boar, that bloody beast,
Which knows no pity, but is still severe.
  Then, gentle shadow—truth I must confess—
  I railed on thee, fearing my love's decease. 150

' 'Tis not my fault: the boar provoked my tongue;
Be wreaked on him, invisible commander;
'Tis he, foul creature, that hath done thee wrong;
I did but act, he's author of thy slander.
  Grief hath two tongues, and never woman yet
  Could rule them both without ten women's wit.'

127 turn their tide] ebb
129 orient] bright
133 hard-believing] believing with difficulty
141 all to nought] vile
143 clepes] names
152 wreaked] revenged

Thus, hoping that Adonis is alive,
Her rash suspect she doth extenuate;
And that his beauty may the better thrive,
With Death she humbly doth insinuate; 160
Tells him of trophies, statues, tombs, and stories
His victories, his triumphs and his glories.

'O Jove,' quoth she, 'how much a fool was I
To be of such a weak and silly mind
To wail his death who lives and must not die
Till mutual overthrow of mortal kind!
For he being dead, with him is Beauty slain,
And, Beauty dead, black Chaos comes again.

'Fie, fie, fond love, thou art as full of fear
As one with treasure laden, hemmed with thieves; 170
Trifles unwitnessed with eye or ear
Thy coward heart with false bethinking grieves.'
Even at this word she hears a merry horn,
Whereat she leaps that was but late forlorn.

As falcons to the lure, away she flies;
The grass stoops not, she treads on it so light;
And in her haste unfortunately spies
The foul boar's conquest on her fair delight;
Which seen, her eyes, as murd'red with the view,
Like stars ashamed of day, themselves withdrew; 180

Or as the snail, whose tender horns being hit,
Shrinks backward in his shelly cave with pain,
And there all smoth'red up in shade doth sit,
Long after fearing to creep forth again;
So at his bloody view her eyes are fled
Into the deep-dark cabins of her head;

Where they resign their office and their light
To the disposing of her troubled brain;
Who bids them still consort with ugly night,
And never wound the heart with looks again; 190
Who, like a king perplexed in his throne,
By their suggestion gives a deadly groan,

158 suspect] suspicion
160 insinuate] ingratiate herself
161 stories] relates
166 mutual] common
174 leaps] jumps for joy
180 ashamed of day] put to shame at the approach of day
191 perplexed] bewildered, deeply troubled

Whereat each tributary subject quakes;
As when the wind, imprisoned in the ground,
Struggling for passage, earth's foundation shakes,
Which with cold terror doth men's minds confound.
  This mutiny each part doth so surprise,
  That from their dark beds once more leap her eyes;

And being opened, threw unwilling light
Upon the wide wound that the boar had trenched 200
In his soft flank; whose wonted lily white
With purple tears that his wound wept was drenched:
  No flower was nigh, no grass, herb, leaf or weed,
  But stole his blood and seemed with him to bleed.

This solemn sympathy poor Venus noteth;
Over one shoulder doth she hang her head;
Dumbly she passions, franticly she doteth;
She thinks he could not die, he is not dead.
  Her voice is stopped, her joints forget to bow;
  Her eyes are mad that they have wept till now. 210

Upon his hurt she looks so steadfastly
That her sight dazzling makes the wound seem three;
And then she reprehends her mangling eye
That makes more gashes where no breach should be:
  His face seems twain, each several limb is doubled;
  For oft the eye mistakes, the brain being troubled.

'My tongue cannot express my grief for one,
And yet,' quoth she, 'behold two Adons dead!
My sighs are blown away, my salt tears gone,
Mine eyes are turned to fire, my heart to lead; 220
  Heavy heart's lead, melt at mine eyes' red fire!
  So shall I die by drops of hot desire.

'Alas, poor world, what treasure hast thou lost!
What face remains alive that's worth the viewing?
Whose tongue is music now? what canst thou boast
Of things long since, or any thing ensuing?
  The flowers are sweet, their colours fresh and trim;
  But true sweet beauty lived and died with him.

200 trenched] cut   207 passions] suffers, grieves

'Bonnet nor veil henceforth no creature wear;
Nor sun nor wind will ever strive to kiss you. 230
Having no fair to lose, you need not fear;
The sun doth scorn you, and the wind doth hiss you.
  But when Adonis lived, sun and sharp air
  Lurked like two thieves to rob him of his fair;

'And therefore would he put his bonnet on,
Under whose brim the gaudy sun would peep;
The wind would blow it off, and, being gone,
Play with his locks. Then would Adonis weep;
  And straight, in pity of his tender years,
  They both would strive who first should dry his tears. 240

'To see his face the lion walked along
Behind some hedge, because he would not fear him;
To recreate himself when he hath sung,
The tiger would be tame and gently hear him;
  If he had spoke, the wolf would leave his prey,
  And never fright the silly lamb that day.

'When he beheld his shadow in the brook,
The fishes spread on it their golden gills;
When he was by, the birds such pleasure took
That some would sing, some other in their bills 250
  Would bring him mulberries and ripe-red cherries;
  He fed them with his sight, they him with berries.

'But this foul, grim, and urchin-snouted boar,
Whose downward eye still looketh for a grave,
Ne'er saw the beauteous livery that he wore;
Witness the entertainment that he gave.
  If he did see his face, why then I know
  He thought to kiss him, and hath killed him so.

' 'Tis true, 'tis true; thus was Adonis slain:
He ran upon the boar with his sharp spear, 260
Who did not whet his teeth at him again,
But by a kiss thought to persuade him there;
  And nuzzling in his flank, the loving swine
  Sheathed unaware the tusk in his soft groin.

231 fair] beauty
236 gaudy] bright
242 fear] frighten
247 shadow] reflection
253 urchin-snouted] snouted like a hedgehog, looking downwards
255 livery] dress, i.e. outward appearance
256 entertainment] reception

'Had I been toothed like him, I must confess,
With kissing him I should have killed him first;
But he is dead, and never did he bless
My youth with his; the more am I accurst.'
  With this, she falleth in the place she stood,
  And stains her face with his congealed blood. 270

She looks upon his lips, and they are pale;
She takes him by the hand, and that is cold;
She whispers in his ears a heavy tale,
As if they heard the woeful words she told;
  She lifts the coffer-lids that close his eyes,
  Where, lo, two lamps, burnt out, in darkness lies;

Two glasses, where herself herself beheld
A thousand times, and now no more, reflect;
Their virtue lost wherein they late excelled,
And every beauty robbed of his effect. 280
  'Wonder of time,' quoth she, 'this is my spite,
  That, thou being dead, the day should yet be light.

'Since thou art dead, lo, here I prophesy
Sorrow on love hereafter shall attend;
It shall be waited on with jealousy,
Find sweet beginning but unsavoury end;
  Ne'er settled equally, but high or low,
  That all love's pleasure shall not match his woe.

'It shall be fickle, false and full of fraud;
Bud, and be blasted, in a breathing while; 290
The bottom poison, and the top o'erstrawed
With sweets that shall the truest sight beguile;
  The strongest body shall it make most weak,
  Strike the wise dumb, and teach the fool to speak.

'It shall be sparing, and too full of riot,
Teaching decrepit age to tread the measures;
The staring ruffian shall it keep in quiet,
Pluck down the rich, enrich the poor with treasures;
  It shall be raging-mad, and silly-mild,
  Make the young old, the old become a child. 300

296 tread the measures] dance 297 staring] truculent

'It shall suspect where is no cause of fear;
It shall not fear where it should most mistrust;
It shall be merciful and too severe,
And most deceiving when it seems most just;
  Perverse it shall be where it shows most toward,
  Put fear to valour, courage to the coward.

'It shall be cause of war and dire events,
And set dissension 'twixt the son and sire;
Subject and servile to all discontents,
As dry combustious matter is to fire. 310
  Sith in his prime death doth my love destroy,
  They that love best their loves shall not enjoy.'

(1593)

from *The Rape of Lucrece*

340 [*Before the Rape*]

WHEN at Collatium this false lord arrived,
Well was he welcomed by the Roman dame,
Within whose face beauty and virtue strived
Which of them both should underprop her fame:
When virtue bragged, beauty would blush for shame;
  When beauty boasted blushes, in despite
  Virtue would stain that o'er with silver white.

But beauty, in that white entituled,
From Venus' doves doth challenge that fair field;
Then virtue claims from beauty beauty's red, 10
Which virtue gave the golden age to gild
Their silver cheeks, and called it then their shield;
  Teaching them thus to use it in the fight,
  When shame assailed, the red should fence the white.

This heraldry in Lucrece' face was seen,
Argued by beauty's red and virtue's white;
Of either's colour was the other queen,
Proving from world's minority their right;
Yet their ambition makes them still to fight,
  The sovereignty of either being so great 20
  That oft they interchange each other's seat.

305 toward] docile

340
8 entituled] having a title
9 challenge] claim
field] (1) battle-field; (2) heraldic 'field' (ground or surface)
14 fence] defend

This silent war of lilies and of roses
Which Tarquin viewed in her fair face's field,
In their pure ranks his traitor eye encloses;
Where, lest between them both it should be killed,
The coward captive vanquished doth yield
  To those two armies that would let him go
  Rather than triumph in so false a foe.

Now thinks he that her husband's shallow tongue,
The niggard prodigal that praised her so, 30
In that high task hath done her beauty wrong,
Which far exceeds his barren skill to show;
Therefore that praise which Collatine doth owe
  Enchanted Tarquin answers with surmise,
  In silent wonder of still-gazing eyes.

This earthly saint, adored by this devil,
Little suspecteth the false worshipper;
For unstained thoughts do seldom dream on evil;
Birds never limed no secret bushes fear.
So guiltless she securely gives good cheer 40
  And reverend welcome to her princely guest,
  Whose inward ill no outward harm expressed;

For that he coloured with his high estate,
Hiding base sin in pleats of majesty;
That nothing in him seemed inordinate,
Save sometime too much wonder of his eye,
Which, having all, all could not satisfy;
  But, poorly rich, so wanteth in his store
  That cloyed with much he pineth still for more.

But she, that never coped with stranger eyes, 50
Could pick no meaning from their parling looks,
Nor read the subtle-shining secrecies
Writ in the glassy margents of such books.
She touched no unknown baits, nor feared no hooks;
  Nor could she moralize his wanton sight,
  More than his eyes were opened to the light.

34 answers] repays
surmise] thought, contemplation
39 limed] caught by bird-lime
40 securely] unsuspectingly
41 reverend] reverent
43 coloured] gave a pleasing appearance
estate] rank
44 pleats] folds
50 coped] dealt
stranger] strangers'
51 parling] speaking
53 margents] margins (referring to marginal glosses)
55 moralize] interpret
56 More than] more than that

He stories to her ears her husband's fame,
Won in the fields of fruitful Italy;
And decks with praises Collatine's high name,
Made glorious by his manly chivalry 60
With bruised arms and wreaths of victory.
  Her joy with heaved-up hand she doth express,
  And wordless so greets heaven for his success.

Far from the purpose of his coming thither,
He makes excuses for his being there.
No cloudy show of stormy blust'ring weather
Doth yet in his fair welkin once appear;
Till sable Night, mother of dread and fear,
  Upon the world dim darkness doth display,
  And in her vaulty prison stows the day. 70

For then is Tarquin brought unto his bed,
Intending weariness with heavy sprite;
For after supper long he questioned
With modest Lucrece, and wore out the night.
Now leaden slumber with life's strength doth fight;
  And every one to rest himself betakes,
  Save thieves and cares and troubled minds that wakes.

As one of which doth Tarquin lie revolving
The sundry dangers of his will's obtaining;
Yet ever to obtain his will resolving, 80
Though weak-built hopes persuade him to abstaining;
Despair to gain doth traffic oft for gaining,
  And when great treasure is the meed proposed,
  Though death be adjunct, there's no death supposed.

Those that much covet are with gain so fond
That what they have not, that which they possess,
They scatter and unloose it from their bond,
And so, by hoping more, they have but less;
Or, gaining more, the profit of excess
  Is but to surfeit, and such griefs sustain 90
  That they prove bankrupt in this poor-rich gain.

67 welkin] sky
69 display] spread
72 Intending] pretending
73 questioned] conversed
82 traffic] trade
83 meed] reward
84 adjunct] joined
85 fond] foolish, infatuated

The aim of all is but to nurse the life
With honour, wealth and ease, in waning age;
And in this aim there is such thwarting strife
That one for all or all for one we gage:
As life for honour in fell battle's rage;
   Honour for wealth; and oft that wealth dost cost
   The death of all, and all together lost.

So that in vent'ring ill we leave to be
The things we are for that which we expect; 100
And this ambitious foul infirmity,
In having much, torments us with defect
Of that we have; so then we do neglect
   The thing we have, and, all for want of wit,
   Make something nothing by augmenting it.

Such hazard now must doting Tarquin make,
Pawning his honour to obtain his lust;
And for himself himself he must forsake:
Then where is truth, if there be no self-trust?
When shall he think to find a stranger just 110
   When he himself himself confounds, betrays
   To sland'rous tongues and wretched hateful days?

Now stole upon the time the dead of night,
When heavy sleep had closed up mortal eyes;
No comfortable star did lend his light,
No noise but owls' and wolves' death-boding cries;
Now serves the season that they may surprise
   The silly lambs. Pure thoughts are dead and still,
   While lust and murder wakes to stain and kill.

(1594)

341 [*Lucrece's Death*]

THUS ebbs and flows the current of her sorrow,
And time doth weary time with her complaining.
She looks for night, and then she longs for morrow,
And both she thinks too long with her remaining.
Short time seems long in sorrow's sharp sustaining;
   Though woe be heavy, yet it seldom sleeps,
   And they that watch see time how slow it creeps.

95 gage] pledge, venture
102 defect] insufficiency
104 wit] intelligence
111 confounds] ruins
115 comfortable] comforting
118 silly] defenceless
341
7 watch] stay awake

Which all this time hath overslipped her thought
That she with painted images hath spent,
Being from the feeling of her own grief brought 10
By deep surmise of others' detriment,
Losing her woes in shows of discontent.
  It easeth some, though none it ever cured,
  To think their dolour others have endured.

But now the mindful messenger come back
Brings home his lord and other company;
Who finds his Lucrece clad in mourning black,
And round about her tear-distained eye
Blue circles streamed, like rainbows in the sky.
  These water-galls in her dim element 20
  Foretell new storms to those already spent.

Which when her sad-beholding husband saw,
Amazedly in her sad face he stares:
Her eyes, though sod in tears, looked red and raw,
Her lively colour killed with deadly cares.
He hath no power to ask her how she fares;
  Both stood, like old acquaintance in a trance,
  Met far from home, wond'ring each other's chance.

At last he takes her by the bloodless hand,
And thus begins: 'What uncouth ill event 30
Hath thee befall'n, that thou dost trembling stand?
Sweet love, what spite hath thy fair colour spent?
Why art thou thus attired in discontent?
  Unmask, dear dear, this moody heaviness,
  And tell thy grief, that we may give redress.'

Three times with sighs she gives her sorrow fire
Ere once she can discharge one word of woe;
At length addressed to answer his desire,
She modestly prepares to let them know
Her honour is ta'en prisoner by the foe; 40
  While Collatine and his consorted lords
  With sad attention long to hear her words.

9 painted images] (the tapestry showing the fall of Troy)
11 surmise] contemplation
12 shows] pictures
20 water-galls] secondary rainbows
element] sky
22 sad-beholding] seriously gazing at
23 Amazedly] perplexedly
24 sod] sodden
28 chance] fortune
30 uncouth] unknown, strange
32 spite] vexation
spent] consumed
34 moody] sorrowful
41 consorted] associated
42 sad] serious

And now this pale swan in her wat'ry nest
Begins the sad dirge of her certain ending.
'Few words', quoth she, 'shall fit the trespass best,
Where no excuse can give the fault amending:
In me moe woes than words are now depending;
  And my laments would be drawn out too long,
  To tell them all with one poor tired tongue.

'Then be this all the task it hath to say: 50
Dear husband, in the interest of thy bed
A stranger came, and on that pillow lay
Where thou wast wont to rest thy weary head;
And what wrong else may be imagined
  By foul enforcement might be done to me,
  From that, alas, thy Lucrece is not free.

'For in the dreadful dead of dark midnight,
With shining falchion in my chamber came
A creeping creature with a flaming light,
And softly cried "Awake, thou Roman dame, 60
And entertain my love; else lasting shame
  On thee and thine this night I will inflict,
  If thou my love's desire do contradict.

'"For some hard-favoured groom of thine," quoth he,
"Unless thou yoke thy liking to my will,
I'll murder straight, and then I'll slaughter thee,
And swear I found you where you did fulfil
The loathsome act of lust, and so did kill
  The lechers in their deed: this act will be
  My fame, and thy perpetual infamy." 70

'With this, I did begin to start and cry,
And then against my heart he set his sword,
Swearing, unless I took all patiently,
I should not live to speak another word;
So should my shame still rest upon record,
  And never be forgot in mighty Rome
  Th'adulterate death of Lucrece and her groom.

47 moe] more
depending] impending
51 in the interest of thy bed] into the bed in which you alone have a rightful interest
58 falchion] sword
61 entertain] receive
64 hard-favoured] ugly, coarse-featured
65 yoke] submit
77 adulterate] adulterous

'Mine enemy was strong, my poor self weak,
And far the weaker with so strong a fear.
My bloody judge forbade my tongue to speak; 80
No rightful plea might plead for justice there.
His scarlet lust came evidence to swear
That my poor beauty had purloined his eyes,
And when the judge is robbed, the prisoner dies.

'O, teach me how to make mine own excuse!
Or, at the least, this refuge let me find:
Though my gross blood be stained with this abuse,
Immaculate and spotless is my mind;
That was not forced; that never was inclined
To accessary yieldings, but still pure 90
Doth in her poisoned closet yet endure.'

Lo, here, the hopeless merchant of this loss,
With head declined, and voice dammed up with woe,
With sad-set eyes and wreathed arms across,
From lips new waxen pale begins to blow
The grief away that stops his answer so;
But, wretched as he is, he strives in vain;
What he breathes out his breath drinks up again.

As through an arch the violent roaring tide
Outruns the eye that doth behold his haste, 100
Yet in the eddy boundeth in his pride
Back to the strait that forced him on so fast,
In rage sent out, recalled in rage, being past;
Even so his sighs, his sorrows, make a saw,
To push grief on and back the same grief draw.

Which speechless woe of his poor she attendeth
And his untimely frenzy thus awaketh:
'Dear lord, thy sorrow to my sorrow lendeth
Another power; no flood by raining slaketh.
My woe too sensible thy passion maketh 110
More feeling-painful. Let it then suffice
To drown one woe, one pair of weeping eyes.

89 forced] ravished
90 accessary] that would make me an accessary
94 wreathed arms across] (a sign of grief)
107 frenzy] distraction, trance-like condition
110 sensible] sensitive

'And for my sake, when I might charm thee so,
For she that was thy Lucrece, now attend me:
Be suddenly revenged on my foe,
Thine, mine, his own; suppose thou dost defend me
From what is past. The help that thou shalt lend me
Comes all too late, yet let the traitor die;
For sparing justice feeds iniquity.

'But ere I name him, you fair lords', quoth she, 120
Speaking to those that came with Collatine,
'Shall plight your honourable faiths to me,
With swift pursuit to venge this wrong of mine;
For 'tis a meritorious fair design
To chase injustice with revengeful arms:
Knights, by their oaths, should right poor ladies' harms.'

At this request, with noble disposition
Each present lord began to promise aid,
As bound in knighthood to her imposition,
Longing to hear the hateful foe bewrayed. 130
But she, that yet her sad task hath not said,
The protestation stops. 'O, speak,' quoth she,
'How may this forced stain be wiped from me?

'What is the quality of my offence,
Being constrained with dreadful circumstance?
May my pure mind with the foul act dispense,
My low-declined honour to advance?
May any terms acquit me from this chance?
The poisoned fountain clears itself again;
And why not I from this compelled stain?' 140

With this, they all at once began to say,
Her body's stain her mind untainted clears;
While with a joyless smile she turns away
The face, that map which deep impression bears
Of hard misfortune, carved in it with tears.
'No, no,' quoth she, 'no dame hereafter living
By my excuse shall claim excuse's giving.'

115 suddenly] immediately 130 bewrayed] revealed

Here with a sigh, as if her heart would break,
She throws forth Tarquin's name: 'He, he,' she says,
But more than 'he' her poor tongue could not speak; 150
Till after many accents and delays,
Untimely breathings, sick and short assays,
She utters this: 'He, he, fair lords, 'tis he,
That guides this hand to give this wound to me.'

Even here she sheathed in her harmless breast
A harmful knife, that thence her soul unsheathed:
That blow did bail it from the deep unrest
Of that polluted prison where it breathed.
Her contrite sighs unto the clouds bequeathed
Her winged sprite and through her wounds doth fly 160
Life's lasting date from cancelled destiny.

Stone-still, astonished with this deadly deed,
Stood Collatine and all his lordly crew;
Till Lucrece' father, that beholds her bleed,
Himself on her self-slaught'red body threw;
And from the purple fountain Brutus drew
The murd'rous knife, and, as it left the place,
Her blood, in poor revenge, held it in chase;

And bubbling from her breast, it doth divide
In two slow rivers, that the crimson blood 170
Circles her body in on every side,
Who like a late-sacked island vastly stood
Bare and unpeopled in this fearful flood.
Some of her blood still pure and red remained,
And some looked black, and that false Tarquin stained.

About the mourning and congealed face
Of that black blood a wat'ry rigol goes,
Which seems to weep upon the tainted place;
And ever since, as pitying Lucrece' woes,
Corrupted blood some watery token shows; 180
And blood untainted still doth red abide,
Blushing at that which is so putrified.

(1594)

151 accents] spoken sounds
152 assays] attempts
157 bail it] buy its release
161 date] duration
162 astonished] stunned
163 crew] company
172 vastly] in desolation
177 rigol] ring, circle (i.e. the serum)

from *The Two Gentlemen of Verona*

342

WHO is Silvia? what is she,
  That all our swains commend her?
Holy, fair, and wise is she;
  The heaven such grace did lend her,
That she might admired be.

Is she kind as she is fair?
  For beauty lives with kindness.
Love doth to her eyes repair,
  To help him of his blindness;
And, being helped, inhabits there. 10

Then to Silvia let us sing,
  That Silvia is excelling;
She excels each mortal thing
  Upon the dull earth dwelling;
To her let us garlands bring.

(Pub. 1623)

from *Love's Labour's Lost*

343

WHEN daisies pied and violets blue
  And lady-smocks all silver-white
And cuckoo-buds of yellow hue
  Do paint the meadows with delight,
The cuckoo then, on every tree,
Mocks married men; for thus sings he,
    'Cuckoo,
Cuckoo, cuckoo'—O, word of fear,
Unpleasing to a married ear!

When shepherds pipe on oaten straws, 10
  And merry larks are ploughmen's clocks,
When turtles tread, and rooks and daws,
  And maidens bleach their summer smocks,
The cuckoo then, on every tree,
Mocks married men; for thus sings he,
    'Cuckoo,
Cuckoo, cuckoo'—O, word of fear,
Unpleasing to a married ear!

343
12 tread] mate

344

WHEN icicles hang by the wall,
    And Dick the shepherd blows his nail,
And Tom bears logs into the hall,
    And milk comes frozen home in pail;
When blood is nipped, and ways be foul,
Then nightly sings the staring owl,
'Tu-whit, tu-who!'—a merry note,
While greasy Joan doth keel the pot.

When all aloud the wind doth blow,
    And coughing drowns the parson's saw, 10
And birds sit brooding in the snow,
    And Marian's nose looks red and raw,
When roasted crabs hiss in the bowl,
Then nightly sings the staring owl,
'Tu-whit, tu-who!'—a merry note,
While greasy Joan doth keel the pot.

(Wr. 1593–5; pub. 1598)

from *A Midsummer Night's Dream*

345

First Fairy. YOU spotted snakes with double tongue,
    Thorny hedgehogs, be not seen;
Newts and blind-worms, do no wrong;
    Come not near our fairy queen.

Chorus. Philomel, with melody,
    Sing in our sweet lullaby;
Lulla, lulla, lullaby; lulla, lulla, lullaby.
        Never harm,
        Nor spell nor charm,
    Come our lovely lady nigh. 10
    So, good night, with lullaby.

First Fairy. Weaving spiders, come not here;
    Hence, you long-legged spinners, hence!
Beetles black, approach not near;
    Worm nor snail, do no offence.

Chorus. Philomel, with melody,
    Sing in our sweet lullaby;
Lulla, lulla, lullaby; lulla, lulla, lullaby.
        Never harm,
        Nor spell nor charm, 20
    Come our lovely lady nigh.
    So, good night, with lullaby.

8 keel] cool   13 crabs] crab-apples

346 *Bottom.* THE ousel cock, so black of hue,
With orange-tawny bill,
The throstle with his note so true,
The wren with little quill—

The finch, the sparrow, and the lark,
The plain-song cuckoo grey,
Whose note full many a man doth mark,
And dares not answer nay—

from *The Merchant of Venice*

347

TELL me where is fancy bred,
Or in the heart or in the head?
How begot, how nourished?
Reply, reply.

It is engendered in the eyes,
With gazing fed; and fancy dies
In the cradle where it lies.
Let us all ring fancy's knell.
I'll begin it. Ding, dong, bell.

All. Ding, dong, bell. 10

(Wr. 1596–7; pub. 1600)

from *Much Ado About Nothing*

348

SIGH no more, ladies, sigh no more,
Men were deceivers ever;
One foot in sea, and one on shore,
To one thing constant never.
Then sigh not so,
But let them go,
And be you blithe and bonny,
Converting all your sounds of woe
Into hey nonny, nonny.

1 ousel cock] cock-blackbird
4 quill] pipe, shrill note
6 plain-song] with its simple melody

Sing no more ditties, sing no mo 10
Of dumps so dull and heavy;
The fraud of men was ever so,
Since summer first was leavy.
Then sigh not so,
But let them go,
And be you blithe and bonny,
Converting all your sounds of woe
Into hey nonny, nonny.

349

PARDON, goddess of the night,
Those that slew thy virgin knight,
For the which, with songs of woe,
Round about her tomb they go.
Midnight, assist our moan;
Help us to sigh and groan,
Heavily, heavily.
Graves, yawn and yield your dead,
Till death be uttered,
Heavily, heavily. 10

(Wr. *c.*1598; pub. 1600)

from *As You Like It*

350

UNDER the greenwood tree
Who loves to lie with me,
And turn his merry note
Unto the sweet bird's throat,
Come hither, come hither, come hither!
Here shall he see
No enemy
But winter and rough weather.

Who doth ambition shun,
And loves to live i' the sun, 10
Seeking the food he eats,
And pleased with what he gets,
Come hither, come hither, come hither!
Here shall he see
No enemy
But winter and rough weather.

11 dumps] sad mood

349
9 uttered] adequately lamented

350
3 turn] adapt

351

BLOW, blow, thou winter wind,
Thou art not so unkind
As man's ingratitude;
Thy tooth is not so keen,
Because thou art not seen,
Although thy breath be rude.
Heigh-ho, sing heigh-ho, unto the green holly:
Most friendship is feigning, most loving mere folly.
Then heigh-ho, the holly,
This life is most jolly. 10

Freeze, freeze, thou bitter sky,
That dost not bite so nigh
As benefits forgot:
Though thou the waters warp,
Thy sting is not so sharp
As friend remembered not.
Heigh-ho, sing heigh-ho, unto the green holly:
Most friendship is feigning, most loving mere folly.
Then heigh-ho, the holly,
This life is most jolly. 20

352

WHAT shall he have that killed the deer?
His leather skin and horns to wear.
Then sing him home.
Take thou no scorn to wear the horn;
It was a crest ere thou wast born:
Thy father's father wore it,
And thy father bore it:
The horn, the horn, the lusty horn
Is not a thing to laugh to scorn.

353

IT was a lover and his lass,
With a hey, and a ho, and a hey nonino,
That o'er the green corn-field did pass,
In spring time, the only pretty ring time,
When birds do sing, hey ding a ding, ding;
Sweet lovers love the spring.

353
4 ring time] (?) time when wedding rings are exchanged or wedding-bells are heard or when ring-dances are danced

Between the acres of the rye,
  With a hey, and a ho, and a hey nonino,
These pretty country folks would lie,
  In spring time, the only pretty ring time, 10
When birds do sing, hey ding a ding, ding;
Sweet lovers love the spring.

This carol they began that hour,
  With a hey, and a ho, and a hey nonino,
How that a life was but a flower
  In spring time, the only pretty ring time,
When birds do sing, hey ding a ding, ding;
Sweet lovers love the spring.

And therefore take the present time,
  With a hey, and a ho, and a hey nonino; 20
For love is crowned with the prime
  In spring time, the only pretty ring time,
When birds do sing, hey ding a ding, ding;
Sweet lovers love the spring.

(Wr. *c.*1599; pub. 1623)

from *Twelfth Night*

354

O MISTRESS mine, where are you roaming?
O, stay and hear; your true love's coming,
  That can sing both high and low.
Trip no further, pretty sweeting;
Journeys end in lovers meeting,
  Every wise man's son doth know.

What is love? 'Tis not hereafter;
Present mirth hath present laughter;
  What's to come is still unsure.
In delay there lies no plenty; 10
Then come kiss me, sweet and twenty;
  Youth's a stuff will not endure.

7 acres] unploughed balks in an open field

21 prime] the best

355

COME away, come away, death,
  And in sad cypress let me be laid.
Fly away, fly away, breath;
  I am slain by a fair cruel maid.
My shroud of white, stuck all with yew,
    O, prepare it!
My part of death, no one so true
    Did share it.

Not a flower, not a flower sweet,
  On my black coffin let there be strown; 10
Not a friend, not a friend greet
  My poor corpse, where my bones shall be thrown.
A thousand thousand sighs to save,
    Lay me, O, where
Sad true lover never find my grave,
    To weep there.

356

WHEN that I was and a little tiny boy,
  With hey, ho, the wind and the rain;
A foolish thing was but a toy,
  For the rain it raineth every day.

But when I came to man's estate,
  With hey, ho, the wind and the rain;
'Gainst knaves and thieves men shut their gate,
  For the rain it raineth every day.

But when I came, alas, to wive,
  With hey, ho, the wind and the rain; 10
By swaggering could I never thrive,
  For the rain it raineth every day.

But when I came unto my beds,
  With hey, ho, the wind and the rain;
With toss-pots still had drunken heads,
  For the rain it raineth every day.

A great while ago the world begun,
  With hey, ho, the wind and the rain;
But that's all one, our play is done,
  And we'll strive to please you every day. 20

(Wr. *c.*1600; pub. 1623)

356
13 came unto my beds] (?) grew old
15 had drunken heads] i.e. I was drunk like the other 'tosspots'

from *Hamlet*

357 *Hamlet.* WHY, let the strucken deer go weep,
The hart ungallèd play;
For some must watch, while some must sleep;
Thus runs the world away.

For thou dost know, O Damon dear,
This realm dismantled was
Of Jove himself; and now reigns here
A very, very—pajock.

358 *Ophelia.* HOW should I your true-love know
From another one?
By his cockle hat and staff,
And his sandal shoon.

He is dead and gone, lady,
He is dead and gone;
At his head a grass-green turf,
At his heels a stone.

White his shroud as the mountain snow,
Larded all with sweet flowers; 10
Which bewept to the grave did not go
With true-love showers.

359 *Ophelia.* TOMORROW is Saint Valentine's day,
All in the morning betime,
And I a maid at your window,
To be your Valentine.

Then up he rose, and donned his clo'es,
And dupped the chamber door;
Let in the maid, that out a maid
Never departed more.

3 watch] stay awake
8 pajock] base contemptible fellow, clown

358
3 cockle hat] hat with a scallop-shell (the mark of a pilgrim)
4 shoon] shoes
10 Larded] strewn

359
6 dupped] opened

By Gis and by Saint Charity,
  Alack, and fie for shame! 10
Young men will do't, if they come to't;
  By Cock, they are to blame.

Quoth she: 'Before you tumbled me,
  You promised me to wed.'
'So would I 'a done, by yonder sun,
  And thou hadst not come to my bed.'

360

Ophelia. AND will 'a not come again?
And will 'a not come again?
  No, no, he is dead.
  Go to thy death-bed,
He never will come again.

His beard was as white as snow,
All flaxen was his poll;
  He is gone, he is gone,
  And we cast away moan:
God-a-mercy on his soul! 10

(Wr. *c.*1600; pub. 1604)

from *Measure for Measure*

361

TAKE, O take those lips away,
  That so sweetly were forsworn,
And those eyes, the break of day,
  Lights that do mislead the morn;
But my kisses bring again, bring again,
Seals of love, but sealed in vain, sealed in vain.

(Wr. *c.*1604; pub. 1623)

from *Antony and Cleopatra*

362

COME, thou monarch of the vine,
Plumpy Bacchus with pink eyne!
In thy fats our cares be drowned,
With thy grapes our hairs be crowned!
  Cup us till the world go round,
  Cup us till the world go round!

(Wr. 1606–7; pub. 1623)

9 Gis] Jesus
12 Cock] (1) God; (2) male organ
360
9 cast away] indulge without purpose

362
2 eyne] eyes
3 fats] vats, casks

from *Cymbeline*

363

HARK, hark, the lark at heaven's gate sings,
And Phoebus gins arise,
His steeds to water at those springs
On chaliced flowers that lies;
And winking Mary-buds begin to ope their golden eyes;
With every thing that pretty is, my lady sweet, arise:
Arise, arise!

364

FEAR no more the heat o' th' sun,
Nor the furious winter's rages,
Thou thy worldly task hast done,
Home art gone, and ta'en thy wages.
Golden lads and girls all must,
As chimney-sweepers, come to dust.

Fear no more the frown o' th' great,
Thou art past the tyrant's stroke;
Care no more to clothe and eat,
To thee the reed is as the oak. 10
The sceptre, learning, physic, must
All follow this, and come to dust.

Fear no more the lightning-flash,
Nor th' all-dreaded thunder-stone.
Fear not slander, censure rash.
Thou hast finished joy and moan.
All lovers young, all lovers must
Consign to thee, and come to dust.

No exorciser harm thee.
Nor no witchcraft charm thee. 20
Ghost unlaid forbear thee.
Nothing ill come near thee.
Quiet consummation have,
And renowned be thy grave.

(Wr. 1609–10; pub. 1623)

363
5 winking] slumbering

from *The Winter's Tale*

365

WHEN daffadils begin to peer,
  With heigh, the doxy over the dale!
Why then comes in the sweet o' th' year,
  For the red blood reigns in the winter's pale.

The white sheet bleaching on the hedge,
  With heigh, the sweet birds, O how they sing!
Doth set my pugging tooth on edge,
  For a quart of ale is a dish for a king.

The lark, that tirra-lirra chants,
  With heigh, with heigh, the thrush and the jay, 10
Are summer songs for me and my aunts,
  While we lie tumbling in the hay.

366

JOG on, jog on, the footpath way,
  And merrily hent the stile-a;
A merry heart goes all the day,
  Your sad tires in a mile-a.

367

LAWN as white as driven snow,
Cypress black as e'er was crow,
Gloves as sweet as damask roses,
Masks for faces and for noses;
Bugle-bracelet, necklace amber,
Perfume for a lady's chamber;
Golden coifs and stomachers
For my lads to give their dears;
Pins and poking-sticks of steel;
What maids lack from head to heel: 10
Come buy of me, come; come buy, come buy;
Buy, lads, or else your lasses cry:
Come buy.

(Wr. 1610–11; pub. 1623)

2 doxy] beggar's wench
4 pale] (1) paleness; (2) domain
7 pugging] thieving
11 aunts] whores

366
2 hent] grasp

367
2 Cypress] filmy crêpe
5 Bugle-bracelet] bracelet made of black beads
7 coifs] tight caps
9 poking-sticks] metal rods used to iron fluted ruffs

from *The Tempest*

368

COME unto these yellow sands,
  And then take hands:
Curtsied when you have, and kissed,
  The wild waves whist:
Foot it featly here and there,
And, sweet sprites, the burden bear.
Hark, hark!
    (*Burden within*) Bow-wow.
The watch-dogs bark!
    (*within*) Bow-wow. 10
Hark, hark, I hear
The strain of strutting Chanticleer:
    (*Cry within*) Cock-a-diddle-dow.

369

FULL fathom five thy father lies;
  Of his bones are coral made:
Those are pearls that were his eyes:
  Nothing of him that doth fade,
But doth suffer a sea-change
Into something rich and strange.
Sea-nymphs hourly ring his knell:
    (*within*) Ding-dong.
Hark now I hear them—ding-dong bell.

370

WHERE the bee sucks, there suck I,
In a cowslip's bell I lie;
There I couch when owls do cry.
On the bat's back I do fly
After summer merrily.
  Merrily, merrily shall I live now
  Under the blossom that hangs on the bough.

(Wr. 1610–11; pub. 1623)

4 whist] being hushed
5 featly] nimbly
6 burden] undersong

from *The Two Noble Kinsmen*

371

ROSES, their sharp spines being gone,
Not royal in their smells alone,
  But in their hue;
Maiden pinks, of odour faint,
Daisies smell-less, yet most quaint,
  And sweet thyme true.

Primrose, first-born child of Ver,
Merry spring-time's harbinger,
  With her bells dim;
Oxlips in their cradles growing, 10
Marigolds on death-beds blowing,
  Larks'-heels trim;

All dear Nature's children sweet,
Lie 'fore bride and bridegroom's feet,
  Blessing their sense;
Not an angel of the air,
Bird melodious, or bird fair,
  Is absent hence.

The crow, the sland'rous cuckoo, nor
The boding raven, nor chough hoar, 20
  Nor chatt'ring pie,
May on our bridehouse perch or sing,
Or with them any discord bring,
  But from it fly.

(Wr. prob. 1613; pub. 1634)

from *Sonnets*

372

WHEN forty winters shall besiege thy brow,
  And dig deep trenches in thy beauty's field,
Thy youth's proud livery, so gazed on now,
  Will be a tattered weed, of small worth held:

5 quaint] pretty, fine
7 Ver] spring
9 dim] pale
11 blowing] blooming
12 Larks'-heels] larkspur
20 chough] crow
21 pie] magpie

372
2 field] (1) meadow; (2) battle-field
4 weed] garment

Then being asked where all thy beauty lies,
Where all the treasure of thy lusty days,
To say, within thine own deep-sunken eyes,
Were an all-eating shame and thriftless praise.
How much more praise deserved thy beauty's use,
If thou couldst answer, 'This fair child of mine 10
Shall sum my count, and make my old excuse',
Proving his beauty by succession thine.
This were to be new made when thou art old,
And see thy blood warm when thou feel'st it cold.

373 WHEN I do count the clock that tells the time,
And see the brave day sunk in hideous night;
When I behold the violet past prime,
And sable curls, all silvered o'er with white;
When lofty trees I see barren of leaves,
Which erst from heat did canopy the herd,
And summer's green all girded up in sheaves,
Borne on the bier with white and bristly beard,
Then of thy beauty do I question make,
That thou among the wastes of time must go, 10
Since sweets and beauties do themselves forsake
And die as fast as they see others grow;
And nothing gainst Time's scythe can make defence
Save breed, to brave him when he takes thee hence.

374 WHEN I consider every thing that grows
Holds in perfection but a little moment,
That this huge stage presenteth nought but shows
Whereon the stars in secret influence comment;
When I perceive that men as plants increase,
Cheered and checked even by the self-same sky,
Vaunt in their youthful sap, at height decrease,
And wear their brave state out of memory;

8 all-eating shame] (1) annihilating shame; (2) shameful greed
thriftless] (1) wasteful; (2) unprofitable
9 use] wise investment
11 sum my count] square my account
make my old excuse] in my old age justify myself

373
1 count] count the strokes of
2 brave] splendid
9 of . . . question make] speculate about
11 do themselves forsake] abandon themselves to decay
14 breed] offspring

374
6 Cheered] encouraged
checked] rebuked; restrained
7 Vaunt] exult; swagger
8 brave state] splendid dress

Then the conceit of this inconstant stay
  Sets you most rich in youth before my sight, 10
Where wasteful Time debateth with Decay,
  To change your day of youth to sullied night;
    And, all in war with Time for love of you,
    As he takes from you, I engraft you new.

375

SHALL I compare thee to a summer's day?
  Thou art more lovely and more temperate:
Rough winds do shake the darling buds of May,
  And summer's lease hath all too short a date:
Sometime too hot the eye of heaven shines,
  And often is his gold complexion dimmed;
And every fair from fair sometime declines,
  By chance, or nature's changing course untrimmed;
But thy eternal summer shall not fade,
  Nor lose possession of that fair thou owest, 10
Nor shall death brag thou wander'st in his shade,
  When in eternal lines to time thou growest;
    So long as men can breathe, or eyes can see,
    So long lives this, and this gives life to thee.

376

DEVOURING Time, blunt thou the lion's paws,
  And make the earth devour her own sweet brood;
Pluck the keen teeth from the fierce tiger's jaws,
  And burn the long-lived phoenix in her blood;
Make glad and sorry seasons as thou fleet'st,
  And do whate'er thou wilt, swift-footed Time,
To the wide world and all her fading sweets;
  But I forbid thee one most heinous crime:
O, carve not with thy hours my Love's fair brow,
  Nor draw no lines there with thine antique pen; 10
Him in thy course untainted do allow
  For beauty's pattern to succeeding men.
    Yet, do thy worst, old Time: despite thy wrong,
    My Love shall in my verse ever live young.

9 conceit] thought
11 debateth] (1) disputes; (2) contends
14 engraft you new] give you new life

375
4 lease] allotted time
too short a date] too brief a duration
8 untrimmed] stripped of its beauty
10 owest] own

376
10 antique] (1) old; (2) capricious, antic, grotesque-making
11 untainted] (1) unsullied; (2) not hit by the lance of his opponent (jousting term)

377 A WOMAN'S face, with Nature's own hand painted,
Hast thou, the master-mistress of my passion;
A woman's gentle heart, but not acquainted
With shifting change, as is false women's fashion;
An eye more bright than theirs, less false in rolling,
Gilding the object whereupon it gazeth;
A man in hue all hues in his controlling,
Which steals men's eyes and women's souls amazeth.
And for a woman wert thou first created,
Till Nature as she wrought thee fell a-doting, 10
And by addition me of thee defeated,
By adding one thing to my purpose nothing.
But since she pricked thee out for women's pleasure,
Mine be thy love, and thy love's use their treasure.

378 As an unperfect actor on the stage,
Who with his fear is put besides his part,
Or some fierce thing replete with too much rage,
Whose strength's abundance weakens his own heart;
So I, for fear of trust, forget to say
The perfect ceremony of love's rite,
And in mine own love's strength seem to decay,
O'ercharged with burden of mine own love's might.
O, let my books be then the eloquence
And dumb presagers of my speaking breast, 10
Who plead for love and look for recompense
More than that tongue that more hath more expressed.
O, learn to read what silent love hath writ;
To hear with eyes belongs to love's fine wit.

5 rolling] roving
6 Gilding] making bright (since eyes were believed to emit beams of light)
7 hue] form, appearance
8 amazeth] overwhelms
11 defeated] deprived
13 pricked thee out] (1) selected you; (2) gave you a penis

378
1 unperfect] who has not properly learned his part
2 put besides] made to forget
3 thing] creature
4 Whose] (refers to 'rage')
5 for fear of trust] (1) afraid to trust myself; (2) afraid that I will not be trusted
6 perfect] properly memorized
10 presagers] messengers, heralds
12 more hath more expressed] has more often said more

379

WEARY with toil, I haste me to my bed,
  The dear repose for limbs with travel tired;
But then begins a journey in my head
  To work my mind, when body's work's expired:
For then my thoughts, from far where I abide,
  Intend a zealous pilgrimage to thee,
And keep my drooping eyelids open wide,
  Looking on darkness which the blind do see:
Save that my soul's imaginary sight
  Presents thy shadow to my sightless view, 10
Which, like a jewel hung in ghastly night,
  Makes black night beauteous and her old face new.
    Lo thus, by day my limbs, by night my mind,
    For thee and for myself no quiet find.

380

WHEN in disgrace with fortune and men's eyes
  I all alone beweep my outcast state,
And trouble deaf heaven with my bootless cries,
  And look upon myself, and curse my fate,
Wishing me like to one more rich in hope,
  Featured like him, like him with friends possessed,
Desiring this man's art, and that man's scope,
  With what I most enjoy contented least;
Yet in these thoughts myself almost despising,
  Haply I think on thee, and then my state, 10
Like to the lark at break of day arising
  From sullen earth, sings hymns at heaven's gate;
    For thy sweet love remembered such wealth brings
    That then I scorn to change my state with kings.

381

WHEN to the sessions of sweet silent thought
  I summon up remembrance of things past,
I sigh the lack of many a thing I sought,
  And with old woes new wail my dear time's waste:
Then can I drown an eye, unused to flow,
  For precious friends hid in death's dateless night,
And weep afresh love's long since cancelled woe,
  And moan the expense of many a vanished sight:

6 Intend] set out on
9 imaginary] imaginative
11 ghastly] terrifying
14 For thee] (1) because of thee; (2) for the sake of

380
1 in disgrace] out of favour
7 scope] range of ability
12 sullen] dull, heavy

381
1 sessions] court sittings
6 dateless] endless
8 expense] loss

Then can I grieve at grievances foregone,
And heavily from woe to woe tell o'er 10
The sad account of fore-bemoaned moan,
Which I new pay as if not paid before.
But if the while I think on thee, dear friend,
All losses are restored and sorrows end.

382

FULL many a glorious morning have I seen
Flatter the mountain-tops with sovereign eye,
Kissing with golden face the meadows green,
Gilding pale streams with heavenly alchemy;
Anon permit the basest clouds to ride
With ugly rack on his celestial face,
And from the forlorn world his visage hide,
Stealing unseen to west with this disgrace:
Even so my sun one early morn did shine,
With all triumphant splendour on my brow; 10
But, out alack, he was but one hour mine,
The region cloud hath masked him from me now.
Yet him for this my love no whit disdaineth;
Suns of the world may stain when heaven's sun staineth.

383

NO more be grieved at that which thou hast done:
Roses have thorns, and silver fountains mud,
Clouds and eclipses stain both moon and sun,
And loathsome canker lives in sweetest bud;
And men make faults, and even I in this,
Authorizing thy trespass with compare,
Myself corrupting salving thy amiss,
Excusing thy sins more than thy sins are;

9 foregone] done with
10 tell] count

382
2 Flatter] (1) caress; (2) encourage (as a king would a courtier)
4 alchemy] (whose aim was to turn base metals to gold)
5 Anon] soon
6 rack] vaporous drift
8 disgrace] disfigurement, loss of beauty
10 triumphant] glorious
12 region] of the upper air
14 Suns of the world] (1) great men; (2) mere mortals ('sons')
stain] grow dim

383
3 stain] dim, obscure the brightness of
4 canker] cankerworm, caterpillar
6 Authorizing . . . with compare] justifying . . . by analogy
7 salving] palliating
amiss] offence

For to thy sensual fault I bring in sense—
Thy adverse party is thy advocate— 10
And gainst myself a lawful plea commence.
Such civil war is in my love and hate
That I an accessory needs must be
To that sweet thief which sourly robs from me.

384

WHAT is your substance, whereof are you made,
That millions of strange shadows on you tend?
Since every one hath, every one, one shade,
And you, but one, can every shadow lend.
Describe Adonis, and the counterfeit
Is poorly imitated after you;
On Helen's cheek all art of beauty set,
And you in Grecian tires are painted new:
Speak of the spring and foison of the year,
The one doth shadow of your beauty show, 10
The other as your bounty doth appear;
And you in every blessed shape we know.
In all external grace you have some part,
But you like none, none you, for constant heart.

385

NOT marble, nor the gilded monuments
Of princes, shall outlive this powerful rhyme;
But you shall shine more bright in these contents
Than unswept stone, besmeared with sluttish time.
When wasteful war shall statues overturn,
And broils root out the work of masonry,
Nor Mars his sword nor war's quick fire shall burn
The living record of your memory.
Gainst death and all oblivious enmity
Shall you pace forth; your praise shall still find room 10
Even in the eyes of all posterity
That wear this world out to the ending doom.
So, till the judgment that yourself arise,
You live in this, and dwell in lovers' eyes.

384
1 substance] (1) material, matter; (2) essence, essential nature
2 shadows] images, reflections
tend] attend
4 lend] supply
5 counterfeit] picture, portrait
8 tires] clothes; headdress
9 foison] harvest

385
3 these contents] this poem; these lines
5 wasteful] destructive
6 broils] tumults
9 all oblivious enmity] all hostile forces which cause everything to be forgotten
12 wear this world out] will last as long as this world
13 judgement] Last Judgement

386

BEING your slave, what should I do but tend
Upon the hours and times of your desire?
I have no precious time at all to spend,
Nor services to do, till you require.
Nor dare I chide the world-without-end hour
Whilst I, my sovereign, watch the clock for you,
Nor think the bitterness of absence sour
When you have bid your servant once adieu.
Nor dare I question with my jealous thought
Where you may be, or your affairs suppose, 10
But like a sad slave stay and think of naught
Save where you are how happy you make those.
So true a fool is love that, in your will
Though you do anything, he thinks no ill.

387

LIKE as the waves make towards the pebbled shore,
So do our minutes hasten to their end;
Each changing place with that which goes before,
In sequent toil all forwards do contend.
Nativity, once in the main of light,
Crawls to maturity, wherewith being crowned,
Crooked eclipses gainst his glory fight,
And Time that gave doth now his gift confound.
Time doth transfix the flourish set on youth
And delves the parallels in beauty's brow, 10
Feeds on the rarities of nature's truth,
And nothing stands but for his scythe to mow.
And yet to times in hope my verse shall stand,
Praising thy worth, despite his cruel hand.

388

WHEN I have seen by Time's fell hand defaced
The rich proud cost of outworn buried age;
When sometime lofty towers I see down razed,
And brass eternal slave to mortal rage;

1–2 tend upon] wait on
5 world-without-end] everlasting
9 jealous] mistrustful
10 your affairs suppose] imagine what you are doing

387
1 Like as] just as
5 Nativity] (1) new-born infant; (2) moment of birth
main] broad expanse
7 Crooked] malignant
8 confound] destroy
9 transfix the flourish] destroy the beauty
10 delves the parallels] digs the wrinkles
11 truth] perfection
13 to times in hope] until future times as yet only imagined

388
1 fell] cruel
3 sometime] once, formerly

When I have seen the hungry ocean gain
Advantage on the kingdom of the shore,
And the firm soil win of the watery main,
Increasing store with loss, and loss with store;
When I have seen such interchange of state,
Or state itself confounded to decay, 10
Ruin hath taught me thus to ruminate,
That Time will come and take my love away.
This thought is as a death, which cannot choose
But weep to have that which it fears to lose.

389

SINCE brass, nor stone, nor earth, nor boundless sea,
But sad mortality o'ersways their power,
How with this rage shall beauty hold a plea,
Whose action is no stronger than a flower?
O, how shall summer's honey breath hold out
Against the wrackful siege of battering days,
When rocks impregnable are not so stout,
Nor gates of steel so strong, but Time decays?
O, fearful meditation! Where, alack,
Shall Time's best jewel from Time's chest lie hid? 10
Or what strong hand can hold his swift foot back?
Or who his spoil of beauty can forbid?
O, none, unless this miracle have might,
That in black ink my love may still shine bright.

390

TIRED with all these, for restful death I cry:
As to behold desert a beggar born,
And needy nothing trimmed in jollity,
And purest faith unhappily forsworn,
And gilded honour shamefully misplaced,
And maiden virtue rudely strumpeted,
And right perfection wrongfully disgraced,
And strength by limping sway disabled,

7 win of] win from
8 store] abundance
9 state] condition
10 state] worldly grandeur

389
1 Since] since there is no
2 o'ersways] overrules
3 hold a plea] win its argument
4 action] (1) law-suit; (2) military action, engagement
6 wrackful] destructive

390
3 jollity] fine clothing
4 unhappily] maliciously; wretchedly
5 honour . . . misplaced] honours given to unworthy persons
6 strumpeted] accused of being a strumpet
7 right] true
8 limping sway] weak or senile rule

And art made tongue-tied by authority,
And folly, doctor-like, controlling skill, 10
And simple truth miscalled simplicity,
And captive good attending captain ill.
Tired with all these, from these would I be gone,
Save that, to die, I leave my love alone.

391

No longer mourn for me when I am dead
Than you shall hear the surly sullen bell
Give warning to the world that I am fled
From this vile world, with vildest worms to dwell:
Nay, if you read this line, remember not
The hand that writ it; for I love you so
That I in your sweet thoughts would be forgot
If thinking on me then should make you woe.
O, if, I say, you look upon this verse,
When I perhaps compounded am with clay, 10
Do not so much as my poor name rehearse,
But let your love even with my life decay;
Lest the wise world should look into your moan,
And mock you with me after I am gone.

392

That time of year thou mayst in me behold
When yellow leaves, or none, or few, do hang
Upon those boughs which shake against the cold,
Bare ruined choirs, where late the sweet birds sang.
In me thou see'st the twilight of such day
As after sunset fadeth in the west;
Which by and by black night doth take away,
Death's second self, that seals up all in rest.
In me thou see'st the glowing of such fire,
That on the ashes of his youth doth lie, 10
As the death-bed whereon it must expire,
Consumed with that which it was nourished by.
This thou perceiv'st, which makes thy love more strong,
To love that well which thou must leave ere long.

10 doctor-like] like a pompous academic
controlling] directing, bossing about
11 truth] honesty
simplicity] simple-mindedness, stupidity
14 to die] if I die
391
4 vildest] vilest
11 rehearse] repeat
392
8 seals up] (1) i.e. encloses (as in a coffin); (2) 'seels up', i.e. stitches up the eyes of a hawk
10 his youth] i.e. its youth
14 leave] give up, forego

393

WHY is my verse so barren of new pride,
 So far from variation or quick change?
Why with the time do I not glance aside
 To new-found methods and to compounds strange?
Why write I still all one, ever the same,
 And keep invention in a noted weed,
That every word doth almost tell my name,
 Showing their birth and where they did proceed?
O, know, sweet love, I always write of you,
 And you and love are still my argument; 10
So all my best is dressing old words new,
 Spending again what is already spent:
 For as the sun is daily new and old,
 So is my love still telling what is told.

394

WAS it the proud full sail of his great verse,
 Bound for the prize of all-too-precious you,
That did my ripe thoughts in my brain inhearse,
 Making their tomb the womb wherein they grew?
Was it his spirit, by spirits taught to write
 Above a mortal pitch, that struck me dead?
No, neither he, nor his compeers by night
 Giving him aid, my verse astonished.
He, nor that affable familiar ghost
 Which nightly gulls him with intelligence, 10
As victors, of my silence cannot boast;
 I was not sick of any fear from thence.
 But when your countenance filled up his line,
 Then lacked I matter; that enfeebled mine.

1 new pride] fashionably novel ornament
2 variation] stylistic variety
3 with the time] in the current mode
glance] turn abruptly
4 compounds strange] striking verbal combinations
6 invention] literary creation
noted weed] the same old dress
8 where] whence
10 still] always
argument] theme
12 spent] (1) paid out; (2) worn out

394
3 ripe] ready for birth
inhearse] entomb, bury
5 spirit] (1) vigorous mind; (2) vivacity; (3) (?) daemon
6 pitch] height
7 compeers] (1) (?) authors or their books; (2) (?) literary associates; (3) (?) supernatural familiars
8 astonished] paralysed
10 intelligence] information
12 sick of] sick from
13 countenance] (1) face; (2) patronage

395 FAREWELL, thou art too dear for my possessing,
And like enough thou know'st thy estimate:
The charter of thy worth gives thee releasing;
My bonds in thee are all determinate.
For how do I hold thee but by thy granting?
And for that riches where is my deserving?
The cause of this fair gift in me is wanting,
And so my patent back again is swerving.
Thyself thou gav'st, thy own worth then not knowing,
Or me, to whom thou gav'st it, else mistaking; 10
So thy great gift, upon misprision growing,
Comes home again, on better judgement making.
Thus have I had thee as a dream doth flatter,
In sleep a king, but waking no such matter.

396 THEN hate me when thou wilt; if ever, now;
Now, while the world is bent my deeds to cross,
Join with the spite of fortune, make me bow,
And do not drop in for an after-loss:
Ah, do not, when my heart hath scaped this sorrow,
Come in the rearward of a conquered woe;
Give not a windy night a rainy morrow,
To linger out a purposed overthrow.
If thou wilt leave me, do not leave me last,
When other petty griefs have done their spite, 10
But in the onset come; so shall I taste
At first the very worst of fortune's might.
And other strains of woe, which now seem woe,
Compared with loss of thee will not seem so.

397 THEY that have power to hurt and will do none,
That do not do the thing they most do show,
Who, moving others, are themselves as stone,
Unmoved, cold, and to temptation slow;

1 dear] (1) costly; (2) beloved; (3) of high rank
2 estimate] value
3 charter] (1) special privilege; (2) the document conferring the privilege
releasing] exemption
4 determinate] ended
8 patent] title of possession
11 upon misprision growing] arising from a misjudgement

396
2 bent] resolved
cross] thwart
4 drop in . . . after-loss] (?) fall upon me in the form of a later, unexpected blow or misfortune
11 in the onset] in the first wave of the attack

397
1 will do] choose to do

They rightly do inherit heaven's graces
And husband nature's riches from expense;
They are the lords and owners of their faces,
Others but stewards of their excellence.
The summer's flower is to the summer sweet,
Though to itself it only live and die; 10
But if that flower with base infection meet,
The basest weed outbraves his dignity:
For sweetest things turn sourest by their deeds;
Lilies that fester smell far worse than weeds.

398 How like a winter hath my absence been
From thee, the pleasure of the fleeting year!
What freezings have I felt, what dark days seen!
What old December's bareness everywhere!
And yet this time removed was summer's time,
The teeming autumn, big with rich increase,
Bearing the wanton burden of the prime,
Like widowed wombs after their lords' decease:
Yet this abundant issue seemed to me
But hope of orphans and unfathered fruit; 10
For summer and his pleasures wait on thee,
And, thou away, the very birds are mute.
Or, if they sing, 'tis with so dull a cheer,
That leaves look pale, dreading the winter's near.

399 From you have I been absent in the spring,
When proud-pied April, dressed in all his trim,
Hath put a spirit of youth in every thing,
That heavy Saturn laughed and leaped with him.
Yet nor the lays of birds, nor the sweet smell
Of different flowers in odour and in hue,
Could make me any summer's story tell,
Or from their proud lap pluck them where they grew:

5 rightly] truly, indeed
6 husband] carefully protect
expense] wasteful spending
10 to itself it only] only to itself it
12 outbraves] makes a finer show than

398
5 time removed] period of separation
7 wanton] (1) frolicsome; (2) luxuriant; (3) amorously sportive
prime] spring
11 his] its
wait on thee] are your servants

399
2 proud-pied] splendidly many-coloured
trim] array
7 summer's story tell] speak (or write) cheerfully
8 proud] showy

Nor did I wonder at the lily's white,
Nor praise the deep vermilion in the rose; 10
They were but sweet, but figures of delight,
Drawn after you, you pattern of all those.
Yet seemed it winter still, and, you away,
As with your shadow I with these did play.

400

To me, fair friend, you never can be old,
For as you were when first your eye I eyed,
Such seems your beauty still. Three winters cold
Have from the forests shook three summers' pride;
Three beauteous springs to yellow autumn turned
In process of the seasons have I seen,
Three April perfumes in three hot Junes burned,
Since first I saw you fresh, which yet are green.
Ah, yet doth beauty, like a dial hand,
Steal from his figure, and no pace perceived; 10
So your sweet hue, which methinks still doth stand,
Hath motion, and mine eye may be deceived.
For fear of which, hear this, thou age unbred;
Ere you were born was beauty's summer dead.

401

When in the chronicle of wasted time
I see descriptions of the fairest wights,
And beauty making beautiful old rhyme,
In praise of ladies dead and lovely knights,
Then, in the blazon of sweet beauty's best,
Of hand, of foot, of lip, of eye, of brow,
I see their antique pen would have expressed
Even such a beauty as you master now.
So all their praises are but prophecies
Of this our time, all you prefiguring; 10
And, for they looked but with divining eyes,
They had not skill enough your worth to sing:
For we, which now behold these present days,
Have eyes to wonder, but lack tongues to praise.

14 shadow] image

400

9 dial] clock, watch

10 Steal from] (1) depart stealthily from; (2) rob

his figure] (1) the number or mark on the 'dial'; (2) the form or appearance of the 'friend'

11 hue] appearance

401

1 wasted] past, gone to ruin, used up

2 wights] men and women

4 lovely] (1) handsome; (2) worthy of love

5 blazon] list of praiseworthy qualities

7 their antique pen] the pens of the old writers

8 master] possess

11 for] because

divining] peering into the future

402 NOT mine own fears, nor the prophetic soul
Of the wide world dreaming on things to come,
Can yet the lease of my true love control,
Supposed as forfeit to a confined doom.
The mortal moon hath her eclipse endured,
And the sad augurs mock their own presage;
Incertainties now crown themselves assured,
And peace proclaims olives of endless age.
Now with the drops of this most balmy time
My love looks fresh, and Death to me subscribes, 10
Since, spite of him, I'll live in this poor rhyme,
While he insults o'er dull and speechless tribes.
And thou in this shalt find thy monument,
When tyrants' crests and tombs of brass are spent.

403 ALAS, 'tis true, I have gone here and there,
And made myself a motley to the view,
Gored mine own thoughts, sold cheap what is most
dear,
Made old offences of affections new.
Most true it is that I have looked on truth
Askance and strangely; but, by all above,
These blenches gave my heart another youth,
And worse essays proved thee my best of love.
Now all is done, have what shall have no end:
Mine appetite I never more will grind 10
On newer proof, to try an older friend,
A god in love, to whom I am confined.
Then give me welcome, next my heaven the best,
Even to thy pure and most most loving breast.

1–2 prophetic soul Of the wide world] speculative forebodings of the world in general
2 dreaming] musing
3 lease] term of duration
4 Supposed . . . doom] thought destined to expire after a limited period
5–8 The mortal moon . . . age] (? Queen Elizabeth's death and the accession of James I)
9 drops] (?) dewdrops
balmy] refreshing, restorative
10 subscribes] submits
12 insults] triumphs
14 spent] wasted away

403
2 motley] fool (fools wore 'motley', parti-coloured clothing)
3 Gored] (1) wounded; (2) furnished with 'gores' (triangular pieces of cloth), dressed in motley; (3) dishonoured (in heraldry a 'gore' was a shaped area between two charges, to serve as a mark of cadency or abatement of honour)
4 Made old . . . new] made new attachments, further instances of the old infidelity
5 truth] constancy, faithfulness
7 blenches] (1) swervings, inconstancies; (2) blemishes; (3) side glances
gave . . . youth] rejuvenated my love
8 worse essays] experiments with inferior materials

404

YOUR love and pity doth th'impression fill
 Which vulgar scandal stamped upon my brow;
For what care I who calls me well or ill,
 So you o'er-green my bad, my good allow?
You are my all the world, and I must strive
 To know my shames and praises from your tongue;
None else to me, nor I to none alive,
 That my steeled sense or changes right or wrong.
In so profound abysm I throw all care
 Of others' voices, that my adder's sense 10
To critic and to flatterer stopped are.
 Mark how with my neglect I do dispense:
 You are so strongly in my purpose bred
 That all the world besides methinks they're dead.

405

LET me not to the marriage of true minds
 Admit impediments. Love is not love
Which alters when it alteration finds,
 Or bends with the remover to remove.
O, no, it is an ever-fixed mark,
 That looks on tempests and is never shaken;
It is the star to every wandering bark,
 Whose worth's unknown, although his height be taken.
Love's not Time's fool, though rosy lips and cheeks
 Within his bending sickle's compass come; 10
Love alters not with his brief hours and weeks,
 But bears it out even to the edge of doom.
 If this be error, and upon me proved,
 I never writ, nor no man ever loved.

1 doth] do
1–2 doth . . . brow] (attempts were often made to fill the hollow scars caused by branding, as in the case of ex-slaves in ancient Rome)
2 vulgar] (1) public; (2) base, common
4 o'er-green] coverup (as in re-turfing)
allow] (1) acknowledge; (2) approve
7–8 None else . . . wrong] there is no one else who exists for me or for whom I exist in such a way that they can change my hardened sensibility either for good or evil
10 adder's sense] deaf ears (cf. 'deaf as an adder')
12 dispense with] excuse
13 in my purpose bred] cherished in my thought

405
1 Let me not] (1) may I never; (2) I will not
5 mark] sea-mark, beacon
8 his] its
10 compass] range, scope
12 bears it out] endures
to the edge of doom] until the day of judgement

406 'TIS better to be vile than vile esteemed
When not to be receives reproach of being,
And the just pleasure lost which is so deemed
Not by our feeling but by others' seeing.
For why should others' false adulterate eyes
Give salutation to my sportive blood?
Or on my frailties why are frailer spies,
Which in their wills count bad what I think good?
No, I am that I am, and they that level
At my abuses reckon up their own; 10
I may be straight though they themselves be bevel.
By their rank thoughts my deeds must not be shown,
Unless this general evil they maintain:
All men are bad and in their badness reign.

407 IF my dear love were but the child of state,
It might for Fortune's bastard be unfathered,
As subject to Time's love, or to Time's hate,
Weeds among weeds, or flowers with flowers gathered.
No, it was builded far from accident;
It suffers not in smiling pomp, nor falls
Under the blow of thralled discontent,
Whereto th'inviting time our fashion calls.
It fears not Policy, that heretic
Which works on leases of short-numbered hours, 10
But all alone stands hugely politic,
That it nor grows with heat, nor drowns with showers.
To this I witness call the fools of Time,
Which die for goodness, who have lived for crime.

3 just] legitimate, innocent
5 adulterate] (1) corrupted; (2) adulterous
6 Give salutation to] greet (as in fellowship or complicitly)
8 which] who
in their wills] (1) in their lusts; (2) arbitrarily, wilfully
9 level] (1) guess at; (2) aim at
11 straight] morally straight
bevel] out of true, crooked
12 rank] lustful, fetid
14 reign] flourish, prosper

407
1 child of state] offspring of circumstance
2 for] as
unfathered] disowned
5 far from accident] out of the reach of chance
6 suffers not] does not deteriorate, endures
in smiling pomp] when favoured by the great
falls] succumbs
7 thralled discontent] embittered people suffering from oppression
8 our fashion] men like us
9 Policy] expediency
10 leases . . . hours] according to a short-term view
11 hugely politic] immensely and massively prudent
12 showers] downpours
13 fools of Time] time-servers, Time's dupes
14 Which] who
Which . . . crime] who, after a life of wrong-doing, renounce their time-serving to die for permanent values

408 WERE'T aught to me I bore the canopy,
  With my extern the outward honouring,
Or laid great bases for eternity,
  Which proves more short than waste or ruining?
Have I not seen dwellers on form and favour
  Lose all and more by paying too much rent,
For compound sweet forgoing simple savour,
  Pitiful thrivers, in their gazing spent?
No, let me be obsequious in thy heart,
  And take thou my oblation, poor but free, 10
Which is not mixed with seconds, knows no art
  But mutual render, only me for thee.
    Hence, thou suborned informer! A true soul
    When most impeached stands least in thy control.

409 TH'EXPENSE of spirit in a waste of shame
  Is lust in action, and till action, lust
Is perjured, murderous, bloody, full of blame,
  Savage, extreme, rude, cruel, not to trust,
Enjoyed no sooner but despised straight,
  Past reason hunted, and no sooner had,
Past reason hated, as a swallowed bait
  On purpose laid to make the taker mad;
Mad in pursuit, and in possession so,
  Had, having, and in quest to have, extreme, 10
A bliss in proof, and proved, a very woe;
  Before, a joy proposed; behind, a dream.
    All this the world well knows, yet none knows well
    To shun the heaven that leads men to this hell.

1 Were't aught to me] would there be any advantage to me if
canopy] a tent-like cloth carried over the head of a dignitary in a procession
2 With my . . . honouring] doing external honour to external worthiness
3 great bases] massive foundations
4 waste or ruining] the forces of decay or destruction
5 dwellers . . . favour] those who attach too much importance to outward show or formality
8 Pitiful] (1) pitiable; (2) contemptible
thrivers] successes, prosperous people
spent] used up, made bankrupt
9 obsequious] (1) servile; (2) dutiful
10 oblation] offering
11 seconds] inferior stuff
12 render] exchange
13 suborned] procured by bribery
14 control] power

409
1 expense] expenditure, lavishing
spirit] (1) vital energy; (2) spiritual essence; (3) semen
waste of shame] (1) shameful orgy; (2) desert of shame; (3) waste='waist'
3 full of blame] utterly culpable
4 extreme] violent
rude] brutal
not to trust] not to be trusted
11 in proof] being experienced
proved] once tested

410

My mistress' eyes are nothing like the sun;
Coral is far more red than her lips' red;
If snow be white, why then her breasts are dun;
If hairs be wires, black wires grow on her head.
I have seen roses damasked, red and white,
But no such roses see I in her cheeks,
And in some perfumes is there more delight
Than in the breath that from my mistress reeks.
I love to hear her speak, yet well I know
That music hath a far more pleasing sound. 10
I grant I never saw a goddess go;
My mistress when she walks treads on the ground.
And yet, by heaven, I think my love as rare
As any she belied with false compare.

411

When my love swears that she is made of truth
I do believe her, though I know she lies,
That she might think me some untutored youth,
Unlearned in the world's false subtleties.
Thus vainly thinking that she thinks me young,
Although she knows my days are past the best,
Simply I credit her false-speaking tongue;
On both sides thus is simple truth suppressed.
But wherefore says she not she is unjust?
And wherefore say not I that I am old? 10
O, love's best habit is in seeming trust,
And age in love loves not to have years told.
Therefore I lie with her, and she with me,
And in our faults by lies we flattered be.

412

Be wise as thou art cruel; do not press
My tongue-tied patience with too much disdain,
Lest sorrow lend me words, and words express
The manner of my pity-wanting pain.

3 dun] dull brown, brownish-grey
5 damasked] dappled, patterned
8 reeks] emanates
11 go] walk

411
1 truth] (1) honesty, fidelity; (2) veracity
2 she lies] (1) she tells an untruth; (2) she sleeps (with someone)
3 That] so that
7 Simply] (1) straightforwardly; (2) naïvely, foolishly
11 love's best habit is in] (1) love looks best dressed in; (2) love's wisest procedure is in
12 told] (1) counted; (2) divulged

412
1 Be . . . cruel] (1) be as wise as you are cruel; (2) be wise since you are cruel
press] (1) assail, harass; (2) torture with weights (prisoners who refused to speak might have heavy weights placed on them)
4 pity-wanting] unpitied

If I might teach thee wit, better it were,
Though not to love, yet, love, to tell me so;
As testy sick men, when their deaths be near,
No news but health from their physicians know.
For if I should despair, I should grow mad,
And in my madness might speak ill of thee. 10
Now this ill-wresting world is grown so bad
Mad sland'rers by mad ears believed be.
That I may not be so, nor thou belied,
Bear thine eyes straight, though thy proud heart go wide.

413

LOVE is too young to know what conscience is;
Yet who knows not conscience is born of love?
Then, gentle cheater, urge not my amiss,
Lest guilty of my faults thy sweet self prove.
For, thou betraying me, I do betray
My nobler part to my gross body's treason:
My soul doth tell my body that he may
Triumph in love; flesh stays no farther reason,
But, rising at thy name, doth point out thee
As his triumphant prize. Proud of this pride, 10
He is contented thy poor drudge to be,
To stand in thy affairs, fall by thy side.
No want of conscience hold it that I call
Her 'love' for whose dear love I rise and fall.

(Pub. 1609)

414

The Phoenix and Turtle

LET the bird of loudest lay
On the sole Arabian tree
Herald sad and trumpet be,
To whose sound chaste wings obey.

5 wit] practical wisdom
7 testy] fretful
8 know] are told
11 ill-wresting] that twists anything to the bad
13 belied] slandered
14 straight . . . wide] (from archery)
go wide] go wrong, range widely

413
1–2 Love . . . love] Cupid . . . the experience
2 conscience] (1) moral sense; (2) sexual activity (*con-science*, knowing the '*con*', the female pudendum)
3 amiss] fault

414
1 lay] song

But thou shrieking harbinger,
  Foul precurrer of the fiend,
  Augur of the fever's end,
To this troop come thou not near.

From this session interdict
  Every fowl of tyrant wing, 10
  Save the eagle, feathered king;
Keep the obsequy so strict.

Let the priest in surplice white
  That defunctive music can,
  Be the death-divining swan,
Lest the requiem lack his right.

And thou treble-dated crow,
  That thy sable gender mak'st
  With the breath thou giv'st and tak'st,
Mongst our mourners shalt thou go. 20

Here the anthem doth commence:
  Love and constancy is dead;
  Phoenix and the turtle fled
In a mutual flame from hence.

So they loved, as love in twain
  Had the essence but in one;
  Two distincts, division none;
Number there in love was slain.

Hearts remote, yet not asunder;
  Distance, and no space was seen 30
  Twixt this turtle and his queen;
But in them it were a wonder.

5 shrieking harbinger] i.e. the screech owl
6 precurrer] forerunner
10 fowl . . . wing] bird of prey
14 defunctive] funereal
can] is skilled in
16 his right] its due
17 treble-dated] long-lived
18–19 thy sable gender . . . tak'st] (there was a belief that crows conceived at the bill)
25 as] that
27 distincts] separate things
28 Number . . . slain] (since 'One is no number')
29 remote] at a distance from each other
32 But in them] except in them

So between them love did shine,
  That the turtle saw his right
  Flaming in the phoenix' sight;
Either was the other's mine.

Property was thus appalled,
  That the self was not the same;
  Single nature's double name
Neither two nor one was called. 40

Reason, in itself confounded,
  Saw division grow together,
  To themselves yet either neither,
Simple were so well compounded:

That it cried, 'How true a twain
  Seemeth this concordant one!
  Love hath reason, reason none,
If what parts can so remain.'

Whereupon it made this threne
  To the phoenix and the dove, 50
  Co-supremes and stars of love,
As chorus to their tragic scene.

THRENOS

Beauty, truth, and rarity,
Grace in all simplicity,
Here enclosed, in cinders lie.

Death is now the phoenix' nest;
And the turtle's loyal breast
To eternity doth rest.

Leaving no posterity,
'Twas not their infirmity, 60
It was married chastity.

Truth may seem, but cannot be;
Beauty brag, but 'tis not she;
Truth and beauty buried be.

34 saw his right] saw what was due to him, a return of love
36 mine] (1) source of wealth; (2) self
37 Property] selfhood; essential quality
44 Simple] single, indivisible
49 threne] threnody, funeral song
51 Co-supremes] joint rulers
53 truth] fidelity

To this urn let those repair
That are either true or fair;
For these dead birds sigh a prayer.

(Pub. 1601)

ANONYMOUS

415 *Crabbed age and youth*

CRABBED age and youth cannot live together:
Youth is full of pleasance, age is full of care;
Youth like summer morn, age like winter weather;
Youth like summer brave, age like winter bare.
Youth is full of sport, age's breath is short;
Youth is nimble, age is lame;
Youth is hot and bold, age is weak and cold;
Youth is wild, and age is tame.
Age, I do abhor thee; youth, I do adore thee;
O my love, my love is young: 10
Age, I do defy thee. O sweet shepherd, hie thee,
For methinks thou stays too long.

(Pub. 1599)

416 *Those eyes which set my fancy on a fire*

THOSE eyes which set my fancy on a fire,
Those crisped hairs which hold my heart in chains,
Those dainty hands which conquered my desire,
That wit which of my thoughts doth hold the reins!
Those eyes, for clearness do the stars surpass,
Those hairs, obscure the brightness of the sun,
Those hands more white than ever ivory was,
That wit, even to the skies hath glory won!
O eyes that pierce our hearts without remorse,
O hairs of right that wear a royal crown, 10
O hands that conquer more than Caesar's force,
O wit that turns huge kingdoms upside down!
Then Love be judge, what heart may thee withstand,
Such eyes, such hair, such wit, and such a hand.

(Pub. 1593)

415
1 Crabbed] cross-tempered, churlish
2 pleasance] pleasantness
4 brave] finely dressed
11 hie thee] hurry

414
2 crisped] curled

417 *A secret murder hath been done of late*

A SECRET murder hath been done of late,
  Unkindness found to be the bloody knife;
And she that did the deed a dame of state,
  Fair, gracious, wise, as any beareth life.

To quit herself this answer did she make:
  'Mistrust', quoth she, 'hath brought him to his end,
Which makes the man so much himself mistake,
  To lay the guilt unto his guiltless friend.'

Lady, not so; not feared I found my death,
  For no desert thus murdered is my mind; 10
And yet before I yield my fainting breath,
  I quit the killer, though I blame the kind.

You kill unkind, I die, and yet am true,
For at your sight my wound doth bleed anew.

(Pub. 1593)

418 *Sought by the world*

SOUGHT by the world, and hath the world disdained,
  Is she, my heart, for whom thou dost endure;
Unto whose grace sith kings have not obtained,
  Sweet is thy choice, though loss of life be sour;
    Yet to the man, whose youth such pains must prove,
    No better end than that which comes by love.

Steer then thy course unto the port of death,
  (Sith thy hard hap no better hap may find,)
Where, when thou shalt unlade thy latest breath,
  Envy herself shall swim, to save thy mind; 10
    Whose body sunk in search to gain that shore
    Where many a prince had perished before.

2 Unkindness] unnatural behaviour, unprovoked hostility
5 quit] acquit
6 Mistrust] lack of trust, suspicion
9 feared] afraid
12 kind] i.e. womankind

And yet, my heart, it might have been foreseen,
Sith skilful medicines mend each kind of grief;
Then in my breast full safely hadst thou been.
But thou, my heart, wouldst never me believe,
Who told thee true when first thou didst aspire,
Death was the end of every such desire.

(Pub. 1593)

419 *The brainsick race that wanton youth ensues*

THE brainsick race that wanton youth ensues
Without regard to grounded wisdom's lore,
As often as I think thereon renews
The fresh remembrance of an ancient sore:
Revoking to my pensive thoughts at last
The worlds of wickedness that I have passed.

And though experience bids me bite on bit,
And champ the bridle of a better smack,
Yet costly is the price of after-wit,
Which brings so cold repentance at her back: 10
And skill that's with so many losses bought
Men say is little better worth than nought.

And yet this fruit I must confess doth grow
Of folly's scourge: that though I now complain
Of error past, yet henceforth I may know
To shun the whip that threats the like again:
For wise men, though they smart a while, had liever
To learn experience at the last, than never.

(Pub. 1593)

420 *Feed still thy self, thou fondling, with belief*

FEED still thy self, thou fondling, with belief,
Go hunt thy hope, that never took effect;
Accuse the wrongs that oft hath wrought thy grief,
And reckon sure where reason would suspect.

419
1 ensues] follows, pursues
8 champ] chew
smack] taste, flavour
17 had liever] had rather

420
1 fondling] foolish person

Dwell in the dreams of wish and vain desire,
  Pursue the faith that flies and seeks to new;
Run after hopes that mock thee with retire,
  And look for love where liking never grew.

Devise conceits to ease thy careful heart,
  Trust upon times and days of grace behind, 10
Presume the rights of promise and desart,
  And measure love by thy believing mind.

Force thy affects that spite doth daily chase,
  Wink at the wrongs with wilful oversight,
See not the soil and stain of thy disgrace,
  Nor reck disdain, to dote on thy delight.

And when thou seest the end of thy reward,
  And these effects ensue of thine assault,
When rashness rues, that reason should regard,
  Yet still accuse thy fortune for the fault. 20

And cry, O love, O death, O vain desire,
When thou complainst the heat, and feeds the fire.

(Pub. 1593)

421 *A Counterlove*

DECLARE, O mind, from fond desires excluded,
That thou didst find erewhile, by Love deluded.

An eye, the plot whereon Love sets his gin;
  Beauty, the trap wherein the heedless fall;
A smile, the train that draws the simple in;
  Sweet words, the wily instrument of all;
    Entreaties, posts; fair promises are charms;
    Writing, the messenger that woos our harms.

9 careful] care-filled
13 affects] inner feelings
14 Wink at] close your eyes to

421
Counterlove] antidote to love
2 That thou didst find] i.e. 'what thou didst find'
3 plot] small area of ground
gin] snare
5 train] pieces of carrion laid in a trail to lure wild animals into a trap
7 posts] couriers, letter-bearers

'Mistress', and 'servant', titles of mischance;
Commandments done, the act of slavery; 10
Their colours worn, a clownish cognisance;
And double duty, petty drudgery;
And when she twines and dallies with thy locks,
Thy freedom then is brought into the stocks.

To touch her hand, her hand binds thy desire;
To wear her ring, her ring is Nessus' gift;
To feel her breast, her breast doth blow the fire;
To see her bare, her bare a baleful drift;
To bait thine eyes thereon, is loss of sight;
To think of it, confounds thy senses quite. 20

Kisses, the keys to sweet consuming sin;
Closings, Cleopatra's adders at thy breast;
Feigned resistance then she will begin,
And yet unsatiable in all the rest;
And when thou dost unto the act proceed,
The bed doth groan and tremble at the deed.

Beauty, a silver dew that falls in May;
Love is an egg-shell, with that humour filled;
Desire, a winged boy, coming that way,
Delights and dallies with it in the field. 30
The fiery sun draws up the shell on high;
Beauty decays, Love dies, Desire doth fly.

Unharmed, give ear: that thing is haply caught
That cost some dear, if thou may'st ha't for nought.

(Pub. 1593)

11 clownish cognisance] badge or insignia of boors or peasants
16 Nessus' gift] i.e. poisoned or fatal (like the poisoned shirt the centaur Nessus gave to Deianira, the wife of Hercules)
18 bare] naked body or part of it, bare skin
19 bait] feed
22 Closings] drawing together, embraces
28 humour] moisture
33 haply] happily

ROBERT DEVEREUX, EARL OF ESSEX
1566–1601

422 *Happy were he*

HAPPY were he could finish forth his fate
  In some unhaunted desert, most obscure
From all societies, from love and hate
  Of worldly folks; then might he sleep secure,
Then wake again and give God ever praise,
  Content with hips and haws and bramble-berry;
In contemplation spending all his days,
  And change of holy thoughts to make him merry;
That, when he dies, his tomb may be a bush,
Where harmless robin dwells with gentle thrush. 10

(Pub. 1853)

ANONYMOUS

423 *Were I a king*

WERE I a king, I could command content;
  Were I obscure, unknown should be my cares;
And were I dead, no thoughts should me torment,
  Nor words, nor wrongs, nor loves, nor hopes, nor fears.
    A doubtful voice, of three things one to crave,
    A kingdom, or a cottage, or a grave.

(Pub. 1594)

1 Happy were he] happy would he be who

2 unhaunted] unfrequented desert] lonely place

THOMAS CAMPION
1567–1620

424 *What fair pomp*

WHAT fair pomp have I spied of glittering ladies,
With locks sparkled abroad, and rosy coronet
On their ivory brows, tracked to the dainty thighs
With robes like Amazons, blue as violet,
With gold aglets adorned, some in a changeable
Pale, with spangs wavering taught to be moveable?

Then those knights, that afar off with dolorous viewing
Cast their eyes hitherward, lo, in an agony,
All unbraced, cry aloud, their heavy state ruing:
Moist cheeks with blubbering, painted as ebony 10
Black; their feltered hair torn with wrathful hand;
And whiles astonied stark in a maze they stand.

But hark, what merry sound, what sudden harmony!
Look, look near the grove, where the ladies do tread
With their knights the measures wayed by the melody.
Wantons, whose traversing make men enamoured;
Now they feign an honour, now by the slender waist
He must lift her aloft, and seal a kiss in haste.

Straight down under a shadow for weariness they lie
With pleasant dalliance, hand knit with arm in arm; 20
Now close, now set aloof, they gaze with an equal eye,
Changing kisses alike; straight with a false alarm,
Mocking kisses alike, pout with a lovely lip.
Thus drowned with jollities their merry days do slip.

1 pomp] procession, parade
2 sparkled] scattered
4 robes . . . Amazons] i.e. with their tunics pulled up to their thighs
5 aglets] metal ends of laces
changeable] showing different colours, shot
6 Pale] cloak
spangs] spangles
11 feltered] matted
12 whiles astonied] sometimes dazed, paralysed
in a maze] in a state of bewilderment
15 wayed] weighed, marked off
17 honour] bow
22 Changing] exchanging

But stay, now I discern they go on a pilgrimage
  Towards Love's holy land, fair Paphos or Cyprus.
Such devotion is meet for a blithesome age;
  With sweet youth it agrees well to be amorous.
    Let old angry fathers lurk in an hermitage.
    Come, we'll associate this jolly pilgrimage. 30

(1591)

425 *My sweetest Lesbia*

My sweetest Lesbia, let us live and love;
And, though the sager sort our deeds reprove,
Let us not weigh them. Heaven's great lamps do dive
Into their west, and straight again revive.
But soon as once set is our little light,
Then must we sleep one ever-during night.

If all would lead their lives in love like me,
Then bloody swords and armour should not be;
No drum nor trumpet peaceful sleeps should move,
Unless alarm came from the camp of Love. 10
But fools do live and waste their little light,
And seek with pain their ever-during night.

When timely death my life and fortune ends,
Let not my hearse be vexed with mourning friends
But let all lovers, rich in triumph, come
And with sweet pastimes grace my happy tomb.
And, Lesbia, close up thou my little light,
And crown with love my ever-during night.

(1601)

426 *I care not for these ladies*

I care not for these ladies
That must be wooed and prayed:
Give me kind Amaryllis,
The wanton country maid.
Nature Art disdaineth;
Her beauty is her own.
  Her when we court and kiss,
  She cries: 'Forsooth, let go!'
  But when we come where comfort is,
  She never will say no. 10

30 associate] join

425
9 move] disturb, trouble

If I love Amaryllis,
She gives me fruit and flowers;
But if we love these ladies,
We must give golden showers.
Give them gold that sell love,
Give me the nut-brown lass,
  Who when we court and kiss,
  She cries: 'Forsooth, let go!'
  But when we come where comfort is,
  She never will say no. 20

These ladies must have pillows
And beds by strangers wrought.
Give me a bower of willows
Of moss and leaves unbought,
And fresh Amaryllis
With milk and honey fed,
  Who when we court and kiss,
  She cries: 'Forsooth, let go!'
  But when we come where comfort is,
  She never will say no. 30

(1601)

427 *Follow thy fair sun*

FOLLOW thy fair sun, unhappy shadow.
  Though thou be black as night,
  And she made all of light,
Yet follow thy fair sun, unhappy shadow.

Follow her whose light thy light depriveth.
  Though here thou livest disgraced,
  And she in heaven is placed,
Yet follow her whose light the world reviveth.

Follow those pure beams whose beauty burneth,
  That so have scorched thee, 10
  As thou still black must be,
Till her kind beams thy black to brightness turneth.

Follow her, while yet her glory shineth.
  There comes a luckless night,
  That will dim all her light;
And this the black unhappy shade divineth.

Follow still, since so thy fates ordained.
The sun must have his shade,
Till both at once do fade,
The sun still proved, the shadow still disdained. 20

(1601)

428 *When to her lute Corinna sings*

WHEN to her lute Corinna sings,
Her voice revives the leaden strings,
And doth in highest notes appear,
As any challenged echo clear.
But when she doth of mourning speak,
E'en with her sighs the strings do break.

And as her lute doth live or die,
Led by her passion, so must I.
For when of pleasure she doth sing,
My thoughts enjoy a sudden spring; 10
But if she doth of sorrow speak,
E'en from my heart the strings do break.

(1601)

429 *Follow your saint*

FOLLOW your saint, follow with accents sweet;
Haste you, sad notes, fall at her flying feet.
There, wrapped in cloud of sorrow, pity move,
And tell the ravisher of my soul I perish for her love.
But if she scorns my never-ceasing pain,
Then burst with sighing in her sight, and ne'er return again.

All that I sung still to her praise did tend.
Still she was first, still she my songs did end.
Yet she my love and music both doth fly,
The music that her echo is, and beauty's sympathy. 10
Then let my notes pursue her scornful flight;
It shall suffice that they were breathed, and died for her delight.

(1601)

20 proved] approved

428
4 challenged] aroused

430 *The man of life upright*

THE man of life upright,
Whose guiltless heart is free
From all dishonest deeds
Or thought of vanity:

The man whose silent days
In harmless joys are spent,
Whom hopes cannot delude,
Nor sorrow discontent:

That man needs neither towers
Nor armour for defence, 10
Nor secret vaults to fly
From thunder's violence.

He only can behold
With unaffrighted eyes
The horrors of the deep
And terrors of the skies.

Thus scorning all the cares
That fate or fortune brings,
He makes the heaven his book,
His wisdom heavenly things, 20

Good thoughts his only friends,
His wealth a well-spent age,
The earth his sober inn
And quiet pilgrimage.

(1601)

431 *Hark, all you ladies that do sleep*

HARK, all you ladies that do sleep,
The fairy queen Proserpina
Bids you awake, and pity them that weep.
You may do in the dark
What the day doth forbid.
Fear not the dogs that bark;
Night will have all hid.

But if you let your lovers moan,
The fairy queen Proserpina
Will send abroad her fairies every one, 10
That shall pinch black and blue
Your white hands and fair arms,
That did not kindly rue
Your paramours' harms.

In myrtle arbours on the downs,
The fairy queen Proserpina,
This night by moonshine, leading merry rounds,
Holds a watch with sweet Love,
Down the dale, up the hill;
No plaints or groans may move 20
Their holy vigil.

All you that will hold watch with Love,
The fairy queen Proserpina
Will make you fairer than Dione's dove.
Roses red, lilies white,
And the clear damask hue,
Shall on your cheeks alight.
Love will adorn you.

All you that love or loved before,
The fairy queen Proserpina 30
Bids you increase that loving humour more.
They that have not yet fed
On delight amorous,
She vows that they shall lead
Apes in Avernus.

(Pr. 1591, 1601)

432 *When thou must home to shades of underground*

WHEN thou must home to shades of underground,
And there arrived, a new admired guest,
The beauteous spirits do engirt thee round,
White Iope, blithe Helen and the rest,
To hear the stories of thy finished love
From that smooth tongue, whose music hell can move:

13 rue] pity
34–5 lead . . . Avernus] leading apes in hell was a proverbial punishment for old maids

Then wilt thou speak of banqueting delights,
  Of masks and revels which sweet youth did make,
Of tourneys and great challenges of knights,
  And all these triumphs for thy beauty's sake. 10
When thou hast told these honours done to thee,
Then tell, O tell, how thou didst murder me.

(1601)

THOMAS NASHE
1567–*c.*1601

from *Summer's Last Will and Testament*

433

FAIR summer droops, droop men and beasts therefore;
So fair a summer look for never more.
All good things vanish, less than in a day,
Peace, plenty, pleasure, suddenly decay.
  Go not yet away, bright soul of the sad year;
  The earth is hell when thou leav'st to appear.

What, shall those flowers, that decked thy garland erst,
Upon thy grave be wastefully dispersed?
O trees, consume your sap in sorrow's source;
Streams, turn to tears your tributary course. 10
  Go not yet hence, bright soul of the sad year;
  The earth is hell when thou leav'st to appear.

434

SPRING, the sweet spring, is the year's pleasant king;
Then blooms each thing, then maids dance in a ring,
Cold doth not sting, the pretty birds do sing:
  Cuckoo, jug-jug, pu-we, to-witta-woo!

The palm and may make country houses gay,
Lambs frisk and play, the shepherds pipe all day,
And we hear aye birds tune this merry lay:
  Cuckoo, jug-jug, pu-we, to-witta-woo!

The fields breathe sweet, the daisies kiss our feet,
Young lovers meet, old wives a-sunning sit;
In every street these tunes our ears do greet:
  Cuckoo, jug-jug, pu-we, to-witta-woo!
    Spring, the sweet spring!

435

ADIEU, farewell earth's bliss,
This world uncertain is;
Fond are life's lustful joys,
Death proves them all but toys,
None from his darts can fly.
I am sick, I must die.
Lord, have mercy on us!

Rich men, trust not in wealth,
Gold cannot buy you health;
Physic himself must fade, 10
All things to end are made.
The plague full swift goes by.
I am sick, I must die.
Lord, have mercy on us!

Beauty is but a flower
Which wrinkles will devour;
Brightness falls from the air,
Queens have died young and fair,
Dust hath closed Helen's eye.
I am sick, I must die. 20
Lord, have mercy on us!

Strength stoops unto the grave,
Worms feed on Hector brave,
Swords may not fight with fate,
Earth still holds ope her gate.
Come! come! the bells do cry.
I am sick, I must die.
Lord, have mercy on us!

Wit with his wantonness
Tasteth death's bitterness; 30
Hell's executioner
Hath no ears for to hear
What vain art can reply.
I am sick, I must die.
Lord, have mercy on us!

Haste, therefore, each degree,
To welcome destiny.
Heaven is our heritage,
Earth but a player's stage;
Mount we unto the sky. 40
I am sick, I must die.
Lord, have mercy on us!

436 AUTUMN hath all the summer's fruitful treasure;
Gone is our sport, fled is poor Croydon's pleasure.
Short days, sharp days, long nights come on apace,
Ah, who shall hide us from the winter's face?
Cold doth increase, the sickness will not cease,
And here we lie, God knows, with little ease.
From winter, plague, and pestilence, good Lord,
deliver us!

London doth mourn, Lambeth is quite forlorn;
Trades cry, woe worth that ever they were born.
The want of term is town and city's harm; 10
Close chambers we do want, to keep us warm.
Long banished must we live from our friends;
This low-built house will bring us to our ends.
From winter, plague, and pestilence, good Lord,
deliver us!

(Wr. 1592; pub. 1600)

ANONYMOUS

[*Madrigals set by Thomas Morley*]

437 CRUEL, you pull away too soon your lips whenas you
kiss me;
But you should hold them still, and then should you
bliss me.
Now or ere I taste them,
Straight away they haste them.
But you perhaps retire them
To move my thoughts thereby the more to fire them.
Alas, such baits you need to find out never:
If you would but let me, I would kiss you ever.

(1593)

438 WHITHER away so fast
From your true love approved?
What haste, I say, what haste,
Tell me, my darling dear beloved?
Then we will try
Who best runs, thou or I.

9 woe worth] cursed be

10 term] the legal term (which brought trade to the city)

Then lo, I come! Dispatch thee!
Hence I say, or else I catch thee.
No, think not thus away to 'scape without me.
But run! You need not doubt me. 10
What, faint you? Of your sweet feet forsaken?
O well I see you mean to mock me.
Run, I say, or else I catch you.
What? You halt? O do you so?
Alack the while! What! Are you down?
Pretty maid, well overtaken!

(1593)

439

WHEN, lo, by break of morning
My love herself adorning
Doth walk the woods so dainty,
Gath'ring sweet violets and cowslips plenty,
The birds enamoured sing and praise my Flora:
Lo, here a new Aurora!

(1595)

440

SWEET nymph, come to thy lover.
Lo here, alone, our loves we may discover,
Where the sweet nightingale with wanton gloses,
Hark, her love too discloses.

(1595)

441

I GO before, my darling.
Follow thou to the bower in the close alley.
There we will together
Sweetly kiss each other,
And like two wantons dally.

(1595)

442

MIRACULOUS love's wounding!
E'en those darts my sweet Phyllis
So fiercely shot against my heart rebounding,
Are turned to roses, violets and lilies,
With odour sweet abounding.

(1595)

440
3 gloses] explanatory remarks (as on an obscure text)

443

NOW is the month of maying,
When merry lads are playing,
Each with his bonny lass
Upon the greeny grass.

The Spring, clad in all gladness,
Doth laugh at Winter's sadness,
And to the bagpipe's sound
The nymphs tread out their ground.

Fie then, why sit we musing,
Youth's sweet delight refusing? 10
Say, dainty nymphs, and speak,
Shall we play barley-break?

(1595)

444

SING we and chant it
While love doth grant it.
Not long youth lasteth,
And old age hasteth.
Now is best leisure
To take our pleasure.

All things invite us
Now to delight us.
Hence, care, be packing!
No mirth be lacking! 10
Let spare no treasure
To live in pleasure.

(1595)

445

LADY, those cherries plenty,
Which grow on your lips dainty,
Ere long will fade and languish.
Then now, while yet they last them,
O let me pull and taste them.

(1595)

12 barley-break] country game played by pairs (one of each sex), in which one pair, being left in a middle den called 'hell', had to catch the others as they ran through it

446

Lo, where with flowery head and hair all brightsome,
Rosy cheeked, crystal eyed, e'en weeping lightsome,
The fresh Aurora springeth!
And wanton Flora flingeth
Amorous odours unto the winds delightsome!
Ah, for pity and anguish
Only my heart doth languish!

(1597)

447

Damon and Phyllis squared,
And to point her the place the nymph him dared.
Her glove she down did cast him,
And to meet her alone she bade him haste him.
Alike their weapons were, alike their smiting,
And little Love came running to the fighting.

(1597)

448

Lady, you think you spite me,
When by the lip you bite me.
But if you think it trouble,
Then let my pain be double,
Ay triple, but you bliss me,
For though you bite, you kiss me,
And with sour sweet delight me.

(1597)

449

You black bright stars, that shine while daylight lasteth,
Ah, why haste you away when night time hasteth?
In darker nights the stars still seem the lighter.
On me shine then a-nights with your beams brighter.
Beams that are cause my heart hath so aspired,
Fire mounts aloft, and they my heart have fired.

(1597)

450

Ladies, you see time flieth,
And beauty too, it dieth.
Then take your pleasure,
While you have leisure.
Nor be so dainty
Of that which you have plenty.

(1597)

447
1 squared] fell out, quarrelled
2 point] appoint, settle

450
5 dainty] sparing

BARNABE BARNES
1569?–1609

from *Parthenophil and Parthenophe*

451

Ode II

LOVELY Maya, Hermes' mother,
Of fair Flora much befriended,
To whom this sweet month is commended,
This month more sweet than any other,
By thy sweet sovereignty defended.

Daisies, cowslips, and primroses,
Fragrant violets, and sweet mynthe,
Matched with purple hyacinth:
Of these, each where, nymphs make trim posies,
Praising their mother Berycinth. 10

Behold, a herd of jolly swains
Go flocking up and down the mead.
A troop of lovely nymphs do tread,
And dearnly dancing on yon plains,
Each doth, in course, her hornpipe lead.

Before the grooms plays Piers the piper.
They bring in hawthorn and sweet brere;
And damask-roses they would bear,
But them they leave till they be riper.
The rest, round morrises dance there. 20

With frisking gambols, and such glee,
Unto the lovely nymphs they haste,
Who there in decent order placed
Expect who shall Queen Flora be
And with the May Crown chiefly graced.

7 mynthe] mint
9 each where] everywhere
14 dearnly] so as to be hidden
15 in course] in order, in turn
17 sweet brere] sweet briar, eglantine (species of rose)
20 morrises] morris-dances
24 Expect] await

The shepherds poopen in their pipe,
  One leads his wench a country round;
  Another sits upon the ground,
  And doth his beard from drivel wipe
  Because he would be handsome found. 30

To see the frisking and the scouping!
  To hear the herdgrooms' wooing speeches!
  Whiles one to dance his girl beseeches.
  The lead-heeled lazy luskins louping,
  Fling out in their new motley breeches.

This done, with jolly cheer and game
  The batch'lor swains and young nymphs met,
  Where in an arbour they were set.
  Thither, to choose a Queen, they came,
  And soon concluded her to fet. 40

There with a garland they did crown
  Parthenophe, my sweet true-love;
  Whose beauty all the nymphs above
  Did put the lovely Graces down.
  The swains with shouts rocks' echoes move.

To see the rounds and morris dances,
  The leaden galliards, for her sake!
  To hear those songs the shepherds make!
  One with his hobby-horse still prances,
  Whiles some with flowers an highway make. 50

There in a mantle of light green,
  Reserved by custom for that day,
  Parthenophe they did array,
  And did create her Summer's Queen
  And Ruler of their merry May.

(1593)

26 poopen] toot, blow
31 scouping] bounding, capering
34 luskins] sluggards
louping] leaping
40 fet] dress, deck

452

Sestina

THEN first with locks dishevelled and bare,
Strait girded, in a cheerful calmy night,
Having a fire made of green cypress wood,
And with male frankincense on altar kindled,
I call on threefold Hecate with tears,
And here, with loud voice, invocate the Furies

For their assistance to me with their furies,
Whilst snowy steeds in coach bright Phoebe bare.
Ay me, Parthenophe smiles at my tears.
I neither take my rest by day or night, 10
Her cruel loves in me such heat have kindled.
Hence, goat, and bring her to me raging wood!

Hecate, tell which way she comes through the wood.
This wine about this altar, to the Furies
I sprinkle, whiles the cypress boughs be kindled,
This brimstone earth within her bowels bare,
And this blue incense, sacred to the night.
This hand, perforce, from this bay this branch tears.

So be she brought which pitied not my tears,
And as it burneth with the cypress wood 20
So burn she with desire, by day and night.
You gods of vengeance, and avengeful Furies,
Revenge, to whom I bend on my knees bare!
Hence, goat, and bring her with love's outrage kindled!

Hecate, make signs, if she with love come kindled;
Think on my passions, Hecate, and my tears.
This rose marine (whose branch she chiefly bare,
And loved best) I cut, both bark and wood,
Broke with this brazen axe, and in love's furies
I tread on it, rejoicing in this night, 30

And saying 'Let her feel such wounds this night!'
About this altar, and rich incense kindled,
This lace and vervain to love's bitter furies
I bind and strew, and with sad sighs and tears
About I bear her image, raging wood.
Hence, goat, and bring her from her bedding bare!

4 male frankincense] frankincense that hangs suspended in a globular drop which in shape resembles testicles

12 wood] mad

16 brimstone] sulphur

27 rose marine] rosemary

Hecate, reveal if she like passions bear.
I knit three true-love knots (this is love's night)
Of three discoloured silks, to make her wood;
But she scorns Venus till her loves be kindled 40
And till she find the grief of sighs and tears.
Sweet Queen of Loves, for mine unpitied furies

Alike torment her with such scalding furies!
And this turtle, when the loss she bare
Of her dear make, in her kind did shed tears,
And mourning did seek him all day and night:
Let such lament in her for me be kindled,
And mourn she still, till she run raging wood.

Hence, goat, and bring her to me raging wood!
These letters, and these verses to the Furies, 50
Which she did write, all in this flame be kindled.
Me with these papers in vain hope she bare
That she to day would turn mine hopeless night.
These, as I rent and burn, so fury tears

Her hardened heart, which pitied not my tears.
The wind-shaked trees make murmur in the wood,
The waters roar at this thrice-sacred night,
The winds come whisking shill to note her furies:
Trees, woods, and winds a part in my plaints bare,
And knew my woes, now joy to see her kindled. 60

See whence she comes, with loves enraged and kindled!
The pitchy clouds in drops send down their tears,
Owls scritch, dogs bark, to see her carried bare;
Wolves yowl and cry, bulls bellow through the wood,
Ravens croap! Now, now, I feel love's fiercest furies!
Seest thou that black goat, brought this silent night

Through empty clouds by th' Daughters of the Night?
See how on him she sits, with love-rage kindled,
Hither perforce brought with avengeful Furies?
Now I wax drowsy, now cease all my tears, 70
Whilst I take rest, and slumber near this wood:
Ah me! Parthenophe naked and bare!

44 turtle] turtle dove kind] nature shill] loudly, shrilly
45 make] mate 58 whisking] rushing 65 croap] croak

Come, blessed goat, that my sweet lady bare!
  Where hast thou been, Parthenophe, this night?
  What, cold? Sleep by this fire of cypress wood,
  Which I much longing for thy sake have kindled.
  Weep not! Come, Loves, and wipe away her tears.
  At length yet, wilt thou take away my furies?

Ay me, embrace me! See those ugly Furies!
  Come to my bed, lest they behold thee bare, 80
  And bear thee hence. They will not pity tears,
  And these still dwell in everlasting night.
  Ah, Loves (sweet love!) sweet fires for us hath kindled,
  But not inflamed with frankincense or wood!

The Furies, they shall hence into the wood,
  Whiles Cupid shall make calmer his hot furies
  And stand appeased at our fires kindled.
  Join, join, Parthenophe! Thyself unbare.
  None can perceive us in the silent night.
  Now will I cease from sighs, laments, and tears. 90

And cease, Parthenophe, sweet, cease thy tears.
  Bear golden apples, thorns in every wood!
  Join heavens, for we conjoin this heavenly night!
  Let alder trees bear apricocks (die, furies!),
  And thistles pears, which prickles lately bare!
  Now both in one with equal flame be kindled!

Die, magic boughs, now die, which late were kindled!
  Here is mine heaven! Loves drop instead of tears!
  It joins, it joins! Ah, both embracing bare!
  Let nettles bring forth roses in each wood! 100
  Last ever-verdant, woods! Hence, former furies!
  O die, live, joy! What! Last continual, night!

Sleep Phoebus still with Thetis! Rule still, night!
  I melt in love! Love's marrow-flame is kindled!
  Here will I be consumed in love's sweet furies!
  I melt, I melt! Watch, Cupid, my love-tears!
  If these be Furies, O let me be wood!
  If all the fiery element I bare,

'Tis now acquitted! Cease your former tears,
  For as she once with rage my body kindled, 110
  So in hers am I buried this night.

(1593)

JOHN DONNE

1572–1631

Satire I

453 *[A London street]*

AWAY thou fondling motley humourist,
Leave me, and in this standing wooden chest,
Consorted with these few books, let me lie
In prison, and here be coffined, when I die;
Here are God's conduits, grave divines; and here
Nature's secretary, the Philosopher;
And jolly statesmen, which teach how to tie
The sinews of a city's mystic body;
Here gathering chroniclers, and by them stand
Giddy fantastic poets of each land. 10
Shall I leave all this constant company,
And follow headlong, wild uncertain thee?
First swear by thy best love in earnest
(If thou which lov'st all, canst love any best)
Thou wilt not leave me in the middle street,
Though some more spruce companion thou dost meet,
Not though a captain do come in thy way
Bright parcel gilt, with forty dead men's pay,
Nor though a brisk perfumed pert courtier
Deign with a nod, thy courtesy to answer. 20
Nor come a velvet Justice with a long
Great train of blue coats, twelve, or fourteen strong,
Wilt thou grin or fawn on him, or prepare
A speech to court his beauteous son and heir.
For better or worse take me, or leave me:
To take, and leave me is adultery.
O monstrous, superstitious puritan,
Of refined manners, yet ceremonial man,

1 fondling motley humourist] foolish changeable zany
2 chest] i.e. his study
5 conduits] channels
6 the Philosopher] Aristotle or perhaps any natural scientist
7 jolly] arrogant
9 gathering] scavenging
15 in the middle street] in the middle of the street
18 parcel gilt] partly gilded
22 blue coats] servants
27 puritan] purist

That when thou meet'st one, with inquiring eyes
Dost search, and like a needy broker prize 30
The silk, and gold he wears, and to that rate
So high or low, dost raise thy formal hat:
That wilt consort none, until thou have known
What lands he hath in hope, or of his own,
As though all thy companions should make thee
Jointures, and marry thy dear company.
Why shouldst thou (that dost not only approve,
But in rank itchy lust, desire, and love
The nakedness and barrenness to enjoy,
Of thy plump muddy whore, or prostitute boy) 40
Hate virtue, though she be naked, and bare?
At birth, and death, our bodies naked are;
And till our souls be unapparelled
Of bodies, they from bliss are banished.
Man's first blessed state was naked, when by sin
He lost that, yet he was clothed but in beast's skin,
And in this coarse attire, which I now wear,
With God, and with the Muses I confer.
But since thou like a contrite penitent,
Charitably warned of thy sins, dost repent 50
These vanities, and giddinesses, lo
I shut my chamber door, and come, let's go.
But sooner may a cheap whore, that hath been
Worn by as many several men in sin,
As are black feathers, or musk-colour hose,
Name her child's right true father, 'mongst all those:
Sooner may one guess, who shall bear away
The Infanta of London, heir to an India;
And sooner may a gulling weather spy
By drawing forth heaven's scheme tell certainly 60
What fashioned hats, or ruffs, or suits next year
Our subtle-witted antic youths will wear;
Than thou, when thou depart'st from me, canst show
Whither, why, when, or with whom thou wouldst go.
But how shall I be pardoned my offence
That thus have sinned against my conscience?
Now we are in the street; he first of all
Improvidently proud, creeps to the wall,
And so imprisoned, and hemmed in by me
Sells for a little state his liberty; 70

30 broker] shopkeeper
prize] appraise
40 muddy] foul, gross
58 Infanta of London] richest heiress in London
59 gulling] cheating
62 antic] fantastic

Yet though he cannot skip forth now to greet
Every fine silken painted fool we meet,
He them to him with amorous smiles allures,
And grins, smacks, shrugs, and such an itch endures,
As prentices, or school-boys which do know
Of some gay sport abroad, yet dare not go.
And as fiddlers stop lowest, at highest sound,
So to the most brave, stoops he nigh'st the ground.
But to a grave man, he doth move no more
Than the wise politic horse would heretofore, 80
Or thou O elephant or ape wilt do,
When any names the King of Spain to you.
Now leaps he upright, jogs me, and cries, 'Do you see
Yonder well-favoured youth?' 'Which?' 'Oh, 'tis he
That dances so divinely'; 'Oh,' said I,
'Stand still, must you dance here for company?'
He drooped, we went, till one (which did excel
Th' Indians, in drinking his tobacco well)
Met us; they talked; I whispered, 'Let us go,
'T may be you smell him not, truly I do.' 90
He hears not me, but, on the other side
A many-coloured peacock having spied,
Leaves him and me; I for my lost sheep stay;
He follows, overtakes, goes on the way,
Saying, 'Him whom I last left, all repute
For his device, in handsoming a suit,
To judge of lace, pink, panes, print, cut, and pleat
Of all the Court, to have the best conceit.'
'Our dull comedians want him, let him go;
But Oh, God strengthen thee, why stoop'st thou so?' 100
'Why? he hath travelled.' 'Long?' 'No, but to me
(Which understand none), he doth seem to be
Perfect French, and Italian'; I replied,
'So is the pox'; he answered not, but spied
More men of sort, of parts, and qualities;
At last his love he in a window spies,
And like light dew exhaled, he flings from me
Violently ravished to his lechery.
Many were there, he could command no more;
He quarrelled, fought, bled; and turned out of door 110
Directly came to me hanging the head,
And constantly a while must keep his bed.

(Wr. 1593; pub. 1633)

78 brave] showily dressed
96 handsoming] embellishing
97 pink] decorative eyehole
panes] decorative strips or slashes
print] crimping of the pleats of a ruff
98 conceit] idea
105 sort] rank

Satire 3

454 [*The search for true religion*]

KIND pity chokes my spleen; brave scorn forbids
Those tears to issue which swell my eye-lids.
I must not laugh, nor weep sins, and be wise,
Can railing then cure these worn maladies?
Is not our mistress fair religion,
As worthy of all our soul's devotion,
As virtue was to the first blinded age?
Are not heaven's joys as valiant to assuage
Lusts, as earth's honour was to them? Alas,
As we do them in means, shall they surpass 10
Us in the end, and shall thy father's spirit
Meet blind philosophers in heaven, whose merit
Of strict life may be imputed faith, and hear
Thee, whom he taught so easy ways and near
To follow, damned? O if thou dar'st, fear this;
This fear great courage, and high valour is.
Dar'st thou aid mutinous Dutch, and dar'st thou lay
Thee in ships' wooden sepulchres, a prey
To leaders' rage, to storms, to shot, to dearth?
Dar'st thou dive seas, and dungeons of the earth? 20
Hast thou courageous fire to thaw the ice
Of frozen north discoveries? and thrice
Colder than salamanders, like divine
Children in th'oven, fires of Spain, and the line,
Whose countries limbecks to our bodies be,
Canst thou for gain bear? and must every he
Which cries not, 'Goddess!' to thy mistress, draw,
Or eat thy poisonous words? courage of straw!
O desperate coward, wilt thou seem bold, and
To thy foes and his (who made thee to stand 30
Sentinel in his world's garrison) thus yield,
And for forbidden wars, leave th'appointed field?
Know thy foes: the foul Devil, he, whom thou
Strivest to please, for hate, not love, would allow

1 spleen] the seat of laughter
3 laugh] laugh to scorn
weep] lament
4 worn] long-standing
7 blinded] i.e. pagan
8 valiant] powerful
12 blind philosophers] pagan moralists
23 salamanders] creatures supposed able to live in fire
24 line] equator
25 limbecks] alembics

Thee fain, his whole realm to be quit; and as
The world's all parts wither away and pass,
So the world's self, thy other loved foe, is
In her decrepit wane, and thou loving this,
Dost love a withered and worn strumpet; last,
Flesh (itself's death) and joys which flesh can taste, 40
Thou lovest; and thy fair goodly soul, which doth
Give this flesh power to taste joy, thou dost loathe.
Seek true religion. O where? Mirreus
Thinking her unhoused here, and fled from us,
Seeks her at Rome, there, because he doth know
That she was there a thousand years ago,
He loves her rags so, as we here obey
The statecloth where the Prince sate yesterday.
Crants to such brave loves will not be enthralled,
But loves her only, who at Geneva is called 50
Religion, plain, simple, sullen, young,
Contemptuous, yet unhandsome; as among
Lecherous humours, there is one that judges
No wenches wholesome, but coarse country drudges.
Graius stays still at home here, and because
Some preachers, vile ambitious bawds, and laws
Still new like fashions, bid him think that she
Which dwells with us, is only perfect, he
Embraceth her, whom his godfathers will
Tender to him, being tender, as wards still 60
Take such wives as their guardians offer, or
Pay values. Careless Phrygius doth abhor
All, because all cannot be good, as one
Knowing some women whores, dares marry none.
Gracchus loves all as one, and thinks that so
As women do in divers countries go
In divers habits, yet are still one kind,
So doth, so is religion; and this blind-
ness too much light breeds; but unmoved thou
Of force must one, and forced but one allow; 70
And the right; ask thy father which is she,
Let him ask his; though truth and falsehood be
Near twins, yet truth a little elder is;
Be busy to seek her, believe me this,
He's not of none, nor worst, that seeks the best.
To adore, or scorn an image, or protest,

49 brave] showy, outwardly splendid
enthralled] enslaved
51 sullen] dismal
60 being tender] because he is young and weak
62 values] fines
70 allow] approve

May all be bad; doubt wisely, in strange way
To stand inquiring right, is not to stray;
To sleep, or run wrong is. On a huge hill,
Cragged, and steep, Truth stands, and he that will 80
Reach her, about must, and about must go;
And what the hill's suddenness resists, win so;
Yet strive so, that before age, death's twilight,
Thy soul rest, for none can work in that night.
To will, implies delay, therefore now do.
Hard deeds, the body's pains; hard knowledge too
The mind's endeavours reach, and mysteries
Are like the sun, dazzling, yet plain to all eyes.
Keep the truth which thou hast found; men do not stand
In so ill case here, that God hath with his hand 90
Signed kings blank-charters to kill whom they hate,
Nor are they vicars, but hangmen to Fate.
Fool and wretch, wilt thou let thy soul be tied
To man's laws, by which she shall not be tried
At the last day? Or will it then boot thee
To say a Philip, or a Gregory,
A Harry, or a Martin taught thee this?
Is not this excuse for mere contraries,
Equally strong; cannot both sides say so?
That thou mayest rightly obey power, her bounds know; 100
Those past, her nature, and name is changed; to be
Then humble to her is idolatry.
As streams are, power is; those blessed flowers that dwell
At the rough stream's calm head, thrive and prove well,
But having left their roots, and themselves given
To the stream's tyrannous rage, alas are driven
Through mills, and rocks, and woods, and at last, almost
Consumed in going, in the sea are lost:
So perish souls, which more choose men's unjust
Power from God claimed, than God himself to trust. 110

(Wr. prob. 1594–5; pub. 1633)

455 *The Perfume*

ONCE, and but once found in thy company,
All thy supposed escapes are laid on me;
And as a thief at bar, is questioned there
By all the men, that have been robbed that year,

92 vicars] proxies
101 Those past] once power oversteps its due limits

455
2 escapes] amorous escapades

So am I, (by this traitorous means surprised)
By thy hydroptic father catechized.
Though he had wont to search with glazed eyes,
As though he came to kill a cockatrice,
Though he have oft sworn, that he would remove
Thy beauty's beauty, and food of our love, 10
Hope of his goods, if I with thee were seen,
Yet close and secret, as our souls, we have been.
Though thy immortal mother which doth lie
Still buried in her bed, yet will not die,
Takes this advantage to sleep out day-light,
And watch thy entries, and returns all night,
And, when she takes thy hand, and would seem kind,
Doth search what rings, and armlets she can find,
And kissing notes the colour of thy face,
And fearing lest thou art swoll'n, doth thee embrace; 20
To try if thou long, doth name strange meats,
And notes thy paleness, blushing, sighs, and sweats;
And politicly will to thee confess
The sins of her own youth's rank lustiness;
Yet love these sorceries did remove, and move
Thee to gull thine own mother for my love.
Thy little brethren, which like faery sprites
Oft skipped into our chamber, those sweet nights,
And kissed, and ingled on thy father's knee,
Were bribed next day, to tell what they did see: 30
The grim eight-foot-high iron-bound serving-man,
That oft names God in oaths, and only then,
He that to bar the first gate, doth as wide
As the great Rhodian Colossus stride,
Which, if in hell no other pains there were,
Makes me fear hell, because he must be there:
Though by thy father he were hired to this,
Could never witness any touch or kiss.
But Oh, too common ill, I brought with me
That, which betrayed me to mine enemy: 40
A loud perfume, which at my entrance cried
Even at thy father's nose, so we were spied.
When, like a tyrant king, that in his bed
Smelt gunpowder, the pale wretch shivered.
Had it been some bad smell, he would have thought
That his own feet, or breath, that smell had wrought.

6 hydroptic] dropsical
23 politicly] cunningly, with an ulterior purpose
26 gull] deceive
29 ingled] fondled

But as we in our isle imprisoned,
Where cattle only, and diverse dogs are bred,
The precious unicorns, strange monsters call,
So thought he good, strange, that had none at all. 50
I taught my silks, their whistling to forbear,
Even my oppressed shoes, dumb and speechless were,
Only, thou bitter sweet, whom I had laid
Next me, me traitorously hast betrayed,
And unsuspected hast invisibly
At once fled unto him, and stayed with me.
Base excrement of earth, which dost confound
Sense, from distinguishing the sick from sound;
By thee the silly amorous sucks his death
By drawing in a leprous harlot's breath; 60
By thee, the greatest stain to man's estate
Falls on us, to be called effeminate;
Though you be much loved in the prince's hall,
There, things that seem, exceed substantial.
Gods, when ye fumed on altars, were pleased well,
Because you were burnt, not that they liked your smell;
You are loathsome all, being taken simply alone,
Shall we love ill things joined, and hate each one?
If you were good, your good doth soon decay;
And you are rare, that takes the good away. 70
All my perfumes, I give most willingly
To embalm thy father's corse. What? will he die?

(Wr. mid-1590s; pub. 1633)

456 *The Bracelet*

Upon the loss of his mistress' chain,
for which he made satisfaction

NOT that in colour it was like thy hair,
For armlets of that thou mayst let me wear;
Nor that thy hand it oft embraced and kissed,
For so it had that good, which oft I missed;
Nor for that silly old morality,
That as those links are tied, our love should be;
Mourn I that I thy sevenfold chain have lost,
Nor for the luck sake; but the bitter cost.

456
7 sevenfold] with seven loops

Oh shall twelve righteous angels, which as yet
No leaven of vile solder did admit, 10
Nor yet by any way have strayed or gone
From the first state of their creation,
Angels, which heaven commanded to provide
All things to me, and be my faithful guide,
To gain new friends, to appease great enemies,
To comfort my soul, when I lie or rise;
Shall these twelve innocents, by thy severe
Sentence (dread judge) my sins' great burden bear?
Shall they be damned, and in the furnace thrown,
And punished for offences not their own? 20
They save not me, they do not ease my pains
When in that hell they are burnt and tied in chains.
Were they but crowns of France, I cared not,
For, most of these, their natural country rot
I think possesseth, they come here to us,
So pale, so lame, so lean, so ruinous.
And howsoe'er French kings most Christian be,
Their crowns are circumcised most Jewishly.
Or were they Spanish stamps, still travelling,
That are become as Catholic as their king, 30
Those unlicked bear-whelps, unfiled pistolets
That, more than cannon shot, avails or lets,
Which, negligently left unrounded, look
Like many-angled figures in the book
Of some great conjurer, that would enforce
Nature, as these do justice, from her course;
Which, as the soul quickens head, feet and heart,
As streams, like veins, run through th' earth's every part,
Visit all countries, and have slily made
Gorgeous France, ruined, ragged and decayed, 40
Scotland, which knew no State, proud in one day,
And mangled seventeen-headed Belgia:
Or were it such gold as that wherewithal
Almighty chemics from each mineral
Having by subtle fire a soul out-pulled,
Are dirtily and desperately gulled:

9 angels] gold coins bearing the figure of an angel
10 solder] metal for repairing coins
17 Shall these . . . innocents] she has demanded that he give twelve 'angels' to be melted down for a new gold bracelet
24 their natural . . . rot] syphilis
31 pistolets] Spanish gold coins
32 avails or lets] opens or bars access
39 Visit all countries] i.e. the corrupting power of Spanish gold in Europe
42 Belgia] the Protestant Netherlands (17 Provinces)
44 chemics] alchemists

I would not spit to quench the fire they were in,
For, they are guilty of much heinous sin.
But, shall my harmless angels perish? Shall
I lose my guard, my ease, my food, my all? 50
Much hope, which they should nourish, will be dead.
Much of my able youth, and lustihead
Will vanish, if thou love let them alone,
For thou wilt love me less when they are gone.
  Oh be content that some loud squeaking crier
Well-pleased with one lean threadbare groat, for hire,
May like a devil roar through every street,
And gall the finder's conscience, if they meet.
Or let me creep to some dread conjurer,
That with fantastic schemes fills full much paper, 60
Which hath divided heaven in tenements,
And with whores, thieves, and murderers stuffed his rents,
So full, that though he pass them all in sin,
He leaves himself no room to enter in.
But if, when all his art and time is spent,
He say 'twill ne'er be found; yet be content;
Receive from him that doom ungrudgingly,
Because he is the mouth of destiny.
  Thou say'st (alas) the gold doth still remain,
Though it be changed, and put into a chain. 70
So in the first fall'n angels, resteth still
Wisdom and knowledge, but, 'tis turned to ill;
As these should do good works, and should provide
Necessities, but now must nurse thy pride.
And they are still bad angels; mine are none,
For form gives being, and their form is gone.
Pity these angels yet; their dignities
Pass Virtues, Powers, and Principalities.
  But thou art resolute; thy will be done.
Yet with such anguish as her only son 80
The mother in the hungry grave doth lay,
Unto the fire these martyrs I betray.
Good souls, for you give life to everything,
Good angels, for good messages you bring,
Destined you might have been to such a one
As would have loved and worshipped you alone,
One that would suffer hunger, nakedness,
Yea death, ere he would make your number less.
But I am guilty of your sad decay,
May your few fellows longer with me stay. 90
  But Oh thou wretched finder whom I hate
So, that I almost pity thy estate;

Gold being the heaviest metal amongst all,
May my most heavy curse upon thee fall.
Here fettered, manacled, and hanged in chains
First mayst thou be, then chained to hellish pains;
Or be with foreign gold bribed to betray
Thy country, and fail both of that and thy pay.
May the next thing thou stoop'st to reach, contain
Poison, whose nimble fume rot thy moist brain; 100
Or libels, or some interdicted thing,
Which negligently kept, thy ruin bring.
Lust-bred diseases rot thee; and dwell with thee
Itchy desire and no ability.
May all the evils that gold ever wrought,
All mischiefs that all devils ever thought,
Want after plenty, poor and gouty age,
The plagues of travellers, love and marriage
Afflict thee, and at thy life's latest moment
May thy swoll'n sins themselves to thee present. 110
But I forgive; repent thee honest man:
Gold is restorative, restore it then.
But if from it thou be'st loth to depart,
Because 'tis cordial, would 'twere at thy heart.

(Wr. early 1590s; pub. 1635)

457 *On his Mistress*

By our first strange and fatal interview,
By all desires which thereof did ensue,
By our long starving hopes, by that remorse
Which my words' masculine persuasive force
Begot in thee, and by the memory
Of hurts, which spies and rivals threatened me,
I calmly beg: but by thy father's wrath,
By all pains, which want and divorcement hath,
I conjure thee; and all the oaths which I
And thou have sworn to seal joint constancy, 10
Here I unswear, and overswear them thus,
Thou shalt not love by ways so dangerous.
Temper, O fair love, love's impetuous rage,
Be my true mistress still, not my feigned page;
I'll go, and, by thy kind leave, leave behind
Thee, only worthy to nurse in my mind

114 would 'twere at thy heart] I wish it would kill you

457
1 strange] when they were strangers
3 remorse] compassion, tender feeling

Thirst to come back; oh, if thou die before,
From other lands my soul towards thee shall soar.
Thy (else almighty) beauty cannot move
Rage from the seas, nor thy love teach them love, 20
Nor tame wild Boreas' harshness; thou hast read
How roughly he in pieces shivered
Fair Orithea, whom he swore he loved.
Fall ill or good, 'tis madness to have proved
Dangers unurged; feed on this flattery,
That absent lovers one in th'other be.
Dissemble nothing, not a boy, nor change
Thy body's habit, nor mind's; be not strange
To thy self only; all will spy in thy face
A blushing womanly discovering grace; 30
Richly clothed apes, are called apes, and as soon
Eclipsed as bright we call the moon the moon.
Men of France, changeable chameleons,
Spitals of diseases, shops of fashions,
Love's fuellers, and the rightest company
Of players, which upon the world's stage be,
Will quickly know thee, and know thee; and alas
Th'indifferent Italian, as we pass
His warm land, well content to think thee page,
Will hunt thee with such lust, and hideous rage, 40
As Lot's fair guests were vexed. But none of these
Nor spongy hydroptic Dutch shall thee displease,
If thou stay here. Oh stay here, for, for thee
England is only a worthy gallery,
To walk in expectation, till from thence
Our greatest King call thee to his presence.
When I am gone, dream me some happiness,
Nor let thy looks our long-hid love confess,
Nor praise, nor dispraise me, nor bless nor curse
Openly love's force, nor in bed fright thy nurse 50
With midnight's startings, crying out, 'Oh, oh
Nurse, O my love is slain, I saw him go
O'er the white Alps alone; I saw him, I,
Assailed, fight, taken, stabbed, bleed, fall, and die.'
Augur me better chance, except dread Jove
Think it enough for me to have had thy love.

(Wr. 1593–6; pub. 1635)

19 move] take away
21 Boreas] north wind
24–25 proved Dangers unurged] undergone dangers when not compelled to
34 Spitals] hospitals
38 indifferent] bisexual
42 hydroptic] soaking up drink
44 only a worthy] the only worthy
55 except] unless

458 *To his Mistress Going to Bed*

COME, Madam, come, all rest my powers defy,
Until I labour, I in labour lie.
The foe oft-times having the foe in sight,
Is tired with standing though they never fight.
Off with that girdle, like heaven's zone glistering,
But a far fairer world encompassing.
Unpin that spangled breastplate which you wear,
That th'eyes of busy fools may be stopped there.
Unlace yourself, for that harmonious chime
Tells me from you, that now 'tis your bed time. 10
Off with that happy busk, which I envy,
That still can be, and still can stand so nigh.
Your gown going off, such beauteous state reveals,
As when from flowery meads th'hill's shadow steals.
Off with that wiry coronet and show
The hairy diadem which on you doth grow;
Now off with those shoes, and then safely tread
In this love's hallowed temple, this soft bed.
In such white robes heaven's angels used to be
Received by men; thou angel bring'st with thee 20
A heaven like Mahomet's paradise; and though
Ill spirits walk in white, we easily know
By this these angels from an evil sprite:
Those set our hairs, but these our flesh upright.
  Licence my roving hands, and let them go
Before, behind, between, above, below.
O my America, my new found land,
My kingdom, safeliest when with one man manned,
My mine of precious stones, my empery,
How blessed am I in this discovering thee! 30
To enter in these bonds, is to be free;
Then where my hand is set, my seal shall be.
  Full nakedness, all joys are due to thee.
As souls unbodied, bodies unclothed must be,
To taste whole joys. Gems which you women use
Are like Atlanta's balls, cast in men's views,
That when a fool's eye lighteth on a gem,
His earthly soul may covet theirs, not them.

1 all rest . . . defy] my (sexual) powers refuse to be inactive
5 heaven's zone] the belt of Orion, or the furthest circle of the universe
7 breastplate] stomacher
9 chime] i.e. she wears a chiming watch
11 busk] corset

Like pictures, or like books' gay coverings made
For laymen, are all women thus arrayed; 40
Themselves are mystic books, which only we
Whom their imputed grace will dignify
Must see revealed. Then since I may know,
As liberally, as to a midwife, show
Thyself: cast all, yea, this white linen hence,
Here is no penance, much less innocence.
To teach thee, I am naked first, why then
What needst thou have more covering than a man.

(Wr. 1593–6; pub. 1633)

459 *The Storm*

To Mr Christopher Brooke

THOU which art I, ('tis nothing to be so)
Thou which art still thyself, by these shalt know
Part of our passage; and, a hand, or eye
By Hilliard drawn, is worth an history,
By a worse painter made; and (without pride)
When by thy judgement they are dignified,
My lines are such: 'tis the pre-eminence
Of friendship only to impute excellence.
England to whom we owe, what we be, and have,
Sad that her sons did seek a foreign grave 10
(For, Fate's, or Fortune's drifts none can soothsay,
Honour and misery have one face and way)
From out her pregnant entrails sighed a wind
Which at th'air's middle marble room did find
Such strong resistance, that itself it threw
Downward again; and so when it did view
How in the port, our fleet dear time did leese,
Withering like prisoners, which lie but for fees,
Mildly it kissed our sails, and, fresh and sweet,
As to a stomach starved, whose insides meet, 20
Meat comes, it came; and swole our sails, when we
So joyed, as Sara her swelling joyed to see.
But 'twas but so kind, as our countrymen,
Which bring friends one day's way, and leave them then.
Then like two mighty kings, which dwelling far
Asunder, meet against a third to war,

459
3 our passage] i.e. the expedition to the Azores under Essex, Howard and Ralegh in 1597.
14 middle . . . room] the (intensely cold) middle region of the air
17 leese] lose

The south and west winds joined, and, as they blew,
Waves like a rolling trench before them threw.
Sooner than you read this line, did the gale,
Like shot, not feared till felt, our sails assail; 30
And what at first was called a gust, the same
Hath now a storm's, anon a tempest's name.
Jonas, I pity thee, and curse those men,
Who when the storm raged most, did wake thee then;
Sleep is pain's easiest salve, and doth fulfil
All offices of death, except to kill.
But when I waked, I saw, that I saw not.
I, and the sun, which should teach me had forgot
East, west, day, night, and I could only say,
If the world had lasted, now it had been day. 40
Thousands our noises were, yet we 'mongst all
Could none by his right name, but thunder call:
Lightning was all our light, and it rained more
Than if the sun had drunk the sea before.
Some coffined in their cabins lie, equally
Grieved that they are not dead, and yet must die.
And as sin-burdened souls from graves will creep,
At the last day, some forth their cabins peep:
And tremblingly ask what news, and do hear so,
Like jealous husbands, what they would not know. 50
Some sitting on the hatches, would seem there,
With hideous gazing to fear away fear.
Then note they the ship's sicknesses, the mast
Shaked with this ague, and the hold and waist
With a salt dropsy clogged, and all our tacklings
Snapping, like too high-stretched treble strings.
And from our tottered sails, rags drop down so,
As from one hanged in chains, a year ago.
Even our ordnance placed for our defence,
Strive to break loose, and 'scape away from thence. 60
Pumping hath tired our men, and what's the gain?
Seas into seas thrown, we suck in again;
Hearing hath deafed our sailors; and if they
Knew how to hear, there's none knows what to say.
Compared to these storms, death is but a qualm,
Hell somewhat lightsome, and the Bermuda calm.
Darkness, light's elder brother, his birth-right
Claims o'er this world, and to heaven hath chased light.
All things are one, and that one none can be,
Since all forms, uniform deformity 70

38 teach] show

Doth cover, so that we, except God say
Another *Fiat*, shall have no more day.
So violent, yet long these furies be,
That though thine absence starve me, I wish not thee.

(Wr. 1597; pub. 1633)

460 *The Calm*

OUR storm is past, and that storm's tyrannous rage,
A stupid calm, but nothing it, doth 'suage.
The fable is inverted, and far more
A block afflicts, now, than a stork before.
Storms chafe, and soon wear out themselves, or us;
In calms, heaven laughs to see us languish thus.
As steady as I can wish, that my thoughts were,
Smooth as thy mistress' glass, or what shines there,
The sea is now. And, as those Isles which we
Seek, when we can move, our ships rooted be. 10
As water did in storms, now pitch runs out
As lead, when a fired church becomes one spout.
And all our beauty, and our trim, decays,
Like courts removing, or like ended plays.
The fighting place now seamen's rags supply;
And all the tackling is a frippery.
No use of lanthorns; and in one place lay
Feathers and dust, today and yesterday.
Earth's hollownesses, which the world's lungs are,
Have no more wind than the upper vault of air. 20
We can nor lost friends nor sought foes recover,
But meteor-like, save that we move not, hover.
Only the calenture together draws
Dear friends, which meet dead in great fishes' jaws:
And on the hatches as on altars lies
Each one, his own priest, and own sacrifice.
Who live, that miracle do multiply
Where walkers in hot ovens do not die.
If in despite of these, we swim, that hath
No more refreshing, than our brimstone bath, 30
But from the sea, into the ship we turn,
Like parboiled wretches, on the coals to burn.

460
16 frippery] old-clothes shop
23 calenture] tropical illness which caused sailors to throw themselves into the sea
31 turn] return

Like Bajazet encaged, the shepherd's scoff,
Or like slack-sinewed Samson, his hair off,
Languish our ships. Now, as a myriad
Of ants, durst th'Emperor's loved snake invade,
The crawling galleys, sea-gaols, finny chips,
Might brave our pinnaces, now bed-rid ships.
Whether a rotten state, and hope of gain,
Or, to disuse me from the queasy pain 40
Of being beloved, and loving, or the thirst
Of honour, or fair death, out pushed me first,
I lose my end: for here as well as I
A desperate may live, and a coward die.
Stag, dog, and all which from, or towards flies,
Is paid with life, or prey, or doing dies.
Fate grudges us all, and doth subtly lay
A scourge, 'gainst which we all forget to pray,
He that at sea prays for more wind, as well
Under the poles may beg cold, heat in hell. 50
What are we then? How little more alas
Is man now, than before he was! he was
Nothing; for us, we are for nothing fit;
Chance, or ourselves still disproportion it.
We have no power, no will, no sense; I lie,
I should not then thus feel this misery.

(Wr. 1597; pub. 1633)

461 *Hero and Leander*

BOTH robbed of air, we both lie in one ground,
Both whom one fire had burnt, one water drowned.

(Pub. 1633)

462 *A Lame Beggar*

I AM unable, yonder beggar cries,
To stand, or move; if he say true, he *lies*.

(Pub. 1607)

33 the shepherd] Tamburlaine, who encaged Bajazeth
36 th'Emperor] Tiberius, whose pet snake was eaten by ants
37 chips] slivers of wood
54 disproportion] mismatch

463 *To Sir Henry Wotton*

SIR, more than kisses, letters mingle souls;
For, thus friends absent speak. This ease controls
The tediousness of my life: but for these
I could ideate nothing which could please,
But I should wither in one day, and pass
To a bottle of hay, that am a lock of grass.
Life is a voyage, and in our life's ways
Countries, courts, towns are rocks, or remoras;
They break or stop all ships, yet our state's such,
That though than pitch they stain worse, we must touch. 10
If in the furnace of the even line,
Or under th'adverse icy poles thou pine,
Thou know'st two temperate regions girded in,
Dwell there: But Oh, what refuge canst thou win
Parched in the Court, and in the country frozen?
Shall cities, built of both extremes, be chosen?
Can dung and garlic be a perfume? or can
A scorpion and torpedo cure a man?
Cities are worst of all three; of all three
(O knotty riddle) each is worst equally. 20
Cities are sepulchres; they who dwell there
Are carcases, as if no such there were.
And Courts are theatres, where some men play
Princes, some slaves, all to one end, and of one clay.
The country is a desert, where no good,
Gained (as habits, not born,) is understood.
There men become beasts, and prone to more evils;
In cities blocks, and in a lewd Court, devils.
As in the first Chaos confusedly
Each element's qualities were in the other three; 30
So pride, lust, covetize, being several
To these three places, yet all are in all,
And mingled thus, their issue incestuous.
Falsehood is denizened. Virtue is barbarous.
Let no man say there, 'Virtue's flinty wall
Shall lock vice in me, I'll do none, but know all.'

2 controls] checks, relieves
4 ideate] conceive
8 remoras] impediments
11 even line] equator
18 torpedo] electric ray (whose sting caused numbness)
28 blocks] dullards
34 denizened] naturalized
barbarous] owes nothing to society

Men are sponges, which to pour out, receive,
Who know false play, rather than lose, deceive.
For in best understandings, sin began,
Angels sinned first, then devils, and then man. 40
Only perchance beasts sin not; wretched we
Are beasts in all, but white integrity.
I think if men, which in these places live
Durst look for themselves, and themselves retrieve,
They would like strangers greet themselves, seeing then
Utopian youth, grown old Italian.
  Be then thine own home, and in thyself dwell;
Inn anywhere, continuance maketh hell.
And seeing the snail, which everywhere doth roam,
Carrying his own house still, still is at home, 50
Follow (for he is easy paced) this snail,
Be thine own palace, or the world's thy gaol.
And in the world's sea, do not like cork sleep
Upon the water's face; nor in the deep
Sink like a lead without a line: but as
Fishes glide, leaving no print where they pass,
Nor making sound, so closely thy course go,
Let men dispute, whether thou breathe, or no.
Only in this one thing, be no Galenist: to make
Courts' hot ambitions wholesome, do not take 60
A dram of country's dullness; do not add
Correctives, but as chemics, purge the bad.
But, Sir, I advise not you, I rather do
Say o'er those lessons, which I learned of you:
Whom, free from German schisms, and lightness
Of France, and fair Italy's faithlessness,
Having from these sucked all they had of worth,
And brought home that faith, which you carried forth,
I throughly love. But if myself, I have won
To know my rules, I have, and you have
Donne.

(Wr. 1597–8; pub. 1633)

42 integrity] innocence
46 Italian] i.e. cunning and corrupt
48 Inn] lodge
59 Galenist] physicians who aimed to cure by contraries
62 chemics] alchemists
69 throughly] entirely

464 *Song*

Go, and catch a falling star,
  Get with child a mandrake root,
Tell me, where all past years are,
  Or who cleft the Devil's foot,
Teach me to hear mermaids singing,
  Or to keep off envy's stinging,
      And find
      What wind
Serves to advance an honest mind.

If thou be'est born to strange sights, 10
  Things invisible to see,
Ride then thousand days and nights,
  Till age snow white hairs on thee,
Thou, when thou return'st, wilt tell me
All strange wonders that befell thee,
      And swear
      No where
Lives a woman true, and fair.

If thou find'st one, let me know,
  Such a pilgrimage were sweet,
Yet do not, I would not go, 20
  Though at next door we might meet,
Though she were true, when you met her,
And last, till you write your letter,
      Yet she
      Will be
False, ere I come, to two, or three.

(Pub. 1633)

465 *Song*

Sweetest love, I do not go,
  For weariness of thee,
Nor in hope the world can show
  A fitter love for me.
    But since that I
Must die at last, 'tis best,
To use my self in jest
  Thus by feigned deaths to die.

Yesternight the sun went hence,
  And yet is here today. 10
He hath no desire nor sense,
  Nor half so short a way.
    Then fear not me,
But believe that I shall make
Speedier journeys, since I take
  More wings and spurs than he.

O how feeble is man's power,
  That if good fortune fall,
Cannot add another hour,
  Nor a lost hour recall! 20
    But come bad chance,
And we join to it our strength,
And we teach it art and length,
  Itself o'er us to advance.

When thou sigh'st, thou sigh'st not wind,
  But sigh'st my soul away,
When thou weep'st, unkindly kind,
  My life's blood doth decay.
    It cannot be
That thou lov'st me, as thou say'st, 30
If in thine my life thou waste,
  Thou art the best of me.

Let not thy divining heart
  Forethink me any ill,
Destiny may take thy part,
  And may thy fears fulfil.
    But think that we
Are but turned aside to sleep;
They who one another keep
  Alive, ne'er parted be. 40

(Pub. 1633)

466 *The Apparition*

When by thy scorn, O murderess, I am dead,
And that thou think'st thee free
From all solicitation from me,
Then shall my ghost come to thy bed,
And thee, feigned vestal, in worse arms shall see;

33 divining] foreseeing

466
2 that thou] when thou

Then thy sick taper will begin to wink,
And he, whose thou art then, being tired before,
Will, if thou stir, or pinch to wake him, think
  Thou call'st for more,
And in false sleep will from thee shrink, 10
And then poor aspen wretch, neglected thou
Bathed in a cold quicksilver sweat wilt lie
  A verier ghost than I;
What I will say, I will not tell thee now,
Lest that preserve thee; and since my love is spent,
I had rather thou shouldst painfully repent,
Than by my threatenings rest still innocent.

(Pub. 1633)

467

The Computation

FOR the first twenty years, since yesterday,
  I scarce believed, thou couldst be gone away,
For forty more, I fed on favours past,
  And forty on hopes, that thou wouldst, they might last.
Tears drowned one hundred, and sighs blew out two,
  A thousand, I did neither think, nor do,
  Or not divide, all being one thought of you;
  Or in a thousand more, forgot that too.
Yet call not this long life; but think that I
Am, by being dead, immortal. Can ghosts die? 10

(Pub. 1633)

468

The Flea

MARK but this flea, and mark in this,
How little that which thou deny'st me is.
Me it sucked first, and now sucks thee,
And in this flea, our two bloods mingled be.
Confess it, this cannot be said
A sin, or shame, or loss of maidenhead,
  Yet this enjoys before it woo,
  And pampered swells with one blood made of two,
  And this, alas, is more than we would do.

Oh stay, three lives in one flea spare, 10
Where we almost, nay more than married are.
This flea is you and I, and this
Our marriage bed, and marriage temple is;

Though parents grudge, and you, we'are met,
And cloistered in these living walls of jet.
Though use make you apt to kill me,
Let not to this, self-murder added be,
And sacrilege, three sins in killing three.

Cruel and sudden, hast thou since
Purpled thy nail, in blood of innocence? 20
In what could this flea guilty be,
Except in that drop which it sucked from thee?
Yet thou triumph'st, and say'st that thou
Find'st not thyself nor me the weaker now.
'Tis true, then learn how false, fears be:
Just so much honour, when thou yield'st to me,
Will waste, as this flea's death took life from thee.

(Pub. 1633)

469 *The Will*

BEFORE I sigh my last gasp, let me breathe,
Great Love, some legacies; here I bequeath
Mine eyes to Argus, if mine eyes can see,
If they be blind, then Love, I give them thee;
My tongue to fame; to ambassadors mine ears;
To women or the sea, my tears.
Thou, Love, hast taught me heretofore
By making me serve her who had twenty more,
That I should give to none, but such, as had too much before.

My constancy I to the planets give; 10
My truth to them, who at the Court do live;
Mine ingenuity and openness,
To Jesuits; to buffoons my pensiveness;
My silence to any, who abroad hath been;
My money to a Capuchin.
Thou Love taught'st me, by appointing me
To love there, where no love received can be,
Only to give to such as have an incapacity.

469
3 Argus] (a giant, whose body was covered with eyes)
5 fame] rumour (who was depicted as being covered with tongues)
10 planets] i.e. 'wandering' or 'errant' stars
12 ingenuity] ingenuousness, candour
13 buffoons] jesters, clowns
15 Capuchin] (order of friars vowed to poverty)

My faith I give to Roman Catholics;
All my good works unto the schismatics 20
Of Amsterdam; my best civility
And courtship, to an university;
My modesty I give to soldiers bare;
My patience let gamesters share.
Thou Love taught'st me, by making me
Love her that holds my love disparity,
Only to give to those that count my gifts indignity.

I give my reputation to those
Which were my friends; mine industry to foes;
To schoolmen I bequeath my doubtfulness; 30
My sickness to physicians, or excess;
To Nature, all that I in rhyme have writ;
And to my company my wit.
Thou Love, by making me adore
Her, who begot this love in me before,
Taught'st me to make, as though I gave, when I did but restore.

To him for whom the passing bell next tolls,
I give my physic books; my written rolls
Of moral counsels, I to Bedlam give;
My brazen medals, unto them which live 40
In want of bread; to them which pass among
All foreigners, mine English tongue.
Thou, Love, by making me love one
Who thinks her friendship a fit portion
For younger lovers, dost my gifts thus disproportion.

Therefore I'll give no more; but I'll undo
The world by dying; because love dies too.
Then all your beauties will be no more worth
Than gold in mines, where none doth draw it forth;
And all your graces no more use shall have 50
Than a sundial in a grave.
Thou Love taught'st me, by making me
Love her, who doth neglect both me and thee,
To invent, and practise this one way, to annihilate all three.

(Pub. 1633)

20–1 schismatics of Amsterdam] i.e. extreme Protestants opposed to 'good works'
23 bare] i.e. without it
26 disparity] beneath her
30 schoolmen] academic metaphysicians
31 excess] i.e. imbalance of the physical constitution
38 physic] medical
40 medals] i.e. old coins, not current money
45 disproportion] make inappropriate

470 *A Lecture upon the Shadow*

STAND still, and I will read to thee
A lecture, love, in love's philosophy.
These three hours that we have spent,
Walking here, two shadows went
Along with us, which we ourselves produced;
But, now the sun is just above our head,
We do those shadows tread;
And to brave clearness all things are reduced.
So whilst our infant loves did grow,
Disguises did, and shadows, flow, 10
From us, and our care; but, now 'tis not so.

That love hath not attained the high'st degree,
Which is still diligent lest others see.

Except our loves at this noon stay,
We shall new shadows make the other way.
As the first were made to blind
Others; these which come behind
Will work upon ourselves, and blind our eyes.
If our loves faint, and westwardly decline;
To me thou, falsely, thine, 20
And I to thee mine actions shall disguise.
The morning shadows wear away,
But these grow longer all the day,
But oh, love's day is short, if love decay.

Love is a growing, or full constant light;
And his first minute, after noon, is night.

(Pub. 1633)

471 *The Anniversary*

ALL kings, and all their favourites,
All glory of honours, beauties, wits,
The sun itself, which makes times, as they pass,
Is elder by a year, now, than it was
When thou and I first one another saw:
All other things, to their destruction draw,
Only our love hath no decay;
This, no tomorrow hath, nor yesterday,
Running it never runs from us away,
But truly keeps his first, last, everlasting day. 10

Two graves must hide thine and my corse,
If one might, death were no divorce.
Alas, as well as other princes, we,
(Who prince enough in one another be,)
Must leave at last in death, these eyes, and ears,
Oft fed with true oaths, and with sweet salt tears;
But souls where nothing dwells but love
(All other thoughts being inmates) then shall prove
This, or a love increased there above,
When bodies to their graves, souls from their graves remove. 20

And then we shall be throughly blessed,
But we no more, than all the rest.
Here upon earth, we are kings, and none but we
Can be such kings, nor of such subjects be;
Who is so safe as we? where none can do
Treason to us, except one of us two.
True and false fears let us refrain,
Let us love nobly, and live, and add again
Years and years unto years, till we attain
To write threescore: this is the second of our reign. 30

(Pub. 1633)

SIR JOHN DAVIES

1569–1626

472 from *Orchestra or A Poem of Dancing*

WHERE lives the man that never yet did hear
Of chaste Penelope, Ulysses' queen?
Who kept her faith unspotted twenty year,
Till he returned, that far away had been,
And many men and many towns had seen;
Ten year at siege of Troy he lingering lay,
And ten year in the midland sea did stray.

17 dwells] has its own home
18 inmates] temporary lodgers
prove] experience
27 refrain] restrain, curb

Homer, to whom the Muses did carouse
A great deep cup with heavenly nectar filled:
The greatest deepest cup in Jove's great house, 10
(For Jove himself had so expressly willed,)
He drank off all, ne let one drop be spilled;
Since when his brain, that had before been dry,
Became the wellspring of all poetry.

Homer doth tell, in his abundant verse,
The long laborious travails of the man,
And of his lady too he doth rehearse,
How she illudes, with all the art she can,
Th'ungrateful love which other lords began;
For of her lord false fame long since had sworn, 20
That Neptune's monsters had his carcass torn.

All this he tells, but one thing he forgot,
One thing most worthy his eternal song;
But he was old and blind and saw it not,
Or else he thought he should Ulysses wrong,
To mingle it his tragic acts among;
Yet was there not, in all the world of things,
A sweeter burden for his Muse's wings.

The courtly love Antinous did make,
Antinous, that fresh and jolly knight, 30
Which of the gallants, that did undertake
To win the widow, had most wealth and might,
Wit to persuade, and beauty to delight:
The courtly love he made unto the queen,
Homer forgot, as if it had not been.

Sing then, Terpsichore, my light Muse, sing
His gentle art and cunning courtesy!
You, lady, can remember everything,
For you are daughter of queen Memory;
But sing a plain and easy melody, 40
For the soft mean that warbleth but the ground
To my rude ear doth yield the sweetest sound.

18 illudes] (1) eludes; (2) deceives
19 ungrateful] unwanted
36 Terpsichore] Muse of dancing

One only night's discourse I can report:
When the great torchbearer of heaven was gone
Down, in a mask, unto the Ocean's court,
To revel it with Tethys all alone;
Antinous, disguised and unknown,
Like to the spring in gaudy ornament,
Unto the castle of the princess went.

The sovereign castle of the rocky isle, 50
Wherein Penelope the princess lay,
Shone with a thousand lamps, which did exile
The shadows dark, and turned the night to day.
Not Jove's blue tent, what time the sunny ray
Behind the bulwark of the earth retires,
Is seen to sparkle with more twinkling fires.

That night the queen came forth from far within,
And in the presence of her court was seen;
For the sweet singer Phemius did begin
To praise the worthies that at Troy had been; 60
Somewhat of her Ulysses she did ween
In his grave hymn the heavenly man would sing,
Or of his wars, or of his wandering.

Pallas that hour, with her sweet breath divine,
Inspired immortal beauty in her eyes,
That with celestial glory she did shine
Brighter than Venus, when she doth arise
Out of the waters to adorn the skies.
The wooers, all amazed, do admire
And check their own presumptuous desire. 70

Only Antinous, when at first he viewed
Her star-bright eyes that with new honour shined,
Was not dismayed; but therewithal renewed
The noblesse and the splendour of his mind;
And as he did fit circumstances find,
Unto the throne he boldly gan advance,
And with fair manners wooed the queen to dance:

61 ween] think

'Goddess of women, sith your heavenliness
  Hath now vouchsafed itself to represent
To our dim eyes, which though they see the less, 80
  Yet are they blest in their astonishment,
  Imitate heaven, whose beauties excellent
    Are in continual motion day and night,
    And move thereby more wonder and delight.

'Let me the mover be, to turn about
  Those glorious ornaments that youth and love
Have fixed in your every part throughout;
  Which if you will in timely measure move,
  Not all those precious gems in heaven above
    Shall yield a sight more pleasing to behold, 90
    With all their turns and tracings manifold.'

With this the modest princess blushed and smiled,
  Like to a clear and rosy eventide,
And softly did return this answer mild:
  'Fair sir, you needs must fairly be denied,
  Where your demand cannot be satisfied.
    My feet, which only nature taught to go,
    Did never yet the art of footing know.

'But why persuade you me to this new rage?
  (For all disorder and misrule is new) 100
For such misgovernment in former age
  Our old divine forefathers never knew;
  Who if they lived, and did the follies view,
    Which their fond nephews make their chief affairs,
    Would hate themselves, that had begot such heirs.'

'Sole heir of virtue, and of beauty both,
  Whence cometh it', Antinous replies,
'That your imperious virtue is so loth
  To grant your beauty her chief exercise?
  Or from what spring doth your opinion rise, 110
    That dancing is a frenzy and a rage,
    First known and used in this new-fangled age?

78 sith] since
97 go] walk
108 imperious] imperial, sovereign

'Dancing, bright lady, then began to be,
When the first seeds whereof the world did spring,
The fire, air, earth, and water, did agree
By Love's persuasion, Nature's mighty king,
To leave their first disordered combating,
And in a dance such measure to observe,
As all the world their motion should preserve.

'Since when they still are carried in a round, 120
And changing come one in another's place;
Yet do they neither mingle nor confound,
But every one doth keep the bounded space
Wherein the dance doth bid it turn or trace.
This wondrous miracle did Love devise,
For dancing is love's proper exercise.

'Like this he framed the gods' eternal bower,
And of a shapeless and confused mass,
By his through-piercing and digesting power,
The turning vault of heaven formed was, 130
Whose starry wheels he hath so made to pass,
As that their movings do a music frame,
And they themselves still dance unto the same.

'Or if this all, which round about we see,
(As idle Morpheus some sick brains hath taught)
Of undivided motes compacted be,
How was this goodly architecture wrought?
Or by what means were they together brought?
They err that say they did concur by chance;
Love made them meet in a well-ordered dance. 140

'As when Amphion with his charming lyre
Begot so sweet a siren of the air,
That, with her rhetoric, made the stones conspire
The ruins of a city to repair,
A work of wit and reason's wise affair;
So Love's smooth tongue the motes such measure taught,
That they joined hands, and so the world was wrought.

'How justly then is dancing termed new,
Which with the world in point of time begun?
Yea, Time itself, whose birth Jove never knew, 150
And which is far more ancient than the sun,
Had not one moment of his age outrun,
When out leapt Dancing from the heap of things
And lightly rode upon his nimble wings.

'Reason hath both their pictures in her treasure;
  Where Time the measure of all moving is,
And Dancing is a moving all in measure.
  Now, if you do resemble that to this,
  And think both one, I think you think amiss;
    But if you judge them twins, together got, 160
    And Time first born, your judgement erreth not.

'Thus doth it equal age with Age enjoy,
  And yet in lusty youth forever flowers;
Like Love, his sire, whom painters make a boy,
  Yet is he eldest of the heavenly powers;
  Or like his brother Time, whose winged hours,
    Going and coming, will not let him die,
    But still preserve him in his infancy.'

This said, the queen, with her sweet lips divine,
  Gently began to move the subtle air, 170
Which gladly yielding, did itself incline
  To take a shape between those rubies fair;
  And being formed, softly did repair,
    With twenty doublings in the empty way,
    Unto Antinous' ears, and thus did say:

'What eye doth see the heaven, but doth admire
  When it the movings of the heavens doth see?
Myself, if I to heaven may once aspire,
  If that be dancing, will a dancer be;
  But as for this, your frantic jollity, 180
    How it began, or whence you did it learn,
    I never could with reason's eye discern.'

Antinous answered: 'Jewel of the earth,
  Worthy you are that heavenly dance to lead;
But for you think our Dancing base of birth,
  And newly born but of a brain-sick head,
  I will forthwith his antique gentry read,
    And, for I love him, will his herald be,
    And blaze his arms, and draw his pedigree.

170 subtle] rarefied   187 gentry] rank by birth

'When Love had shaped this world, this great fair wight, 190
That all wights else in this wide womb contains,
And had instructed it to dance aright
A thousand measures, with a thousand strains,
Which it should practise with delightful pains,
Until that fatal instant should revolve,
When all to nothing should again resolve;

'The comely order and proportion fair
On every side did please his wandering eye;
Till, glancing through the thin transparent air,
A rude disordered rout he did espy 200
Of men and women, that most spitefully
Did one another throng and crowd so sore,
That his kind eye, in pity, wept therefore.

'And swifter than the lightning down he came,
Another shapeless chaos to digest;
He will begin another world to frame,
For Love, till all be well, will never rest.
Then with such words as cannot be expressed
He cuts the troops, that all asunder fling,
And ere they wist he casts them in a ring. 210

'Then did he rarefy the element,
And in the centre of the ring appear;
The beams that from his forehead spreading went
Begot a horror and religious fear
In all the souls that round about him were,
Which in their ears attentiveness procures,
While he, with such like sounds, their minds allures:

' "How doth Confusion's mother, headlong Chance,
Put Reason's noble squadron to the rout?
Or how should you, that have the governance 220
Of Nature's children, heaven and earth throughout,
Prescribe them rules, and live yourselves without?
Why should your fellowship a trouble be,
Since man's chief pleasure is society?

190 wight] creature 205 digest] reduce to order

'If sense hath not yet taught you, learn of me
A comely moderation and discreet,
That your assemblies may well ordered be
When my uniting power shall make you meet;
With heavenly tunes it shall be tempered sweet,
And be the model of the world's great frame, 230
And you, earth's children, Dancing shall it name.

'Behold the world, how it is whirled round!
And for it is so whirled, is named so;
In whose large volume many rules are found
Of this new art, which it doth fairly show.
For your quick eyes in wandering to and fro,
From east to west, on no one thing can glance,
But, if you mark it well, it seems to dance.

'First you see fixed in this huge mirror blue
Of trembling lights a number numberless; 240
Fixed, they are named, but with a name untrue;
For they all move and in a dance express
The great long year that doth contain no less
Than threescore hundreds of those years in all,
Which the sun makes with his course natural.

'What if to you these sparks disordered seem,
As if by chance they had been scattered there?
The gods a solemn measure do it deem
And see a just proportion everywhere,
And know the points whence first their movings were, 250
To which first points when all return again,
The axletree of heaven shall break in twain.

'Under that spangled sky five wandering flames,
Besides the king of day and queen of night,
Are wheeled around, all in their sundry frames,
And all in sundry measures do delight;
Yet altogether keep no measure right;
For by itself each doth itself advance,
And by itself each doth a galliard dance.

230 model] copy 259 galliard] lively dance

'Venus, the mother of that bastard Love, 260
Which doth usurp the world's great marshal's name,
Just with the sun her dainty feet doth move,
And unto him doth all her gestures frame;
Now after, now afore, the flattering dame
With divers cunning passages doth err,
Still him respecting that respects not her.

'For that brave sun, the father of the day,
Doth love this earth, the mother of the night;
And, like a reveller in rich array,
Doth dance his galliard in his leman's sight, 270
Both back and forth and sideways passing light.
His gallant grace doth so the gods amaze,
That all stand still and at his beauty gaze.

'But see the earth when she approacheth near,
How she for joy doth spring and sweetly smile;
But see again her sad and heavy cheer,
When changing places he retires a while;
But those black clouds he shortly will exile,
And make them all before his presence fly,
As mists consumed before his cheerful eye. 280

'Who doth not see the measures of the moon?
Which thirteen times she danceth every year,
And ends her pavan thirteen times as soon
As doth her brother, of whose golden hair
She borroweth part, and proudly doth it wear.
Then doth she coyly turn her face aside,
That half her cheek is scarce sometimes descried.

'Next her, the pure, subtile, and cleansing fire
Is swiftly carried in a circle even,
Though Vulcan be pronounced by many a liar 290
The only halting god that dwells in heaven;
But that foul name may be more fitly given
To your false fire, that far from heaven is fall,
And doth consume, waste, spoil, disorder all.

270 leman's] sweetheart's
283 pavan] slow, stately processional dance, in the course of which the partners turn their faces away from each other
291 halting] limping

'And now behold your tender nurse, the air,
And common neighbour that aye runs around;
How many pictures and impressions fair
Within her empty regions are there found,
Which to your senses dancing do propound?
For what are breath, speech, echoes, music, winds, 300
But dancings of the air, in sundry kinds?

'For, when you breathe, the air in order moves,
Now in, now out, in time and measure true,
And when you speak, so well she dancing loves,
That doubling oft and oft redoubling new
With thousand forms she doth herself endue;
For all the words that from your lips repair
Are nought but tricks and turnings of the air.

'Hence is her prattling daughter, Echo, born,
That dances to all voices she can hear. 310
There is no sound so harsh that she doth scorn,
Nor any time wherein she will forbear
The airy pavement with her feet to wear;
And yet her hearing sense is nothing quick,
For after time she endeth every trick.

'And thou, sweet music, dancing's only life,
The ear's sole happiness, the air's best speech,
Lodestone of fellowship, charming rod of strife,
The soft mind's paradise, the sick mind's leech,
With thine own tongue thou trees and stones canst teach, 320
That when the air doth dance her finest measure,
Then art thou born, the gods' and men's sweet pleasure.

'Lastly, where keep the winds their revelry,
Their violent turnings and wild whirling hays,
But in the air's tralucent gallery?
Where she herself is turned a hundred ways,
While with those maskers wantonly she plays.
Yet in this misrule they such rule embrace
As two, at once, encumber not the place.

319 leech] physician, healer
324 hays] country dances
325 tralucent] translucent

'If then fire, air, wandering and fixed lights, 330
In every province of th'imperial sky,
Yield perfect forms of dancing to your sights,
In vain I teach the ear that which the eye,
With certain view, already doth descry;
But for your eyes perceive not all they see,
In this I will your senses' master be.

'For lo, the sea that fleets about the land,
And like a girdle clips her solid waist,
Music and measure both doth understand;
For his great crystal eye is always cast 340
Up to the moon, and on her fixed fast;
And as she danceth in her pallid sphere,
So danceth he about the centre here.

'Sometimes his proud green waves in order set,
One after other, flow unto the shore;
Which when they have with many kisses wet,
They ebb away in order, as before;
And to make known his courtly love the more,
He oft doth lay aside his three-forked mace,
And with his arms the timorous earth embrace. 350

'Only the earth doth stand forever still:
Her rocks remove not, nor her mountains meet,
Although some wits enriched with learning's skill
Say heaven stands firm and that the earth doth fleet,
And swiftly turneth underneath their feet;
Yet, though the earth is ever steadfast seen,
On her broad breast hath dancing ever been.

'For those blue veins that through her body spread,
Those sapphire streams which from great hills do spring,
(The earth's great dugs, for every wight is fed 360
With sweet fresh moisture from them issuing)
Observe a dance in their wild wandering;
And still their dance begets a murmur sweet,
And still the murmur with the dance doth meet.

338 clips] embraces

'Of all their ways, I love Meander's path,
Which, to the tunes of dying swans, doth dance;
Such winding sleights, such turns and tricks he hath,
Such creeks, such wrenches, and such dalliance,
That, whether it be hap or heedless chance,
In his indented course and wriggling play, 370
He seems to dance a perfect cunning hay.

'But wherefore do these streams forever run?
To keep themselves forever sweet and clear;
For let their everlasting course be done,
They straight corrupt and foul with mud appear.
O ye sweet nymphs, that beauty's loss do fear,
Contemn the drugs that physic doth devise,
And learn of Love this dainty exercise.

'See how those flowers, that have sweet beauty too,
The only jewels that the earth doth wear, 380
When the young sun in bravery her doth woo,
As oft as they the whistling wind do hear,
Do wave their tender bodies here and there;
And though their dance no perfect measure is,
Yet oftentimes their music makes them kiss.

'What makes the vine about the elm to dance
With turnings, windings, and embracements round?
What makes the lodestone to the north advance
His subtle point, as if from thence he found
His chief attractive virtue to redound? 390
Kind nature first doth cause all things to love;
Love makes them dance, and in just order move.

'Hark how the birds do sing, and mark then how,
Jump with the modulation of their lays,
They lightly leap and skip from bough to bough;
Yet do the cranes deserve a greater praise,
Which keep such measure in their airy ways,
As when they all in order ranked are,
They make a perfect form triangular.

368 creeks] turnings wrenches] sharp turns
381 bravery] splendid appearance
394 Jump with] in exact time with

'In the chief angle flies the watchful guide; 400
And all the followers their heads do lay
On their foregoers' backs, on either side;
But, for the captain hath no rest to stay
His head, forwearied with the windy way,
He back retires; and then the next behind,
As his lieutenant, leads them through the wind.

'But why relate I every singular?
Since all the world's great fortunes and affairs
Forward and backward rapt and whirled are,
According to the music of the spheres; 410
And Chance herself her nimble feet upbears
On a round slippery wheel, that rolleth aye,
And turns all states with her imperious sway;

'Learn then to dance, you that are princes born,
And lawful lords of earthly creatures all;
Imitate them, and thereof take no scorn,
(For this new art to them is natural)
And imitate the stars celestial.
For when pale death your vital twist shall sever,
Your better parts must dance with them forever." ' 420

(Wr. 1594; pub. 1596)

from *Gulling Sonnets*

473

As when the bright cerulian firmament
Hath not his glory with black clouds defaced,
So were my thoughts void of all discontent
And with no mist of passions overcast.
They all were pure and clear, till at the last
An idle careless thought forth wand'ring went,
And of that poisonous beauty took a taste
Which does the hearts of lovers so torment.
Then as it chanceth in a flock of sheep
When some contagious ill breeds first in one, 10
Daily it spreads, and secretly doth creep
Till all the silly troop be overgone;
So by close neighbourhood within my breast
One scurvy thought infecteth all the rest.

474

WHAT eagle can behold her sunbright eye,
Her sunbright eye that lights the world with love,
The world of love wherein I live and die,
I live and die and divers changes prove;
I changes prove, yet still the same am I,
The same am I and never will remove,
Never remove until my soul doth fly,
My soul doth fly and I surcease to move;
I cease to move which now am moved by you,
Am moved by you that move all mortal hearts, 10
All mortal hearts whose eyes your eyes doth view,
Your eyes doth view whence Cupid shoots his darts,
Whence Cupid shoots his darts and woundeth those
That honour you, and never were his foes.

475

THE sacred muse that first made Love divine
Hath made him naked and without attire;
But I will clothe him with this pen of mine
That all the world his fashion shall admire:
His hat of hope, his band of beauty fine,
His cloak of craft, his doublet of desire;
Grief for a girdle shall about him twine;
His points of pride, his ilet-holes of ire,
His hose of hate, his codpiece of conceit,
His stockings of stern strife, his shirt of shame; 10
His garters of vain-glory, gay and slight,
His pantofles of passions I will frame;
Pumps of presumption shall adorn his feet,
And socks of sullenness exceeding sweet.

476

MY case is this, I love Zepheria bright.
Of her I hold my heart by fealty
Which I discharge to her perpetually,
Yet she thereof will never me acquit.
For now supposing I withhold her right,
She hath distrained my heart to satisfy
The duty which I never did deny,
And far away impounds it with despite.

475
5 band] collar
8 points] ties
ilet-holes] i.e. eyelet holes
9 hose] breeches
12 pantofles] over-shoes
13 Pumps] soft shoes

476
6 distrained] (legal) seized
8 impounds] (legal) takes possession of

I labour therefore justly to repleve
  My heart which she unjustly doth impound, 10
But quick conceit which now is love's high shrieve
  Returns it as esloined, not to be found;
    Then, which the law affords, I only crave
    Her heart for mine in withernam to have.

(Wr. prob. 1594; pub. 1873)

from *Epigrams*

477

TITUS the brave and valorous gallant
Three years together in this town hath been,
Yet my Lord Chancellor's tomb he hath not seen,
Nor the new water work, nor the elephant.
  I cannot tell the cause without a smile:
  He hath been in the Counter all this while.

478

COSMUS hath more discoursing in his head
Than Jove when Pallas issued from his brain,
And still he strives to be delivered
Of all his thoughts at once, but all in vain.
For as we see at all the play-house doors,
When ended is the play, the dance and song,
A thousand townsmen, gentlemen, and whores,
Porters and serving-men together throng,
So thoughts of drinking, thriving, wenching, war,
And borrowing money, raging in his mind, 10
To issue all at once so forward are
As none at all can perfect passage find.

479

THE fine youth Ciprius is more terse and neat
Than the new garden of th'old Temple is,
And still the newest fashion he doth get,
And with the time doth change from that to this.

9 repleve] (legal) bail out
11 shrieve] sheriff
12 esloined] eloigned, removed out of the jurisdiction of the court or sheriff
14 withernam] (legal) taking other goods in place of those eloigned

477
6 Counter] a debtor's prison

479
1 terse] spruce, trim

He wears a hat now of the flat-crown block,
The treble ruffs, long cloak and doublet French.
He takes tobacco, and doth wear a lock,
And wastes more time in dressing than a wench.
  Yet this new-fangled youth, made for these times,
  Doth above all praise old George Gascoigne's rhymes. 10

480

AMONGST the poets Dacus numbered is,
Yet could he never make an English rhyme;
But some prose speeches I have heard of his,
Which have been spoken many a hundred time.
  The man that keeps the elephant hath one,
  Wherein he tells the wonders of the beast.
  Another Banks pronounced long agone,
  When he his curtal's qualities expressed.
He first taught him that keeps the monuments
At Westminster his formal tale to say, 10
And also him which puppets represents,
And also him which with the ape doth play.
  Though all his poetry be like to this,
  Amongst the poets Dacus numbered is.

481

PHILO the gentleman, the fortune teller,
The schoolmaster, the midwife and the bawd,
The conjurer, the buyer and the seller
Of painting which with breathing will be thawed,
  Doth practice physic, and his credit grows
  As doth the ballad-singer's auditory,
  Which hath at Temple Bar his standing chose,
  And to the vulgar sings an ale-house story.
First stands a porter, then an oyster-wife
Doth stint her cry and stays her steps to hear him, 10
Then comes a cutpurse ready with his knife,
And then a country client presseth near him.

5 block] style, fashion (of hat)
7 lock] lovelock
10 Gascoigne] (pronounced 'Gaskin')

480
7 Banks] (the owner of a famous performing horse called Morocco)
8 curtal] horse with its tail cut short or docked
10 Westminster] i.e. Westminster Abbey
11 represents] presents, exhibits

481
5 physic] medicine
6 auditory] audience
10 stint] stop, check

There stands the constable, there stands the whore,
And hearkening to the song mark not each other.
There by the sergeant stands the debtor poor,
And doth no more mistrust him than his brother.
Thus Orpheus to such hearers giveth music,
And Philo to such patients giveth physic.

(Pub. without date, prob. 1595 or 1596)

from *Nosce Teipsum*

482 *Of Human Knowledge*

WHY did my parents send me to the schools,
That I with knowledge might enrich my mind?
Since the desire to know first made men fools,
And did corrupt the root of all mankind.

For when God's hand had written in the hearts
Of the first Parents all the rules of good,
So that their skill infused did pass all arts
That ever were, before or since the Flood,

And when their reason's eye was sharp and clear,
And (as an eagle can behold the sun) 10
Could have approached the Eternal Light as near
As th'intellectual angels could have done:

Even then to them the Spirit of Lies suggests
That they were blind because they saw not ill,
And breathes into their incorrupted breasts
A curious wish, which did corrupt their will.

For that same ill they straight desired to know:
Which ill, being nought but a defect of good,
In all God's works the Devil could not show
While Man their lord in his perfection stood. 20

So that themselves were first to do the ill,
Ere they thereof the knowledge could attain,
Like him that knew not poison's power to kill,
Until (by tasting it) himself was slain.

15 sergeant] law-officer with the duty of arresting offenders

482 *Nosce Teipsum*] know thyself

Even so, by tasting of that fruit forbid,
Where they sought knowledge they did error find:
Ill they desired to know, and ill they did,
And to give Passion eyes made Reason blind.

For then their minds did first in Passion see
Those wretched shapes of Misery and Woe, 30
Of Nakedness, of Shame, of Poverty,
Which then their own experience made them know.

But then grew Reason dark, that she no more
Could the fair forms of Good and Truth discern:
Bats they became that eagles were before,
And this they got by their desire to learn.

But we their wretched offspring, what do we?
Do we not still taste of the fruit forbid
Whiles with fond fruitless curiosity,
In books profane we seek for knowledge hid? 40

What is this knowledge but the sky-stoln fire
For which the thief still chain'd in ice doth sit,
And which the poor rude Satyr did admire,
And needs would kiss but burnt his lips with it?

What is it but the cloud empty of rain
Which when Jove's guest embraced, he monsters got?
Or the false pails which, oft being filled with pain,
Received the water but retained it not?

Shortly, what is it but the fiery coach
Which the youth sought, and sought his death withal? 50
Or the boy's wings which, when he did approach
The sun's hot beams, did melt and let him fall?

And yet alas, when all our lamps are burned,
Our bodies wasted and our spirits spent,
When we have all the learned volumes turned
Which yield men's wits both help and ornament,

What can we know? or what can we discern?
When Error chokes the windows of the mind,
The divers forms of things how can we learn
That have been ever from our birthday blind? 60

When Reason's lamp, which like the sun in sky
Throughout Man's little world her beams did spread,
Is now become a sparkle, which doth lie
Under the ashes, half extinct and dead,

How can we hope that through the eye and ear
This dying sparkle in this cloudy place
Can re-collect those beams of knowledge clear
Which were infused in the first minds by grace?

So might the heir, whose father hath in play
Wasted a thousand pounds of ancient rent, 70
By painful earning of one groat a day
Hope to restore the patrimony spent.

The wits that dived most deep and soared most high,
Seeking Man's powers, have found his weakness such:
'Skill comes so slow, and life so fast doth fly,
We learn so little and forget so much.'

For this the wisest of all moral men
Said he knew nought but that he nought did know:
And the great mocking-master mocked not then,
When he said truth was buried deep below. 80

For how may we to other things attain
When none of us his own soul understands?
For which the Devil mocks our curious brain
When 'Know thyself' his oracle commands.

For why should we the busy Soul believe
When boldly she concludes of that and this,
When of herself she can no judgment give,
Nor how, nor whence, nor where, nor what she is?

All things without, which round about we see,
We seek to know and have therewith to do; 90
But that whereby we reason, live, and be,
Within ourselves, we strangers are thereto.

We seek to know the moving of each sphere,
And the strange cause of th'ebbs and floods of Nile;
But of that clock which in our breasts we bear,
The subtle motions, we forget the while.

77 the wisest] Socrates

79 the great mocking-master] Democritus

We that acquaint ourselves with every Zone
And pass both Tropics and behold both Poles,
When we come home are to ourselves unknown,
And unacquainted still with our own Souls. 100

We study Speech, but others we persuade;
We leech-craft learn, but others cure with it;
We'interpret laws, which other men have made,
But read not those which in our hearts are writ.

Is it because the mind is like the eye,
Through which it gathers knowledge by degrees,
Whose rays reflect not, but spread outwardly,
Not seeing itself when other things it sees?

No, doubtless; for the mind can backward cast
Upon herself her understanding light; 110
But she is so corrupt, and so defaced,
As her own image doth herself affright.

As in the fable of that Lady fair
Which for her lust was turned into a cow,
When thirsty to a stream she did repair
And saw herself transformed (she wist not how),

At first she startles, then she stands amazed,
At last with terror she from thence doth fly;
And loathes the watery glass wherein she gazed,
And shuns it still though she for thirst do die: 120

Even so Man's Soul which did God's image bear,
And was at first fair, good, and spotless pure,
Since with her sins her beauties blotted were,
Doth of all sights her own sight least endure:

For even at first reflection she espies
Such strange chimeras and such monsters there;
Such toys, such antics, and such vanities,
As she retires, and shrinks for shame and fear.

And as the man loves least at home to be
That hath a sluttish house haunted with sprites, 130
So she, impatient her own faults to see,
Turns from herself and in strange things delights.

102 leech-craft] medicine
116 wist] knew
127 toys] fantastic notions
antics] grotesqueries

For this, few know themselves: for merchants broke
View their estate with discontent and pain;
And seas are troubled when they do revoke
Their flowing waves into themselves again.

And while the face of outward things we find
Pleasing and fair, agreeable and sweet,
These things transport and carry out the mind,
That with herself her self can never meet. 140

Yet if Affliction once her wars begin,
And threat the feebler Sense with sword and fire,
The Mind contracts herself and shrinketh in,
And to herself she gladly doth retire—

As spiders, touched, seek their webs' inmost part;
As bees in storms unto their hives return;
As blood, in danger, gathers to the heart;
As men seek towns when foes the country burn.

If aught can teach us aught, Affliction's looks,
Making us look into ourselves so near, 150
Teach us to know ourselves beyond all books,
Or all the learned Schools that ever were.

This mistress lately plucked me by the ear,
And many a golden lesson hath me taught;
Hath made my Senses quick, and Reason clear,
Reformed my Will and rectified my Thought.

So do the winds and thunders cleanse the air;
So working seas settle and purge the wine;
So lopped and pruned trees do flourish fair;
So doth the fire the drossy gold refine. 160

Neither Minerva nor the learned Muse,
Nor rules of art nor precepts of the wise,
Could in my brain those beams of skill infuse
As but the glance of this Dame's angry eyes.

She within lists my ranging mind hath brought
That now beyond myself I list not go:
Myself am centre of my circling thought,
Only myself I study, learn, and know.

I know my body's of so frail a kind
As force without, fevers within, can kill; 170
I know the heavenly nature of my mind,
But 'tis corrupted both in wit and will.

I know my Soul hath power to know all things,
Yet is she blind and ignorant in all;
I know I'am one of Nature's little kings,
Yet to the least and vilest things am thrall.

I know my life's a pain and but a span,
I know my Sense is mocked with every thing:
And, to conclude, I know myself a MAN,
Which is a proud, and yet a wretched thing. 180

(Wr. 1598; pub. 1599)

ANONYMOUS

483 [*Things forbidden*]

THE man that hath a handsome wife
  And keeps her as a treasure,
It is my chiefest joy of life
  To have her to my pleasure.

But if that man regardless were
  As though he cared not for her,
Though she were like to Venus fair,
  In faith I would abhor her.

If to do good I were restrained,
  And to do evil bidden, 10
I would be Puritan, I swear,
  For I love the thing forbidden.

It is the care that makes the theft;
  None loves the thing forsaken;
The bold and willing whore is left
  When the modest wench is taken.

172 wit] intellect

She dull is that's too forwards bent;
Not good, but want, is reason.
Fish at a feast, and flesh in Lent,
Are never out of season. 20

(Pub. 1867)

GEORGE PEELE
1556–1596

from *The Old Wive's Tale*

484

THREE merry men, and three merry men,
And three merry men be we;
I in the wood, and thou on the ground,
And Jack sleeps in the tree.

485

WHEN as the rye reach to the chin,
And chopcherry, chopcherry ripe within,
Strawberries swimming in the cream,
And school-boys playing in the stream;
Then O, then O, then O my true love said,
Till that time come again,
She could not live a maid.

486

SPREAD, table, spread,
Meat, drink, and bread,
Ever may I have
What I ever crave,
When I am spread,
For meat for my black cock,
And meat for my red.

487

[*Voices from the Well of Life*]

GENTLY dip, but not too deep,
For fear you make the golden beard to weep.
Fair maiden, white and red,
Comb me smooth, and stroke my head,
And thou shalt have some cockle-bread.

Gently dip, but not too deep,
For fear thou make the golden beard to weep.
Fair maiden, white and red,
Comb me smooth, and stroke my head,
And every hair a sheave shall be,
And every sheave a golden tree.

(Pub. 1595)

from *David and Fair Bethsabe*

488

[*Bethsabe's Song*]

HOT sun, cool fire, tempered with sweet air,
Black shade, fair nurse, shadow my white hair:
Shine, sun; burn, fire; breathe, air, and ease me;
Black shade, fair nurse, shroud me and please me:
Shadow, my sweet nurse, keep me from burning,
Make not my glad cause cause of mourning.
    Let not my beauty's fire
    Inflame unstaid desire,
    Nor pierce any bright eye
    That wandereth lightly.

(Pub. 1599)

RICHARD BARNFIELD
1574–1627

489

The Affectionate Shepherd

SCARCE had the morning star hid from the light
  Heaven's crimson canopy with stars bespangled,
But I began to rue th'unhappy sight
  Of that fair boy that had my heart entangled;
    Cursing the time, the place, the sense, the sin;
    I came, I saw, I viewed, I slipped in.

If it be sin to love a sweet-faced boy
  (Whose amber locks trussed up in golden trammels
Dangle adown his lovely cheeks with joy,
  When pearl and flowers his fair hair enamels)   10
    If it be sin to love a lovely lad,
    Oh then sin I, for whom my soul is sad.

His ivory-white and alabaster skin
  Is stained throughout with rare vermilion red,
Whose twinkling starry lights do never blin
  To shine on lovely Venus, beauty's bed;
    But as the lily and the blushing rose,
    So white and red on him in order grows.

Upon a time the nymphs bestirred themselves
  To try who could his beauty soonest win; 20
But he accounted them but all as elves,
  Except it were the fair Queen Gwendolen:
    Her he embraced, of her he was beloved,
    With plaints he proved, and with tears he moved.

But her an old man had been suitor to,
  That in his age began to dote again.
Her would he often pray, and often woo,
  When through old age enfeebled was his brain.
    But she before had loved a lusty youth
    That now was dead, the cause of all her ruth. 30

And thus it happened. Death and Cupid met
  Upon a time at swilling Bacchus' house,
Where dainty cates upon the board were set
  And goblets full of wine to drink carouse:
    Where Love and Death did love the liquor so
    That out they fall and to the fray they go.

And having both their quivers at their back
  Filled full of arrows; th'one of fatal steel,
The other all of gold; Death's shaft was black,
  But Love's was yellow: Fortune turned her wheel; 40
    And from Death's quiver fell a fatal shaft,
    That under Cupid by the wind was waft.

And at the same time by ill hap there fell
  Another arrow out of Cupid's quiver;
The which was carried by the wind at will,
  And under Death the amorous shaft did shiver.
    They being parted, Love took up Death's dart,
    And Death took up Love's arrow, for his part.

15 blin] cease
21 elves] imps, i.e. mischievous and spiteful creatures
34 drink carouse] drink to the bottom, drink a full can to someone's health

Thus as they wandered both about the world,
  At last Death met with one of feeble age; 50
Wherewith he drew a shaft and at him hurled
  The unknown arrow, with a furious rage,
    Thinking to strike him dead with Death's black dart,
    But he (alas) with Love did wound his heart.

This was the doting fool, this was the man
  That loved fair Gwendolena Queen of Beauty.
She cannot shake him off, do what she can,
  For he hath vowed to her his soul's last duty,
    Making him trim upon the holy-days,
    And crowns his love with garlands made of bays. 60

Now doth he stroke his beard, and now (again)
  He wipes the drivel from his filthy chin;
Now offers he a kiss; but high disdain
  Will not permit her heart to pity him:
    Her heart more hard than adamant or steel,
    Her heart more changeable than Fortune's wheel.

But leave we him in love (up to the ears)
  And tell how Love behaved himself abroad;
Who seeing one that mourned still in tears
  (A young man groaning under love's great load) 70
    Thinking to ease his burden, rid his pains:
    For men have grief as long as life remains.

Alas the while, that unawares he drew
  The fatal shaft that Death had dropped before;
By which deceit great harm did then issue,
  Staining his face with blood and filthy gore.
    His face, that was to Gwendolen more dear
    Than love of lords, of any lordly peer.

This was that fair and beautiful young man
  Whom Gwendolena so lamented for; 80
This is that love whom she doth curse and ban,
  Because she doth that dismal chance abhor;
    And if it were not for his mother's sake,
    Even Ganymede himself she would forsake.

Oh would she would forsake my Ganymede,
  Whose sugared love is full of sweet delight,
Upon whose forehead you may plainly read
  Love's pleasure, graved in ivory tablets bright;
    In whose fair eye-balls you may clearly see
    Base love still stained with foul indignity. 90

Oh would to God he would but pity me,
  That love him more than any mortal wight:
Then he and I with love would soon agree,
  That now cannot abide his suitors' sight.
    O would to God (so I might have my fee)
    My lips were honey, and thy mouth a bee.

Then shouldst thou suck my sweet and my fair flower
  That now is ripe and full of honey-berries;
Then would I lead thee to my pleasant bower
  Filled full of grapes, of mulberries, and cherries; 100
    Then shouldst thou be my wasp or else my bee,
    I would thy hive, and thou my honey be.

I would put amber bracelets on thy wrests,
  Crownets of pearl about thy naked arms;
And when thou sit'st at swilling Bacchus' feasts,
  My lips with charms should save thee from all harms;
    And when in sleep thou took'st thy chiefest pleasure,
    Mine eyes should gaze upon thine eye-lids' treasure.

And every morn by dawning of the day,
  When Phoebus riseth with a blushing face, 110
Silvanus' chapel-clerks shall chaunt a lay,
  And play thee hunts-up in thy resting place;
    My cote thy chamber, my bosom thy bed,
    Shall be appointed for thy sleepy head.

And when it pleaseth thee to walk abroad
  (Abroad into the fields to take fresh air),
The meads with Flora's treasure should be strowed
  (The mantled meadows and the fields so fair),
    And by a silver well, with golden sands,
    I'll sit me down, and wash thine ivory hands. 120

And in the sweltering heat of summer time,
  I would make cabinets for thee, my love:
Sweet-smelling arbours made of eglantine
  Should be thy shrine, and I would be thy dove.
    Cool cabinets of fresh green laurel boughs
    Should shadow us, o'er-set with thick-set yews.

113 cote] cottage

122 cabinets] bowers, summer-houses

Or if thou list to bathe thy naked limbs
Within the crystal of a pearl-bright brook,
Paved with the dainty pebbles to the brims,
Or clear, wherein thyself thyself mayst look, 130
We'll go to Ladon, whose still trickling noise
Will lull thee fast sleep amidst thy joys.

Or if thou'lt go unto the river side
To angle for the sweet fresh-water fish,
Armed with thy implements that will abide
(Thy rod, hook, line) to take a dainty dish;
Thy rods shall be of cane, thy lines of silk,
Thy hooks of silver, and thy baits of milk.

Or if thou lov'st to hear sweet melody,
Or pipe a round upon an oaten reed, 140
Or make thyself glad with some mirthful glee,
Or play them music whilst thy flock doth feed;
To Pan's own pipe I'll help my lovely lad,
Pan's golden pipe which he of Syrinx had.

Or if thou dar'st to climb the highest trees
For apples, cherries, medlars, pears, or plums,
Nuts, walnuts, filberts, chestnuts, services,
The hoary peach, when snowy winter comes;
I have fine orchards full of mellowed fruit,
Which I will give thee to obtain my suit. 150

Not proud Alcinous himself can vaunt
Of goodlier orchards or of braver trees
Than I have planted; yet thou wilt not grant
My simple suit; but like the honey bees
Thou suck'st the flower till all the sweet be gone,
And lov'st me for my coin till I have none.

Leave Gwendolen (sweet-heart). Though she is fair
Yet is she light; not light in virtue shining,
But light in her behaviour, to impair
Her honour in her chastity's declining. 160
Trust not her tears, for they can wantonise,
When tears in pearl are trickling from her eyes.

147 filberts] hazelnuts services] pear-shaped fruit

If thou wilt come and dwell with me at home,
  My sheep-cote shall be strowed with new green rushes;
We'll haunt the trembling prickets as they roam
  About the fields, along the hawthorn bushes.
    I have a piebald cur to hunt the hare:
    So we will live with dainty forest fare.

Nay more than this, I have a garden-plot,
  Wherein there wants nor herbs, nor roots, nor flowers 170
(Flowers to smell, roots to eat, herbs for the pot),
  And dainty shelters when the welkin lowers:
    Sweet-smelling beds of lilies and of roses,
    Which rosemary banks and lavender encloses.

There grows the gillyflower, the mint, the daisy
  (Both red and white), the blue-veined violet;
The purple hyacinth, the spike to please thee;
  The scarlet-dyed carnation bleeding yet;
    The sage, the savory, and sweet marjoram,
    Hyssop, thyme, and eye-bright, good for the blind and dumb. 180

The pink, the primrose, cowslip, and daffadilly,
  The harebell blue, the crimson columbine,
Sage, lettuce, parsley, and the milk-white lily,
  The rose, and speckled flowers called sops-in-wine,
    Fine pretty king-cups, and the yellow boots
    That grows by rivers and by shallow brooks.

And many thousand moe I cannot name
  Of herbs and flowers that in gardens grow
I have for thee; and coneys that be tame,
  Young rabbits, white as swan and black as crow, 190
    Some speckled here and there with dainty spots;
    And more I have two milch and milk-white goats.

All these, and more, I'll give thee for thy love,
  If these, and more, may tice thy love away.
I have a pigeon-house, in it a dove,
  Which I love more than mortal tongue can say.
    And last of all, I'll give thee a little lamb
    To play withal, new-weaned from her dam.

165 prickets] bucks in their second year
177 spike] French lavender
179 savory] species of herb
180 Hyssop] aromatic herb
  eye-bright] i.e. euphrasy
185 boots] marsh-marigolds
187 moe] more
189 coneys] rabbits
194 tice] entice

But if thou wilt not pity my complaint,
  My tears, nor vows, nor oaths, made to thy beauty, 200
What shall I do? But languish, die, or faint,
  Since thou dost scorn my tears and my soul's duty;
    And tears contemned, vows and oaths must fail,
    For where tears cannot, nothing cannot prevail.

Compare the love of fair Queen Gwendolin
  With mine, and thou shalt see how she doth love thee:
I love thee for thy qualities divine,
  But she doth love another swain above thee.
    I love thee for thy gifts, she for her pleasure;
    I for thy virtue, she for beauty's treasure. 210

And always (I am sure) it cannot last,
  But sometime Nature will deny those dimples:
Instead of beauty (when thy blossom's past)
  Thy face will be deformed, full of wrinkles.
    Then she that loved thee for thy beauty's sake,
    When age draws on, thy love will soon forsake.

But I that loved thee for thy gifts divine,
  In the December of thy beauty's waning,
Will still admire, with joy, those lovely eyne,
  That now behold me with their beauties baning. 220
    Though January will never come again,
    Yet April years will come in showers of rain.

When will my May come, that I may embrace thee?
  When will the hour be of my soul's joying?
Why dost thou seek in mirth still to disgrace me?
  Whose mirth's my health, whose grief's my heart's annoying.
    Thy bane my bale, thy bliss my blessedness,
    Thy ill my hell, thy weal my welfare is.

Thus do I honour thee that love thee so,
  And love thee so, that so do honour thee 230
Much more than any mortal man doth know
  Or can discern by love or jealousy.
    But if that thou disdain'st my loving ever,
    Oh happy I, if I had loved never.

(1594)

219 eyne] eyes   220 baning] killing, poisoning

490

Man's life

MAN'S life is well compared to a feast,
Furnished with choice of all variety;
To it comes Time; and as a bidden guest
He sets him down, in pomp and majesty;
The three-fold Age of man the waiters be.
Then with an earthen voider (made of clay)
Comes Death, and takes the table clean away.

(1598)

GEORGE CHAPMAN
1559?–1634

from *Ovid's Banquet of Sense*

491

[*'The Ears' Delight'*]

WHILE this was singing, Ovid young in love
With her perfections, never proving yet
How merciful a mistress she would prove,
Boldly embraced the power he could not let,
And like a fiery exhalation
Followed the sun he wished might never set;
Trusting herein his constellation
Ruled by love's beams, which Julia's eyes erected,
Whose beauty was the star his life directed.

And having drenched his ankles in those seas, 10
He needs would swim, and cared not if he drowned.
Love's feet are in his eyes; for if he please
The depth of beauty's gulfy flood to sound,
He goes upon his eyes, and up to them
At the first step he is. No shader ground
Could Ovid find; but in love's holy stream
Was past his eyes, and now did wet his ears,
For his high sovereign's silver voice he hears.

6 voider] tray, basket or other vessel in which dirty dishes or utensils, fragments of broken food etc., are placed in clearing the table or during a meal

491
1 While this was singing] while this was being sung (i.e. the song of Julia, Ovid's beloved)
4 let] prevent
15 shader] more sloping

Whereat his wit assumed fiery wings,
  Soaring above the temper of his soul, 20
And he the purifying rapture sings
  Of his ears' sense; takes full the Thespian bowl
And it carouseth to his mistress' health,
  Whose sprightful verdure did dull flesh control;
And his conceit he crowneth with the wealth
    Of all the Muses in his pleased senses,
    When with the ears' delight he thus commences:

'Now Muses, come, repair your broken wings
  (Plucked and prophaned by rustic ignorance)
With feathers of these notes my mistress sings; 30
  And let quick verse her drooping head advance
From dungeons of contempt to smite the stars.
  In Julia's tunes, led forth by furious trance
A thousand Muses come to bid you wars,
    Dive to your spring, and hide you from the stroke,
    All poets' furies will her tunes invoke.

Never was any sense so set on fire
  With an immortal ardour, as mine ears;
Her fingers to the strings doth speech inspire
  And numbered laughter, that the descant bears 40
To her sweet voice; whose species through my sense
  My spirits to their highest function rears;
To which, impressed with ceaseless confluence,
    It useth them as proper to her power,
    Marries my soul, and makes itself her dower.

Methinks her tunes fly gilt like Attic bees
  To my ears' hives, with honey tried to air:
My brain is but the comb, the wax, the lees,
  My soul the drone, that lives by their affair.
O, so it sweets, refines, and ravisheth, 50
  And with what sport they sting in their repair!
Rise then in swarms, and sting me thus to death,
    Or turn me into swound; possess me whole,
    Soul to my life and essence to my soul!

24 verdure] freshness
control] challenge, expose to criticism
25 conceit] conception, idea
31 advance] raise
40 numbered] rhythmical
41 species] emanation
47 tried] purified, refined
49 affair] activity
51 repair] dwelling-place

Say, gentle air, O does it not thee good
  Thus to be smit with her correcting voice?
Why dance ye not, ye daughters of the wood?
  Wither for ever, if not now rejoice.
Rise stones, and build a city with her notes,
  And notes infuse with your most Cynthian noise 60
To all the trees, sweet flowers, and crystal floats
    That crown and make this cheerful garden quick,
    Virtue, that every touch may make such music.

O that as man is called a little world
  The world might shrink into a little man
To hear the notes about this garden hurled,
  That skill dispersed in tunes so Orphean
Might not be lost in smiting stocks and trees
  That have no ears, but grown as it began
Spread their renowns as far as Phoebus sees 70
    Through earth's dull veins; that she like heaven might move
    In ceaseless music, and be filled with love.

In precious incense of her holy breath
  My love doth offer hecatombs of notes
To all the gods, who now despise the death
  Of oxen, heifers, wethers, swine and goats.
A sonnet in her breathing sacrificed
  Delights them more than all beasts' bellowing throats,
As much with heaven as with my hearing prized.
    And as gilt atoms in the sun appear, 80
    So greet these sounds the gristles of mine ear,

Whose pores do open wide to their regreet,
  And my implanted air that air embraceth
Which they impress. I feel their nimble feet
  Tread my ears' labyrinth; their sport amazeth
They keep such measure; play themselves and dance.
  And now my soul in Cupid's furnace blazeth,
Wrought into fury with their dalliance;
    And as the fire the parched stubble burns,
    So fades my flesh, and into spirit turns. 90

56 correcting] harmonizing, bringing to order
61 floats] waves
63 touch] note, sound
74 hecatombs] sacrifices of many victims, i.e. a large number
81 gristles] cartilages
82 regreet] returned greeting

Sweet tunes, brave issue, that from Julia come,
  Shook from her brain, armed like the Queen of Ire;
For first conceived in her mental womb,
  And nourished with her soul's discursive fire
They grew into the power of her thought.
  She gave them downy plumes from her attire,
And them to strong imagination brought:
    That, to her voice; wherein most movingly
    She (blessing them with kisses) lets them fly.

Who fly rejoicing; but (like noblest minds) 100
  In giving others life themselves do die,
Not able to endure earth's rude unkinds,
  Bred in my sovereign's parts too tenderly.
O that as intellects themselves transite
  To each intelligible quality,
My life might pass into my love's conceit,
    Thus to be formed in words, her tunes, and breath,
    And with her kisses sing itself to death.

This life were wholly sweet, this only bliss;
  Thus would I live to die, thus sense were feasted; 120
My life that in my flesh a chaos is
  Should to a golden world be thus digested.
Thus should I rule her face's monarchy,
  Whose looks in several empires are invested
Crowned now with smiles, and then with modesty.
    Thus in her tunes' division I should reign,
    For her conceit does all, in every vein.

My life then turned to that, t'each note and word
  Should I consort her look; which sweeter sings
Where songs of solid harmony accord, 130
  Ruled with love's rule and pricked with all his stings.
Thus should I be her notes before they be;
  While in her blood they sit with fiery wings
Not vapoured in her voice's stillery.
    Nought are these notes her breast so sweetly frames
    But motions, fled out of her spirit's flames.

92 Queen of Ire] i.e. Minerva
102 unkinds] unkindnesses
104 transite] pass over or through
122 digested] reduced, assimilated
126 division] rapid series of notes, descant
129 consort] accompany, escort
134 vapoured] evaporated
stillery] still, distillery

For as when steel and flint together smit
  With violent action spit forth sparks of fire
And make the tender tinder burn with it,
  So my love's soul doth lighten her desire 140
Upon her spirits in her notes' pretence;
  And they convey them (for distinct attire)
To use the wardrobe of the common sense;
    From whence in veils of her rich breath they fly
    And feast the ear with this felicity.

Methinks they raise me from the heavy ground
  And move me swimming in the yielding air,
As zephyrs' flowery blast do toss a sound;
  Upon their wings will I to heaven repair,
And sing them so, gods shall descend and hear. 150
  Ladies must be adored that are but fair,
But apt besides with art to tempt the ear
    In notes of nature, is a goddess' part,
    Though oft men's nature's notes please more than art.

But here are art and nature both confined,
  Art casting nature in so deep a trance
That both seem dead, because they be divined;
  Buried is heaven in earthly ignorance:
Why break not men then strumpet folly's bounds,
  To learn at this pure virgin's utterance? 160
No; none but Ovid's ears can sound these sounds,
    Where sing the hearts of love and poesy
    Which make my Muse so strong she works too high.'

Now in his glowing ears her tunes did sleep;
  And as a silver bell, with violent blow
Of steel or iron, when his sounds most deep
  Do from his sides and airs' soft bosom flow,
A great while after murmurs at the stroke,
  Letting the hearers' ears his hardness know,
So chid the air to be no longer broke, 170
    And left the accents panting in his ear,
    Which in this banquet his first service were.

(1595)

from *Hero and Leander*

492 NEW light gives new directions, fortunes new,
To fashion our endeavours that ensue;
More harsh (at least more hard), more grave and high
Our subject runs, and our stern muse must fly;
Love's edge is taken off, and that light flame,
Those thoughts, joys, longings, that before became
High unexperienced blood, and maids' sharp plights,
Must now grow staid, and censure the delights,
That being enjoyed ask judgement; now we praise,
As having parted: evenings crown the days. 10
And now ye wanton loves and young desires
Pied vanity, the mint of strange attires,
Ye lisping flatteries and obsequious glances,
Relentful musics and attractive dances,
And you detested charms constraining love,
Shun love's stol'n sports by that these lovers prove.
By this the sovereign of heaven's golden fires,
And young Leander, lord of his desires,
Together from their lovers' arms arose:
Leander into Hellespontus throws 20
His Hero-handled body, whose delight
Made him disdain each other epithet.
And as amidst th'enamoured waves he swims,
The god of gold of purpose gilt his limbs,
That this word gilt including double sense,
The double guilt of his incontinence
Might be expressed, that had no stay t'employ
The treasure which the love-god let him joy
In his dear Hero, with such sacred thrift
As had beseemed so sanctified a gift; 30
But like a greedy vulgar prodigal
Would on the stock dispend, and rudely fall
Before his time, to that unblessed blessing,
Which for lust's plague doth perish with possessing.
Joy graven in sense, like snow in water, wastes;
Without preserve of virtue nothing lasts.
What man is he that with a wealthy eye
Enjoys a beauty richer than the sky,
Through whose white skin, softer than soundest sleep,
With damask eyes the ruby blood doth peep, 40

8 staid] grave, serious
16 prove] experience
32 on the stock dispend] trench on his capital

And runs in branches through her azure veins,
Whose mixture and first fire his love attains;
Whose both hands limit both love's deities,
And sweeten human thoughts like paradise;
Whose disposition silken is and kind,
Directed with an earth-exempted mind—
Who thinks not heaven with such a love is given?
And who like earth would spend that dower of heaven,
With rank desire to joy it all at first?
What simply kills our hunger, quencheth thirst, 50
Clothes but our nakedness, and makes us live,
Praise doth not any of her favours give:
But what doth plentifully minister
Beauteous apparel and delicious cheer,
So ordered that it still excites desire,
And still gives pleasure freeness to aspire,
The palm of bounty ever moist preserving:
To love's sweet life this is the courtly carving.
Thus Time, and all-states-ordering Ceremony
Had banished all offence: Time's golden thigh 60
Upholds the flowery body of the earth
In sacred harmony, and every birth
Of men and actions makes legitimate,
Being used aright. *The use of time is Fate.*
  Yet did the gentle flood transfer once more
This prize of love home to his father's shore,
Where he unlades himself of that false wealth
That makes few rich, treasures composed by stealth;
And to his sister, kind Hermione
(Who on the shore kneeled, praying to the sea 70
For his return), he all love's goods did show,
In Hero seised for him, in him for Hero.
His most kind sister all his secrets knew,
And to her singing like a shower he flew,
Sprinkling the earth, that to their tombs took in
Streams dead for love to leave his ivory skin,
Which yet a snowy foam did leave above,
As soul to the dead water that did love;
And from thence did the first white roses spring
(For love is sweet and fair in every thing) 80
And all the sweetened shore as he did go,
Was crowned with od'rous roses white as snow.
Love-blest Leander was with love so filled,
That love to all that touched him he instilled.

42 mixture] sexual intercourse
43 limit] confine, contain
57 moist] fresh
72 seised for] legally possessed by

And as the colours of all things we see
To our sight's powers communicated be,
So to all objects that in compass came
Of any sense he had, his senses' flame
Flowed from his parts with force so virtual,
It fired with sense things mere insensual. 90
Now (with warm baths and odours comforted)
When he lay down he kindly kissed his bed,
As consecrating it to Hero's right,
And vowed thereafter that whatever sight
Put him in mind of Hero, or her bliss,
Should be her altar to prefer a kiss.
Then laid he forth his late enriched arms,
In whose white circle Love writ all his charms,
And made his characters sweet Hero's limbs,
When on his breast's warm sea she sidling swims. 100
And as those arms (held up in circle) met,
He said: 'See, sister, Hero's carcanet,
Which she had rather wear about her neck,
Than all the jewels that do Juno deck.'
But as he shook with passionate desire
To put in flame his other secret fire,
A music so divine did pierce his ear,
As never yet his ravished sense did hear:
When suddenly a light of twenty hues
Brake through the roof, and like the rainbow views 110
Amazed Leander; in whose beam came down
The goddess Ceremony, with a crown
Of all the stars, and heaven with her descended;
Her flaming hair to her bright feet extended,
By which hung all the bench of deities,
And in a chain, compact of ears and eyes,
She led Religion. All her body was
Clear and transparent as the purest glass:
For she was all presented to the sense:
Devotion, Order, State, and Reverence 120
Her shadows were; Society, Memory;
All which her sight made live, her absence die.
A rich disparent pentacle she wears,
Drawn full of circles and strange characters;
Her face was changeable to every eye,
One way looked ill, another graciously;

89 virtual] powerful
90 insensual] insensible
96 prefer] offer
99 characters] letters
102 carcanet] necklace
123 disparent] diverse

Which while men viewed, they cheerful were and holy,
But looking off, vicious and melancholy.
The snaky paths to each observed law
Did Policy in her broad bosom draw; 130
One hand a mathematic crystal sways,
Which gathering in one line a thousand rays
From her bright eyes, Confusion burns to death,
And all estates of men distinguisheth.
By it Morality and Comeliness
Themselves in all their sightly figures dress.
Her other hand a laurel rod applies,
To beat back Barbarism and Avarice
That followed, eating earth and excrement
And human limbs, and would make proud ascent 140
To seats of gods, were Ceremony slain.
The Hours and Graces bore her glorious train,
And all the sweets of our society
Were sphered and treasured in her bounteous eye.
Thus she appeared, and sharply did reprove
Leander's bluntness in his violent love;
Told him how poor was substance without rites,
Like bills unsigned, desires without delights;
Like meats unseasoned; like rank corn that grows
On cottages, that none or reaps or sows; 150
Not being with civil forms confirmed and bounded,
For human dignities and comforts founded,
But loose and secret, all their glories hide;
Fear fills the chamber, darkness decks the bride.
She vanished, leaving pierced Leander's heart
With sense of his unceremonious part,
In which with plain neglect of nuptial rites,
He close and flatly fell to his delights;
And instantly he vowed to celebrate
All rites pertaining to his married state. 160
So up he gets, and to his father goes,
To whose glad ears he doth his vows disclose.
The nuptials are resolved with utmost power,
And he at night would swim to Hero's tower,
From whence he meant to Sestos' forked bay
To bring her covertly, where ships must stay,
Sent by his father, throughly rigged and manned,
To waft her safely to Abydos' strand.
There leave we him, and with fresh wing pursue
Astonished Hero, whose most wished view 170
I thus long have forborne, because I left her
So out of count'nance, and her spirits bereft her.

To look on one abashed is impudence,
When of slight faults he hath too deep a sense.
Her blushing het her chamber; she looked out,
And all the air she purpled round about;
And after it a foul black day befell,
Which ever since a red morn doth foretell,
And still renews our woes for Hero's woe.
And foul it proved, because it figured so 180
The next night's horror, which prepare to hear:
I fail, if it profane your daintiest ear.
  Then thou most strangely-intellectual fire,
That proper to my soul hast power t'inspire
Her burning faculties, and with the wings
Of thy unsphered flame visit'st the springs
Of spirits immortal; now (as swift as Time
Doth follow Motion) find th'eternal clime
Of his free soul, whose living subject stood
Up to the chin in the Pierian flood, 190
And drunk to me half this Musaean story,
Inscribing it to deathless memory:
Confer with it, and make my pledge as deep,
That neither's draught be consecrate to sleep.
Tell it how much his late desires I tender
(If yet it know not), and to light surrender
My soul's dark offspring, willing it should die
To loves, to passions, and society.

493 THIS told, strange Teras touched her lute, and sung
This ditty that the torchy evening sprung.

Epithalamium Teratos

Come, come, dear Night, Love's mart of kisses,
Sweet close of his ambitious line,
The fruitful summer of his blisses,
Love's glory doth in darkness shine.
O come, soft rest of cares, come Night,
Come naked Virtue's only tire,
The reaped harvest of the light,
Bound up in sheaves of sacred fire. 10

175 het] heated
189 his free soul] i.e. Marlowe's, whose poem Chapman is offering to complete

493
Epithalamium Teratos] Teras's wedding song
8 tire] attire

Love calls to war,
Sighs his alarms,
Lips his swords are,
The field his arms.
Come, Night, and lay thy velvet hand
On glorious Day's outfacing face,
And all thy crowned flames command
For torches to our nuptial grace.
Love calls to war,
Sighs his alarms, 20
Lips his swords are,
The field his arms.
No need have we of factious Day,
To cast in envy of thy peace
Her balls of discord in thy way:
Here Beauty's day doth never cease;
Day is abstracted here,
And varied in a triple sphere.
Hero, Alcmane, Mya so outshine thee,
Ere thou come here let Thetis thrice refine thee. 30
Love calls to war,
Sighs his alarms,
Lips his swords are,
The field his arms.
The evening star I see:
Rise, youths, the evening star
Helps Love to summon war;
Both now embracing be.
Rise, youths, Love's right claims more than banquets, rise.
Now the bright marigolds that deck the skies, 40
Phoebus' celestial flowers, that (contrary
To his flowers here) ope when he shuts his eye,
And shut when he doth open, crown your sports.
Now Love in Night, and Night in Love exhorts
Courtship and dances. All your parts employ,
And suit Night's rich expansure with your joy.
Love paints his longings in sweet virgins' eyes:
Rise, youths, Love's right claims more than banquets, rise.
Rise, virgins, let fair nuptial loves enfold
Your fruitless breasts: the maidenheads ye hold 50
Are not your own alone, but parted are;
Part in disposing them your parents share,
And that a third part is, so must ye save
Your loves a third, and you your thirds must have.

Love paints his longings in sweet virgins' eyes:
Rise, youths, Love's right claims more than banquets,
rise.

(1598)

from *Achilles' Shield*

[Part of Book 18 of Homer's *Iliad*]

494 [*Thetis asks Vulcan to make a shield for Achilles*]

BRIGHT-FOOTED Thetis did the sphere aspire
(Amongst th'Immortals) of the god of fire,
Starry, incorruptible, and had the frame
Of ruddy brass, right shaped by the lame.
She found him at his swelling bellows sweating
And twenty tripods seriously beating,
To stand and beautify his royal hall
For chairs of honour round about the wall.
And to the feet he fixed of every one
Wheels of man-making gold to run alone 10
To the gods' temples—to the which they were
Religious ornaments; when, standing there
Till sacrifice were done, they would retire
To Vulcan's house, which all eyes did admire.
Yet the Daedalean handles to hold by
Were unimposed, which straight he did apply.
These while he fashioned with miraculous art,
The fair white-footed dame appeared apart
To Charis with the rich-attired head,
Whose heavenly beauties strowed the nuptial bed 20
Of that illustrate smith. She took her hand
And entertained her with this kind demand:
'What makes the goddess with the ample train,
Reverend and friendly Thetis, entertain
Conceit to honour us with her repair,
That never yet was kind in that affair?
But enter further, that so wished a guest
May be received with hospitable feast.'
Thus led she Thetis to a chair of state,
Rich and exceedingly elaborate, 30

494
1 aspire] go up to
6 tripods] three-legged stools
seriously] in due order, one after another
14 admire] wonder at
15 Daedalean] ingenious, skilfully made
24–25 entertain Conceit] have the idea
25 repair] visit

And set a footstool at her silver feet.
Then called her famous smith: 'Vulcan my sweet,
Thetis in some use needs thy fiery hand.'
He answered: 'Thetis hath a strong command
Of all my powers, who gave my life defence,
Cast by my mother's wilful impudence
Out of Olympus, who would have obscured
My native lameness. Then had I endured
Unhelped griefs if on her shining breast
Hospitious Thetis had not let me rest, 40
And bright Eurynome, my guardian,
Fair daughter of the labouring Ocean,
With whom nine years I wrought up divers things—
Buttons and bracelets, whistles, chains and rings—
In concluse of a cave; and over us
The swelling waves of old Oceanus
With foamy murmur flowed, and not a god
Nor any mortal knew my close abode
But Thetis and divine Eurynome,
Who succoured me. And now from gulfy sea 50
To our steep house hath Thetis made ascent,
To whom requital more than competent
It fits me much my safety should repay.
Charis, do thou some sumptuous feast purvey
Whiles I my airy bellows may lay by
And all my tools of heavenly ferrary.'
Thus from his anvil the huge monster rose,
And with distorted knees he limping goes
To a bright chest of silver-ore composed,
Where all his wonder-working tools were closed, 60
And took his sighing bellows from the fire.
Then with a sponge his breast with hairs like wire,
His brawned neck, his hard hands and his face
He cleansed; put on his robe, assumed his mace,
And halted forth; and on his steps attended
Handmaids of gold that with strong paces wended
Like dames in flower of life, in whom were minds
Furnished with wisdom, knowing all the kinds
Of the god's powers; from whom did voices fly,
In whom were strengths and motions voluntary. 70
These at his elbow ever ministered,
And these (drawing after him his legs) he led

40 Hospitious] hospitable
45 concluse] enclosure, confines
52 competent] merely sufficient, only moderate
56 ferrary] farriery (the art of the shoeing-smith)
65 halted] limped

To Thetis seated in a shining throne,
Whose hand he shook, and asked this question:
  'What wished occasion brings the sea's bright queen
To Vulcan's house, that ever yet hath been
So great a stranger? Show thy reverend will,
Which mine of choice commands me to fulfil,
If in the reach of all mine art it lie
Or it be possible to satisfy.' 80
  Thetis poured out this sad reply in tears:
'O Vulcan, is there any goddess bears
(Of all the deities that deck the sky)
So much of mortal wretchedness as I,
Whom Jove past all deprives of heavenly peace?
Myself of all the blue Nereides
He hath subjected to a mortal's bed,
Which I against my will have suffered
To Peleus, surnamed Æacides,
Who in his court lies slain with the disease 90
Of woeful age; and now with new infortunes
He all my joys to discontents importunes
In giving me a son, chief in renown
Of all Heroes, who hath palm-like grown
Set in a fruitful soil; and, when my care
Had nursed him to a form so singular,
I sent him in the Grecians' crook-sterned fleet
To Ilion, with the swiftness of his feet
And dreadful strength that his choice limbs indued
To fight against the Troyan fortitude; 100
And him I never shall receive, retired
To Peleus' court, but while he lives inspired
With human breath and sees the sun's clear light,
He must live sad and moody as the night.
Nor can I cheer him, since his valure's prize,
Resigned by all the Grecians' compromise,
Atrides forced into his fortune's part,
For which consumption tires upon his heart.
Yet, since the Troyans all the Greeks conclude
Within their fort, the peers of Greece have sued 110
With worthiness of gifts and humble prayers
To win his hand to hearten their affairs;
Which he denied; but to appease their harms
He decked his dear Patroclus in his arms
And sent him with his bands to those debates.
All day they fought before the Scæan gates

105 valure's] valour's   109 conclude] enclose, confine   115 debates] (military) conflicts

And well might have expugned, by that black light,
The Ilian city, if Apollo's spite,
Thirsting the blood of good Menœtius' son,
Had not in face of all the fight foredone 120
His faultless life and authored the renown
On Hector's prowess, making th'act his own.
Since, therefore, to revenge the timeless death
Of his true friend my son determineth
T'embrue the field, for want whereof he lies
Buried in dust and drowned in miseries,
Here at thy knees I sue, that the short date
Prefixed his life by power of envious fate
Thou wilt with heavenly arms grace and maintain,
Since his are lost with his Patroclus slain.' 130
He answered: 'Be assured, nor let the care
Of these desires thy firmest hopes impair.
Would God as far from lamentable death,
When heavy fates shall see it with his breath,
I could reserve him, as unequalled arms
Shall be found near t'avert all instant harms—
Such arms as all worlds shall for art admire,
That by their eyes their excellence aspire.'
This said, the smith did to his bellows go,
Set them to fire and made his Cyclops blow. 140
Full twenty pair breathed through his furnace holes
All sorts of blasts t'enflame his tempered coals,
Now blustered hard and now did contrarise
As Vulcan would, and, as his exercise
Might with perfection serve the dame's desire,
Hard brass and tin he cast into the fire,
High-prized gold and silver, and did set
Within the stock an anvil bright and great.
His massy hammer then his right hand held,
His other hand his gasping tongs compelled. 150

(1598)

117 expugned] captured, taken by storm
120 in face of] in defiance of
foredone] put an end to
123 timeless] untimely
125 embrue] stain with blood
127 date] duration
128 envious] hostile, malicious
143 contrarise] do the opposite

BARTHOLOMEW GRIFFIN

d. 1602

495 *Venus and Adonis*

VENUS, and young Adonis sitting by her,
Under a myrtle shade began to woo him:
She told the youngling how god Mars did try her,
And as he fell to her, so fell she to him.
'Even thus', quoth she, 'the wanton god embraced me'
(And then she clasped Adonis in her arms);
'Even thus', quoth she, 'the warlike god unlaced me',
As if the boy should use like loving charms.
But he, a wayward boy, refused her offer,
And ran away, the beauteous queen neglecting, 10
Showing both folly to abuse her proffer,
And all his sex of cowardice detecting.
O that I had my mistress at that bay,
To kiss and clip me till I ran away!

(1596)

496 *Care-charmer sleep*

CARE-CHARMER sleep, sweet ease in restless misery,
The captive's liberty, and his freedom's song,
Balm of the bruised heart, man's chief felicity,
Brother of quiet death, when life is too, too long!
A comedy it is, and now an history.
What is not sleep unto the feeble mind?
It easeth him that toils and him that's sorry,
It makes the deaf to hear, to see the blind.
Ungentle sleep, thou helpest all but me,
For when I sleep my soul is vexed most. 10
It is Fidessa that doth master thee;
If she approach, alas, thy power is lost.
But here she is. See, how he runs amain!
I fear at night he will not come again.

(1596)

12 detecting] accusing

SIR FRANCIS BACON
1561–1626

497 [*The life of man*]

THE world's a bubble, and the life of man
 Less than a span,
In his conception wretched, from the womb,
 So to the tomb;
Curst from the cradle, and brought up to years,
 With cares and fears.
Who then to frail mortality shall trust,
But limns on water, or but writes in dust.

Yet since with sorrow here we live oppressed,
 What life is best? 10
Courts are but only superficial schools
 To dandle fools.
The rural parts are turned into a den
 Of savage men.
And where's a city from all vice so free,
But may be termed the worst of all the three?

Domestic cares afflict the husband's bed,
 Or pains his head.
Those that live single take it for a curse,
 Or do things worse. 20
Some would have children; those that have them none,
 Or wish them gone.
What is it then to have or have no wife,
But single thraldom, or a double strife?

Our own affections still at home to please
 Is a disease;
To cross the sea to any foreign soil,
 Perils and toil.
Wars with their noise affright us; when they cease,
 W'are worse in peace. 30
What then remains, but that we still should cry,
Not to be born, or being born to die?

(Probably wr. 1592–8; pub. 1629)

2 span] short space of time; the distance from the tip of the thumb to the tip of the little finger, when the hand is fully extended

ROBERT SIDNEY
1563–1626

498 *Alas, why say you I am rich*

ALAS, why say you I am rich? when I
  Do beg, and begging scant a life sustain.
  Why do you say that I am well?—when pain
Louder than on the rack, in me doth cry.
O let me know myself! My poverty
  With whitening rotten walls no stay doth gain,
  And these small hopes you tell, keep but in vain
Life with hot drinks, in one laid down to die.
If in my face my wants and sores so great
  Do not appear, a canker (think) unseen   10
The apple's heart, though sound without, doth eat;
  Or if on me from my fair heaven are seen
    Some scattered beams, know such heat gives their light
    As frosty morning's sun, as moonshine night.

(Pub. 1975)

499 *Ah dearest limbs, my life's best joy and stay*

'AH dearest limbs, my life's best joy and stay,
  How must I thus let you be cut from me,
  And losing you, myself unuseful see,
And keeping you, cast life and all away?':
Full of dead gangrenes doth the sickman say,
  Whose death of part, health of the rest must be.
  Alas my love, from no infections free,
Like law doth give of it or my decay.
My love, more dear to me than hands or eyes,
  Nearer to me than what with me was born,   10
  Delayed, betrayed, cast under change and scorn,
Sick past all help or hope, or kills or dies;
    While all the blood it sheds my heart doth bleed
    And with my bowels I his cancers feed.

(Pub. 1981)

500 *Forsaken woods, trees with sharp storms oppressed*

FORSAKEN woods, trees with sharp storms oppressed,
  Whose leaves once hid the sun, now strew the ground,
  Once bred delight, now scorn, late used to sound
Of sweetest birds, now of hoarse crows the nest;
Gardens, which once in thousand colours dressed
  Showed nature's pride, now in dead sticks abound,
  In whom proud summer's treasure late was found
Now but the rags of winter's torn coat rest;
Meadows whose sides late fair brooks kissed, now slime
  Embraced holds; fields whose youth green and brave   10
  Promised long life, now frosts lay in the grave:
Say all, and I with them, 'What doth not Time!'
    But they, who knew Time, Time will find again;
    I that fair times lost, on Time call in vain.

(Pub. 1975)

501 *The sun is set, and masked night*

THE sun is set, and masked night
  Veils heaven's fair eyes:
Ah what trust is there to a light
  That so swift flies?

A new world doth his flames enjoy,
  New hearts rejoice:
In other eyes is now his joy,
  In other choice.

(Pub. 1975)

JOSEPH HALL

1574–1636

from *Virgidemiae. Toothless Satires*

502 [*Advertisement for a Chaplain*]

A GENTLE squire would gladly entertain
Into his house some trencher-chaplain,
Some willing man that might instruct his sons,
And that would stand to good conditions.
First, that he lie upon the truckle-bed,
Whiles his young master lieth o'er his head.
Secondly, that he doe, on no default,
Ever presume to sit above the salt.
Third, that he never change his trencher twice.
Fourth, that he use all comely courtesies: 10
Sit bare at meals, and one half rise and wait.
Last, that he never his young master beat,
But he must ask his mother to define
How many jerks she would his breech should line.
All those observed, he could contented be
To give five marks, and winter livery.

(1597)

503 [*A Drunkard arrives in Hades*]

WHEN Gullion died (who knows not Gullion?)
And his dry soul arrived at Acheron,
He fair besought the ferryman of hell
That he might drink to dead Pantagruel.
Charon was 'fraid lest thirsty Gullion
Would have drunk dry the river Acheron;
Yet last consented for a little hire,
And down he dips his chops deep in the mire,
And drinks, and drinks, and swallows in the stream
Until the shallow shores all naked seem. 10
Yet still he drinks, nor can the boatman's cries,
Nor crabbed oars, nor prayers make him rise.

2 chaplain] (pronounced with three syllables)
5 truckle-bed] low bed on wheels
14 jerks] strokes (of the birch)

So long he drinks till the black caravel
Stands still fast gravelled on the mud of hell.
There stand they still, nor can go, nor retire,
Though greedy ghosts quick passage did require.
Yet stand they still, as though they lay at road,
Till Gullion his bladder would unload.
They stand, and wait, and pray for that good hour;
Which when it came, they sailed to the shore. 20
But never since dareth the ferryman
Once entertain the ghost of Gullion.
Drink on, dry soul, and pledge Sir Gullion:
Drink to all healths, but drink not to thine own.

(1597)

from *Virgidemiae. Biting Satires*

504 [*Landlords and Tenants*]

PARDON, ye glowing ears; needs will it out,
Though brazen walls compassed my tongue about,
As thick as wealthy Scrobio's quick-set rows
In the wide common that he did enclose.
Pull out mine eyes, if I shall see no vice,
Or let me see it with detesting eyes.
Renowned Aquine, now I follow thee
Far as I may for fear of jeopardy,
And to thy hand yield up the ivy-mace
From crabbed Persius and more smooth Horace, 10
Or from that shrew, the Roman poetess,
That taught her gossips learned bitterness;
Or Lucile's Muse whom thou didst imitate,
Or Menips old, or Pasquillers of late.
Yet name I not Mutius, or Tigilline,
Though they deserve a keener style than mine;
Nor mean to ransack up the quiet grave,
Nor burn dead bones, as he example gave.

13 caravel] light ship
17 road] harbour

504
7 Aquine] Juvenal
9 ivy-mace] the thyrsus with which Bacchus controlled his satyrs
11 Roman poetess] Sulpicia, believed to have written a satirical poem
13 Lucile] Lucilius, first writer of formal satire
14 Menips] Menippus (supposed originator of Menippean satire) Pasquillers] writers of lampoons in 16th-cent. Rome
15 Mutius, Tigilline] (mentioned by Juvenal)

I tax the living, let dead ashes rest,
Whose faults are dead, and nailed in their chest; 20
Who can refrain, that's guiltless of their crime,
Whiles yet he lives in such a cruel time?
  When Titius' grounds that in his grandsire's days
But one pound fine, one penny rent did raise,
A summer snowball, or a winter rose,
Is grown to thousands as the world now goes.
So thrift and time sets other things on float,
That now his son swoops in a silken coat.
Whose grandsire haply a poor hungry swain
Begged some cast abbey in the Church's wane, 30
And but for that, whatever he may vaunt,
Who now's a monk, had been a mendicant;
While freezing Matho, that for one lean fee
Wont term each term the Term of Hilary,
May now instead of those his simple fees
Get the fee-simples of fair manneries.
  What, did he counterfeit his Prince's hand,
For some strave lordship of concealed land?
Or on each Michael, and Lady Day,
Took he deep forfeits for an hour's delay? 40
And gained no less by such injurious brawl
Than Gamius by his sixth wife's burial?
Or hath he won some wider interest,
By hoary charters from his grandsire's chest,
Which late some bribed scribe for slender wage
Writ in the characters of another age
That Plowden self might stammer to rehearse,
Whose date o'erlooks three centuries of years?
  Who ever yet the tracks of weal so tried
But there hath been one beaten way beside? 50
He, when he lets a lease for life, or years,
(As never he doth until the date expires;
For when the full state in his fist doth lie,
He may take vantage of the vacancy),

24 fine] fee paid by tenant to the lord on the transfer of the tenant right
28 swoops] sweeps along proudly
30 wane] decline
32 Who now's a monk] who now lives in what was once a monastery
34 Hilary] i.e. the coldest time of year
36 fee-simples] absolute possession without restriction
manneries] manor-houses
38 strave] stray (?)
concealed land] land illicitly annexed
39 Michael . . . Lady Day] Quarter days, on which debts were settled
40 forfeits] penal fines
46 characters] lettering
47 Plowden] the great 16th-cent. lawyer
48 Whose date] i.e. of the charters
49 tracks of weal] ways of getting wealth

His fine affords so many trebled pounds
As he agreeth years to lease his grounds:
His rent in fair respondence must arise
To double trebles of his one year's price.
Of one bay's breadth, God wot, a silly cote,
Whose thatched spars are furred with sluttish soot 60
A whole inch thick, shining like black-moor's brows
Through smoke that down the headless barrel blows.
At his bed's-feet feeden his stalled team,
His swine beneath, his pullen o'er the beam:
A starved tenement, such, as I guess,
Stand straggling in the wastes of Holderness,
Or such as shiver on a Peak hill-side,
When March's lungs beat on their turf-clad hide;
Such as nice Lipsius would grudge to see
Above his lodging in wild Westphaly; 70
Or as the Saxon king his court might make
When his sides plained of the neat-herd's cake.
Yet must he haunt his greedy landlord's hall
With often presents at each festival;
With crammed capons every New Year's morn,
Or with green cheeses when his sheep are shorn,
Or many maunds-full of his mellow fruit
To make some way to win his weighty suit.
Whom cannot gifts at last cause to relent,
Or to win favour, or flee punishment? 80
When gripple patrons turn their sturdy steel
To wax, when they the golden flame do feel;
When grand Maecenas casts a glavering eye
On the cold present of a poesy,
And lest he might more frankly take than give,
Gropes for a French crown in his empty sleeve:
Thence Clodius hopes to set his shoulders free
From the light burden of his napery.
The smiling landlord shows a sunshine face,
Feigning that he will grant him further grace, 90

59 bay] division of a building; space under one gable
silly] simple
64 pullen] poultry
66 Holderness] district north of the Humber estuary
69 nice] fastidious
Lipsius] Flemish scholar (who had criticized German inns)
71 Saxon king] Alfred
72 plained] complained
neat-herd] cow-herd
76 green cheeses] new or fresh cheeses
77 maund] wicker basket
81 gripple] niggardly
83 Maecenas] the Patron
glavering] flattering
87–8 hopes . . . napery] i.e. he is making his landlord a present of household linen (?)

And leers like Æsop's fox upon a crane,
Whose neck he craves for his chirurgeon;
So lingers off the lease until the last—
What recks he then of pains, or promise past?
Was ever feather, or fond woman's mind,
More light than words; the blasts of idle wind?
What's sib or sire, to take the gentle slip,
And in th'Exchequer rot for surety-ship?
Or thence thy starved brother live and die
Within the cold Coal-harbour sanctuary? 100
Will one from Scot's-bank bid but one groat more,
My old tenant may be turned out of door,
Though much he spent in th'rotten roof's repair
In hope to have it left unto his heir,
Though many a load of marl and manure led,
Revived his barren leas, that erst lay dead.
Were he as Furius, he would defy
Such pilf'ring slips of petty landlordry,
And might dislodge whole colonies of poor,
And lay their roof quite level with their floor, 110
Whiles yet he gives as to a yielding fence
Their bag and baggage to his citizens,
And ships them to the new-named Virgin-lond,
Or wilder Wales, where never wight yet wonned.
Would it not vex thee, where thy sires did keep,
To see the dunged folds of dag-tailed sheep,
And ruined house where holy things were said,
Whose free-stone walls the thatched roof upbraid,
Whose shrill saint's-bell hangs on his louvery,
While the rest are damned to the plumbery? 120
Yet pure devotion lets the steeple stand,
And idle battlements on either hand,
Lest that, perhaps, were all those relics gone,
Furius's sacrilege could not be known.

(1598)

92 chirurgeon] surgeon
95 fond] foolish
97 sib] blood-relation
take . . . slip] take your false word and suffer for it
98 for surety-ship] for going surety for you
100 Coal-harbour] sanctuary for debtors
112 citizens] townsmen (as opposed to country-men)
113 Virgin-lond] Virginia
114 wight] human being
wonned] lived
115 sires] forefathers
116 dag-tailed] with tails matted into clotted locks
118 thatched roof] (which has replaced the lead roof)
119 saint's-bell] Sanctus-bell
louvery] turret-like erection on roof to let out smoke or let in light
120 plumbery] lead-works, smelting furnace

WILLIAM ALABASTER
1567–1640

505 *A Divine Sonnet*

JESU, thy love within me is so main,
And my poor heart so narrow of content,
That with thy love my heart wellnigh is rent,
And yet I love to bear such loving pain.
O take thy Cross and nails and therewith strain
My heart's desire unto his full extent,
That thy dear love may not therein be pent,
But thoughts may have free scope thy love to explain.
O now my heart more paineth than before,
Because it can receive and hath no more. 10
O fill this emptiness or else I die.
Now stretch my heart again and now supply;
Now I want space, now grace. To end this smart,
Since my heart holds not thee, hold thou my heart.

(Wr. 1597–8; pub. 1959)

506 *Upon the Ensigns of Christ's Crucifying*

[*The Sponge*]

O SWEET and bitter monuments of pain,
Bitter to Christ who all the pain endured,
But sweet to me whose death my life procured,
How shall I full express such loss, such gain?
My tongue shall be my pen, mine eyes shall rain
Tears for my ink, the place where I was cured
Shall be my book, where, having all abjured,
And calling heavens to record in that plain,
Thus plainly will I write: no sin like mine.
When I have done, do thou, Jesu divine, 10
Take up the tart sponge of thy Passion
And blot it forth; then be thy spirit the quill,
Thy blood the ink, and with compassion
Write thus upon my soul: thy Jesu still.

(Wr. 1597–8; pub. 1959)

1 main] great, strong
506
6 the place . . . cured] i.e. his soul
12 blot . . . forth] wipe out, efface
14 thy Jesu still] thy Saviour always

507 *Of the Reed that the Jews Set in Our Saviour's Hand*

LONG time hath Christ, long time I must confess,
Held me a hollow reed within his hand,
That merited in hell to make a brand,
Had not his grace supplied mine emptiness.
Oft time with languor and newfangleness,
Had I been borne away like sifted sand,
When sin and Satan got the upper hand,
But that his steadfast mercy did me bless.
Still let me grow upon that living land,
Within that wound which iron did impress, 10
And made a spring of blood flow from thy hand.
Then will I gather sap and rise and stand,
That all that see this wonder may express,
Upon this ground how well grows barrenness.

(Wr. 1597–8; pub. 1959)

508 *Upon the Crucifix*

NOW I have found thee, I will evermore
Embrace this standard where thou sitst above.
Feed greedy eyes and from hence never rove,
Suck hungry soul of this eternal store,
Issue my heart from thy two-leaved door,
And let my lips from kissing not remove.
O that I were transformed into love,
And as a plant might spring upon this flower;
Like wandering ivy or sweet honeysuckle,
How would I with my twine about it buckle, 10
And kiss his feet with my ambitious boughs,
And climb along upon his sacred breast,
And make a garland for his wounded brows.
Lord, so I am if here my thoughts might rest.

(Wr. 1597–8; pub. 1959)

4 supplied] made good, compensated for
508
2 this standard] i.e. the Cross
5 two-leaved door] i.e. the mouth
14 so I am] so I am transformed

509 *To the Blessed Virgin*

HAIL graceful morning of eternal day,
The period of Judah's throned right,
And latest minute of the legal night,
Whom wakeful prophets spied far away,
Chasing the night from the world's eastern bay;
Within whose pudent lap and roseal plight,
Conceived was the Son of unborn light,
Whose light gave being to the world's array;
Unspotted morning whom no mist of sin,
Nor cloud of human mixture did obscure, 10
Strange morning that since day hath entered in,
Before and after doth alike endure.
And well it seems a day that never wasteth,
Should have a morning that for ever lasteth.

(Wr. 1597–8; pub. 1959)

510 *To Christ*

LO here I am, lord, whither wilt thou send me?
To which part of my soul, which region?
Whether the palace of my whole dominion,
My mind? which doth not rightly apprehend thee,
And needs more light of knowledge to amend me;
Or to the parliamental session,
My will? that doth design all action,
And doth not as it ought attend thee,
But suffers sin and pleasures, which offend thee,
Within thy kingdom to continue faction; 10
Or to my heart's great lordship shall I bend me,
Where love, the steward of affection,
On vain and barren pleasures doth dispend me?
Lord I am here, O give me thy commission.

(Wr. 1597–8; pub. 1959)

1 graceful] full of divine grace
2 period] termination
Judah's throned right] the Kingdom of Judah, i.e. the old dispensation of the law
6 pudent] chaste
roseal plight] rosy womb
8 array] disposition, order
13 wasteth] passes away

510
13 dispend] expend

511 *Incarnatio est Maximum Dei Donum*

LIKE as the fountain of all light created,
Doth pour out streams of brightness undefined,
Through all the conduits of transparent kind,
That heaven and air are both illuminated,
And yet his light is not thereby abated:
So God's eternal bounty ever shined
The beams of being, moving, life, sense, mind,
And to all things himself communicated.
But see the violent diffusive pleasure
Of goodness, that left not till God had spent 10
Himself by giving us himself, his treasure,
In making man a God omnipotent.
How might this goodness draw our souls above,
Which drew down God with such attractive love!

(Wr. 1597–8; pub. 1959)

512 *Away, fear, with thy projects*

AWAY, fear, with thy projects, no false fire
Which thou dost make can aught my courage quail,
Or cause me leeward run or strike my sail.
What if the world do frown at my retire,
What if denial dash my wished desire,
And purblind pity do my state bewail,
And wonder cross itself and free speech rail,
And greatness take it not and death show nigher!
Tell them, my soul, the fears that make me quake:
The smouldering brimstone and the burning lake, 10
Life feeding death, death ever life devouring,
Torments not moved, unheard, yet still roaring,
God lost, hell found,—ever, never begun.
Now bid me into flame from smoke to run!

(Wr. 1597–8; pub. 1959)

Incarnatio . . . Donum] the Incarnation is God's greatest gift
2 undefined] unlimited

512
1 projects] speculations, notions
false fire] blank discharge of fire-arms
3 leeward] away from the wind (i.e. seeking shelter)
strike sail] lower sails
6 purblind] having impaired vision
7 cross itself] make the sign of the cross
free] frank, uninhibited
rail] speak abusively
8 greatness] persons of high rank
12 not moved] everlasting

513

Exaltatio Humanae Naturae

HUMANITY, the field of miseries,
Nature's abortive table of mischance,
Stage of complaint, the fair that doth enhance
The price of error and of vanities,
Whither? who seeth it? whither doth it rise?
Or do I see, or am I in a trance?
I see it far above the clouds advance,
And under it to tread the starry skies.
My dazzling thoughts do hold this sight for pain.
Vouchsafe me, Christ, to look: see, now again 10
Above the angels it hath distance won,
And left the winged cherubins behind,
And is within God's secret curtain gone,
And still it soareth: gaze no more my mind!

(Wr. 1597–8; pub. 1959)

THOMAS BASTARD
1566–1618

514

De Puero Balbutiente

METHINKS 'tis pretty sport to hear a child,
Rocking a word in mouth yet undefiled;
The tender racquet rudely plays the sound,
Which, weakly bandied, cannot back rebound;
And the soft air the softer roof doth kiss
With a sweet dying and a pretty miss,
Which hears no answer yet from the white rank
Of teeth, not risen from their coral bank.
The alphabet is searched for letters soft,
To try a word before it can be wrought, 10
And when it slideth forth, it goes as nice
As when a man doth walk upon the ice.

(1598)

Exaltatio Humanae Naturae] the raising up of human nature (in the Incarnation)
2 abortive] fruitless, sterile
table] picture (?)
3 complaint] lamentation
fair] place where buyers and sellers periodically assemble
enhance] raise
7 advance] rise

514
De Puero Balbutiente] on a child stammering, or learning to talk
11 nice] carefully, fastidiously

JOSUAH SYLVESTER
1562 or 1563–1618

from *The Divine Weeks and Works of Guillaume de Saluste Sieur Du Bartas*

515 [*The Tower of Babel*]

O HAPPY people, where good princes reign,
Who tender public more than private gain;
Who, virtue's patrons and the plagues of vice,
Hate parasites, and hearken to the wise;
Who, self-commanders, rather sin suppress
By self-examples than by rigorousness;
Whose inward-humble, outward majesty
With subjects' love is guarded loyally;
Who idol not their pearly sceptres' glory
But know themselves set on a lofty storey 10
For all the world to see, and censure too:
So, not their lust, but what is just, they do.
But 'tis a hell, in hateful vassalage
Under a tyrant to consume one's age:
A self-shav'n Dennis, or a Nero fell,
Whose cursed courts with blood and incest swell;
An owl, that flies the light of parliaments
And state assemblies; jealous of th'intents
Of private tongues; who, for a pastime, sets
His peers at odds, and on their fury whets; 20
Who neither faith, honour, nor right respects;
Who every day new offices erects;
Who brooks no learned, wise, nor valiant subjects,
But daily crops such vice-upbraiding objects;
Who, worse than beasts or savage monsters been,
Spares neither mother, brother, kiff nor kin;
Who, though round-fenced with guard of armed knights,
A many moe he fears than he affrights;

2 tender] cherish, regard favourably
9 idol] idolize, worship
10 storey] platform, stage
11 censure] judge
15 self-shav'n Dennis] (Dionysius of Syracuse feared assassination so much that he refused to let a barber shave him)
fell] savage
18 jealous] suspicious
22 offices] official posts
23 brooks] endures
26 kiff] kith
28 moe] more

Who taxes strange extorts; and cannibal
Gnaws to the bones his wretched subjects all. 30
  Print (O heaven's King) in our Kings' hearts a zeal,
First of thy laws, then of their public weal.
And if our courtiers' now-Po-poisoned phrase
Or now-contagion of corrupted days
Leave any tract of Nimrodising there,
O cancel it, that they may everywhere
Instead of Babel build Jerusalem,
That loud my Muse may echo under them.
  Ere Nimrod had attained to twice six years,
He tyrannised among his stripling peers, 40
Outstripped his equals, and in happy hour
Laid the foundations of his after-power;
And bearing reeds for sceptres, first he reigns
In prentice-princedom over shepherd swains.
Then knowing well that whoso aims illustre
At fancied bliss of empire's awful lustre
In valiant acts must pass the vulgar sort,
Or mask (at least) in lovely virtue's port,
He spends not night on beds of down or feathers,
Nor day in tents, but hardens to all weathers 50
His youthful limbs, and takes ambitiously
A rock for pillow, heaven for canopy.
Instead of softlings' jests and jollities,
He joys in jousts and manly exercise;
His dainty cates a fat kid's trembling flesh,
Scarce fully slain, luke-warm, and bleeding fresh.
  Then, with one breath, he striveth to attain
A mountain's top, that overpeers the plain;
Against the stream to cleave the rolling ridges
Of nymph-strong floods, that have borne down their bridges, 60
Running unreined with swift rebounding sallies
Across the rocks within the narrow valleys,
To overtake the dart himself did throw
And in plain course to catch the hind or roe.
  But, when five lustres of his age expired,
Feeling his stomach and his strength aspired

33 Po-poisoned phrase] deadly flattery (i.e. words as lethal as Italian poisons)
35 tract of Nimrodising] trace of tyrannous aspiration
45 illustre] illustrious (i.e. illustriously)
47 pass] surpass
48 mask] be disguised with
port] demeanour
53 softlings] weaklings, effeminate persons
60 nymph-strong] (the rising of the Hyades, five stars in Taurus, ushers in the spring rains)
64 in plain course] at full speed
65 five lustres] i.e. 25 years
66 stomach] courage, spirit

To worthier wars, perceived he anywhere
Boar, leopard, lion, tiger, ounce or bear,
Him dreadless combats; and in combat foils,
And rears high trophies of his bloody spoils. 70
  The people, seeing by his warlike deed
From thieves and robbers every passage freed,
From hideous yells the deserts round about,
From fear their flocks; this monster-master stout,
This Hercules, this hammer-ill, they tender,
And call him all their father and defender.
  Then Nimrod snatching fortune by the tresses
Strikes the hot steel; sues, soothes, importunes, presses
Now these, now those; and hast'ning his good hap
Leaves hunting beasts and hunteth men to trap. 80
For like as he in former quests did use
Calls, pit-falls, toils, springes, and baits, and glues,
And in the end against the wilder game
Clubs, darts, and shafts, and swords, their rage to tame,
So, some he wins with promise-full entreats,
With presents some, and some with rougher threats,
And boldly breaking bounds of equity,
Usurps the child-world's maiden monarchy;
Whereas before, each kindred had for guide
Their proper chief, ere that the youthful pride 90
Of upstart state, ambitious, boiling, fickle,
Did thrust as now in others' corn his sickle.
  Enthronised thus, this tyrant gan devise
To perpetrate a thousand cruelties,
Pell-mell subverting for his appetite
God's, Man's, and Nature's triple-sacred right.
He braves th'Almighty, lifting to his nose
His flowering sceptre; and for fear he lose
The people's awe, who, idle, in the end
Might slip their yoke, he subtly makes them spend, 100
Draws dry their wealth, and busies them to build
A lofty tower, or rather Atlas wild.
'W'have lived', quoth he, 'too long like pilgrim grooms.
Leave we these rolling tents and wand'ring rooms.
Let's raise a palace, whose proud front and feet
With heaven and hell may in an instant meet;

68 ounce] lynx
78 Strikes the hot steel] strikes while the iron is hot
82 Calls] bird-calls
springes] snares
glues] bird-limes
85 entreats] negotiations, entreaties
102 Atlas wild] pillar rising unchecked to the sky
103 pilgrim grooms] wandering herdsmen

A sure asylum and a safe retreat,
If th'ireful storm of yet more floods should threat.
Let's found a city, and united there
Under a king let's lead our lives, for fear 110
Lest severed thus in princes and in tents
We be dispersed o'er all the regiments
That in his course the day's bright champion eyes,
Mightless ourselves to succour or advise.
But if the fire of some intestine war
Or other mischief should divide us far,
Brethren, at least let's leave memorials
Of our great names on these cloud-neighbouring walls.'
  Now, as a spark that shepherds unespied
Have fallen by chance upon a forest-side 120
Among dry leaves, a while in secret shrouds,
Lifting aloft small, smoky-waving clouds,
Till fanned by the fawning winds it blushes
With angry rage; and rising through the bushes,
Climbs fragrant hawthorns, thence the oak, and than
The pine and fir, that bridge the ocean,
It still gets ground; and running doth augment,
And never leaves till all near woods be brent;
So this sweet speech, first broached by certain minions,
Is soon applauded mong the light opinions, 130
And by degrees from hand to hand renewed,
To all the base confused multitude,
Who longing now to see this castle reared
Them night and day in differing crafts bestirred.
  Some fall to felling with a thousand strokes
Adventurous alders, ashes, long-lived oaks,
Degrading forests, that the sun might view
Fields that before his bright rays never knew.
  Ha'ye seen a town exposed to spoil and slaughter,
At victor's pleasure, where laments and laughter 140
Mixedly resound; some carry, some convey,
Some lug, some load; gainst soldiers seeking prey
No place is sure; and ere a day be done
Out at her gate the ransacked town doth run:
So in a trice, these carpenters disrobe
Th'Assyrian hills of all their leafy robe,

112 regiments] kingdoms, places under a particular rule
120 fallen] let fall
125 and than] and then
128 brent] burnt
136 Adventurous] (alder-wood was much used in shipbuilding)
137 Degrading] laying low

Strip the steep mountains of their ghastly shades,
And poll the broad plains of their branchy glades:
Carts, sleds, and mules, thick-justling meet abroad,
And bending axles groan beneath their load. 150
 Here, for hard cement, heap they night and day
The gummy slime of chalky waters grey;
There, busy kil-men ply their occupations
For brick and tile; there, for their firm foundations
They dig to hell; and damned ghosts again
(Past hope) behold the sun's bright glorious wain.
Their hammers' noise, through heaven's rebounding brim,
Affrights the fish that in fair Tigris swim.
These ruddy walls in height and compass grow,
They cast long shadow, and far-off do show. 160
All swarms with workmen, that (poor sots) surmise
Even the first day to touch the very skies.
 Which God perceiving, bending wrathful frowns,
And with a noise that roaring thunder drowns,
Mid cloudy fields hills by the roots he rakes,
And th'unmoved hinges of the heavens he shakes.
'See, see', quoth He, 'these dust-spawn, feeble dwarfs!
See their huge castles, walls, and counter-scarfs!
O strength-full piece, impregnable and sure,
All my just anger's batteries to endure! 170
I swore to them, the fruitful earth no more
Henceforth should fear the raging ocean's roar,
Yet build they towers. I willed that, scattered wide,
They should go man the world, and lo they bide
Self-prisoned here. I meant to be their master
Myself alone, their law, their prince, and pastor:
And they for lord a tyrant fell have ta'en them,
Who, to their cost, will roughly curb and rein them,
Who scorns mine arm, and with these braving towers
Attempts to scale this crystal throne of ours. 180
Come, come, let's dash their drift; and sith, combined
As well in voice, as blood, and law, and mind,
In ill they harden, and with language bold
Encourage-on themselves their works to hold,
Let's cast a let gainst their quick diligence,
Let's strike them straight with spirit of difference,

147 ghastly] fear-inducing
151 cement] (pronounced with stress on first syllable)
153 kil-men] kiln-men
161 surmise] expect
168 counter-scarfs] fortified outer walls
177 fell] savage, fierce
181 sith] since
185 let] obstruction, impediment

Let's all-confound their speech, let's make the brother,
The sire, and son, not understand each other.'
  This said, as soon confusedly did bound
Through all the work I wot not what strange sound, 190
A jangling noise not much unlike the rumours
Of Bacchus' swains amid their drunken humours.
Some speak between the teeth, some in the nose,
Some in the throat their words do ill dispose,
Some howl, some hallow, some do stut and strain;
Each hath his gibberish, and all strive in vain
To find again their known beloved tongue
That with their milk they sucked in cradle young.
  Arise betimes, while th'opal-coloured morn
In golden pomp doth May-day's door adorn, 200
And patient hear th'all-differing voices sweet
Of painted singers, that in groves do greet
Their love-Bon-jours, each in his phrase and fashion
From trembling perch utt'ring his earnest passion;
And so thou may'st conceit what mingle-mangle
Among this people everywhere did jangle.
  'Bring me', quoth one, 'a trowel, quickly, quick':
One brings him up a hammer. 'Hew this brick',
Another bids, and then they cleave a tree.
'Make fast this rope', and then they let it flee. 210
One calls for planks, another mortar lacks:
They bear the first a stone, the last an axe.
One would have spikes, and him a spade they give.
Another asks a saw, and gets a sieve.
Thus crossly-crossed, they prate and point in vain;
What one hath made, another mars again.
Nigh breathless all with their confused yawling,
In bootless labour, now begins appalling.
  In brief, as those that in some channel deep
Begin to build a bridge with arches steep, 220
Perceiving once in thousand streams extending
The course-changed river from the hills descending
With watery mountains bearing down their bay
As if it scorned such bondage to obey,
Abandon quickly all their work begun,
And here and there for swifter safety run,
These masons so, seeing the storm arrived
Of God's just wrath, all weak and heart-deprived,

190 I wot not] I know not
191 rumours] uproar, tumult
195 hallow] shout
202–3 greet Their love-Bon-jours] say good-morning to their loves
205 conceit] conceive

Forsake their purpose, and like frantic fools
Scatter their stuff and tumble down their tools. 230
O proud revolt! O traitorous felony!
See in what sort the Lord hath punished thee
By this confusion! Ah that language sweet,
Sure bond of cities, friendship's mastic meet,
Strong curb of anger, erst united, now
In thousand dry brooks strays; I wot not how
That rare-rich gold, that charm-grief fancy mover,
That calm-rage heart's-thief, quell-pride conjure-lover,
That purest coin, then current in each coast,
Now mingled, hath sound, weight and colour lost: 240
'Tis counterfeit, and over every shore
The confused fall of Babel yet doth roar.
Then Finland-folk might visit Africa,
The Spaniard Inde, and ours America
Without a truchman; now, the banks that bound
Our towns about our tongues do also mound.
For, who from home but half a furlong goes,
As dumb (alas) his reason's tool doth lose;
Or if we talk but with our near confines,
We borrow mouths, or else we work by signs. 250
Untoiled, untutored, sucking tender food,
We learned a language all men understood;
And seven-years-old, in glass-dust did commence
To draw the round earth's fair circumference.
To cipher well, and climbing art by art,
We reached betimes that castle's highest part,
Where th'encyclopedy her darlings crowns,
In sign of conquest, with eterne renowns.
Now, ever boys, we wax old while we seek
The Hebrew tongue, the Latin, and the Greek. 260
We can but babble, and for knowledge whole
Of nature's secrets and of th'Essence sole,
Which essence gives to all, we tire our mind
To vary verbs, and finest words to find,

234 mastic] cement, sticking substance (hence bond, link)
meet] fit, proper
238 conjure-lover] (?) bewitching enchanter
245 truchman] interpreter
246 mound] enclose
253 glass-dust] (?) glass enamel (i.e. children traced geographical outlines on globes)
256 betimes] early
257 encyclopedy] general learning (personified)
258 eterne] eternal
261 for knowledge] i.e. instead of knowledge
262 Essence] Being
264 vary] express in grammatically different ways

Our letters and our syllables to weigh.
At tutors' lips we hang with heads all grey,
Who teach us yet to read, and give us raw
An A.B.C. for great Justinian's law,
Hippocrates, or that diviner lore
Where God appears to whom Him right adore. 270
What shall I more say? Then, all spake the speech
Of God Himself, th'old sacred Idiom rich,
Rich perfect language where's no point, nor sign,
But hides some rare deep mystery divine.
But since that pride, each people hath apart
A bastard gibberish, harsh and overthwart,
Which, daily changed and losing light, well-near
Nothing retains of that first language clear.

(1598)

BEN JONSON
1572–1637

516 [*To Thomas Palmer, on his book 'The Sprite of Trees and Herbs'*]

WHEN late, grave Palmer, these thy grafts and flowers,
So well disposed by thy auspicious hand,
Were made the objects to my weaker powers,
I could not but in admiration stand.
First, thy success did strike my sense with wonder,
That 'mongst so many plants transplanted hither
Not one but thrives, in spite of storms and thunder,
Unseasoned frosts, or the most envious weather.
Then I admired the rare and precious use
Thy skill hath made of rank despised weeds, 10
Whilst other souls convert to base abuse
The sweetest simples, and most sovereign seeds.
Next, that which rapt me was, I might behold
How, like the carbuncle in Aaron's breast,
The seven-fold flower of art, more rich than gold,
Did sparkle forth in centre of the rest;

267 raw] unskilled
273 point] mark used to indicate the vowel in Hebrew
277 well-near] almost

516
3 made the objects] made visible
8 Unseasoned] unseasonable
envious] harshly inclement
12 simples] herbs
14 carbuncle] cf. Exod. 28:17, 39:10

Thus, as a ponderous thing in water cast
Extendeth circles into infinites,
Still making that the greatest that is last,
Till the one hath drowned the other in our sights: 20
So in my brain the strong impression
Of thy rich labours worlds of thoughts created,
Which thoughts being circumvolved in gyre-like motion
Were spent with wonder as they were dilated,
Till giddy with amazement I fell down
In a deep trance; *****
***** when, lo! to crown thy worth
I struggled with this passion that did drown
My abler faculties; and thus brake forth:
Palmer, thy travails well become thy name, 30
And thou in them shalt live as long as fame.

Dignum laude virum musa vetat mori.

(Wr. 1598–9; pub. 1895)

JOHN MARSTON
1576–1634

from *The Scourge of Villainy*

517 *To* Detraction *I present my* Poesy

FOUL canker of fair virtuous action,
Vile blaster of the freshest blooms on earth,
Envy's abhorred child, Detraction,
I here expose, to thy all-tainting breath,
The issue of my brain: snarl, rail, bark, bite,
Know that my spirit scorns Detraction's spite.

Know that the Genius, which attendeth on
And guides my powers intellectual,
Holds in all vile repute Detraction;
My soul an essence metaphysical, 10
That in the basest sort scorns critics' rage
Because he knows his sacred parentage.

23 circumvolved] turned around
25 amazement] mental stupefaction, frenzy
30 travails] (1) labours; (2) travels
thy name] Palmer (meaning pilgrim)
32 *Dignum . . . mori*] the Muse forbids that the hero worthy of fame should perish

517
7 Genius] the rational principle lodged in the body

My spirit is not puft up with fat fume
Of slimy ale, nor Bacchus' heating grape.
My mind disdains the dungy muddy scum
Of abject thoughts and Envy's raging hate.
  True judgment slight regards Opinion,
  A spritely wit disdains Detraction.

A partial praise shall never elevate
My settled censure of my own esteem; 20
A cankered verdict of malignant hate
Shall ne'er provoke me worse myself to deem.
  Spite of despite and rancour's villainy,
  I am myself, so is my poesy.

(1598)

from *Satire VII*

518

A Cynic Satire

'A MAN, a man, a kingdom for a man!'
'Why, how now, currish, mad Athenian?
Thou Cynic dog, see'st not the streets do swarm
With troops of men?' 'No, no: for Circe's charm
Hath turn'd them all to swine. I never shall
Think those same Samian saws authentical:
But rather, I dare swear, the souls of swine
Do live in men. For that same radiant shine—
That lustre wherewith Nature's nature decked
Our intellectual part—that gloss is soiled 10
With staining spots of vile impiety,
And muddy dirt of sensuality.
These are no men, but apparitions,
Ignes fatui, glowworms, fictions,
Meteors, rats of Nilus, fantasies,
Colosses, pictures, shades, resemblances.
  Ho, Lynceus!
Seest thou yon gallant in the sumptuous clothes,
How brisk, how spruce, how gorgeously he shows?

17 Opinion] the ill-grounded views of the multitude
20 censure] judgement

518
Cynic Satire] in the manner of Diogenes the Cynic
1 A man, a man] cf. *Richard III* 5.4.7, 13
6 Samian saws] teachings of Pythagoras the Samian concerning the transmigration of human souls into animals
14 Ignes fatui] will-o'-the-wisps

Note his French herring-bones: but note no more, 20
Unless thou spy his fair appendant whore,
That lackies him. Mark nothing but his clothes,
His new-stamped compliment, his cannon oaths;
Mark those: for naught but such lewd viciousness
E'er graced him, save Sodom beastliness.
Is this a man? Nay, an incarnate devil,
That struts in vice and glorieth in evil.
  A man, a man!' 'Peace, Cynic, yon is one:
A complete soul of all perfection.'
'What, mean'st thou him that walks all open-breasted, 30
Drawn through the ear, with ribands, plumy-crested;
He that doth snort in fat-fed luxury,
And gapes for some grinding monopoly;
He that in effeminate invention,
In beastly source of all pollution,
In riot, lust, and fleshly seeming sweetness,
Sleeps sound, secure, under the shade of greatness?
Mean'st thou that senseless, sensual epicure—
That sink of filth, that guzzle most impure—
What, he? Lynceus, on my word thus presume, 40
He's nought but clothes, and scenting sweet perfume;
His very soul, assure thee, Lynceus,
Is not so big as is an atomus:
Nay, he is spriteless, sense or soul hath none,
Since last Medusa turned him to a stone.
A man, a man!' 'Lo, yonder I espy
The shade of Nestor in sad gravity.'
'Since old Silenus brake his ass's back,
He now is forced his paunch and guts to pack
In a fair tumbrel.' 'Why, sour satirist, 50
Canst thou unman him? Here I dare insist
And soothly say, he is a perfect soul,
Eats nectar, drinks ambrosia, sans control;
An inundation of felicity
Fats him with honour and huge treasury.'
'Canst thou not, Lynceus, cast thy searching eye,
And spy his imminent catastrophe?
He's but a sponge, and shortly needs must leese

20 French herring-bones] silk or velvet from Lyons, with a herring-bone pattern
39 guzzle] gutter, drain
43 atomus] atom
44 spriteless] without spirit
47 Nestor] i.e. type of sobriety and wisdom
sad] serious
48 Silenus] i.e. type of self-indulgence
50 tumbrel] dung-cart, i.e. coach
52 soothly] truly
58 leese] lose

His wrong-got juice, when greatness' fist shall squeeze
His liquor out. Would not some shallow head, 60
That is with seeming shadows only fed,
Swear yon same damask-coat, yon garded man,
Were some grave sober Cato Utican?
When, let him but in judgment's sight uncase,
He's naught but budge, old gards, brown fox-fur face;
He hath no soul the which the Stagyrite
Termed rational: for beastly appetite,
Base dunghill thoughts, and sensual action,
Hath made him lose that fair creation.
And now no man, since Circe's magic charm 70
Hath turned him to a maggot that doth swarm
In tainted flesh, whose foul corruption
Is his fair food: whose generation
Another's ruin. O Canaan's dread curse,
To live in people's sins! Nay, far more worse,
To muck rank hate! But, sirrah Lynceus,
Seest thou that troop that now affronteth us?
They are naught but eels, that never will appear
Till that tempestuous winds or thunder tear
Their slimy beds. But prithee stay a while; 80
Look, yon comes John-a-Noke and John-a-Stile;
They are nought but slow-paced, dilatory pleas,
Demure demurrers, still striving to appease
Hot zealous love. The language that they speak
Is the pure barbarous blacksaunt of the Gete;
Their only skill rests in collusions,
Abatements, stoppels, inhibitions.
Heavy-paced jades, dull-pated jobbernowls,
Quick in delays, checking with vain controls
Fair Justice' course; vile necessary evils, 90
Smooth seem-saints, yet damned incarnate devils.
Far be it from my sharp satiric muse,
Those grave and reverent legists to abuse,

62 damask] rich patterned material (e.g. silk and wool)
garded] with ornamental trimmings
64 uncase] take off his outer garment
65 budge] shabby fur
gards] borders, trimmings
66 the Stagyrite] Aristotle
74 curse] cf. Genesis 9.25–7
75 in people's sins] i.e. as a usurer
76 muck] manure, help to grow
77 that troop] i.e. a group of lawyers
78–9 eels . . . thunder] (thunder was supposed to rouse eels from the mud)
81 John-a-Noke and John-a-Stile] (type-names for parties in a legal action)
83 demurrers] a form of legal pleading
85 blacksaunt] blacksanctus, i.e. a burlesque hymn
86–7 collusions . . . inhibitions] (more legal terms)
88 jades] worn-out horses
jobbernowls] blockheads
93 reverent] reverend

That aid Astræa, that do further right;
But these Megaeras that inflame despite,
That broach deep rancour, that study still
To ruin right, that they their paunch may fill
With Irus' blood—these furies I do mean,
These hedgehogs, that disturb Astrea's scene.
  A man, a man!' 'Peace, Cynic, yon's a man; 100
Behold yon sprightly dread Mavortian;
With him I stop thy currish barking chops.'—
'What, mean'st thou him that in his swaggering slops
Wallows unbraced, all along the street;
He that salutes each gallant he doth meet
With "Farewell, sweet captain, kind heart, adieu;"
He that last night, tumbling thou didst view
From out the great man's head, and thinking still
He had been sentinel of warlike Brill,
Cries out, "Que va la? zounds, que?" and out doth draw 110
His transform'd poniard, to a syringe straw,
And stabs the drawer? What, that ringo-root!
Mean'st thou that wasted leg, puff bumbast boot;
What, he that's drawn and quartered with lace;
That Westphalian gammon clove-stuck face?
Why, he is nought but huge blaspheming oaths,
Swart snout, big looks, misshapen Switzers' clothes;
Weak meagre lust hath now consumed quite,
And wasted clean away his martial sprite;
Enfeebling riot, all vices' confluence, 120
Hath eaten out that sacred influence
Which made him man.
That divine part is soaked away in sin,
In sensual lust, and midnight bezzling,
Rank inundation of luxuriousness
Have tainted him with such gross beastliness,
That now the seat of that celestial essence
Is all possessed with Naples' pestilence.

94 Astræa] Equity
right] justice
95 Megaeras] Megaera was one of the Furies
96 broach] introduce
98 Irus] type of poor man
99 hedgehogs] i.e. heedless of other people's feelings and rights
101 Mavortian] man of Mars, soldier
103 slops] wide baggy breeches
108 great man's head] an inn (perhaps 'The Saracen's Head')
109 Brill] (in Low Countries)
111 straw] needle
112 ringo] eringo, sea-holly (an aphrodisiac)
113 bumbast boot] padded boot
115 clove-stuck] full of blackheads
124 bezzling] guzzling, riotous living
125 luxuriousness] gross sensuality
128 Naples' pestilence] syphilis

Fat peace, and dissolute impiety,
Have lulled him in such security, 130
That now, let whirlwinds and confusion tear
The centre of our state; let giants rear
Hill upon hill; let western Termagant
Shake heaven's vault: he, with his occupant,
Are cling'd so close, like dew-worms in the morn,
That he'll not stir till out his guts are torn
With eating filth. Tubrio, snort on, snort on,
Till thou art waked with sad confusion.
  Now rail no more at my sharp cynic sound,
Thou brutish world, that in all vileness drowned 140
Hast lost thy soul: for nought but shades I see—
Resemblances of men inhabit thee.'

(1598)

from *Satire XI*

519 *Humours*

SLEEP, grim Reproof; my jocund muse doth sing
In other keys, to nimbler fingering.
Dull-sprighted Melancholy, leave my brain—
To hell, Cimmerian night! In lively vein
I strive to paint, then hence all dark intent
And sullen frowns! Come, sporting Merriment,
Cheek-dimpling Laughter, crown my very soul
With jouisance, whilst mirthful jests control
The gouty humours of these pride-swollen days,
Which I do long until my pen displays. 10
O, I am great with Mirth! Some midwif'ry,
Or I shall break my sides at vanity!
Room for a capering mouth, whose lips ne'er stir
But in discoursing of the graceful slur.
Who ever heard spruce skipping Curio
E'er prate of ought but of the whirl on toe,
The turn-above-ground, Robrus' sprawling kicks,
Fabius' caper, Harry's tossing tricks?

130 security] heedlessness
133 western Termagant] i.e. Jupiter
134 occupant] harlot
135 dew-worms] earth-worms
137 eating filth] filthy disease which eats away the flesh
138 confusion] ruin

519
Humours] whims, caprices, inclinations, habits
8 control] check, restrain
10 long] long for (like a pregnant woman)
14 slur] gliding dance-movement
15 Curio] Sir John Davies

Did ever any ear e'er hear him speak
Unless his tongue of cross-points did entreat? 20
His teeth do caper whilst he eats his meat,
His heels do caper whilst he takes his seat;
His very soul, his intellectual
Is nothing but a mincing capreal.
He dreams of toe-turns; each gallant he doth meet
He fronts him with a traverse in the street.
Praise but *Orchestra*, and the skipping art,
You shall command him, faith you have his heart
Even cap'ring in your fist. A hall, a hall!
Room for the spheres, the orbs celestial 30
Will dance Kemp's jig: they'll revel with neat jumps;
A worthy poet hath put on their pumps.
O wit's quick traverse, but *sance ceo's* slow;
Good faith 'tis hard for nimble Curio.
"Ye gracious orbs, keep the old measuring;
All's spoiled if once ye fall to capering."
Luscus, what's played to-day? Faith now I know
I set thy lips abroach, from whence doth flow
Naught but pure Juliet and Romeo.
Say who acts best? Drusus or Roscio? 40
Now I have him, that ne'er of ought did speak
But when of plays or players he did treat—
Hath made a common-place book out of plays,
And speaks in print: at least what e'er he says
Is warranted by Curtain plaudities.
If e'er you heard him courting Lesbia's eyes,
Say (courteous sir), speaks he not movingly,
From out some new pathetic tragedy?
He writes, he rails, he jests, he courts (what not?),
And all from out his huge long-scraped stock 50
Of well-penned plays.
O for a humour, look, who yon doth go,
The meagre lecher, lewd Luxurio!

20 cross-points] dance-steps used in the galliard
entreat] treat
24 capreal] capriole; a leap or caper
27 *Orchestra*] Davies's poem
29 A hall, a hall!] 'make room for a dance!'
31 Kemp's jig] the jig danced by William Kemp, the famous clown. A jig was a farce sung and danced to ballad measure
32 pumps] dancing shoes
33 *sance ceo's*] (Law Latin for 'sans cela') on the contrary
35 measuring] (1) rhythm, movement; (2) a grave and stately dance
37 played] acted (in the theatres)
38 set thy lips abroach] made your mouth water
45 Curtain plaudities] rounds of applause at the Curtain playhouse

'Tis he that hath the sole monopoly,
By patent, of the suburb lechery;
No new edition of drabs comes out,
But seen and allowed by Luxurio's snout.
Did ever any man e'er hear him talk,
But of Pick-hatch, or of some Shoreditch balk,
Aretine's filth, or of his wand'ring whore; 60
Of some Cinædian, or of Tacedore;
Of Ruscus' nasty, loathsome brothel rhyme,
That stinks like Ajax' froth, or muck-pit slime?
The news he tells you is of some new flesh,
Lately broke up, span new, hot piping fresh.
The courtesy he shows you is some morn
To give you Venus 'fore her smock be on.
His eyes, his tongue, his soul, his all, is lust,
Which vengeance and confusion follow must.
Out on this salt humour, letcher's dropsy, 70
Fie! it doth soil my chaster poesy!
  O spruce! How now, Piso, Aurelius' ape,
What strange disguise, what new deformed shape,
Doth hold thy thoughts in contemplation?
Faith say, what fashion art thou thinking on?
A stitch'd taffeta cloak, a pair of slops
Of Spanish leather? O, who heard his chops
E'er chew of ought but of some strange disguise?
This fashion-monger, each morn 'fore he rise,
Contemplates suit-shapes, and once from out his bed, 80
He hath them straight full lively portrayed.
And then he chucks, and is as proud of this
As Taphus when he got his neighbour's bliss.
All fashions, since the first year of this queen,
May in his study fairly drawn be seen;
And all that shall be to his day of doom
You may peruse within that little room;
For not a fashion once dare show his face,
But from neat Piso first must take his grace:
The long fool's coat, the huge slop, the lugged boot, 90
From mimic Piso all do claim their root.

55 suburb] (the brothels were in the suburbs of London)
56 drabs] prostitutes
57 allowed] given formal permission
59 balk] projecting lower part of a building
60 wand'ring whore] *Puttana Errante*, a poem (wrongly) attributed to Aretino
61 Cinædian] homosexual
Tacedore] (unexplained)
63 Ajax] (punning on 'a jakes', privy)
65 broke up] (term used for the cutting up of the body of a deer or fowl)
70 salt] lecherous
76 slops] wide breeches or thigh-boots
82 chucks] chuckles
90 lugged boot] boot with a loop or tag on the heel

O that the boundless power of the soul
Should be cooped up in fashioning some roll!
  But O, Suffenus! (that doth hug, embrace
His proper self, admires his own sweet face;
Praiseth his own fair limbs' proportion,
Kisseth his shade, recounteth all alone
His own good parts) who envies him? Not I,
For well he may, without all rivalry.
  Fie! whither's fled my sprite's alacrity? 100
How dull I vent this humorous poesy!
In faith I am sad, I am possessed with ruth,
To see the vainness of fair Albion's youth;
To see their richest time even wholly spent
In that which is but gentry's ornament;
Which, being meanly done, becomes them well;
But when with dear time's loss they do excel,
How ill they do things well! To dance and sing,
To vault, to fence, and fairly trot a ring
With good grace, meanly done, O what repute 110
They do beget! But being absolute,
It argues too much time, too much regard
Employed in that which might be better spared
Than substance should be lost. If one should sue
For Lesbia's love, having two days to woo,
And not one more, and should employ those twain
The favour of her waiting-wench to gain,
Were he not mad? Your apprehension,
Your wits are quick in application.
Gallants, 120
Methinks your souls should grudge and inly scorn
To be made slaves to humours that are born
In slime of filthy sensuality.
That part not subject to mortality
(Boundless, discursive apprehension
Giving it wings to act his function),
Methinks should murmur when you stop his course,
And soil his beauties in some beastly source
Of brutish pleasures; but it is so poor,
So weak, so hunger-bitten, evermore 130
Kept from his food, meagre for want of meat,
Scorned and rejected, thrust from out his seat,

93 roll] roll of cloth
101 humorous poesy] poetry about humours
106 meanly] i.e. neither too little nor too much
109 trot a ring] exercise a horse in the manège
111 absolute] perfect

Upbraid by capons' grease, consumed quite
By eating stews, that waste the better sprite,
Snibbed by his baser parts, that now poor soul
(Thus peasanted to each lewd thought's control)
Hath lost all heart, bearing all injuries,
The utmost spite and rank'st indignities,
With forced willingness; taking great joy,
If you will deign his faculties employ 140
But in the mean'st ingenious quality
(How proud he'll be of any dignity!).
Put it to music, dancing, fencing-school,
Lord, how I laugh to hear the pretty fool,
How it will prate! His tongue shall never lie,
But still discourse of his spruce quality,
Egging his master to proceed from this,
And get the substance of celestial bliss.
His lord straight calls his parliament of sense;
But still the sensual have pre-eminence. 150
The poor soul's better part so feeble is,
So cold and dead is his Synderesis,
That shadows, by odd chance, sometimes are got;
But O the substance is respected not!
Here ends my rage. Though angry brow was bent,
Yet I have sung in sporting merriment.

(1598)

520 *To Everlasting Oblivion*

THOU mighty gulf, insatiate cormorant,
Deride me not, though I seem petulant
To fall into thy chops. Let others pray
Forever their fair poems flourish may.
But as for me, hungry Oblivion,
Devour me quick, accept my orison,
  My earnest prayers, which do importune thee,
  With gloomy shade of thy still empery
  To veil both me and my rude poesy.

Far worthier lines in silence of thy state 10
Do sleep securely, free from love or hate,

133 Upbraid] made uneasy with indigestion
135 Snibbed] checked, reproved
152 Synderesis] that part of conscience that serves as a guide for conduct

520
2 petulant] peevishly eager

From which this living ne'er can be exempt,
But whilst it breathes will hate and fury tempt.
Then close his eyes with thy all-dimming hand,
Which not right glorious actions can withstand.
Peace, hateful tongues, I now in silence pace;
Unless some hound do wake me from my place,
  I with this sharp, yet well-meant poesy,
  Will sleep secure, right free from injury
  Of cankered hate or rankest villainy. 20

(1598)

THOMAS DEKKER

*c.*1570–*c.*1632

from *The Shoemaker's Holiday*

521

O THE month of May, the merry month of May,
  So frolic, so gay, and so green, so green, so green!
O and then did I unto my true love say,
  Sweet Peg, thou shalt be my Summer's Queen.

Now the nightingale, the pretty nightingale,
  The sweetest singer in all the forest's choir,
Entreats thee, sweet Peggy, to hear thy true love's tale:
  Lo, yonder she sitteth, her breast against a briar.

But O I spy the cuckoo, the cuckoo, the cuckoo;
  See where she sitteth; come away, my joy: 10
Come away, I prithee, I do not like the cuckoo
  Should sing where my Peggy and I kiss and toy.

O the month of May, the merry month of May,
  So frolic, so gay, and so green, so green, so green!
And then did I unto my true love say,
  Sweet Peg, thou shalt be my Summer's Queen.

522

COLD'S the wind, and wet's the rain,
  Saint Hugh be our good speed!
Ill is the weather that bringeth no gain,
  Nor helps good hearts in need.

Troll the bowl, the jolly nut-brown bowl,
And here, kind mate, to thee!
Let's sing a dirge for Saint Hugh's soul,
And down it merrily.

Down-a-down, hey, down-a-down,
Hey derry derry down-a-down!
Ho! well done, to me let come,
Ring compass, gentle joy!

Cold's the wind, and wet's the rain,
Saint Hugh be our good speed!
Ill is the weather that bringeth no gain,
Nor helps good hearts in need.

(1600)

from *Old Fortunatus*

523

FORTUNE smiles, cry holy day!
Dimples on her cheeks do dwell.
Fortune frowns, cry well-a-day!
Her love is heaven, her hate is hell.
Since heaven and hell obey her power,
Tremble when her eyes do lour;
Since heaven and hell her power obey,
When she smiles cry holy day!
Holy day with joy we cry,
And bend, and bend, and merrily 10
Sing hymns to Fortune's deity,
Sing hymns to Fortune's deity.

Let us sing merrily, merrily, merrily!
With our song let heaven resound,
Fortune's hands our heads have crowned;
Let us sing merrily, merrily, merrily!

524

VIRTUE'S branches wither, virtue pines,
O pity, pity, and alack the time!
Vice doth flourish, vice in glory shines,
Her gilded boughs above the cedar climb.

Vice hath golden cheeks, O pity, pity!
  She in every land doth monarchize.
Virtue is exiled from every city,
  Virtue is a fool, vice only wise.

O pity, pity! virtue weeping dies.
  Vice laughs to see her faint, alack the time! 10
This sinks; with painted wings the other flies.
  Alack, that best should fall, and bad should climb!

O pity, pity, pity! mourn, not sing!
Vice is a saint, virtue an underling.
Vice doth flourish, vice in glory shines,
Virtue's branches wither, virtue pines.

(1600)

from *Patient Grissil*

525

ART thou poor, yet hast thou golden slumbers?
    O sweet content!
Art thou rich, yet is thy mind perplexed?
    O punishment!
Dost thou laugh to see how fools are vexed
To add to golden numbers, golden numbers?
O sweet content! O sweet content!
   Work apace, apace, apace, apace;
   Honest labour bears a lovely face;
  Then hey nonny nonny, hey nonny nonny! 10

Canst drink the waters of the crisped spring?
    O sweet content!
Swim'st thou in wealth, yet sink'st in thine own tears?
    O punishment!
Then he that patiently want's burden bears
No burden bears, but is a king, a king!
O sweet content! O sweet content!
   Work apace, apace, apace, apace;
   Honest labour bears a lovely face;
  Then hey nonny nonny, hey nonny nonny! 20

526

GOLDEN slumbers kiss your eyes,
Smiles awake you when you rise.
Sleep, pretty wantons, do not cry,
And I will sing a lullaby:
Rock them, rock them, lullaby.

Care is heavy, therefore sleep you;
You are care, and care must keep you.
Sleep, pretty wantons, do not cry,
And I will sing a lullaby:
Rock them, rock them, lullaby. 10

(Wr. 1599; pub. 1603)

ANONYMOUS

[*Lute Songs set by John Dowland*]

527

COME away, come, sweet love,
The golden morning breaks;
All the earth, all the air
Of love and pleasure speaks.
Teach thine arms then to embrace,
And sweet, rosy lips to kiss,
And mix our souls in mutual bliss.
Eyes were made for beauty's grace,
Viewing, ruing love's long pain,
Procured by beauty's rude disdain. 10

Come away, come, sweet love,
The golden morning wastes,
While the sun from his sphere
His fiery arrows casts,
Making all the shadows fly,
Playing, staying in the grove
To entertain the stealth of love.
Thither, sweet love, let us hie,
Flying, dying in desire,
Winged with sweet hopes and heavenly fire. 20

Come away, come, sweet love,
Do not in vain adorn
Beauty's grace, that should rise
Like to the naked morn.

527
18 hie] hasten

Lilies on the river's side,
And fair, Cyprian flowers new-blown,
Desire no beauties but their own;
Ornament is nurse of pride,
Pleasure, measure love's delight.
Haste then, sweet love, our wished flight. 30

(1597)

528

DIE not before thy day, poor man condemned,
But lift thy low looks from the humble earth.
Kiss not Despair and see sweet Hope contemned;
The hag hath no delight but moan for mirth.
O fie, poor fondling, fie! Be willing
To preserve thyself from killing.
Hope, thy keeper, glad to free thee,
Bids thee go and will not see thee.
Hie thee quickly from thy wrong!
So she ends her willing song. 10

(1600)

529

I SAW my lady weep,
And Sorrow proud to be advanced so
In those fair eyes where all perfections keep.
Her face was full of woe;
But such a woe, believe me, as wins more hearts
Than Mirth can do with her enticing parts.

Sorrow was there made fair,
And Passion wise, tears a delightful thing;
Silence beyond all speech a wisdom rare.
She made her sighs to sing, 10
And all things with so sweet a sadness move
As made my heart at once both grieve and love.

O fairer than aught else
The world can show, leave off in time to grieve.
Enough, enough your joyful looks excels;
Tears kills the heart, believe.
O strive not to be excellent in woe,
Which only breeds your beauty's overthrow.

(1600)

528
5 fondling] foolish person

530 FINE knacks for ladies, cheap, choice, brave and new!
Good pennyworths! but money cannot move.
I keep a fair but for the fair to view;
A beggar may be liberal of love.
Though all my wares be trash, the heart is true,
The heart is true,
The heart is true.

Great gifts are guiles and look for gifts again;
My trifles come as treasures from my mind.
It is a precious jewel to be plain; 10
Sometimes in shell th'orient's pearls we find.
Of others take a sheaf, of me a grain,
Of me a grain,
Of me a grain.

Within this pack pins, points, laces, and gloves,
And divers toys fitting a country fair.
But in my heart, where duty serves and loves,
Turtles and twins, court's brood, a heavenly pair.
Happy the heart that thinks of no removes,
Of no removes, 20
Of no removes.

(1600)

531 TOSS not my soul, O Love, 'twixt hope and fear.
Show me some ground where I may firmly stand
Or surely fall; I care not which appear,
So one will close me in a certain band.

Take me, Assurance, to thy blissful hold,
Or thou, Despair, unto thy darkest cell.
Each hath full rest, the one in joys enrolled,
Th'other, in that he fears no more, is well.

(1600)

532 WEEP you no more, sad fountains;
What need you flow so fast?
Look how the snowy mountains
Heaven's sun doth gently waste.

19 removes] removals, departures

But my sun's heavenly eyes
View not your weeping,
That now lies sleeping
Softly, now softly lies
Sleeping.

Sleep is a reconciling, 10
A rest that peace begets.
Doth not the sun rise smiling
When fair at ev'n he sets?
Rest you then, rest, sad eyes,
Melt not in weeping
While she lies sleeping
Softly, now softly lies
Sleeping.

(1603)

[*Madrigal set by John Farmer*]

533

TAKE time while Time doth last;
Mark how Fair fadeth fast;
Beware if Envy reign;
Take heed of proud Disdain.
Hold fast now in thy youth;
Regard thy vowed Truth;
Lest when thou waxeth old
Friends fail and Love grow cold.

(1599)

[*Madrigal set by John Bennet*]

534

THYRSIS, sleepest thou? Holla! Let not sorrow stay us.
Hold up thy head, man, said the gentle Meliboeus.
See Summer comes again, the country's pride adorning,
Hark how the cuckoo singeth this fair April morning.
O, said the shepherd, and sighed as one all undone,
Let me alone, alas, and drive him back to London.

(1599)

[*Madrigals set by Thomas Weelkes*]

535 LIKE two proud armies marching in the field,
Joining a thundering fight each scorns to yield,
So in my heart your Beauty and my Reason,
The one claims the crown, the other says 'tis treason.
But O, your Beauty shineth as the sun,
And dazzled Reason yields as quite undone.

(1600)

536 THULE, the period of cosmography,
Doth vaunt of Hecla, whose sulphurious fire
Doth melt the frozen clime and thaw the sky;
Trinacrian Ætna's flames ascend not higher.
These things seem wondrous, yet more wondrous I,
Whose heart with fear doth freeze, with love doth fry.

The Andalusian merchant, that returns
Laden with cochineal and China dishes,
Reports in Spain how strangely Fogo burns
Amidst an ocean full of flying fishes. 10
These things seem wondrous, yet more wondrous I,
Whose heart with fear doth freeze, with love doth fry.

(1600)

[*Lute Song set by Robert Jones*]

537 FAREWELL, dear love, since thou wilt needs be gone.
Mine eyes do show my life is almost done.
Nay, I will never die,
So long as I can spy.
There be many moe
Though that she do go.
There be many moe, I fear not.
Why then, let her go, I care not.

536
1 Thule] prob. Iceland
period of cosmography] limit of the known world
4 Trinacrian] Sicilian
8 cochineal] scarlet dye
China dishes] dishes from China
9 Fogo] Terra del Fuego (at the southernmost tip of South America)

Farewell, farewell, since this I find is true,
I will not spend more time in wooing you. 10
    But I will seek elsewhere
    If I may find her there.
    Shall I bid her go?
    What and if I do?
  Shall I bid her go, and spare not?
  O no, no, no, no, I dare not.

Ten thousand times farewell! Yet stay awhile!
Sweet, kiss me once; sweet kisses time beguile.
    I have no power to move.
    How now, am I in love? 20
    Wilt thou needs be gone?
    Go then, all is one.
  Wilt thou needs be gone? O hie thee!
  Nay, stay, and do no more deny me.

Once more farewell! I see loth to depart
Bids oft adieu to her that holds my heart.
    But seeing I must lose
    Thy love which I did choose,
    Go thy ways for me
    Since it may not be. 30
  Go thy ways for me. But whither?
  Go, O but where I may come thither.

What shall I do? My love is now departed.
She is as fair as she is cruel-hearted.
    She would not be entreated
    With prayers oft repeated.
    If she come no more
    Shall I die therefore?
  If she come no more, what care I?
  Faith, let her go, or come, or tarry. 40

(1600)

EDMUND BOLTON
1575?–1633?

538 *A Palinode*

As withereth the primrose by the river,
  As fadeth summer's sun from gliding fountains,
As vanisheth the light-blown bubble ever,
  As melteth snow upon the mossy mountains:
So melts, so vanisheth, so fades, so withers
  The rose, the shine, the bubble, and the snow
Of praise, pomp, glory, joy (which short life gathers),
  Fair praise, vain pomp, sweet glory, brittle joy.
The withered primrose by the mourning river,
  The faded summer's sun from weeping fountains, 10
The light-blown bubble vanished for ever,
  The molten snow upon the naked mountains,
Are emblems that the treasures we up-lay
Soon wither, vanish, fade, and melt away.

For as the snow, whose lawn did overspread
  Th'ambitious hills, which giant-like did threat
To pierce the heaven with their aspiring head,
  Naked and bare doth leave their craggy seat;
Whenas the bubble, which did empty fly
  The dalliance of the undiscerned wind, 20
On whose calm rolling waves it did rely,
  Hath shipwreck made, where it did dalliance find;
And when the sunshine which dissolved the snow,
  Coloured the bubble with a pleasant vary,
And made the rathe and timely primrose grow,
  Swarth clouds withdrawn (which longer time do tarry)
O, what is praise, pomp, glory, joy, but so
As shine by fountains, bubbles, flowers, or snow?

(1600)

539 *As to the blooming prime*

As to the blooming prime,
  Bleak winter being fled
From compass of the clime,
  When nature lay as dead,

25 rathe] early

539
1 prime] spring

The rivers dulled with time,
The green leaves withered,
Fresh Zephiri (the western brethren) be;
So th' honour of your favour is to me.
For as the plains revive,
And put on youthful green; 10
As plants begin to thrive,
That disattired had been,
And arbours now alive
In former pomp are seen;
So if my spring had any flowers before,
Your breaths, Favonius, hath increased the store.

(1600)

SAMUEL ROWLANDS

1570?–1630?

540 ### *Boreas*

'HANG him, base gull; I'll stab him, by the Lord,
If he presume to speak but half a word!
I'll paunch the villain with my rapier's point,
Or hew him with my falchion joint by joint.
Through both his cheeks my poniard he shall have,
Or mincepie-like I'll mangle out the slave.
Ask who I am, you whoreson frieze-gown patch!
Call me before the constable, or watch!
Cannot a captain walk the Queen's highway?
Swounds, who de speak to? Know, ye villains, ha? 10
You drunken peasants, runs your tongues on wheels?
Long you to see your guts about your heels?
Dost love me, Tom? Let go my rapier, then,
Persuade me not from killing nine or ten.
I care no more to kill them in bravado
Than for to drink a pipe of Trinidado.
Thy mind to patience never will restore me
Until their blood do gush in streams before me!'
Thus doth Sir Launcelot in his drunken stagger
Swear, curse and rail, threaten, protest, and swagger. 20
But being next day to sober answer brought,
He's not the man can breed so base a thought.

(1600)

16 Favonius] the west wind

540
3 paunch] stab
7 patch] fool
16 drink] smoke

541

Thraso

WHEN Thraso meets his friend, he swears by God
Unto his chamber he shall welcome be:
Not that he'll cloy him there with roast or sod,
Such vulgar diet with cooks' shops agree.
But he'll present, most kind, exceeding frank,
The best tobacco that he ever drank:

Such as himself did make a voyage for
And with his own hands gathered from the ground.
All that which other fetch, he doth abhor:
His grew upon an island never found. 10
O rare compound, a dying horse to choke,
Of English fire and of India smoke!

(1600)

542

Sir Revel

'SPEAK, gentlemen, what shall we do today?
Drink some brave health upon the Dutch carouse?
Or shall we to the Globe and see a play?
Or visit Shoreditch for a bawdy-house?
Let's call for cards or dice, and have a game:
To sit thus idle is both sin and shame.'

This speaks Sir Revel, furnished out with fashion,
From dish-crowned hat unto the shoes' square toe,
That haunts a whore-house but for recreation,
Plays but at dice to coney-catch or so; 10
Drinks drunk in kindness, for good fellowship,
Or to the play goes but some purse to nip.

(1600)

3 sod] boiled (meat) 5 frank] generous 9 other] others

EDWARD FAIRFAX
d. 1635

from *Godfrey of Bulloigne* (Tasso's *Gerusalemme Liberata*)

543 [*Erminia among the shepherds*]

ERMINIA'S steed this while his mistress bore
Through forests thick among the shady treen,
Her feeble hand the bridle reins forlore,
Half in a swoon she was, for fear I ween;
But her fleet courser spared ne'er the more,
To bear her through the desert woods unseen
Of her strong foes, that chased her through the plain,
And still pursued, but still pursued in vain.

Like as the weary hounds at last retire,
Windless, displeased, from the fruitless chase, 10
When the sly beast tapished in bush and brier,
No art nor pains can rouse out of his place:
The Christian knights so full of shame and ire
Returned back, with faint and weary pace:
Yet still the fearful dame fled swift as wind,
Nor ever stayed, nor ever looked behind.

Through thick and thin, all night, all day, she drived,
Withouten comfort, company, or guide,
Her plaints and tears with every thought revived,
She heard and saw her griefs, but nought beside: 20
But when the sun his burning chariot dived
In Thetis' wave, and weary team untied,
On Jordan's sandy banks her course she stayed
At last, there down she light, and down she laid.

Her tears, her drink; her food, her sorrowings,
This was her diet that unhappy night:
But sleep, that sweet repose and quiet brings,
To ease the griefs of discontented wight,
Spread forth his tender, soft, and nimble wings,
In his dull arms folding the virgin bright; 30
And Love, his mother, and the Graces kept
Strong watch and ward, while this fair lady slept.

11 tapished] hidden

The birds awaked her with their morning song,
  Their warbling music pierced her tender ear,
The murmuring brooks and whistling winds among
  The rattling boughs and leaves, their parts did bear;
Her eyes unclosed beheld the groves along
  Of swains and shepherd grooms that dwellings were;
    And that sweet noise, birds, winds and waters sent,
    Provoked again the virgin to lament. 40

Her plaints were interrupted with a sound,
  That seemed from thickest bushes to proceed,
Some jolly shepherd sung a lusty round,
  And to his voice he tuned his oaten reed;
Thither she went, an old man there she found,
  At whose right hand his little flock did feed,
    Sat making baskets, his three sons among,
    That learned their father's art, and learned his song.

Beholding one in shining arms appear,
  The silly man and his were sore dismayed; 50
But sweet Erminia comforted their fear,
  Her vental up, her visage open laid.
'You happy folk, of heaven beloved dear,
  Work on,' quoth she, 'upon your harmless trade.
    These dreadful arms I bear, no warfare bring
    To your sweet toil, nor those sweet tunes you sing.

'But father, since this land, these towns and towers
  Destroyed are with sword, with fire and spoil,
How may it be unhurt that you and yours
  In safety thus apply your harmless toil?' 60
'My son,' quoth he, 'this poor estate of ours
  Is ever safe from storm of warlike broil;
    This wilderness doth us in safety keep,
    No thundering drum, no trumpet breaks our sleep.

'Haply just Heaven's defence and shield of right
  Doth love the innocence of simple swains;
The thunderbolts on highest mountains light,
  And seld or never strike the lower plains;
So kings have cause to fear Bellona's might,
  Not they whose sweat and toil their dinner gains, 70
    Nor ever greedy soldier was enticed
    By poverty, neglected and despised.

'O poverty, chief of the heavenly brood,
  Dearer to me than wealth or kingly crown:
No wish for honour, thirst of others' good,
  Can move my heart, contented with mine own:
We quench our thirst with water of this flood,
  Nor fear we poison should therein be thrown;
    These little flocks of sheep and tender goats
    Give milk for food, and wool to make us coats. 80

'We little wish, we need but little wealth,
  From cold and hunger us to clothe and feed;
These are my sons, their care preserves from stealth
  Their father's flocks, nor servants moe I need:
Amid these groves I walk oft for my health,
  And to the fishes, birds, and beasts give heed,
    How they are fed, in forest, spring and lake,
    And their contentment for ensample take.

'Time was—for each one hath his doting time,
  These silver locks were golden tresses then— 90
That country life I hated as a crime,
  And from the forest's sweet contentment ran.
To Memphis' stately palace would I climb,
  And there became the mighty Caliph's man,
    And though I but a simple gardener were,
    Yet could I mark abuses, see and hear.

'Enticed on with hope of future gain,
  I suffered long what did my soul displease;
But when my youth was spent, my hope was vain,
  I felt my native strength at last decrease; 100
I gan my loss of lusty years complain,
  And wished I had enjoyed the country's peace;
    I bade the court farewell, and with content
    My latter age here have I quiet spent.'

While thus he spake, Erminia hushed and still
  His wise discourses heard, with great attention;
His speeches grave those idle fancies kill
  Which in her troubled soul bred such dissension;
After much thought reformed was her will;
  Within those woods to dwell was her intention, 110
    Till Fortune should occasion new afford,
    To turn her home to her desired lord.

She said therefore, 'O shepherd fortunate!
  That troubles some didst whilom feel and prove,
Yet livest now in this contented state,
  Let my mishap thy thoughts to pity move,
To entertain me as a willing mate
  In shepherd's life, which I admire and love;
    Within these pleasant groves perchance my heart,
    Of her discomforts, may unload some part. 120

'If gold or wealth, of most esteemed dear,
  If jewels rich thou diddest hold in prize,
Such store thereof, such plenty have I here,
  As to a greedy mind might well suffice:'
With that down trickled many a silver tear,
  Two crystal streams fell from her watery eyes;
    Part of her sad misfortunes then she told,
    And wept, and with her wept that shepherd old.

With speeches kind, he gan the virgin dear
  Towards his cottage gently home to guide; 130
His aged wife there made her homely cheer,
  Yet welcomed her, and placed her by her side.
The princess donned a poor pastora's gear,
  A kerchief coarse upon her head she tied;
    But yet her gestures and her looks, I guess,
    Were such as ill beseemed a shepherdess.

Not those rude garments could obscure and hide
  The heavenly beauty of her angel's face,
Nor was her princely offspring damnified
  Or aught disparaged by those labours base; 140
Her little flocks to pasture would she guide,
  And milk her goats, and in their folds them place;
    Both cheese and butter could she make, and frame
    Herself to please the shepherd and his dame.

But oft, when underneath the greenwood shade
  Her flocks lay hid from Phœbus' scorching rays,
Unto her knight she songs and sonnets made,
  And them engraved in bark of beech and bays;
She told how Cupid did her first invade,
  How conquered her, and ends with Tancred's praise: 150
    And when her passion's writ she over read,
    Again she mourned, again salt tears she shed.

114 whilom] formerly
133 pastora] shepherdess
139 damnified] injured
140 disparaged] dishonoured

'You happy trees for ever keep,' quoth she,
'This woeful story in your tender rind;
Another day under your shade maybe
Will come to rest again some lover kind;
Who if these trophies of my griefs he see,
Shall feel dear pity pierce his gentle mind;'
With that she sighed and said, 'Too late I prove
There is no troth in fortune, trust in love. 160

'Yet may it be, if gracious heavens attend
The earnest suit of a distressed wight,
At my entreat they will vouchsafe to send
To these huge deserts that unthankful knight,
That when to earth the man his eyes shall bend,
And sees my grave, my tomb, and ashes light,
My woeful death his stubborn heart may move,
With tears and sorrows to reward my love.

'So, though my life hath most unhappy been,
At least yet shall my spirit dead be blest, 170
My ashes cold shall, buried on this green,
Enjoy that good this body ne'er possessed.'
Thus she complained to the senseless treen,
Floods in her eyes, and fires were in her breast;
But he for whom these streams of tears she shed,
Wandered far off, alas, as chance him led.

544 [*The garden of Armida*]

THE palace great is builded rich and round,
And in the centre of the inmost hold
There lies a garden sweet, on fertile ground,
Fairer than that where grew the trees of gold:
The cunning sprites had buildings reared around
With doors and entries false a thousandfold;
A labyrinth they made that fortress brave,
Like Dædal's prison, or Porsenna's grave.

The knights passed through the castle's largest gate,
Though round about an hundred ports there shine. 10
The door-leaves framed of carved silver-plate,
Upon their golden hinges turn and twine.
They stayed to view this work of wit and state;
The workmanship excelled the substance fine,
For all the shapes in that rich metal wrought,
Save speech, of living bodies wanted nought.

Alcides there sat telling tales, and spun
Among the feeble troops of damsels mild,
He that the fiery gates of hell had won
And heaven upheld; false Love stood by and smiled: 20
Armed with his club fair Iole forth run,
His club with blood of monsters foul defiled,
And on her back his lion's skin had she,
Too rough a bark for such a tender tree.

Beyond was made a sea, whose azure flood
The hoary froth crushed from the surges blue,
Wherein two navies great well ranged stood
Of warlike ships; fire from their arms outflew,
The waters burned about their vessels good,
Such flames the gold therein enchased threw; 30
Cæsar his Romans hence, the Asian kings
Thence Antony and Indian princes brings.

The Cyclades seemed to swim amid the main,
And hill gainst hill, and mount gainst mountain smote,
With such great fury met those armies twain;
Here burnt a ship, there sunk a bark or boat,
Here darts and wild-fire flew, there drowned or slain
Of princes dead the bodies fleet and float;
Here Cæsar wins, and yonder conquered been
The Eastern ships, there fled the Egyptian queen: 40

Antonius eke himself to flight betook,
The empire lost to which he would aspire,
Yet fled not he nor fight for fear forsook,
But followed her, drawn on by fond desire:
Well might you see within his troubled look,
Strive and contend, love, courage, shame and ire;
Oft looked he back, oft gazed he on the fight,
But oftener on his mistress and her flight.

Then in the secret creeks of fruitful Nile,
Cast in her lap, he would sad death await, 50
And in the pleasure of her lovely smile
Sweeten the bitter stroke of cursed fate:
All this did art with curious hand compile
In the rich metal of that princely gate.
The knights these stories viewed first and last,
Which seen, they forward pressed, and in they passed:

As through his channel crookt Meander glides
  With turns and twines, and rolls now to, now fro,
Whose streams run forth there to the salt sea sides,
  Here back return and to their springward go: 60
Such crooked paths, such ways this palace hides;
  Yet all the maze their map described so,
    That through the labyrinth they got in fine,
    As Theseus did by Ariadne's line.

When they had passed all those troubled ways,
  The garden sweet spread forth her green to show,
The moving crystal from the fountains plays,
  Fair trees, high plants, strange herbs and flowerets new,
Sunshiny hills, dales hid from Phœbus' rays,
  Groves, arbours, mossy caves, at once they view, 70
    And that which beauty most, most wonder brought,
    Nowhere appeared the art which all this wrought.

So with the rude the polished mingled was
  That natural seemed all and every part;
Nature would craft in counterfeiting pass,
  And imitate her imitator art:
Mild was the air, the skies were clear as glass,
  The trees no whirlwind felt, nor tempest smart,
    But ere the fruit drop off, the blossom comes,
    This springs, that falls, that ripeneth and this blooms. 80

The leaves upon the self-same bough did hide
  Beside the young the old and ripened fig;
Here fruit was green, there ripe with vermeil side,
  The apples new and old grew on one twig;
The fruitful vine her arms spread high and wide
  That bended underneath their clusters big,
    The grapes were tender here, hard, young and sour,
    There purple ripe, and nectar sweet forth pour.

The joyous birds, hid under greenwood shade,
  Sung merry notes on every branch and bough; 90
The wind that in the leaves and waters played
  With murmur sweet, now sung, and whistled now;
Ceased the birds, the wind loud answer made,
  And while they sung, it rumbled soft and low;
    Thus were it hap or cunning, chance or art,
    The wind in this strange music bore his part.

With party-coloured plumes and purple bill,
  A wondrous bird among the rest there flew,
That in plain speech sung love-lays loud and shrill,
  Her leden was like human language true; 100
So much she talked, and with such wit and skill,
  That strange it seemed how much good she knew:
    Her feathered fellows all stood hush to hear,
    Dumb was the wind, the waters silent were.

'The gently budding rose,' quoth she, 'behold,
  That first scant peeping forth with virgin beams,
Half ope, half shut, her beauties doth upfold
  In their dear leaves, and less seen, fairer seems,
And after spreads them forth more broad and bold,
  Then languisheth and dies in last extremes, 110
    Nor seems the same, that decked bed and bower
    Of many a lady late, and paramour;

'So, in the passing of a day, doth pass
  The bud and blossom of the life of man,
Nor e'er doth flourish more, but like the grass
  Cut down, becometh withered, pale and wan:
O gather then the rose while time thou has.
  Short is the day, done when it scant began.
    Gather the rose of love, while yet thou mayest,
    Loving, be loved; embracing, be embraced.' 120

She ceased, and as approving all she spoke,
  The choir of birds their heavenly tunes renew;
The turtles sighed, and sighs with kisses broke,
  The fowls to shades unseen by pairs withdrew.
It seemed the laurel chaste, and stubborn oak,
  And all the gentle trees on earth that grew,
    It seemed the land, the sea, and heaven above,
    All breathed out fancy sweet, and sighed out love.

Through all this music rare, and strong consent
  Of strange allurements, sweet 'bove mean and measure, 130
Severe, firm, constant, still the knights forth went,
  Hardening their hearts gainst false enticing pleasure.
'Twixt leaf and leaf their sight before they sent,
  And after crept themselves at ease and leisure,
    Till they beheld the queen, set with their knight
    Besides the lake, shaded with boughs from sight.

100 leden] language

Her breasts were naked, for the day was hot;
  Her locks unbound waved in the wanton wind;
Somedeal she sweat, tired with the game you wot,
  Her sweat-drops bright, white, round, like pearls of Inde; 140
Her humid eyes a fiery smile forth shot
  That like sunbeams in silver fountains shined;
    O'er him her looks she hung, and her soft breast
    The pillow was, where he and love took rest.

His hungry eyes upon her face he fed,
  And feeding them, so, pined himself away;
And she, declining often down her head,
  His lips, his cheeks, his eyes kissed, as he lay,
Wherewith he sighed, as if his soul had fled
  From his frail breast to hers, and there would stay 150
    With her beloved sprite: the armed pair
    These follies all beheld and this hot fare.

Down by the lovers' side there pendent was
  A crystal mirror, bright, pure, smooth, and neat.
He rose, and to his mistress held the glass,
  A noble page, graced with that service great;
She, with glad looks, he with inflamed, alas,
  Beauty and love beheld, both in one seat;
    Yet them in sundry objects each espies,
    She, in the glass, he saw them in her eyes: 160

Her, to command; to serve, it pleased the knight;
  He proud of bondage; of her empire, she;
'My dear,' he said, 'that blessest with thy sight
  Even blessed angels, turn thine eyes to me,
For painted in my heart and portrayed right
  Thy worth, thy beauties and perfections be,
    Of which the form, the shape and fashion best,
    Not in this glass is seen, but in my breast.

'And if thou me disdain, yet be content
  At least so to behold thy lovely hue, 170
That while thereon thy looks are fixed and bent
  Thy happy eyes themselves may see and view;
So rare a shape no crystal can present,
  No glass contain that heaven of beauties true;
    O let the skies thy worthy mirror be!
    And in clear stars thy shape and image see.'

139 Somedeal] somewhat

And with that word she smiled, and ne'ertheless
  Her love-toys still she used, and pleasures bold.
Her hair, that done, she twisted up in tress,
  And looser locks in silken laces rolled; 180
Her curles garlandwise she did up-dress,
  Wherein, like rich enamel laid on gold,
    The twisted flowers smiled, and her white breast
    The lilies there that spring with roses dressed.

The jolly peacock spreads not half so fair
  The eyed feathers of his pompous train;
Nor golden Iris so bends in the air
  Her twenty-coloured bow, through clouds of rain;
Yet all her ornaments, strange, rich and rare,
  Her girdle did in price and beauty stain; 190
    Not that, with scorn, which Tuscan Guilla lost,
    Nor Venus' ceston, could match this for cost.

Of mild denays, of tender scorns, of sweet
  Repulses, war, peace, hope, despair, joy, fear,
Of smiles, jests, mirth, woe, grief, and sad regreet,
  Sighs, sorrows, tears, embracements, kisses dear,
That mixed first by weight and measure meet,
  Then at an easy fire attempered were,
    This wondrous girdle did Armida frame,
    And, when she would be loved, wore the same. 200

But when her wooing fit was brought to end,
  She congee took, kissed him, and went her way;
For once she used every day to wend
  Bout her affairs, her spells and charms to say:
The youth remained, yet had no power to bend
  One step from thence, but used there to stray
    Mongst the sweet birds, through every walk and grove
    Alone, save for an hermit false called Love.

(1600)

190 stain] eclipse
192 ceston] girdle
193 denays] denials
202 congee] leave

NOTES AND REFERENCES

1–10. John Skelton. Texts from *John Skelton*, ed. John Scattergood (1983). **5.** *Philip Sparrow.* Lines 1–142, 386–587 are included in this extract.

11. Anonymous. Text from William A. Ringler, Jr., in *English Literary Renaissance*, i (1971).

12–13. Stephen Hawes. **12.** Text from *The Pastime of Pleasure*, ed. W. E. Mead (1928). **13.** Text from P. J. Frankis, in *Notes and Queries*, NS, xiv (1967).

14–15. Anonymous. Texts from Royal MSS, Appendix 58, ed. E. Flügel in *Anglia*, xii (1889).

16. Heath. MS. Harleian 7578.

17–19. Attributed to King Henry VIII. Texts from Add. MS. 31922, ed. Flügel in *Anglia*, xii (1889).

20. William Cornish. Text from Add. MS. 31922, ed. Flügel in *Anglia*, xii (1889).

21. Anonymous. Text from Balliol MS. 354, ed. Roman Dyboski (1908). I have followed E. K. Chambers and F. Sidgwick, *Early English Lyrics* (1907), in dividing the text into stanzas of five lines.

22–3. Sir Thomas More. Texts from *English Works*, ed. W. E. Campbell *et al.* (1931).

24. Alexander Barclay. Text from *The Eclogues of Alexander Barclay*, ed. Beatrice White (1928).

25. Anonymous. Text from *Scotish Feilde and Flodden Feilde. Two Flodden Poems*, ed. Ian F. Baird (1982). I have divided Baird's unmodernized text into paragraphs.

26–53. Sir Thomas Wyatt. Texts from *Poems*, ed. R. A. Rebholz (1978). **39.** In lines 8, 11, 13, 14, and 16, emendations have been adopted from H. A. Mason, *Sir Thomas Wyatt: A Literary Portrait* (1986). **41.** For their alleged sexual involvement with Anne Boleyn, Lord Rochford, Sir Henry Norris, Sir Francis Weston, Sir William Brereton, and Mark Smeaton were executed. **44.** The poem is usually thought to have been occasioned by the execution on 28 July 1540 of Wyatt's patron Thomas Cromwell, Earl of Essex.

54. Attributed to Sir Thomas Wyatt. Text from *Poems*, ed. Rebholz.

55. Anonymous. Text from *Chaucerian and Other Pieces*, ed. Walter W. Skeat (1897).

56–67. Henry Howard, Earl of Surrey. Texts from *Poems*, ed. Emrys Jones (1964). **67.** Thomas Clere, Surrey's squire and companion, died on 14 April 1545 and was buried in the Howard chapel in St Mary's Church, Lambeth. Clere's cousin was Anne Boleyn, his 'love' Mary Shelton.

68. Robert Copland. Text from *Early Popular Poetry of England*, ed. W. Carew Hazlitt (1866).

69–72. John Harington. **69.** Text from *John Harington of Stepney: Tudor Gentleman. His Life and Works* by Ruth Hughey (1971). **70–2.** Texts from *The Arundel Harington Manuscript of Tudor Poetry*, ed. Ruth Hughey (1960). **70–1.** These are probably versions of Latin poems published later in Walter Haddon's *Poemata* (1567). **72.**

Sir Thomas Seymour (Baron Seymour of Sudeley) was executed for treason in 1549. Cf. Seymour's poem 'Forgetting God' (no. 75).

73. Anonymous. Text from *The Arundel Harington Manuscript of Tudor Poetry*, ed. Hughey (1960).

74. Anne Askew. Text from *The lattre examinacion of Anne Askew, lately martyred in Smythfelde, by the wicked Synagoge of Antichrist, with the Elucydacyon of Johan Bale* (no date). *DNB* states that it was printed in 'Marburg, in Hesse, January 1547'. Anne Askew, a Protestant martyr, was tortured on the rack and burned at the stake for denying transubstantiation.

75. Sir Thomas Seymour (Baron Seymour of Sudeley). Text from *The Arundel Harington Manuscript of Tudor Poetry*, ed. Hughey (1960). In *Nugae Antiquae* (1769), the poem is given the heading: *'Verses found written by the Lord Admiral* SEYMOUR *the Week before he was beheaded*, 1549.'

76. John Heywood. Text from *John Heywood's 'Works' and Miscellaneous Short Poems*, ed. Burton A. Milligan (1956).

77. Nicholas Grimald. Text from *Tottel's Miscellany (1557–1587)*, ed. Hyder Edward Rollins (1965). The poem translates Beza's epigram 'Descriptio Virtutis' in *Poemata* (1548).

78–81. Thomas, Lord Vaux. **78.** Text from *Tottel's Miscellany (1557–1587)*, ed. Hyder Edward Rollins (1965). **79, 80.** Texts from *A Paradise of Dainty Delights (1576–1606)*, ed. Rollins (1927). **81.** Text from *The Arundel Harington Manuscript of Tudor Poetry*, ed. Hughey (1960). The poem is given to Vaux in *A Paradise of Dainty Delights*, but it also occurs in the *Arundel Harington Manuscript*, in a better text; Ruth Hughey argues that it could have been written by either Vaux or John Harington (the Elder).

82. George Cavendish. Text from *Metrical Visions*, ed. A. S. G. Edwards (1980).

83. Thomas Phaer. *The nine first books of the Eneidos* (1562).

84–6. Barnaby Googe. Texts from *Eglogs, Epytaphes and Sonettes*, ed. Edward Arber (1868).

87. Thomas Sackville, Earl of Dorset. Text from *The Mirror for Magistrates*, ed. Lily B. Campbell (1938).

88. Anonymous. Text from *Old English Ballads 1553–1625*, ed. Rollins (1920).

89–95. Edward de Vere, Earl of Oxford. Texts from L. G. Black, unpublished Oxford D.Phil. thesis 'Studies in Some Related Manuscript Poetic Miscellanies of the 1580s' (1970). **93.** Black places this among 'doubtful poems'; it may have been written by Anne Vavasour, Oxford's mistress.

96. Attributed to Edward de Vere, Earl of Oxford. Text from Black.

97. Anonymous. Text from *A Handful of Pleasant Delights (1584)*, ed. Rollins (1924).

98. Arthur Golding. Text from *Ovid's 'Metamorphoses'. The Arthur Golding Translation*, ed. John Frederick Nims (1965).

99–101. John Pikeryng. Texts from *Three Tudor Classical Interludes*, ed. Marie Axton (1982).

102. Anonymous. Text from *A Handful of Pleasant Delights (1584)*, ed. Rollins (1924).

103–6. George Turberville. **103–5.** Texts from *Epitaphs, Epigrams, Songs and Sonnets 1567*, ed. J. P. Collier (1867). **106.** *Tragical Tales* (1587).

107–9. Queen Elizabeth I. **107–8.** Texts from *The Poems of Queen Elizabeth I*, ed. Leicester Bradner (1964). **109.** Text from Black, unpublished Oxford D.Phil. thesis 'Studies in Some Related Manuscript Poetic Miscellanies of the 1580s' (1970). There is a lacuna in the MS. after line 15.

110. Anonymous. Text from *The Poems of Queen Elizabeth I*, ed. Leicester Bradner (1964). The poem has traditionally been attributed to Queen Elizabeth I, but she is unlikely to have written it.

111–12. Thomas Tusser. Texts from *Five Hundred Points of Good Husbandry*, eds. W. Payne and S. J. Herrtage (1876); reprinted 1984.

113. Isabella Whitney. Text from 'The "Wyll and Testament" of Isabella Whitney', ed. Betty Travitsky, in *English Literary History*, x (1980).

114–20. George Gascoigne. **114–19.** Texts from *A Hundreth Sundrie Flowres*, ed. C. T. Prouty (1942). **120.** Text from *The Posies*, ed. John W. Cunliffe (1907).

121. Bewe. Text from *A Gorgeous Gallery of Gallant Inventions (1578)*, ed. Rollins (1926). Bewe's first name is not known.

122. Thomas Proctor. Text from *A Gorgeous Gallery of Gallant Inventions (1578)*, ed. Rollins (1926).

123. Thomas Churchyard. Text from *The firste parte of Churchyardes chippes*, ed. Collier (1867).

124–8. Timothy Kendall. Texts from *Flowers of Epigrammes (1577)*, (Spenser Society Reprint, 1874).

129–34. Nicholas Breton. **129, 133–4.** Texts from *Complete Works*, i, ed. A. B. Grosart (1879). **130.** Text from 'The Honourable Entertainment at Elvetham' (1591), in *The Complete Works of John Lyly*, i, ed. R. Warwick Bond (1902). **131.** Text from *The Phoenix Nest 1593*, ed. Rollins (1931). **132.** Text from *England's Helicon 1600, 1614*, ed. Rollins (1935).

135–47. Edmund Spenser. **135, 136, 142–7.** Texts from *Minor Poems*, ed. Ernest de Selincourt (1910). **137–41.** Texts from *The Faerie Queene*, ed. J. C. Smith (1909). Also consulted: *The Faerie Queene*, ed. Thomas P. Roche, Jr. (1978).

148–85. Sir Philip Sidney. Texts from *Poems*, ed. Ringler (1962).

186. Sir Edward Dyer. Text from MS. Rawl. Poet. 85, in Black, unpublished Oxford D.Phil. thesis 'Studies in Some Related Manuscript Poetic Miscellanies of the 1580s' (1970). In line 12, for 'rest and run' I have adopted the reading 'run and rest' which occurs in the text of the poem in Sidney's *Arcadia* (1598), *England's Helicon* (1600), and MS. Harl. 6910.

187. Attributed to Sir Edward Dyer. Text from Steven W. May, whose copy-text is in Inner Temple MS. Petyt 538. 10. See 'The Authorship of "My Mind to me a Kingdom Is" ', in *Review of English Studies*, NS, xxvi (1975). The poem is attributed to Dyer in Bodleian MS. Rawl. Poet. 85. May puts forward what seem to him the slightly stronger claims of Edward de Vere, Earl of Oxford, but admits that the evidence is inconclusive.

188. Anonymous. Text from Black, unpublished Oxford D.Phil. thesis 'Studies in Some Related Manuscript Poetic Miscellanies of the 1580s' (1970). I have departed from Black's copy-text (MS. Malone 19) in lines 5 and 13, where it reads 'course' and 'floods'; the adopted readings are found in other contemporary texts of this poem.

189–90. Humphrey Gifford. Texts from *Poems*, ed. Grosart (1870).

191. Richard Stanyhurst. Text from *The First Foure Bookes of Virgil his Æneis*, ed. Arber (1880).

192. Thomas Watson. Text from *Hekatompathia or The Passionate Centurie of Love 1582* (Spenser Society Reprint, 1869).

193–4. Anonymous. Texts from *The Arundel Harington Manuscript of Tudor Poetry*, ed. Hughey (1960). **193.** This poem has traditionally been attributed to Henry Walpole, though there is no contemporary evidence. Ruth Hughey prefers the claims of Stephen Valenger.

195. Thomas Gilbart. Text from *Old English Ballads 1553–1625*, ed. Rollins (1920).

196–7. Anonymous. Texts from *A Handful of Pleasant Delights (1584)*, ed. Rollins (1924).

198–208. John Lyly. Texts from *Complete Works*, ed. Bond (1902).

209–22. Fulke Greville, Lord Brooke. Texts from *Poems and Dramas*, ed. Geoffrey Bullough (1939).

223–42. Sir Walter Ralegh. **223–33, 235–42.** Texts from *Poems*, ed. Agnes M. C. Latham (1951). **234.** Text from Black, unpublished Oxford D.Phil. thesis 'Studies in Some Related Manuscript Poetic Miscellanies of the 1580s' (1970). **229.** This poem was evidently written as a reply to Marlowe's 'The passionate shepherd to his love'. **233.** Variants found in contemporary manuscripts include 'knows' for 'shows' (line 9) and 'a faction' or 'their faction' for 'affection' (line 16).

243. Sir Arthur Gorges. Text from *Poems*, ed. H. E. Sandison (1953).

244. Chidiock Tichborne. Text from *Verses of Praise and Joy* (1586), reprinted in *The Works of Thomas Kyd*, ed. F. S. Boas (1901).

245–53. Robert Southwell SJ. Texts from *Poems*, eds. James E. McDonald and Nancy Pollard Brown (1967).

254. Anonymous. Text from *The Poems of Robert Southwell SJ*, eds. McDonald and Pollard Brown (1967). Although traditionally attributed to Southwell, this poem is unlikely to be his.

255–7. Anonymous. Texts from E. H. Fellowes, *English Madrigal Verse 1588–1632*, revised and enlarged by Frederick W. Sternfeld and David Greer (1967).

258–60. Anonymous. Texts from *An English Garner*, ed. Arber (1896).

261. Lodowick Bryskett (Lodovico Bruschetto). Text from *Works*, ed. J. H. P. Pafford (1972).

262. Anonymous. Text from *The Poems of Sir Walter Ralegh*, ed. Agnes M. C. Latham (1951).

263–71. Robert Greene. Texts from *Life and Complete Works*, ed. Grosart (1881–6). **263** is from *Perimedes*, **264** from *Menaphon*, **265–6** from *Greene's Mourning Garment*, **267** from *Orpharion*, **268–9** from *Never too Late*, **270** from *The Groatsworth of Wit*, **271** from *Greene's Vision*. Cf. **281** for another version of this Anacreontic subject.

272. William Warner. *Albions England* (1589).

273. Sir Henry Lee. Text from Thomas Clayton, ' "Sir Henry Lee's Farewell to the Court": The Texts and Authorship of "His Golden Locks Time Hath to Silver Turned" ', in *English Literary Renaissance*, iv (1974). This poem, which was first printed at the back of George Peele's *Polyhymnia* (1590), has usually been attributed to Peele, but Clayton argues convincingly for Lee's authorship. The poem was written for a formal court occasion (17 November 1590) at which, in the presence of the Queen, Lee handed over his office of Queen's Champion to the Earl of Cumberland.

274–8. Thomas Lodge. **274–6, 278.** Texts from *Complete Works*, ed. Hunterian Club (1883). **277.** Text from *The Phoenix Next 1593*, ed. Rollins (1931); the title is taken from *England's Helicon* (1600). **274** is from *Rosalynde*, **275–7** from *Phillis*, **278** from *A Fig for Momus*.

279–81. Anonymous. Texts from *A Poetical Rhapsody 1602–1621*, ed. Rollins (1931). **279.** This is one of the poems in *A Poetical Rhapsody* attributed by its editor Francis Davison to 'A. W.'. Davison thought that these poems had been written 'almost twenty years since' (i.e. 20 years before 1602). Rollins argues, plausibly I think, that 'A. W.' refers not to a single poet but to a group of poets whose identities were unknown to Davison. **281.** Cf. **267** for a free version of Anacreon's poem by Greene.

282. Francis Tregian. MS. Tregian-Yate, Oscott College, Sutton Coldfield. Tregian's verse-letter forms a poem of 168 lines. Included here are lines 1–12, 40–4, 53–64, 125–44, and 157–60. For further information about Tregian, see Louise Imogen Guiney, *Recusant Poets* (1938).

283–8. Sir John Harington. **283–4.** Texts from *Ludovico Ariosto's 'Orlando Furioso'*, ed. Robert McNulty (1972). **285–8.** Texts from *Letters and Epigrams*, ed. N. E. McClure (1930).

289. Anonymous. Text from *The Poems of Sir Walter Ralegh*, ed. Latham (1951). Though traditionally attributed to Ralegh, the poem cannot be his, as Philip Edwards has shown. See 'Who Wrote "The Passionate Man's Pilgrimage"?', in *English Literary Renaissance*, iv (1974). Like **338**, it may have been written by a recusant towards the end of the sixteenth century.

290–9. Henry Constable. Texts from *Poems*, ed. Joan Grundy (1960).

300–5. Mary Herbert, Countess of Pembroke. Texts from *The Psalms of Sir Philip Sidney and the Countess of Pembroke*, ed. J. C. A. Rathmell (1963).

306–313. Christopher Marlowe. Texts from *Poems*, ed. Roma Gill (1987). **310.** The text printed here is from *England's Helicon* (1600). A different version in only four stanzas appeared in *The Passionate Pilgrim* (1599).

314–15. Sir Henry Wotton. **314.** Text from *Poems*, ed. John Hannah (1845). **315.** Text from H. J. C. Grierson, in *Modern Language Review*, vi (1911), p. 155.

316–29. Samuel Daniel. **316–24, 326–9.** Texts from *Poems and 'A Defence of Rhyme'*, ed. Arthur Colby Sprague (1930). **325.** Text from *Works*, ed. Grosart (1883).

330–7. Michael Drayton. Texts from *Works*, ed. J. William Hebel, Kathleen Tillotson and Bernard H. Newdigate (1961). **337.** I have followed the Hebel–Tillotson–Newdigate edition in using the latest editions corrected by Drayton, i.e. until 1637. **337(ii).** Croggen. 'In the voyage (i.e. expedition) that Henry the second made against the Welshmen, as his soldiers passed Offa's Ditch at Croggen Castle, they were overthrown by the Welshmen: which word *Croggen* hath since been used to the Welshmen's disgrace, which was at first begun with their honour' (Drayton).

338. Anonymous. Bodleian MS. Rawlinson Poet. 219.

339–414. William Shakespeare. **339–41** and **414.** Texts from *Poems*, ed. J. C. Maxwell (1966). **342–71.** Texts from *The Riverside Shakespeare*, ed. G. Blakemore Evans (1974). **372–413.** Texts from *Sonnets*, ed. Stephen Booth (1977). Editions by W. G. Ingrams and Theodore Redpath (1964) and John Kerrigan (1986) were also consulted.

415. Anonymous. Text from *Shakespeare: The Poems*, ed. Maxwell (1966).

416–21. Anonymous. Texts from *The Poems of Sir Walter Ralegh*, ed. Latham (1951). These poems from *The Phoenix Nest* (1593) form part of what Miss Latham calls 'a conjectural "Ralegh Group" '. They are possibly by Ralegh.

422. Robert Devereux, Earl of Essex. Bodleian MS. Ashmole 781, which reads 'give God praise' (line 5) and 'Where when' (line 9); the readings adopted in the text are introduced from B.L. MS. Chetham 8012.

423. Anonymous. Text from Black, unpublished Oxford D.Phil. thesis 'Studies in Some Related Manuscript Poetic Miscellanies of the 1580s' (1970). Black's copy-text is B.L. MS. Add. 22583, where the poem occurs on a page of writings by William Gager. Black infers that Gager (1555/1560–1621) may well have been the author. In the later Farmer Chetham MS. 8012, the poem is said to be 'By the Earl of Oxford'.

424–32. Thomas Campion. Texts from *Poems*, ed. Walter R. Davis (1969).

433–6. Thomas Nashe. Texts from *Works*, ed. R. B. McKerrow, revised F. P. Wilson (1958).

437–50. Anonymous. Texts from *English Madrigal Verse 1588–1632* (3rd edn.), ed. Fellowes, revised and enlarged by Sternfeld and Greer (1967).

451–2. Barnabe Barnes. Texts from *Parthenophil and Parthenophe*, ed. Victor A. Doyno (1971).

453–71. John Donne. Texts from *The Complete English Poems*, ed. A. J. Smith (1971).

472–82. Sir John Davies. Texts from *Poems*, ed. Robert Krueger (1973).

483. Anonymous. Text from *Bishop Percy's Folio Manuscript. Loose and Humorous Songs*, ed. John W. Hales and Frederick J. Furnivall (1867).

484–8. George Peele. Texts from *Dramatic Works*, ed. R. Mark Benbow, Elmer Blistein, and Frank S. Hook (1970). **484–7** are from *The Old Wive's Tale*, **488** from *David and Fair Bethsabe*.

489–90. Richard Barnfield. **489.** Text from *Some Longer Elizabethan Poems*, ed. A. H. Bullen (1903). **490.** *Poems in Divers Humours* (1598).

491–4. George Chapman. **491–3.** Texts from *Poems*, ed. Phyllis Brooks Bartlett (1941). **494.** Text from *Chapman's Homer* I, ed. Allardyce Nicoll (1957).

495–6. Bartholomew Griffin. Texts from *Poems*, ed. Grosart (1876).

497. Sir Francis Bacon. Thomas Farnaby, *Florilegium Epigrammatum Graecorum* (1629).

498–501. Robert Sidney. Texts from *Poems*, ed. P. J. Croft (1984).

502–4. Joseph Hall. Texts from *Collected Poems*, ed. Arnold Davenport (1949).

505–13. William Alabaster. Texts from *The Sonnets of William Alabaster*, ed. G. M. Story and Helen Gardner (1959).

514. Thomas Bastard. Text from *Chrestoleros. Seven Bookes of Epigrames* (Spenser Society Reprint, 1888).

515. Josuah Sylvester. Text from *The Divine Weeks and Works of Guillaume de Saluste, Sieur du Bartas. Translated by Josuah Sylvester*, ed. Susan Snyder (1979).

516. Ben Jonson. Text from *Poems*, ed. Ian Donaldson (1975).

517–20. John Marston. Texts from *Poems*, ed. Davenport (1961).

521–6. Thomas Dekker. Texts from *Dramatic Works*, ed. Fredson Bowers (1953). **521–2** are from *The Shoemaker's Holiday*, **523–4** from *Old Fortunatus*, **525–6** from *Patient Grissil.*

527–37. Anonymous. Texts from *English Madrigal Verse 1588–1632* (3rd edn.), ed. Fellowes, revised and enlarged by Sternfeld and Greer (1967).

538–9. Edmund Bolton. Texts from *England's Helicon 1600, 1614*, ed. Rollins (1935).

540–2. Samuel Rowlands. Texts from *Complete Works* (Hunterian Club, 1880).

543–4. Edward Fairfax. Texts from *Godfrey of Bulloigne*, ed. Kathleen M. Lea and T. M. Gang (1981).

INDEX OF FIRST LINES

The references are to the numbers of the poems

INDEX OF FIRST LINES

INDEX OF FIRST LINES

INDEX OF AUTHORS

The references are to the numbers of the poems